Job of Edessa's Book of Treasures

Syriac Studies Library

46

The Syriac Studies Library brings back to active circulation major reference works in the field of Syriac studies, including dictionaries, grammars, text editions, manuscript catalogues, and monographs. The books were reproduced from originals at The Catholic University of America, one of the largest collections of Eastern Christianity in North America. The project is a collaboration between CUA, Beth Mardutho: The Syriac Institute, and Brigham Young University.

Job of Edessa's Book of Treasures

Encyclopaedia of Philosophical and Natural Sciences as taught in Baghdad A.D. 817

Edited and Translated by

Alphonse Mingana

2010
gorgias press

Gorgias Press LLC, 954 River Road, Piscataway, NJ, 08854, USA

www.gorgiaspress.com

Originally published in 1935

2010 ܐ

ISBN 978-1-60724-909-2

Reprinted from the 1935 Cambridge edition.

Digitized by Brigham Young University. Printed in the United States of America.

Series Foreword

This series provides reference works in Syriac studies from original books digitized at the ICOR library of The Catholic University of America under the supervision of Monica Blanchard, ICOR's librarian. The project was carried out by Beth Mardutho: The Syriac Institute and Brigham Young University. About 675 books were digitized, most of which will appear in this series.

Our aim is to present the volumes as they have been digitized, preserving images of the covers, front matter, and back matter (if any). Marks by patrons, which may shed some light on the history of the library and its users, have been retained. In some cases, even inserts have been digitized and appear here in the location where they were found.

The books digitized by Brigham Young University are in color, even when the original text is not. These have been produced here in grayscale for economic reasons. The grayscale images retain original colors in the form of gray shades. The books digitized by Beth Mardutho and black on white.

We are grateful to the head librarian at CUA, Adele R. Chwalek, who was kind enough to permit this project. "We are custodians, not owners of this collection," she generously said at a small gathering that celebrated the completion of the project. We are also grateful to Sidney Griffith who supported the project.

BOOK OF TREASURES

Published for the Trustees of
THE WOODBROOKE SETTLEMENT, SELLY OAK, BIRMINGHAM
by
W. HEFFER & SONS LIMITED
CAMBRIDGE

ENCYCLOPÆDIA OF PHILOSOPHICAL AND NATURAL SCIENCES
AS TAUGHT IN BAGHDAD ABOUT A.D. 817

OR

BOOK OF TREASURES

BY

JOB OF EDESSA

SYRIAC TEXT EDITED AND TRANSLATED WITH A CRITICAL APPARATUS

BY

A. MINGANA

[Volume I of Woodbrooke Scientific Publications]

CAMBRIDGE:
W. HEFFER & SONS LIMITED
1935

22256

PRINTED IN GREAT BRITAIN
BY THE ABERDEEN UNIVERSITY PRESS LIMITED

PREFATORY NOTE.

THE present volume is the first of a series of scientific publications, the contents of which are drawn from MSS. in my collection. My previous series, comprising seven volumes and entitled *Woodbrooke Studies: Christian Documents in Syriac, Arabic and Garshūni* (1927-1934), came to an end last year with the volume containing *Early Christian Mystics.*

This first volume of the new series gives the text and the translation of an important work, by Job of Edessa, on philosophical and natural sciences, as known and taught in Baghdad in about A.D. 817. No complete work dealing with such a variety of subjects has come down to us in any language from that early period.

The cost of the publication of the present volume has been defrayed, as in the case of the volumes of the previous series, by Mr. Edward Cadbury, to whom I wish here to express my sincerest thanks. If, as our author affirms in the introduction to his work and elsewhere, "the knowledge of the causes of the created beings is the knowledge of God, and the knowledge of God is life, and life is the kingdom of heaven," to help to spread such knowledge is to help to spread the kingdom of heaven.

A word of thanks is also due to the Aberdeen University Press for the care they have displayed in printing the volume, and to my secretary, Miss N. K. Garnett, for the pains she has taken in its production.

A. MINGANA.

SELLY OAK COLLEGES LIBRARY,
BIRMINGHAM,
9th April, 1935.

CONTENTS.

INTRODUCTION.

I.

I GIVE in the following pages the text and the translation of a work entitled "Book of Treasures," by the Syrian writer Job of Edessa. It is written in the form of an encyclopædia, embracing almost all the natural and philosophical sciences as known and taught in Baghdad about A.D. 817. Although dealing with the branches of science the foundations of which had been laid centuries earlier by Aristotle, one of the greatest thinkers of all times, the author displays much ingenuity in discussing the "cause" or the elemental origin of biological, physiological, chemical, physical and astronomical facts known to him.

The end of the 8th and the beginning of the 9th centuries were characterised by a great revival of the study of exact sciences in the new capital of the 'Abbasid empire. Much credit for this revival was undoubtedly due to the enlightened attitude of the Caliphs Manṣūr, Mahdi, Hārūn, Amīn and especially Ma'mūn (A.D. 754-833), and considering that Baghdad itself was only founded by Manṣūr in A.D. 762, it is true to say that this revival coincided with the beginning of the 'Abbasid dynasty. Greek manuscripts of early masters of medical and natural sciences were eagerly sought after, and brought to Baghdad from Syria, Egypt and Asia Minor,[1] and gradually translated by Christian Arab authors. Hospitals and astronomical observatories were erected later, to facilitate researches. An imposing list of the early writers who adorned the crown of the empire of the sons of 'Abbās, and whose works are mostly lost in our days, is found in the *Fihrist*,[2] in the history of

[1] *Fihrist*, pp. 339-340 (ed. Cairo, A.H. 1348). [2] *Ibid.*, pp. 339-421.

philosophy and medicine by Ibn al-Qifṭi [1] and i. a. Uṣaibi'ah,[2] and in other historical and bibliographical lucubrations.

The end of the 10th century marks the apogee of this Arabic Hellenistic culture. At its dawn history makes mention, so far as medicine is concerned, of the East Syrian Arab writer Ibn Sahda or Sahdé from Karkh (near Baghdad), of the beginning of the 9th century, who, according to the *Fihrist*,[3] and i. a. Uṣaibi'ah [4] translated from Syriac into Arabic some works of Hippocrates. According to Ḥunain ibn Isḥāq, he also translated into Syriac the works of Galen, *De Sectis, De Partibus Artis Medicativae* and *De Pulsibus ad Tirones*.[5]

A second Christian Arab author is Abu Yaḥya al-Baṭrīq, who died about A.D. 805, and who was employed by the Caliph Manṣūr. According to i. a. Uṣaibi'ah,[6] he translated many works of Hippocrates and Galen, and he is given also as the translator of Ptolemy's *Quadripartitum*.[7]

To Yaḥya ibn Baṭrīq, the son of the above writer, who flourished in the first quarter of the 9th century, is ascribed the translation of Hippocrates' book *Signs of Death*, some works of Aristotle, and the *De Theriaca ad Pisonem* of Galen.[8] To him is also attributed, although without much probability, the famous *Secretum Secretorum*, which has lately been so well studied by R. Steele.[9]

We will be satisfied here with the mention of the above three writers, as our aim is not to furnish detailed references concerning physicians who did not leave any translations of Greek medical works, such as Simon of Ṭaibūtheh, and the two most eminent members of the Bokhtīsho' family, about

[1] *Akhbār al-Ḥukamā* (ed. Cairo, 1326 A.H.).

[2] *Ṭabaqāt al-aṭibbā* (Wahb Press, 1882).

[3] *Fihrist*, p. 341.

[4] *Ṭabaqāt*, i. 204.

[5] *Ḥunain ibn Isḥāq, Über die Syrischen und Arabischen Galen-übersetzungen* (edit. G. Bergsträsser, 1925), pp. 4, 6.

[6] *Ṭabaqāt*, p. 205.

[7] Sarton, *Introduction to the History of Science*, Vol. I, p. 537 (Carnegie Institution of Washington, 1927).

[8] *Ḥunain ibn Isḥāq* (*ibid.*), p. 39; *Fihrist*, pp. 351 and 405. *Cf.* Meyerhof, *Isis*, VIII, 705.

[9] *Secretum Secretorum, cum glossis et notulis*, Oxford, 1920.

whom see *Fihrist*,[1] Ibn al-Qifṭi[2] and i. a. Uṣaibi'ah,[3] nor to enumerate the various writers who flourished in the decades that followed the death of our author, and so do not come within the scope of our enquiry.

After medicine, the branch of science in which the writers contemporary with our author seem to have evinced most interest was astronomy, but since Ptolemy was not translated into Syriac or Arabic in the first years of the revival of the Hellenistic culture mentioned above, they generally based their conclusions upon the half astrological aberrations found in early Indian and Persian astronomical books,[4] the contents of which had little in common with the solid data found in Ptolemy's work. An early but unsatisfactory translation of the *Almagest* of Ptolemy was made for Yaḥya b. Khālid b. Barmak,[5] towards the beginning of the 9th century[6] but the most reliable and trustworthy translation was that made by Ḥajjāj b. Maṭar for the Caliph Ma'mūn, in A.D. 827-828 (and not in 829-830).[7] With the exception of a few insignificant dissentients Ptolemy's great work became the main authority of the Arab astronomers of later generations, as is testified by such independent writers as Ibn Rabban,[8] who died about A.D. 855, Jāḥiẓ,[9] who died in A.D. 869, and especially Muḥammad al-Khwārizmī, who died about A.D. 850.[10] The same method was followed by the astronomical observers or astrolabe makers Ḥabash, Sanad, 'Ali ibn'Īsa,

[1] *Fihrist*, p. 413.
[2] *Akhbār*, pp. 71-73.
[3] *Ṭabaqāt*, I, pp. 125-144.
[4] The names of many such Indian and Persian writers are given in the *Fihrist*, pp. 331-342, Ibn al-Qifti, 148-150, etc. See on them H. Suter, *Die Mathematiker und Astronomen der Araber* (1900), and Sarton, *ibid.*, pp. 530-533.
[5] *Fihrist*, p. 374.
[6] The information found in Sarton (*Introduction*, p. 565) that Rabban aṭ-Ṭabari, the father of Ibn Rabban aṭ-Ṭabari, the author of *Firdausu'l-Ḥikmat*, translated the *Almagest* into Arabic does not seem to me to be correct. Sarton (*ibid.*, p. 574) states also that 'Ali ibn Rabban aṭ-Ṭabari was the son of a Jewish physician, but as has been convincingly proved by my edition of his *Book of Religion and Empire* and by the publication of his *Firdausu'l-Ḥikmat*, his father was a Christian and not a Jew.
[7] *Fihrist*, p. 341, and Nallino, *Encyclopædia of Islam*, I, 498.
[8] *Firdausu'l Ḥikmat*, pp. 547-551 (ed. Siddiqi).
[9] *Three Essays of Jāḥiẓ*, p. 16 (ed. J. Finkel).
[10] *Fihrist*, p. 383, and Suter, *ibid.*, Nachträge (158-160), 1902.

Yaḥya i. a. Manṣūr, Marwarrūdhi and the like, mentioned by the author of the *Fihrist*, by Suter, and by Sarton.[1] This, however, cannot be said of the astronomers who flourished before A.D. 828, such as Fazārī, Ibn Ṭāriq, Māshāllāh, and Naubakht.[2]

In the realm of chemistry, a word must be said about the celebrated Jābir ibn Ḥayyān who flourished in Kūfa about A.D. 776, and to whom are ascribed scores of works, many of which never existed. His chemistry seems to have been largely connected with the ancient alchemy. It is not probable that he was the author of the philosophical works attributed to him which are under the influence of the peripatetic teaching, as the works of Aristotle were not yet translated into Arabic in his time, and there is no reason to believe that he could read them in the original.[3] Some of the opinions held at present about him may be modified by the forthcoming work of P. Kraus, *Jābir ibn Ḥayyān, essai sur l'histoire des idées scientifiques dans l'Islam,* which is to appear in three volumes.

So far as the philosophical works of the masters of Greek philosophy—Plato and Aristotle—are concerned, they began to be systematically translated into Arabic at the beginning of the 9th century. The man who more than any other was responsible for spreading the knowledge of Greek learning in the 'Abbasid capital was undoubtedly the Christian Arab writer Ḥunain ibn Isḥāq, who died in A.D. 876. Long pages of the *Fihrist,*[4] of Ibn al-Qifṭi,[5] and of i. a. Uṣaibi'ah,[6] are devoted to the enumeration of the translators of the works of the two above-named coryphæi of Greek philosophy.

In the forefront of the phalanx of the translators of Aristotle and Galen, our author stands as one of the earliest figures. No importance should be attached to the statement of Ibn al-Qifṭi[7] that an earlier writer 'Abdallah b. Muqaffa' translated into Arabic the *Categories* and the *Analytica* of Aristotle, since

[1] *Introduction*, pp. 565-569.
[2] *Ibid.*, pp. 530-532.
[3] About him and his works see *Fihrist*, pp. 498-503, Berthelot, *La Chimie au Moyen Age*, Vol. 3, and Sarton, *Introduction*, pp. 532-533.
[4] *Fihrist*, pp. 345-352.
[5] *Akhbār*, pp. 13-40.
[6] *Ṭabaqāt*, Vol. I, pp. 54-69, etc.
[7] *Akhbār*, pp. 148-149.

there is no reason to believe that Ibn al-Muqaffaʿ knew any Greek or Syriac.

What marks the present work as one of the best that have come down to us is its completeness. All the scientific books written in the time of Ma'mūn, if not mere translations of Greek works, are cast into the mould of short monographs on particular subjects. This point will be more fully discussed in the following sections.

II.

The author of the present work is Job of Edessa, or *Ayyūb ar-Ruhāwi*, as the Arab writers called him. He was, as his name implies, born in Edessa, possibly about 760 A.D. The author of the *Fihrist* mentions him as a translator of Greek works,[1] and in the very same line names a Job who, together with a writer called Simʿān, translated the *Zīj*, or astronomical tables of Ptolemy for the nobleman Muḥammad, son of Barmak. No one, however, who reads with care the astronomical data of the present work can induce himself to believe that their author had read Ptolemy on the subject, much less translated him. If this Job is to be identified with our author, it may be presumed that he translated the above *Zīj* after the composition of the present work.

The great Ḥunain ibn Isḥāq[2] also mentions our author, and attributes to him the translation of thirty-six different works of Galen, especially the translation into Syriac of his famous "Book on Simple Drugs," or *De Simplicium Medicamentorum Temperamentis et Facultatibus*.[3] It is a noteworthy fact that Ḥunain himself used the translation into Syriac of some Galenic works made by our author, such as the *Anatomicæ Administrationes*, or ἀνατομικαὶ ἐγχειρήσεις.[4] The full list of the

[1] *Fihrist*, p. 341.

[2] In the work referred to above: *Ḥunain ibn Isḥāq, über die Syrischen und Arabischen Galen-Übersetzungen*, edit. by G. Bergsträsser (1925), pp. 3-50.

[3] *Cf.* Meyerhof and Sobhy, *Abridged Version of "The Book of Simple Drugs" of Ghāfiqi by Barhebræus*, p. 8 (Cairo, 1932).

[4] *Cf. The Book of the Ten Treatises on the Eye Ascribed to Ḥunain ibn Isḥāq*, edit. by Meyerhof (1928), p. xli.

works of Galen translated by our author, according to Ḥunain ibn Isḥāq,[1] is the following:

(1) *Πίναξ* (p. 3).

(2) *Τέχνη ἰατρική* (p. 6).

(3) The last seven discourses of the *Περὶ τῶν σφυγμῶν πραγματεία*, translated for Gabriel b. Bokhtīsho' (p. 15).

(4) *Ἀνατομικαὶ ἐγχειρήσεις*, translated for Gabriel b. Bokhtīsho' (p. 20).

(5) *Περὶ τῆς ἀνατομικῆς διαφωνίας* (p 21).

(6) *Περὶ τεθνεώτων ἀνατομῆς* (p. 21).

(7) *Περὶ ζώντων ἀνατομῆς* (p. 21).

(8) *Περὶ τῆς Ἱπποκράτους ἀνατομῆς* (p. 21).

(9) *Περὶ μήτρας ἀνατομῆς* (p. 22).

(10) On the Anatomy of the Eye (apocryphal), (p. 23).

(11) *Περὶ τῆς ἀναπνοῆς αἰτιῶν* (p. 24).

(12) *Περὶ τῆς τῶν καθαιρόντων φαρμάκων δυνάμεως* (p. 26).

(13) *Περὶ τῶν Ἱπποκράτους και Πλάτωνος δογμάτων* (p. 26).

(14) On the Difficult and Unknown Movements (p.27).

(15) *Περὶ ἀνωμάλου δυσκρασίας* (p. 29).

(16) The first part, composed of five discourses, of the *Περὶ ἁπλῶν φαρμάκων* (p. 30).

(17) *Περὶ τῶν ἐν τοῖς νόσοις καιρῶν* (p. 30).

(18) *Περὶ τῶν παρὰ φύσιν ὄγκων* (p. 31).

(19) *Περὶ τῶν προκαταρκτικῶν αἰτιῶν* (p. 31).

(20) *Περὶ μελαίνης χολῆς*, translated for Bokhtīsho' son of Gabriel (p. 33).

(21) *Περὶ δυσπνοίας* (p. 34).

(22) *Περὶ προγινώσκειν πρὸς Ἐπιγένην* (p. 34).

(23) *Περὶ μαρασμοῦ* (p. 35).

(24) *Περὶ τροφῶν δυνάμεων* (p. 35).

(25) *Πρὸς Πίσωνα περὶ τῆς θηριακῆς* (p. 39).

(26) Galen's Commentary on the *ἀφορισμοί* of Hippocrates (p. 40).

(27) Galen's Commentary on the *περὶ ἀγμῶν* of Hippocrates (p. 40).

[1] References are to the pages of Ḥunain's Arabic text in Bergsträsser's edition. I have reconstructed the original Greek titles as found in Bergsträsser's edition instead of the Arabic ones given in Ḥunain's text.

(28) Galen's Commentary on the *περὶ διαίτης ὀξέων* of Hippocrates (p. 41).

(29) Galen's Commentary on the *περὶ τῶν ἐν κεφαλῇ τρωμάτων* of Hippocrates (p. 41).

(30) Galen's Commentary on the *ἐπιδημίαι* of Hippocrates (p. 41).

(31) *Ὅτι ὁ ἄριστος ἰατρὸς καὶ φιλόσοφος* (p. 44).

(32) *Περὶ τῶν ἰδίων δοξάντων* (p. 46).

(33) *Περὶ ἀποδείξεως* (p. 47).

(34) *Περὶ ἠθῶν* (p. 49).

(35) *Περὶ ἀλυπίας* (p. 49).

(36) *Ὅτι ταῖς τοῦ σώματος κράσεσιν αἱ τῆς ψυχῆς δυνάμεις ἕπονται* (p. 50).

Our Job is further spoken of by Barhebraeus [1] in the following terms :

"And in the time (of Timothy I) lived Job of Edessa, a philosopher who followed the doctrine of the Nestorians."

Ibn a. Uṣaibi'ah [2] devotes also a special section to him, and considers him to be a good translator, versed in languages, but adds that he was more versed in Syriac than in Arabic. He is inaccurate, however, in distinguishing him from Job *al-Abrash*, or "The Spotted," to whom he has devoted another section.[3] Ḥunain ibn Isḥāq [4] clearly identifies our Job of Edessa with Job the Spotted. Another paragraph is devoted by i. a. Uṣaibi'ah [5] to a son of our author called Abraham, in connection with the Caliph Mutawakkil and other high personages.

Our author is also mentioned by Yāqūt,[6] with reference to an anecdote told of the Caliph Ma'mūn, as one of the greatest physicians of his day. This anecdote is important because it shows that our author was still alive in A.H. 217 (A.D. 832) when Ma'mūn appointed 'Abdallah b. Ṭāhir governor of the Persian province of Khurāsān.

[1] *Chron. Eccl.*, II, 181.
[2] *Ṭabaqāt*, I, 204.
[3] *Ibid.*, p. 204.
[4] Bergsträsser's edition, p. 3 (of the text).
[5] *Ibid.*, i. 170-171.
[6] *Dictionary of Learned Men* (ed. of D. S. Margoliouth), Vol. I, p. 122 of the second edition.

We have no precise information as to the date of our author's death, but we may presume that he did not survive long the Caliph Ma'mūn, and that he died about A.D. 835.

As I stated in the first volume of the catalogue of the MSS. of my collection,[1] we may infer from the above sentence of Barhebræus that the author belonged by birth either to the Melchite or to the West Syrian or Jacobite community, which he left in order to join the East Syrian or Nestorian Church.

Job seems to have been a prolific writer, both in Arabic and in Syriac. None of his Arabic works, however, has come down to posterity, nor is any of them mentioned by name in the authorities referred to above. So far as his Syriac books are concerned, on fol. 36b of the MS. from which the present work is derived, he enumerates some of them, in the following terms:

"After having completed our book *On the Causes of Fevers*, and our other book *On the Soul*, which is divided into twenty chapters, and after having composed the book *On the Causes of the Coming into Existence of the Universe from the Elements*, we wrote our other book *On Urine*, and you request us now, O brother, to write to you a treatise *On Canine Hydrophobia*."

Four of these five works—(a) *On the Soul*, (b) *On the Causes of Fevers*, (c) *On Canine Hydrophobia*, and (d) *On Urine*, are also mentioned in the present work.[2]

Another work of the author was entitled *Book on Faith*. From the terms which he uses to describe it,[3] we are entitled to believe that it dealt with the Trinity and the Incarnation, and included other points of Christian dogma, such as the Holy Communion and the worship towards the east.

A seventh book by our author is mentioned in the same chapter,[4] with the title *Book of Ten Syllogisms*, in which, among other things, he demonstrated that Christ was both God and man.

An eighth book by him, entitled *On the Five Senses*, and a ninth entitled *On Essences*, are referred to in other passages.[5]

[1] *Catalogue of the Mingana Collection of MSS.*, Vol. I, p. 1036. By inadvertence I stated on this page that Job lived towards the *end*, instead of the *beginning*, of the 9th century. [2] Pp. 25, 94, 136, 191, 257, 279, 280.
[3] See here p. 279. [4] See here p. 278. [5] See here pp. 126, 154.

Of all his books only two, that *On Canine Hydrophobia* and the present *Book of Treasures*, have come down to us. Both are found in the MS. from which the present edition is derived, the former on ff. 36b-39b, and the latter on ff. 40a-125b. The book called *On the Causes of the Coming into Existence of the Universe from the Elements* does not seem to have been identical with the *Book of Treasures*. The latter contains many more points than the title of the former implies. The lost book seems to have dealt exclusively with the emanation of created things from the elements, while the *Book of Treasures* has a much wider scope. This hypothesis is rendered probable by the fact that the books *On Urine*, and *On Canine Hydrophobia*, which followed that *On the Coming into Existence of the Universe from the Elements*, are both quoted in the present work.

III.

It is very difficult to fix on a precise year for the composition of the present work. Two independent considerations induce me to name a date about A.D. 817. On the one hand the author states in Chapter XVI of the 3rd Discourse of his book[1] that he wrote at a time of great tribulations and wars " such as were not heard of since the beginning of the world." These tribulations and wars were so catastrophic that he was unable to predict whether he would be alive to finish the work upon which he was engaged. On the other hand, as we remarked above, the author did not know of the existence of Ptolemy's astronomical work which was translated in A.D. 828, and consequently we have to fix on a year preceding this date. Further, since all his works of which we have any record were written before the present *Book of Treasures*, he must have been a fairly old man when he wrote it, and in this case we are not at liberty to name a date much earlier than 828, especially as we know, from the sentence of Yāqūt quoted above, that he was still alive in A.D. 832. The question arises now whether in the annals of the history of Baghdad we can point to a period which would square with the calamitous times so vividly described

[1] See here p. 154.

by the author. Although hardly a year passed in the stormy weather of Baghdad politics without wars or rumours of wars in some of the outlying provinces of the Caliphate, yet the years 816-817 seem to have experienced a particularly severe recrudescence of disturbances, which for a time endangered the life of Ma'mūn himself.

In A.D. 816 the famous general Harthama had died in prison as a result of injuries received through the machinations of the treacherous Faḍl ibn Sahl, the vizier of Ma'mūn. His cruel and unmerited death caused great excitement in Baghdad. The Caliph himself was in Merw, and the loyal troops rose against the viceroy Ḥasan, the brother of Faḍl, and after three days' battle drove him from the city to Madāin, and eventually to Wāsiṭ. While the troops were pursuing Ḥasan, great confusion prevailed in Baghdad itself, which was for a long time at the mercy of adventurers of the robber class, who committed in it all kinds of cruelties, and introduced into it an era of the worst and crudest type of Oriental anarchy.[1] In order to defend their lives and property, the better class of citizens banded themselves together, held in check the excesses of the populace, and offered the throne to Manṣūr, son of Mahdi, who wisely declined it, but promised to conduct the affairs of the government in the name of the rightful Caliph Ma'mūn.

The years which immediately preceded and followed the above disturbances may also be counted among the most disastrous that the capital of the 'Abbasids experienced in the early years of its foundation, and the historian Ibn 'Imād [2] had strong reasons for writing: "The distress of the inhabitants of Baghdad reached such a pitch that they believed that the end of the city had come, and many of them left it on account of robberies, plunders, famine and the destruction of their homes."

The question of the sources of the author is more easily determined. He refers several times to Aristotle, twice to Galen, and once to Hippocrates. He refers also to early Indian and Persian sages, but without mentioning any proper names.

[1] See the vivid account of these disturbances in the historians Ṭabari, *Annales*, Vol. X, pp. 238-243 (edit. Cairo), and Ibn Athīr, *Kāmil*, Vol. VI, pp. 109-110 (edit. Cairo).

[2] *Shadharāt* II, 2.

So far as the Indian philosophers or physicians are concerned, they must have been Charaka Samhita and Susruta Samhita, so often quoted by Ibn Rabban, a contemporary of the author, who survived him by more than twenty years.[1]

The author's statements concerning the "functional origin" of the different organs of the animal body are largely dependent on Galen, while his borrowings from Aristotle are very extensive. One might almost say that he was saturated with Aristotle, many references to whose works will be found in the footnotes. These references, as well as those to Galen, could have been easily multiplied, but as my aim was not to trace every single line of the work to its source, I felt compelled to set a limit to them, and to confine myself to the main points of comparison. To have done otherwise would have entailed an abnormal multiplication of footnotes without appreciable advantage. No wider scope, therefore, should be given to the *apparatus criticus* found in the book than that which it is strictly meant to convey. Almost all the scientific works of Aristotle are pressed into service, especially his *Meteorologica*, *Physica*, *De Mundo*, *De Anima*, *De Juventute et Senectute*, *De Sensu*, *De Somno et Vigilia*, *De Cælo*, *De Generatione et Corruptione*, and *Metaphysica*.

The author's inspirations from Aristotle do not imply that he was a servile plagiarist. On the contrary, he often adds fresh data to this master's investigations, and often also expressly contradicts him, as on the subject of matter, the existence of a fifth element in the heavenly bodies, the divinity of the stars, the eternity of the movement of the first elements, the nature of the ether, and the doctrine of the "dry evaporation" found in the earth.

I have also noticed in the author some borrowings from Hippocrates, more especially in his treatment of the "cause" of the procreation of males and females, and the influence of environment on physical characteristics.

The author lays special stress on the fact that the main thesis of his work, the emanation of the genera and species from the elements, is original. This originality applies in a particular

[1] See his *Firdausu'l-Ḥikmat*, pp. 567, 573, 576, 578, 579 for Charaka, and pp. 558, 560-562, 578 for Susruta.

way to the formation of the various animal organs: head, hands, feet, toes, inner organs, organs of the five senses, hair, nails, etc. He emphatically asserts in Chapter VIII of the Second Discourse,[1] that he had not found this subject discussed in any book written by the Indian, Persian, Greek or Syrian scholars who had preceded him, and on p. 279 he asserts that in this book "he has not troubled anybody with the labours of others."

Until all the works written by the Arab and Syrian investigators of the 9th and 10th centuries are published and scientifically studied, we shall not be able to measure the influence exercised by our author on the physicians and philosophers who came after him. I have noticed, however, statements by two different writers which seem to be derived from the present work. The first writer who seems to have borrowed from our Job in a rather systematic way is the often-mentioned Ibn Rabban aṭ-Ṭabari, the author of *Firdausu'l-Ḥikmat*.[2] The second writer is the lexicographer Bar 'Ali, whose borrowing was detected in a rather unusual way. He had read the Greek word μελωδός, as *mehloros*[3] owing to an oversight of a copyist who had wrongly transliterated it in this way in the present work, and so the mistake was perpetuated in Bar 'Ali's lexicon. What renders certain the hypothesis that Bar 'Ali borrowed this word from our author is the fact that he translates the Greek vocable by exactly the same Syriac words as those used by our author in Chapter VIII of his Third Discourse.[4]

The influence exercised by our author may possibly be extended to a wider sphere. He was the first to develop in detail, through a deductive method of reasoning based on natural phenomena, the idea of the elemental origin of the universe and of the different bodies comprising it. We have noted above his definite statement that he was the first in the field of the concrete application of the elemental principle to the physical bodies. Ibn Rabban aṭ-Ṭabari, who wrote some years after him, distinctly borrows some of his conclusions from him. After many vicissitudes, the conclusions first drawn by our author

[1] See here p. 77. [2] See references here pp. 13, 49, 107, 108, 131, 174.

[3] In Syriac the letters *d* and *r* are graphically identical and distinguished only by an extraneous dot over the *r*. [4] See here p. 143.

reached Cordova, in Spain, another Muslim centre of learning, and from it they radiated to the *foyers* of knowledge in Italy and France. The philosophy and physics of Averroes (or Ibn Rushd), Avicenna (or Ibn Sīna), Alpharabius (or Al-Farābī) and many others were translated into Latin, and exercised great influence on the scientific teaching of the Middle Ages. The western scientists of that period, having lost sight of the works of Aristotle and his followers in their original Greek, were mostly dependent on the above Arab writers, and their development of the idea of the elemental origin of bodies was not more advanced than that worked out by our author at the beginning of the 9th century. Our author may, therefore, be rightly considered as the father of the concrete development of the theory of the elemental origin of bodies, a conception which became firmly rooted in the minds of the scientific investigators of later generations.

IV.

There is no need to dilate here on the contents of the *Book of Treasures*. The translation of the text and the detailed index of chapters at the beginning of the volume will help the reader to form his own judgment on the author's method of reasoning, but to facilitate his task, it would be useful to draw attention to some outstanding points discussed in the book.

(a) *Metaphysics.*

The author follows Aristotle in dividing everything that exists into essence and accident. The word "substance" might often have been substituted for "essence" which I have generally used, but I deemed it advisable to make use of one word only, in order to throw into relief more clearly the divisional differences of ontological notions which the author intended to convey.

According to the author, there are two kinds of essences in the universe. The first and universal essence is that of the simple and abstract elements, heat, cold, humidity and dryness, even in their composed state as found in the compound elements, earth, air, fire and water, heat being the strongest and most important in its action. The second kind of essence is that of every individual body and is sometimes called by him *substratum*. This essence is not universal as the first is, and remains only as long as the body which it constitutes stands in its own individual composition; it is, as it were, a relative essence, and when the body which it forms ceases to exist, and the parts which compose it suffer complete disintegration, this essence reverts to the universal type of essence of the simple elements. In this way everything referring to the quality or quantity of an object, such as taste, colour, dimensions, smoothness, roughness, etc., is an accident, since a thing existing not in itself but in another and liable to modifications, cannot be called an essence, which by its very definition means something self-existing, or subsisting in itself.

The primordial essences of the simple elements cannot be infinite and eternal, because an infinite being is one which cannot be limited and circumscribed. In the mere fact that the simple elements are limited and circumscribed they cannot be infinite. How can they be infinite, when they do not possess uniformity, but undergo distinctions and quantitative changes, brought about by the inherent weakness of their finite nature? When one element in its abstract conception increases or decreases, we have to introduce for its right working one or more of the remaining elements, and in doing so we create mentally limit or space between them, and thus remove from them every idea of infinity.

This leads to the demonstration of the existence of God, which is attested by the coming together of the two antagonistic forces of heat and cold. Since these powers are mutually antipathetic, and would never have come together naturally to form the material bodies, we must assume the action of a pre-existing and infinite Being who brought them together for the purpose of carrying out the work of creation. That Being we call "God."

The existence of the primordial and *intelligible* matter on which, according to Aristotle, the two antagonistic forces, heat-cold and dryness-humidity, worked for the formation of the universe, is rejected by the author, on the ground that

that matter, not being endowed with any quality whereby it may be conceived, cannot possibly receive things that are dissimilar to itself, namely the two antagonistic forces. The only kind of matter in which the author believes is the *concrete* matter, synonymous with a material body, endowed with the quantity and quality possessed by all bodies found in the universe.

The above universal essences of heat, cold, humidity and dryness, were created by God from nothing, and they did not emanate from His essence. He created them and imparted to them the power to produce everything found in the universe. It is they that by their different combinations with one another gave birth to all material objects, and all changing accidents, found in the universe. They first formed the compound elements simultaneously with their own creation, then the bodies, with special reference to the humours in their bearing upon the formation of animal bodies. The author mentions in this connection the six days of the creation, and asserts that since all things done by God are performed immediately He wills them, if the method of the six days of the creation is to possess any reality, it must be presumed that God made use of it in order to make His work understandable to the weakness of the human mind. The real fact of the creation is to be conceived in the notion that God created the simple elements and allowed them to carry out the complicated work of building up the creation into its innumerable genera and species.

The author follows Aristotle in his law of contraries, and holds, like him, that the creation came into being through the antagonistic movement found in the two active elements of heat and cold coming into mutual contact through the passive elements of humidity and dryness. Without contraries nothing comes into being. When, therefore, the elements cease to strive and struggle, the generating activity of the universe will automatically come to an end. The activity of the elements working together in balanced proportions brings also about the health and well-being of bodies, while their lack of balanced proportions brings about their ill-health and corruption. "A balance of parts preserves and does not injure, while a non-balance of parts injures and causes deterioration."

(b) *Theology.*

The *Book of Treasures*, not being strictly theological, discusses only those theological questions which have a direct bearing upon science. The points that are somewhat extensively touched upon by the author are the existence of God, spoken of above, angelology and eschatology. The existence of angels is proved by metaphysical reasoning. Man is composed of two principles: the corporeal and the incorporeal. Since the corporeal principle is represented in a complete form in the lower orders of the creation, and the incorporeal principle in an imperfect way in man, there must be another order of beings who would represent the incorporeal principle in a complete way. The angels fill up this gap in the creation, and constitute the perfect type of the incorporeal or spiritual principle so imperfectly represented in man. Further methods of reasoning of a somewhat recondite and abstract character are resorted to by the author, especially in the case of the strange method, based on the perfection of the number three, which he adopts to prove rationally the existence of the three hierarchies and the nine orders of the angels.

He argues that the present world will have an end, owing to the weakness that will affect the activity of the elements spoken of above. In process of time everything in the world will suffer from this eventual weakness of the elements, which, no longer able to generate, will degenerate. This degeneration is called the end of the world. The end of this world and the beginning of the next are alike explained by this change which the elements will undergo in their inner working. Being less active, they will have less antagonism and more affinity with one another, and, animated by the harmonious "circular movement" of the peripatetic thought, they will give birth to a totally different world, devoid of the generation and corruption, increase and decrease, which characterise the present world, itself the result of the working of the "straight movement."

The author does not believe in the resurrection of material bodies. He thinks that the "resurrection of the body" found in the Creed is to be explained by the "spiritual body" spoken of by St. Paul. This spiritual body consists of the individual parts of the first or simple elements, by the antagonistic action of which a particular body was composed while on earth. The soul, by its long union with such a body, will distinguish, in the next world, the parts of the elements of which it was composed but which now will possess a harmonious action.

Heaven and hell are explained by the author in a rather ingenious way. The essential function of the soul, says he, is the acquisition of the knowledge of God, sciences and arts. "In the same way as a hungry man is urged by his body to eat, so also human nature is urged forcibly by the soul to enquire into things and to know their causes." The above constituents of the soul will remain as long as the soul itself remains, as without them there is strictly speaking no soul. The happiness of the soul would naturally consist in its possession of its essential constituents, and its unhappiness in its deprivation of them. The greatest torment, thinks the author, that a being can undergo, is to be deprived of acts pertaining to its essential function, and its greatest happiness is to have full and harmonius use of this function. "When the soul wishes to understand, in accordance with its nature, the things pertaining to God and the things pertaining to the world, and cannot comprehend them, it becomes greatly distressed, even when the body is satisfied. . . . When we comprehend a difficult point concerning God or concerning His providence towards the created beings, we rejoice and jubilate greatly, in such a way that we would die for it, in order not to give it the lie or deny it."

There are no material fire and no tormenting demons in what we call hell: "The kingdom of heaven will, therefore, be meditation on God, and the knowledge of the nature of the created beings, while lack of them will be hell." Nothing pertaining to the body on this earth will rise with the body, because, as stated above, it will be a spiritual and not a material body: "Because the things pertaining to the body will not rise with it—as the soul will overcome these things when it is reunited with the body and works in it the things pertaining to itself—there will be no happiness emanating from food and drink; and since this happiness emanating from food and drink will not be found in the next world, neither will the torment emanating from them be found in it. Indeed where there is no food, there is no deprivation of food, and where there is no antagonism to give rise to fight, there are no afflictions, suffering and corruption; and if these are not found there, what will be the happiness (of the just) and the torment of the wicked which will be found there, except meditation on God and on the created beings, and ignorance of them respectively?"

"As to the torment, it consists in a complete deprivation of the knowledge of God and of the created beings, a thing contrary to the light of the kingdom (of heaven), and this is the outer darkness which stands before the knowledge of the soul and impedes it from rejoicing in the glorious and ineffable light of the Holy Trinity, one God. The sins and evil deeds which the wicked have committed will stand before the eye of the soul, and will blind it; and it will in consequence be affected with ἀμβλυωπία, that is to say, dimness of vision, which is an incurable and hopeless disease which acts like the thick matter of walls that impedes light from penetrating and from illuminating those who are imprisoned in a house. This is the outer and the extreme darkness."

In the author's opinion it is we who make ourselves fit for heaven or for hell: "He who does good," says he, "is good, and he who does not do good is not good."

(c) *Psychology.*

The author believes that the soul is totally different from the "animal spirit" found in the material organs of the brain. Although permeating this animal spirit, it works independently of it, through the nerves of the brain. This it is able to do by reason of the subtile affinity which it possesses with the "thin parts" of the animal spirit. The function of the animal spirit is confined to the nutrition, growth and welfare of the body, while that of the soul consists in the higher work of thinking and reasoning in the domain of religion, science and art. The soul is not the immediate source of the emotions of the body, such as love, fear, greed, generosity, courage, pusillanimity, etc. Since the animals also possess them, their origin must be traced to a physical principle, based on the different combinations of the first elements. The soul is endowed with free-will, while the animal spirit is not.

The existence of the soul is proved by metaphysical arguments, of which the following two seem to be the most important: first, the ability of man to plan and visualise objects before they are produced, and second, his power to check the strong desires connected with the law of self-preservation.

The union of the soul with the body is explained in the following terms:

"The soul is inside the body, but not in the way in which a body is inside another body, for instance water or oil in a skin; but it is in it as an integral yet composite entity; and it is not mixed with it, because, like the body which keeps its entity intact, it also keeps its entity without change. It is in the body according to its size, whether large or small, while itself is neither large nor small. This is illustrated by a piece of iron or other metal, in which fire is seen. The greatness or smallness of the fire is seen according to the greatness and shape of this object, although fire itself has neither greatness nor smallness."

The upright stature of man is explained as being due to the action of his soul mixing with his animal spirit, and drawing him upwards through its extreme expansiveness, which, combining with the expansiveness of the light and hot elements, tends to increase the strength of their movement upwards towards the cranium, where the soul has its seat.

The author explains the capacity of the brain to hold images in the following way:

"How can the small space of the brain hold the images of towns and countries and, which is more wonderful, of all the world?"—The soul, being simple, and having no antagonistic forces in it, "imparts its unlimitedness to the limited body of the brain. . . . I mean by the unlimitedness of the soul that by means of which it is in an instant everywhere, through the swift movement of the imagination, immediately it wills it, without any intervention of time which bodies have to endure."

(d) *Biology, Anatomy and Physiology.*

Many biological points find their way into the author's scheme of discussion. The distinctions of the genera and species of animals and plants are clearly defined, and as usual traced back to the elements. According to him, the principal divisions which characterise the life of existing things were fixed at the beginning, when the combination of the first elements brought them into being. Having, however, preceded Pasteur and his bacteriological experiments by more than a thousand years, he believes that the lower forms of life, as represented by fleas, lice, tape-worms and such like, are derived from the action of the elements working in corrupted bodies, and that such lower forms of life do not procreate.

The cause of individual differentiations within the species is clearly set forth, as may be seen from the following sentences: "Accidental variations also occur in individuals of each species, not, however, through that variation which perfected this species in its essential existence and constituted all the genera and species, but through the accidental variation that occurred to each individual at the beginning of its composition from the semen. This resulted either from addition to food, or from climate, or from any other external accident that affected individuals of the species and not all the species. In this way, one became red-haired, another snub-nosed, and another tall. In this way also this one may be said to be the son of that one, and both of them are the children of another. . . . Genera and species are always and invariably the very same, while individuals may or may not be alike."

Many interesting details are discussed by the author, including the reason why birds are oviparous and the terrestrial animals viviparous, and why some animals are carnivorous and others herbivorous.

He explains the origin of the animal bodies by the process of seething that took place in the primordial liquid matter, or "chyle," which was the result of the meeting together of the simple elements: "When there was heat, the parts (of the primordial matter) that were thin and light departed from it and thick parts were formed, and by their mutual composition a body that was harder and more earthy than it came into being. In the same way as when we cook a body, such as barley, etc., the first thing that is produced is a liquid and watery chyle, and when we cook afresh this chyle so that its humid and thin part rises up, it becomes itself a body thick, hard, and non-liquid —in this same way the humours became bodies in the process of seething."

He shows much ingenuity in a realm which may not be very attractive to some adherents of modern positive science. He explains how the head, feet, hands, fingers, toes, hair, nails, organs of the senses, and inner organs of man and other animals came into being; how it is that man alone of all animals has an upright posture; how it is that the lower animals have tails, to the exclusion of man; and how it is that nerves and veins are composed of one whole structure, while bones are formed of separate parts joined together. Evidently early scientists wished to find a reason for natural phenomena which modern science leaves unexplored for lack of positive data. Modern science is satisfied with the elucidation of known facts of the creation, while early scientists wished to know the ultimate origin of these facts. They had worked out a philosophical system of "reasons of things," within the framework of which they included all biological phenomena. It is in this domain of the physical explanation of biological facts that the author chiefly displays the originality to which we have drawn attention above. As regards the functional explanation of the organs of bodies, he is much under the influence of Galen, whose main translator he was, and it is noteworthy that although in many places he contradicts Aristotle, he never runs counter, in an explicit way, to the physiological and anatomical doctrine of Galen, as exposed in his great work *περι χρείας τῶν ἀνθρώπου σώματος μορίων*.[1] His general knowledge in connection with the three classes of animals, "the terrestrial, the aerial and the aquatic," and his insight into the method of growth and fruition of trees and plants are often remarkable.

Weeping is physiologically explained as due to the superfluity of hot and humid matter which man possesses more than the other animals owing to the diversity of his food. The tendency

[1] For my references to this work of Galen, I have made use of the edition of Daremberg, *Oeuvres Anatomiques, Physiologiques et Médicales de Galien*, Paris, 1854-1856.

of heat being to rise upwards, and man having an upright posture, this hot superfluity rises to his head, and being checked in its rising by the cranium, it follows the "path of the factors of vision," at the end of which it finds an outlet.

Sleep in man takes place in the brain, and is physiologically accounted for by the fact that labour causes displacement of its natural humidity, and thus renders it needful of rest in order to become humid. It is asserted in this connection that "When there is no sleep we moisten the brain until it comes." During sleep the natural humidity found in the body rises, as usual, and reaches the brain, which it moistens. The sleep which animals experience is not comparable to that of man. Animals simply *rest*, while man truly *sleeps*, the reason being that animals do not labour with their brain in the same degree as man.

The subject of growth from the seminal stage to that of full maturity in both plants and animals is explained in the following manner :

"The winter was made so that through cold it might contract, collect and condense the humidity which the species receive from rain, from air, and from other causes. If this did not take place, there would be no budding and no increase in the trees (and plants), since the humidity, before being contracted and condensed, would evaporate outside, and no fruit or corn would grow. This happens also to the seed in the womb, because if it is not caught at a propitious moment, it will not increase little by little, and perfect the fruit which is the embryo. Since, however, the humidity is imprisoned, it increases and condenses, and being invigorated from outside by the earthiness, airiness and fieriness in which it is imprisoned it gives rise to budding.

"As to the summer, it expands, through its heat, the humidity which the bodies of animals, plants and corn have received, draws it out, and causes it to evaporate outside. (This humidity) first causes in plants green leaves, and then blossoms ; but when dryness overcomes humidity little by little, it hardens the blossoms and produces fruits ; and after they have ripened through heat and humidity—and this is even more true of animals—humidity diminishes, and dryness predominates and greatly contracts this humidity inwards."

"The earth does not produce much grass when rain falls on it and makes it too humid, till the sun shines on it and heats this humidity, and an equilibrium is established between them, and humidity rises up according to its nature. The same thing happens to the stage of childhood, in which no hair grows, until it reaches the stage of puberty, in which, through the addition of the strength of heat, hair grows."

Contrary to the opinion of ancient medical authorities to the effect that the young emanate exclusively from the semen of the male, our author seems to imply that both the male and the female combine to produce the embryo, which thus takes its qualities from both of them. In this respect he is far removed from the notion that the mother is only, as Aeschylus makes Apollo say, "the nurse of the embryo," a notion which had such a far-reaching influence on the social position of women.

(e) *Medicine.*

In spite of the fact that the author was considered as one of the greatest physicians of his day, the therapeutics and general pathology of his work are not very extensive, owing, no doubt, to his having already written many books on the subject. The anatomical and physiological aspects of medicine are, however, fairly fully dealt with, and are summarised in the previous section. Red bile, black bile, blood and phlegm, the "humours" of the fathers of ancient medicine, Hippocrates and Galen, are referred to at some length, and, after the simple elements, are regarded as the immediate originators of the physical body.

According to the author, good health is acquired through the equilibrium that should exist between the elements in and around our bodies. Illness is caused by an undue predominance of one element over the others, and may be remedied by administering drugs which would re-establish the former harmony by creating opposite effects.

The purgative and vomitive action of certain drugs is scientifically explained by the degree of their active power in comparison with the power of resistance of the stomach. If their active power is stronger than the power of the stomach, they force it to yield, and purgation takes place, but if the power of resistance of the stomach is stronger than the active power of the drugs, it forces them upwards, and vomiting takes place. This is explained in the following quotation :

"The drugs that contain more active powers have a stronger movement, and those that contain less have a weaker movement. When they are drunk, and go down to the stomach, they become its enemies and fight against it ; and those of them that are stronger in the active power win a victory over the power of resistance of the stomach which weakens ; and then they cause a downward and cleansing movement ; and those of them that are weak in active power, when they go down to the stomach fight against it, and the power of resistance of the stomach prevails, and it repels them backwards ; and in the fight that ensues they get hold of the matter (found in the stomach) and seize it ; and after they have been repelled they drag that matter with them upwards."

"As to voluntary vomiting," says the author, "it takes place when we introduce our finger to the root of the tongue and to the head of the gullet, and tickle this head of the gullet, which communicates with the stomach through its lower part. When this titillation takes place, this lower part rises to help the upper part, and in doing so it drags with it, through a sudden upward spasm, the food that it had brought to the stomach, and drags it through the expelling power to where it emerges outside."

(f) *Chemistry and Physics.*

The minerals are, according to the author, formed of different proportions of the simple elements found in the earth. No notion of an atomic or molecular theory is found in his book. Minerals could not be an exception to the general principle that the origin of everything is to be sought in the antagonism between the simple elements. The reason why some metals, such as iron, rust, while others, such as gold, do not, is also referred to these different proportions of each element found in them, and likewise their hardness or softness, and their relative resistance to heat. There are, therefore, no simple chemical bodies in the modern sense of the term, since even these simple bodies may be reduced to the simple elements of heat, cold, humidity and dryness.

The reason why some springs of water are hot is explained by the fact that the earth in that particular place contains hot minerals, such as sulphur.

The author holds that the hot metals such as sulphur, yellow orpiment and bitumen leave nothing behind when they are burnt because of the predominance in them of airy and fiery parts; these parts being affinitive to air and fire are taken up by them in the process of burning. In the case of other objects, such as pieces of wood, "which contain a thick, earthy, heavy and compact part, when fire is kindled in them it takes from them that which is affinitive to it—as does also the air—and what is left of them remains in their place."

An ingenious theory is set forth concerning the different tastes found in bodies. Pleasant tastes are produced by the well-balanced parts of bodies coming into contact with correspondingly well-balanced parts of our tasting organs. "All things," says the author, "that have well-balanced parts in all species are agreeable and are tasted with pleasure. . . . Life itself is sweet and pleasant through the harmonious balance between the elements." Unpleasant tastes are caused by the lack of the above balance of parts.

The following quotation deals with the sweet taste:

"Sweetness comes about when an even quantity of heat permeates the humidity (of a body), and through its predominance softens its contracted and hard parts, or impedes them from forming, . . . Sweetness causes pleasantness to the palate in the following way: sweetness comes into contact and joins with the humidity of the mouth, which itself consists naturally of a well-balanced and even humidity mixed with heat. Because the humidity and heat found in that sweetness are affinitive to the humidity and heat of the mouth, the latter does not show aversion to the sweetness, but it receives it and presents it to the palate, which takes pleasure from it on account of the general balance of proportions which is not impaired by any increase or decrease of other tastes that might cause repulsion to the palate."

The other varieties of tastes are similarly explained.

In the author's opinion there are two principal colours, white and black, of which all others are formed in different proportions according to the predominance of one element over another.

The perception of colours and shapes is explained in the following way: The eye possesses in itself the two principal colours, with the addition of grey, or some other colour. These colours in the eye create a certain affinity between it and the outside colours, which enables it to receive them. The brightness of the eye is a factor for the reception of the light of the sun from outside, and the perceiving light of the soul from inside: "The eye, on account of its brightness and clearness, receives the bright and perceiving power of the soul from inside, and in addition it receives the light of the sun from outside. Thus in it the perceiving and bright power of the soul is united with the light of the sun. This perceiving power is carried by the light of the sun till it reaches the reflected bodies; and it perceives their shapes and colours, and first draws them to the eye from outside, and then presents them to the imagination from inside. This explains why, when we close our eyes immediately after having perceived colours, we see them in our imagination as if with our eyes."

The other objects of the senses are explained by the author on similar lines. "Sound," says he, "is the movement of air when it comes into contact with bodies and strikes them and then, after being driven by them, brings that striking to the ears, through the instrumentality of which it reaches the brain." Sound is sub-divided into two kinds: that caused by concussion and that caused by the friction of air. "Sound is either breathy or non-breathy. The non-breathy sound occurs, for instance, when two stones come into contact with each other and strike each other, and from the concussion thus caused between them, the air leaps out and carries their striking and brings it to the ears. The breathy sound occurs when an animal inhales the air by means of the indrawing power inside the lungs, and exhales it through the expelling power, and in its exit it strikes the larynx and the trachea, and then goes out and reaches the ear." He is aware of the fact that sound travels more slowly than light: "It is well known that what the eye receives comes to it first, and what the ear receives comes to it afterwards, from the fact that when we see from far a tree being chopped, the eye receives (the impression of) the movement when looking at it, before the ear receives the sound." This phenomenon he explains by the fact that sight is fiery, while hearing is aerial, and fire is more rapid and penetrating than air.

The meteorological doctrine of the author is, on the whole, sound, and, considering the early period in which he wrote, compares favourably with that of other scholars. He follows Aristotle closely in many points, but contradicts him or completes him in many others. There is no need to stress here his views on the origin of winds, hurricanes, snow, hail, rain, thunder and lightning, earthquakes, thunderbolts, shooting stars, rainbows, and the halos of the sun and the moon. The reader will find these points clearly explained in the book. It would be useful, however, to state that he believes that rain leaves the clouds in the form of a downpour or cascade, which gradually becomes broken up into drops by the resistance of the air which it meets in falling.

His doctrine regarding physical geography may be summarised in a few sentences. The origin of mountains and springs is explained by gases trying to escape from the earth, when it was in process of formation through the hot seething produced by the elements coming together to give birth to it. "We affirm," says he, "that in the upheaval which took place through fire and air going upwards from the inside of the earth, and through which mountains came into existence, some places were found to be weak and soft, and through them air leapt upwards and in doing so gave rise to slits and clefts formed in an upward direction inside the mountains at their top and at their side. In this way it opened up fissures and escaped. This very air drew also the water after it, through the affinity which its humidity possessed with water, and springs came into being at the extremities of the mountains, from which water flowed." Having no instruments to measure the temperature of water, and working only through the natural sense of touch, he believes that the water of springs is essentially warmer in the winter than in the summer. This is part of his doctrine that in the winter the inside of the earth is warmer than in the summer, owing to the fact that the summer heat finds a refuge there from the onslaught of the winter cold which affects the upper surface.

He asserts that mountains have more trees than plains, "because of the contraction of their parts which imprison within themselves heat and humidity, which give rise to many trees and plants."

The salinity of sea water is explained accurately as due to the effect of the sun and the admixture of some parts of the earth: "The water of the sea has been heated by the sun from the beginning of the world up till now, and in this way the heat of the sun acts upon it and draws from it, through the air, the thin, volatile and expanded parts which it contains, and which are affinitive to itself. The water that remains consists of the thick parts which collect together and bear in themselves the result of the scorching of heat which changed them to saltness. They further receive from the nature of the upper surface of the earth a certain dryness and thickness."

(g) *Music.*

The main musical notes are four: first, the high or thin, second, the low or thick, and third and fourth, two intermediate ones between these. Each one of the four notes is divided into three relatively high or low notes. "From these twelve notes," says the author, "is derived the number of all melodies, in which is included all the myriads of innumerable tunes found among all peoples."

The physiological formation of high and low notes is explained as follows: "The voice of men is deep for two reasons: first because heat predominates in the throat, and thus expands its contracted parts and loosens them little by little, according to its own nature. In this way it differentiates the throat of men from that of children, and expands (the vocal cords) outward, according to its nature, thus widening the throat. . . . As to the voices of children, women and eunuchs, they are unbroken because the parts which come into contact (with the air) are tighter, on account of cold; and when the voice comes out, it harps on those tight parts, and they give a high note. . . . When we strike two bodies, such as brass and wood, the brass gives a high sound and the wood a low sound, the former because of the contraction of its parts, and the latter because of the expansion of its parts."

Melody takes place "when the thin notes and the thick notes are mutually balanced, and none of them is in great excess over its neighbour, but all of them are formed in harmony with one another. This is illustrated by the strings of a guitar, in which when the vibration of the thin part is made to balance with that of the thick part, and the high note with the low note, they are moved in harmony with one another." Lack of harmony between parts causes a discordant sound.

The art of composing music is not an easy one: "All people possess naturally in themselves the faculty of composing tunes, in the same way as every one of them possesses, from natural thought, the power of building a house; but the question of an *artistic* production belongs to the artist. Anyone can compose tunes and believe that he is composing in a masterly way, while only those of a *musician* are so."

In this connection we will quote the following remarkable passage: "When the voice of animals and of man comes out in a natural way, it is not delightful to the ear, but is, on the contrary, jarring to it; only the one which is divided into melodies is delightful, although the natural voice, being one and evenly-balanced, should have delighted the hearing, on account of this balance. But because like is pleasing to like, and because the body is composed of varieties of different parts, and its pleasure is caused by the harmony between them, which, in creating an even balance and an even temperament causes a happy life, it follows that the body is delighted only by sounds which are formed with one another, through the diversity of their parts, in a balanced harmony. When no balanced harmony of the elements exists in the body, it does not derive any happiness out of life. In this same way, when there is no even harmony of sounds, no pleasure is produced. When the voice comes out naturally, and reaches the hearing, it does not delight it, because it is not composed, like the body, of many parts, but penetrates abruptly and does not enter the hearing in parts which are harmoniously composed, and so does not cause it any pleasure."

(h) *Mathematics.*

The author often speaks of geometry, with its angles, points, sides, triangles and circles. He also mentions the three dimensions of bodies. He is not, however, very charitable towards those who juggle with numbers, and whom he calls "arithmeticians." There is evidently with him a difference between a "mathematician" and an "arithmetician." In spite of this hostility to "arithmeticians," he himself sometimes juggles with numbers: with the number three in connection with its completeness and with the Trinity; with the number seven in connection with the seven planets and the seven days of the week; with the number nine in connection with the nine orders of angels and the nine numbers before the cipher; with the number twelve in connection with the twelve signs of the Zodiac, the twelve months of the year, and the twelve musical notes; and with the number thirty-two in connection with the thirty-two teeth and the thirty-two years of the full growth of man.

We must draw attention here to an interesting historical point dealing with the origin of the cipher. This important arithmetical sign is historically explained by the fact that numbering in general stopped at the number nine, and then turned back to the number one, after which it began again to mount up. Early mathematicians considered the number nine as being complete in every respect. They considered the number three to be also complete, but not to the same degree as the number nine. The number three contains only the number one added to itself three times, and the number one is not in itself a complete number; but the number nine contains the number three added to itself three times, and this makes it doubly complete.

We have it on the authority of our Job that early mathematicians made use of the fingers for counting the above numbers, and that after having reached the ninth finger, which would be the forefinger, they began again with the first finger, which would be the little finger of the other hand: but in order to find a link to connect number nine with number one, they formed the cipher, by linking the forefinger and the thumb together in a circle. This explanation of the origin of the cipher is all the more important because the author is the earliest writer to make mention of the cipher, as it was in his time that this indispensable arithmetical sign reached Baghdad.

Four points are to be considered in connection with the right understanding of the evolution of the cipher :

(1) The cipher is a circular link between number nine and number one, causing numbering in general to proceed in a succession of cycles between one and nine.

(2) The figure of the cipher is a circle made by linking the forefinger and the thumb together.

(3) The cipher, the very name of which means "emptiness" has no arithmetical value in itself, but gives value to another number with which it is linked. In written numbering this linking of the cipher with another number always takes place from right to left, that is to say, the cipher is placed *before* the other number,[1] as in counting with the fingers the cipher must be formed immediately after the number nine, and immediately before the number one.

(4) The very figure of the cipher represents both the emptiness and the linking spoken of above.

We will quote here a passage dealing with numbering :

"At the beginning of all numbering, the number one moves first towards the number three, which is composed of uneven and even numbers. Indeed numbering does not begin until the number two, which is even, follows the number one (making three), and it continues thus until the number three reaches the number nine, where the movement of mounting-up stops, according to the angelic division. It is after this that, in the composed beings, on account of suffering, an addition begins towards the number ten, as if the number nine wished to make a kind of link, in order to strengthen the numbering on account

[1] The Arabs, from whom we took the cipher, read numbers from right to left. We wrongly read 1935 as : one thousand nine hundred and thirty-five, while they rightly read it as : five and thirty and nine hundred and one thousand.

of its weakness, in accordance with the order required by the composed beings that these beings should be linked in all their compositions ; and so that it might be linked, it reverted towards the number one in a circular way. The movement of numbering is thus completed in a kind of cycle. It is for this reason that the ancients invented, as a first sign for this number (ten) the (empty) space between the forefinger and the thumb, formed in a circular way. Indeed when the numbers which we have with us reach a denary state they stop, and then turn back and mount up indefinitely."

(i) *Astronomy.*

Since the author had no intention of writing a systematic treatise on cosmography, it is sometimes difficult to define clearly his opinion on certain astronomical problems. He believes that the stars are bodies composed of the same elements as those which we have in this world, the only difference being that the elements which constitute the stars are endowed principally with a "circular movement," unlimited in its action, while the earth is animated chiefly by a "straight movement." These two movements, which are well studied in Aristotle's works, are of paramount importance in the elaboration of the author's system, and should be studied with care by anyone wishing to understand him rightly. He adduces many mathematical and metaphysical reasons to demonstrate the nature of these two movements.

Since the course of the circular movement has no end, because it has neither increase nor decrease in its essential functioning, the stars will by necessity have no end ; but the earth, with its straight movement constantly affected by increases and decreases, will by necessity come to an end.

Influenced by the first chapter of Genesis, the author believes that what the ancients called "the firmament" is a real body, and quotes with disapproval an opinion to the effect that this firmament is an optical illusion created by distance. He holds that the earth is a spherical body, but is very far from the doctrine first enunciated by Copernicus and Galileo that it is a moving planet. He upholds the opinion that the stars move in a circle, with the earth as its centre. Being unaware of Ptolemy's mathematical researches as to the size of the planets, and basing his calculations on the apparent size of the different stars, he thinks that the sun and the moon are the largest stellar bodies. To be fair to him, however, we must at once state that he clearly affirms also that the moon appears larger than it really is, because of its nearness to us. He devotes a special chapter to the Galaxy, which he describes, against Aristotle's doctrine, as a thick mass of stars, the blurred appearance of which is accounted for by their great distance.

He attacks vigorously the opinion of the ancients that the stars are reasonable beings endowed with a rational soul, and has not enough strong words with which to inveigh against the theory that the planetary system has anything to do with the prognostication of future events affecting the lives of men. All types, therefore, of astrological calculations dealing with the horoscope are severely criticised and ridiculed. The true master of our destiny, says he, can only be God, and the stars, being bodies composed of the same elements as our earth, cannot possibly be endowed with a power of prognosis not possessed by our earth. He is, therefore, far from accepting the opinion of Hippocrates, the father of scientific and positive medicine: "Astronomy renders no small service to medicine. On the contrary it is of great help to it."[1]

V.

The MS. reproduced here in facsimile is Mingana Syriac 559,[2] of which no other copy is found in any European library. The Colophon found at the end of the present edition shows that it is derived from an original dated April 1532 of the Greeks (A.D. 1221) and written in the town of Caesarea by the deacon Basil, son of the Notary Public of Melitene. A year later it was carefully collated with the MS. from which it had been copied, by the physician Abu'l-Ḥasan.

[1] *Περὶ ἀέρων ὑδάτων τόπων*, §2, Vol. II, p. 14 of Kühn's edition.

[2] For a description of it see my *Catalogue of the Mingana Collection of MSS.*, Vol. I, pp. 1034-1039.

The last facsimile exhibits another colophon written by the present Syrian Patriarch of Antioch, Ignatius Ephrem I, who testifies to the accuracy of the text of the MS., which he kindly compared for me with another copy found in the town of Ḥoms (Emesa) in North Syria. As, however, the work is intrinsically difficult, many grammatical and lexicographical mistakes have crept into the structure of its text. In the footnotes I have corrected those which affect the meaning of the sentences, but, in order not to swell unnecessarily the number of these footnotes, I have refrained from drawing attention to all the diacritical and other minor grammatical and lexicographical inaccuracies which can easily be detected by an intelligent reader.

The author was certainly no stylist. I have referred above to a sentence of Ibn a. Uṣaibi'ah, to the effect that he was more versed in Syriac than in Arabic. If this Arab writer had known any Syriac at all, he would have added that, strictly speaking, Job was not what we call a *littérateur* either in Syriac or in Arabic. He seems, however, to have possessed a perfect knowledge of Greek, as is attested by his translation of the works of Aristotle and Galen.

In my translation I have kept to the original whenever this could be done without great injury to the right understanding of the meaning of a passage ; but in order to render the work more readable I felt occasionally compelled to deviate from the peculiar method of expression adopted by the author.

The numbers found on the margins of the translation refer to the pages (and not the folios) of the MS. reproduced at the end of the volume.

TRANSLATION.

We begin to write the "Book of Treasures," composed by the philosopher of the spirit and of the body, Mar Job of Edessa, the well-doer (εὐποιός) *and the head of the physicians, on the knowledge of the created things: on the fact that their division into the genera, species and individua of this world originated from the elements, through a natural process; on the nature of the heaven and the stars, and on the fact that they do not constitute a fifth element, but originated from the four elements; on the knowledge of the beginning and end of the world; on the resurrection; and on the heavenly kingdom and the fact that, together with the next world, it will have no end. (All these things are proved) by natural demonstrations* (ἀποδείξεις) *taken from the universe.* 77

The question whether the genera and species of the heaven and of the earth emanated from the first principles,[1] and whether or not their nature resembles that of these first principles, was subjected by the ancients to innumerable researches, with the result that contradictions (ἀντίθεσις) [2] increased among them. Some of them, in giving their opinion, answered in the negative, and some others in the affirmative, the question of the emanation of the created beings and the composition of their species from the moving course of the elements.[3] It is for this reason that we, who are devoid of wisdom, have raised our steps that walk on the earth, to the high mountain of the investigation of knowledge, with the hope that a certain weight upon the mountain [4] will make it come down to the level of our intelligence. We have trusted the One who bestowed reason on the rational beings, to impart through us, to the hearing sense (of our listeners), an unerring discourse dealing with the causes of the

[1] *I.e.*, the simple elements.

[2] Correct text accordingly.

[3] On the movements of the elements, see below, First Discourse, Chapters VI, XI, etc.

[4] Read *ṭūra*.

things which constitute the beauty of the universe. A detailed description of each species and of the characteristics with which it is or it is not endowed concerns others and not us. We only give the *αἰτία αἱρετικά*, that is to say, the origin of the division of the created beings into their different genera and species from the first principles—not that these are in themselves [1] the cause of the universe, as it is not permissible to the Adamic race (to think this), but rather that they are themselves from the One who is ever existing, and not from one who never existed—together with the functional ground for their existence, the latter not in every case, but only here and there. In doing this, in the measure that is possible for a created nature, we impart a fresh reality to that intelligible Being whom we shall behold in the hereafter openly and face to face.

It has been said: "The word of the wise man will delight you,"[2] and "Why did He then separate the wise from the unwise, as He separated light from darkness?"[3] The knowledge of the divisions of the genera and species, and the understanding of the universe, are, therefore, light, and ignorance of them is darkness.

In this world we know a few things out of many, while the next world, which will defy numbering by days and all that this involves, being one and the same, without change, and perfect—because there is no other world by comparison with which its imperfections might be rectified, as those of this world are rectified by comparison with the next—will impart a perfect light of knowledge, which will not suffer any increase or decrease, and will grant to those in the high stage (of knowledge) to shine with all the saints in the kingdom of the Father.

If a wise man is likened to light, and an ignorant man to darkness—as it is said: "Cast him into outer darkness"[4]—
it follows that the knowledge of the created beings is light, and
78 ignorance of them is darkness. We see the kingdom of heaven
as in a mirror,[5] as also we see that life of which it is written: "This is the true life, that they might know Thee the only God";[6] and because it is by His works that God is known to

[1] Remove the *Dalath* before *mān*.
[2] *Cf.* Eccles. x. 12.
[3] *Cf.* Eccles. ii. 13.
[4] Matt. xxii. 13.
[5] 1 Co. xiii. 12.
[6] John xvii. 3.

be God, and the created beings are His works, it follows that the knowledge of the origin of the created beings is the knowledge of God.[1]

If then the knowledge of God is life, and life is the kingdom of heaven—as it is said: "And they will go into life eternal"[2] —the knowledge of the origin of the created beings is therefore of paramount importance, and it is incumbent upon us to learn it. In the measure of our relation to it we shall become related to God. Let us therefore strive and endeavour to walk in the high path of the knowledge of the origin of the created beings, so that philosophy may be acknowledged to be not a shadow which prepares empty barns for those who are remote from knowledge, but, on the contrary, to be the knowledge of the existing beings as they exist in their relation to both God and man.

[1] *Cf.* Rom. i. 20.

[2] Matt. xxv. 46.

FIRST DISCOURSE.

CHAPTER I.

On the origin of the simple and the compound elements.[1]

EVERYTHING that exists either falls under the senses, or is perceived mentally. The simple elements, of which we shall speak first, as they are the first principles, are perceived mentally, while the compound elements fall under the senses.

The simple elements are heat, cold, dryness and humidity; and the compound elements are fire, water, earth and air. The compound ones are in no need of reasoning to be perceived, and because of this there is no question and answer about them, as their existence is self-evident. Earth is in the first layer, water is above it, air is above the water, and fire is above the air. The position of their layers testifies to us that they exist, and that they are four, neither more nor less. Indeed, the knowledge of the senses is in no need of witnesses. As to the simple elements, because they do not fall under the senses, the question may arise whether they exist or not, but we affirm that they do exist, and we prove it in this way: when we climb, as on a ladder, from the compound to the simple elements, we find that the compound elements are composed of two parts—fire, for instance, emanates from heat and dryness, water from cold and humidity, and the two others emanate from the very same things. If fire were only hot, we should not have said that it was compound, and if water were only cold, we should not have said that it was compound, and so also is the case with earth and air; but because we notice changes in them, we call them compound. When, however, we proceed to the parts of which

[1] On the simple and compound elements, see Aristotle, *De Generatione et Corruptione*, Book II, ch. 1-7.

they are composed, and see that these parts do not brook changes, we know that they are simple, because a simple element is one which possesses in the whole of itself the very same quality without change; while a compound element is one which possesses what it possesses through another. Heat being simply hot, and cold simply cold—and so also is the case with dryness and humidity—they have no change in them; they are simple
79 elements, and there is no other element before them. Indeed, if they were changeable in their essence, they would have emanated from other unchangeables;[1] but since there is nothing in this universe but themselves, the compound elements composed of them, and bodies composed of the compound elements, there can therefore be nothing else beyond themselves.

Since there was a time when the parts of the compound elements approached one another—as heat and dryness with regard to fire, and cold and humidity with regard to water—it is necessary to suppose the existence of simple elements, to the number of four, which in their essence do not brook any change. If one says that they do change, let him name us their changes, in the same way as we name the changes in the compound elements. As it is not possible for him to name a thing that does not exist, simple as well as compound elements therefore exist.

The characteristics of the simple elements are as follows:[2] heat is a simple element that expands, kindles, and is antipathetic[3] to cold; and cold is a simple element that condenses, congeals, and is contrary to heat. As to dryness, it is a simple element which hardens, contracts, and is antipathetic to humidity; while humidity is a simple element which is dilative and antipathetic to dryness. The same thing applies to the compound elements. Fire is an element that is both hot and dry, and water is an element that is both cold and humid, while earth is both dry and cold, and air is both hot and humid. The reverse of these definitions is also true, as follows: that which is anti-

[1] Text: "changeables."

[2] *Cf.* Aristotle, *De Generatione et Corruptione*, Book II, ch. 2-3.

[3] The word ܣܘܩܒܠܐ and the verbs ܕܠܘܩܒܠ and ܣܪܒ ܥܡ are often interchangeable. We have rendered them by "contrary," "antagonistic" and "antipathetic," according to the context. All express the Aristotelian doctrine of *contraries*.

pathetic to cold, which kindles, and which is expansive, has as its first element, heat. Heat therefore, is a simple element that expands, kindles, and is antipathetic to cold. In the same way the definitions of the remaining elements revert on themselves. These elements therefore exist.

If someone says that since the simple elements precede [1] the compound ones, therefore they existed before the latter were composed—because if they approached one another, they must have been remote, and if they did not approach one another, how did composition take place?—and that consequently the simple as well as the compound elements have a separate [2] existence:—

We will answer that composition occurs in two ways: one through priority of time, and the other not through priority of time, but through priority of order only. Priority of time affects all compositions of parts which existed, one before the other, in time, all of which parts were composed from other compounds which preceded them in time. This may be illustrated by the building of a house, which is composed of different parts, such as bricks and other things. These component parts precede the composition of the house in time, and they themselves were composed [3] of other things, such as earth and straw, which themselves were composed of yet other things. The same may be said of all other compositions in the category of priority of time which resemble this.

The simple elements, the composition of which into the compound elements involved only priority of order, and not of time, did not exist before the compound ones, as if they were themselves compound elements generated from other simple ones, because there is no other thing requiring time, from which the simple elements could be composed. If such things existed, where were they when there was no "where"? If you name
the "where," name a "where" comparable to that which we 80
name in the case of the nails and wood that precede a door.

In the same way as a line is composed of points [4] which exist simultaneously, and one does not precede another, and

[1] Read *ḳaddīmīn*.
[2] Read *lau* instead of *au*.
[3] Remove the final *Wāu*.
[4] Text has: "composed of four points."

they come together, and there is no question of priority or posteriority of time between the point and the line, as there is no preceding thing from which the points originated—and the points give birth to the line, and the line to the triangle—in this same way we may speak of the simple and compound elements. Their simple state does not precede their compound state, and there is no time between them, but in spite of that we give them priority of order.

CHAPTER II.

On matter.[1]

Having proceeded from the compound elements to the simple elements, and shown that the latter exist only in the mind, let us now speak of matter. How do we know matter? Is it through the senses? Some people have said, "Not at all, because it does not possess likeness, shape, quality, taste, tangibility, or any other things which are known through the senses." They assert that it is known through the mind only, (but we answer that) it is evident that the mind perceives the first principles and the things which are related to them by means of the senses. It is indeed possible for us to know through reasoning (συλλογισμός) the existence of a thing that does not fall under the senses. We have already shown that it is from the compound elements that fall under the senses that we know of the existence of the simple elements that are only perceived mentally. Since we see the working of heat and dryness in fire, and of cold and humidity in water, we infer their existence.

Now by what working is matter known? They say that it precedes the simple elements, as these precede the compound ones, for Aristotle taught [2] that there are three principles in the universe: matter, and two antagonistic forces, matter receiving

[1] The author is speaking here of the Aristotelian "matter" in its intelligible and abstract sense. Concrete "matter," designating a given object, is called by him "a body," and is spoken of at the end of the chapter. See Aristotle, *Metaphysica*, Books VII-VIII, etc., *Physica*, Book I, ch. 7, etc.

[2] See Aristotle, *Physica*, Book I, ch. 6-7.

these two antagonistic forces, which are heat-cold and dryness-humidity. These two antagonistic forces we perceive mentally, and as we infer the simple [1] elements from the compound ones, so it is necessary that matter should be inferred from the simple elements.[2] But how can we infer matter from the simple elements? What working does it give us? Not one, either through the senses or mentally, since it is neither hot nor cold, dry nor humid. We cannot therefore know it through the simple elements, in the same way as we know the simple elements through the compound ones. Indeed it possesses nothing that resembles them, nor has it any name to indicate its identity. From what then (can we deduce it)? From nothing? Absolutely impossible, because nothing has neither likeness, quantity, shape, quality, nor any other *summum genus* from among the ten categories. This applies also to matter. How then do we know it? From the fact that, as they say, it receives the antagonistic forces which are in it? We see in the universe that like is affinitive to like, and unlike is antipathetic to unlike; how, therefore, can matter receive things that are dissimilar to itself? [3] It should indeed flee from them; for if it is true to say that nothing receives antipathetic things repugnant to itself, then 81
matter should follow this rule. As, therefore, I cannot assert that a non-existent thing exists, I can apply the same with regard to matter.

Everything that is known is known either mentally or through the senses: the compound elements are known through the senses, and the simple elements are known mentally through the compound elements; as to matter, it is known neither mentally nor through the senses. The simple and compound elements exist therefore, while matter does not exist.

If you say that this name "matter" has then been used by men to no purpose, lo, matter [4] consists in anything, such as gold, silver, brass, iron, stones and wood, which receives different likenesses, shapes and accidents.

[1] Text "compound."

[2] This last sentence has been transferred, for the sake of clearness.

[3] *I.e.*, how can a thing that itself has no quality receive antagonistic forces which have definite and contrary qualities?

[4] *I.e.*, concrete matter.

We know that there are five terms used in connection with every created being: essence, matter, nature, species, and the particular quantities of heat, cold, humidity and dryness which combine to produce the semen. The matter of this semen, that is to say, its body, is something that possesses forms and likenesses, and its nature, called in Greek "φύσις," is the quality of these incorporeal forms and likenesses which, when belonging exclusively to an individual being, distinguish it as an individuum. This is how I understand the above five terms of which men make use.

We affirm, therefore, that this name (matter) denotes something which falls under the senses, and is a body apt to receive changes, such as gold, which receives the figure of a king, and may also be melted and then receive the figure of a crown, and innumerable other figures. The same is true of all other bodies. As regards the *substratum*,[1] the species, and the individua let us give as an illustration, man. He is an essence, because he is in himself and not in another, and his essence is one composition (namely) a body capable of receiving changes. We shall speak later of his nature, which little by little, through the essential increase of heat, *et cetera*, which he possesses, builds up for him constituent forms which distinguish him from other animals, and makes him a species having uprightness of stature, roundness of head, rationality, *et cetera*. We have written these things with reference to man's body, but the same process may be applied to all other animals and plants, of whose different genera and species we will treat.

CHAPTER III.

What are the elements—essences or accidents?

Having shown that there are simple and compound elements, and that matter does not exist, let us now, in a natural order of things, show what the elements are: essences or accidents.

We may state that all that exists is either essence or accident, and as we have attributed existence to the elements, they are

[1] I.e., substance or essence. For the meaning of this word see Aristotle, *Metaphysica*, Book V, ch. VIII; *Physica*, Book II, ch. I, etc.

therefore either essences or accidents. We may proceed to show that they are essences, from the definition of essence and accident. An essence is that which is what it is in itself, and does not change; and an accident is that which is what it is in another, and sometimes exists and sometimes not. Let us see now which definition is more congruous to the elements—that of essence or that of accident.

We see that heat is constituted of [1] itself, and does not change 82
or cease to exist; and so also is the case with cold, dryness and humidity. How do we prove this? We affirm that heat is not constituted of cold—on the contrary it is antipathetic to it—nor is cold constituted of heat, nor dryness of humidity, nor humidity of dryness. They are indeed antipathetic one to another, and because they are not constituted of one another, and there is no fifth element of which they could be constituted, it is necessary to affirm that they are constituted of themselves, and that they are, therefore, essences. If they are essences in their simple state, how much more will they be so when they are mixed with one another, and give rise to fire, air, water and earth, and thus fall under the senses? In this way fire possesses an essential identity, and so also do the remaining elements.

If you say: "Since the (elements) are liable to classification under the category of quality, and quality is an accident, because constituted of an essence, how then do you say that the elements are essences? Would you make an essence an accident, or an accident an essence?" We answer that the word "accident" refers to the way in which the elements are united in the body. When, for instance, we say that this body—that is to say, essence—has received heat or cold *et cetera*, it is not in the sense that it had not in it heat or cold, as it is constituted of them, but in the sense that these have been given to it in certain proportions. We call "accident" not the necessary cause of their addition (to the body), but the addition itself. When, for example, to a vessel [2] containing a hundred litres of water we add ten more litres, we call this addition an accident, because

[1] This expression may be translated more literally here and in the following sentences by changing "is constituted of" to "stands in." *Cf.* Aristotle, *Metaphysica*, Book VII, ch. 4.

[2] Read *māna*.

this same addition did not previously exist, and may be diminished. As to the water, since it was previously water, it was not an accident, because it was water—an essence—both before and after the addition. So also is the case with the elements from which water was composed. Now, how does water possess its essence? It possesses it from the fact that it took it from the simple elements from which it emanates, and which are essences. In the same way, heat is not an accident from the fact that it is found in an object that is hot; what is an accident is its quantity. It is in the same way as that in which the body, which is an essence, falls under the division of quantity, that the elements, heat, *et cetera*, which are also essences, fall under the division of quality.

(You may object): "The simple elements are not essences. Remove (from your thought) their first congealment into the four (compound) elements,[1] and their second congealment into the bodies of animals *et cetera*, which are from them, and show us the essence (of these simple elements), either mentally or through the senses. Consider their composition and decomposition.[2] The compound elements were composed of the simple elements, and the bodies of the compound elements. After the bodies have been dissolved into the compound elements, such as fire *et cetera*, where is the essence[3] of these bodies? After their
83 reversion to the compound elements, fire, *et cetera*, it became an accident and ceased to exist."

(We will answer): Lo, the elements are found both before and after the composition, and they did not come into being at that moment, as did the body which was composed[4] of them. They remain what they are both before and after the composition. After the composition, fire is still fire, and air is still air, *et cetera*. The elements only underwent a displacement, while their essence did not cease to exist as it is the very substance of their identity, although composition happened to the body by their coming together, and (decomposition) by their separation from one another. Essence is therefore found in the first principles before the composition of the bodies.[5]

[1] Remove the *Wāu*. [2] Read *sherāya*. [3] *I.e.*, its simple elements.
[4] Read *itherkheb*. [5] Change the first *Baith* into *Dalath*.

As to the nine classes of accidents which happen through increase, decrease, contraction, expansion, composition, dissolution, and others, and from which originate colours, tastes, sounds, smells, and other changes, we will speak of the modality of their coming into existence when discussing the different species of animals.

If someone asks what differentiates the heat and the cold *et cetera* which a body receives, from accidents such as blackness, whiteness, sweetness, bitterness, *et cetera* which the body sometimes receives, and which cease to exist in it at other times :—

We will answer that heat, cold, humidity and dryness only displace themselves, while colours, tastes, smells *et cetera* vanish.[1] Lo, bodies become hot by fire, and cold by water, and before a body received these they were self-existing ; and they only displaced themselves. Fire existed in an upper sphere, and water in a lower sphere. The bodies receive from them, but they, when they cease to exist in the bodies, revert from the bodies to their first state as essences.[2] Colours, however, tastes, and the rest, do not exist before they are received by the body. If they existed, where was their place and their position, in which they were self-existing like the four elements ?[3] There is no such place. Further, after they have displaced themselves, to what first essential state do they revert, as heat and dryness to fire, and cold and humidity to water ? Since they were not self-existing before and after the body received them, they are therefore accidents, and they vanish ; while the elements of the body remain, and only suffer displacement ; and they are consequently essences. Moreover, the body does not receive them in the same way as it receives accidents. This is known by the fact that if one of them suffers decrease or increase beyond the measure required by the body, the latter becomes sick from heat, or from cold, or from dryness, or from humidity, and we can heal it by making it either hotter or colder, *et cetera*

[1] Read *buṭṭāla.*

[2] Lit. "to the first principles." The author refers to the compound state of the simple elements.

[3] A similar argument against the theory that colours, etc., are essences is found in *Firdausu'l-Ḥikmat* of Ibn Rabban, p. 82.

(as the case may be). This is not so in the case of colour, voice, taste, or any of the remaining accidents.

If you say: "Water, fire, earth and air are not the only four elements, but there are many, because earth is both cold and dry, and possesses quantity, length, expansion and density"—and you would call all these essences, O new philosopher!—

(We will answer): If each one of these characteristics were an essence, why did they not acquire an individual existence in
84 themselves and possess a particular position, like earth, water, air and fire, every one of which possesses a particular position? Or why did not each of them become one element? And when they separated, why did they not give rise to as many elements as there are characteristics, but only to four? (The fact that they did not do this) shows that they are not essences, but they became divided according to the number of the essences. As to the four (elements) they remain always in their four-fold quality, and exist always. Lo, earth never changes from being earth, either in its nature or in its position, and this applies also to the remaining elements. Indeed earth becomes sometimes black, and some other times white; some of it is sometimes round, and some of it long; sometimes it is contracted, and some other times it is expanded, but it never ceases from being earth. It is never sometimes earth and sometimes not. This shows that unless a man wishes to be recalcitrant, like the one who calls night day, or a man an ox, he will not deny that earth is an essence; and the same applies to the remaining elements. The fact that earth is sometimes black and sometimes contracted, expanded, spherical, or long, is an accident, as we will show later at some length.

CHAPTER IV.

On the cause [1] *of the coming together of the elements.*

After having demonstrated the existence of the simple and compound elements, and the non-existence of matter, and the fact that the former are essences, let us now speak of their coming together. How can they, while they are antipathetic to one another—as heat is antipathetic to cold, and humidity to dryness—combine and become compound elements from which originate bodies, and the bodies themselves dissolve into those very elements, while all this happens by necessity? We will answer that there are two causes for any coming together of the elements, one concerned with function, and the other with existence. The functional cause of the coming together of the elements was that they might produce the genera and the species of all this material world, according to the will of the Creator. (The elements) have thus within their scope both the beginning and the end of the world; the number of years, months, days and hours; the changes of times; life and death; sickness and good health; and the length of time of this world is according to the measure which pleased the will of the Maker, our God. So far as the physical cause of their coming together is concerned, it is different, and it acts in this way: every happening that affects existence is brought about either by necessity or not. What is brought about [2] by necessity is effected by an agent, and what is not brought about by necessity is effected spontaneously. Now the happening that affects the elements is by necessity: heat does not come together spontaneously with cold, nor humidity with dryness; therefore this is brought about by another. It takes place by necessity, not of themselves, because if left to themselves, they would not have been induced to do things against their nature. There is therefore another Being beyond the elements, and that Being is God.

[1] Or "origin." The Syriac word means both "origin" and "cause," especially the latter. With the author both meanings seem to be interchangeable, although its translation by the word "origin" seems to be more appropriate in some sentences.

[2] Add a *Tāu* before the *Nūn*.

CHAPTER V.

On the fact that God exists, and that He is one and infinite.

God does not resemble the elements, which are four and
85 finite, but is infinite. And because He is infinite, He is one; and there is no limit and no end to the One who is infinite. And since He has no end and no limit, this applies to the whole of Him, and to every single attribute in Him. He is one, and has always been one, because anyone who becomes one, or became one, or will become one, is not truly one, but only relatively. If there is another one with the infinite, the latter will not be infinite, but because God was the only One from eternity, He is infinite.

After having shown in short terms that the first cause of the coming together of the elements was not from themselves but from another, who is God, and that God is one, let us now show that they are not without a beginning, but that rather they began to exist through Him, and that He created them from nothing.

CHAPTER VI.

On the fact that the elements were created from nothing, that they had a beginning, and that they are not infinite.

Everything that has a limit has also an end, and everything that has an end has also a beginning; and everything that has no limit has no end, and everything that does not possess an end did not possess a beginning.[1] It follows that since the elements have a limit towards one another, and are also circumscribed, they have an end; and since they have an end they had also a beginning. They are not therefore divine, without a beginning,

[1] See Discourse VI, ch. V, p. 271, where the author, speaking of the next world, asserts that although having a beginning it will have no end. This is explained to be possible owing to the particular working given to the next world by God. Compare this with Aristotle's doctrine of the divine nature of heaven. See Aristotle *De Caelo*, Book II, ch. 3, etc.

but they had a beginning ; and if they had a beginning, they are under time, and are circumscribed. Further, if their nature is limited—and by necessity it is limited—they possess the partition, limit and division of a limited nature ; and since division has a beginning, the nature of a divided thing has also a beginning. The elements, therefore, are not without a beginning.

The proof that they came into existence not from anything, but from nothing, is as follows : there is nothing in existence but God the Creator, the created beings—and, if you wish, also the intelligible natures.[1] The elements began, therefore, to exist, but their beginning was not from (the nature of) God,[2] who being uncircumscribed Himself does not give birth to a circumscribed being. Indeed nothing [3] from His nature extends to other natures, nor from other natures to His nature, because He does not receive any increase or decrease. If the elements came from Him, they would therefore be Gods, and He would be circumscribed like them. Because this [4] conduces to impossibility, and because there was no other thing from which they could have come into existence, they came into existence, therefore, from nothing, and they were consequently created.

We affirm that the functional cause of their coming together for the purpose of functioning is God, who created the world by His will, and who placed in them the cause of its [5] existence, which is their movement towards one another. How can their movement be the cause of composition ? We answer that two of the four elements are active, and two others are passive. The active elements are heat and cold, and the passive ones are humidity and dryness. The active elements have two movements, one contrary and the other not. The same thing 86
applies to the passive elements. Heat, for instance, has a contrary movement whereby it flees from cold, but it has also an affinitive movement towards humidity and dryness, that is to say, it is sympathetic to them and inclines towards them. In the same way, cold has one movement towards heat, and another

[1] *I.e.*, angels.

[2] There is in the text a small blank which should probably be filled by the words ܠܐ ܗܘܐ .

[3] Read *lā* instead of *illa*.

[4] Read *d-hādhé.*

[5] Text : their.

towards humidity and dryness, the latter being sympathetic and affinitive. Likewise, in the case of the passive elements, dryness is antipathetic to humidity, but sympathetic to heat and cold; and humidity is antipathetic to dryness, but sympathetic to heat and cold.

Concomitantly with their creation from nothing, the elements acquired an antipathetic and a sympathetic movement in their nature. This characteristic of their composition came to them simultaneously with their natural existence. As a sequel to what we have said (we may state that) the beginning of the two contrary forces consists in their fleeing from each other through the contrary movement, and in their combining [1] through the two sympathetic movements. If there were no contrary movements, there would be no existing beings and no composition.[2] Without these contrary movements there would be no dissolution and no corruption, and without composition, generation, dissolution and corruption, this world would not have taken its natural course. There would not even have been a world. God therefore made a wise arrangement in implanting two movements in each of the four elements, from which the composition of the compound elements originated; and from the latter the different bodies, genera and species of the universe have emanated.

CHAPTER VII.

On the reason why the movement of the simple elements did not produce one composition only, but four: earth, water, air and fire.

We affirm that composition took place through the contrary and through the sympathetic and affinitive movements. In the process of composition, what had affinity was drawn more closely towards its like, and what had no affinity was repelled

[1] Add an *Alaph* after *Tāu*.

[2] For the principle of *contraries*, see Aristotle, *Physica*, Book I, ch. 5, Book V, ch. 5; *De Mundo*, ch. 5; *De Longitudine et Brevitate Vitae*, ch. 3; *De Generatione et Corruptione*, Book II, ch. 2-3; and especially *De Caelo*, Book I, ch. 3.

more strongly, even when a balance between them and their affinitive objects was obtained. And because cold was drawn to cold, and heat to heat, they acquired a certain limit in space.[1] And because expansion, which applies to heat, is swifter, it lifted itself above and stood in a higher sphere. As to cold, because its contracted state is heavier, it sank down and established itself in a lower sphere. Their contrary movement, however, did not continue to separate them indefinitely, on account of the affinitive and sympathetic movement which they also possess, and which binds them to one another. This is the reason why they are placed in the figure found below in a kind of diametrically opposite position. In that figure (it is seen) how heat, which is at the top, is joined with both dryness and humidity, and also how cold, which is at the bottom, is joined with dryness and humidity. In this way two compositions take place between heat and cold, one of which is dry and the other humid, and between themselves they give rise to the four compositions : heat and dryness, which
is fire, cold and humidity, which is water, dryness and cold, which 87
is earth, and heat and humidity which is air. And there is an equal balance between heat and cold, and between humidity and dryness, so that heat may not lift itself unduly high, through excess of heat, nor cold sink unduly low through excess of cold, nor move unduly hither and thither through excess of dryness and humidity. They are joined together in one complete harmony, by the incomprehensible power of the Maker. Earth is not placed on anything, nor does fire hang on anything, but He, by His infinity, keeps their two movements, through the cause which He Himself implanted in them, according to the sign of His power, as in a kind of equilibrium, if we can use the term "equilibrium" in connection with His infinity and His incomprehensibility. While the elements possess both antagonistic and sympathetic and affinitive movements, there is also between them an even and well-balanced movement which takes from both (active) principles equally, and which has no excess of either of them. It has therefore neither increase nor decrease, and consequently neither corruption nor change. It is one single and perpetual movement, which is in itself a complete equilibrium.

[1] Remove *Wāu* before *Meddaim*.

This affects also the circle of the heaven, that is to say, of the stars which it contains, because, if the Creator vouchsafes life to us, we will speak also of it, after we have completed the discussion of the causes of the composition of this world of ours. We will discourse on why it will not change, why it is not without a beginning, as some people believe, how it is composed of the four elements, and why, in spite of the fact that it is composed of the four elements, it has in it no generation or corruption, and no increase or decrease, as it happens to-day in this world.

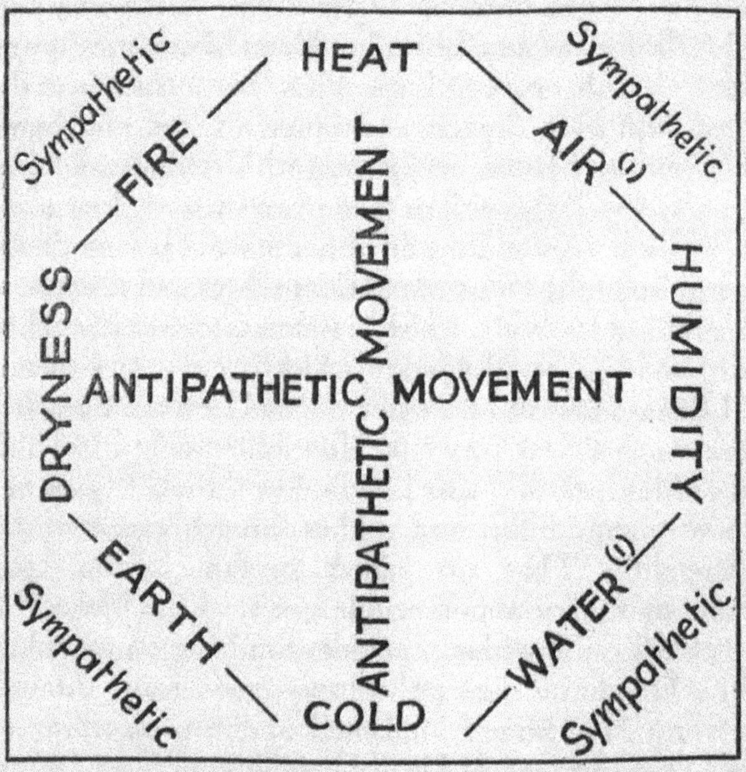

[1] In the original diagram, the Syriac word "water" is found in the place of the word "air," and vice versa, against the indications found in the body of the text.

CHAPTER VIII.

On the reason why the simple elements, which are only conceived mentally, give rise to compound elements, which fall under the senses, and have three dimensions.

We affirm that because the simple elements have expansive parts, they do not fall under the senses, nor do they possess three dimensions ; that they flee from one another by nature, and the more they flee from one another, the more expansive they become ; and that the more expansive they become, the more remote they become from tangibility, and the less do they possess the three dimensions. Take the example of dust and chaff of wheat, which, because they have expansive parts, flee from our vision when they mount up in the air, and also from the three dimensions. When, however, we collect them, bring their parts together and condense them or mix them with water, so that through the humidity of water their parts may stick together, they fall under the senses and receive a form. The same thing happens in the case of the simple elements, because of their characteristic of expansion. When their parts come together, through the sympathetic and affinitive movement, to form a
composition, they become condensed and fall under the senses, 88
and by falling under the senses they acquire the three dimensions.

Take a demonstration from the compound elements. The one which is more condensed, and the parts of which are more contracted, acquires to a greater degree the three dimensions, and falls also to a greater degree under the senses. Earth [1] is more visible because its parts are more dense, and it possesses the three dimensions in itself. After it, water possesses the three dimensions, not in itself, but in an earthen object, such as a vessel. As to air, because it is more expansive than water, it is not visible. As to fire, because it is more expansive than all, it is contained in the form of power [2] in other bodies, and this quality is not possessed by the other elements. In fleeing, in the form of power, from one object to another, it receives the three dimensions in itself, while the dimensions of water and

[1] Read *ar'a* (earth).

[2] Or : potentially.

air are limited by another object. In this way, it is not tangible except in the mind only. If this applies to fire, which is a compound element, how much more will it not apply to the simple elements?

If you agree with me [1] that the cause of the three dimensions is density, and the cause of density is composition, and the cause of composition is the two antagonistic movements, and the cause of these two movements is the simple elements (you will also agree that) there is in the created beings no cause for the First Cause, as there is no other being above Him. The cause of the created beings is their Creator, not in the same way as they are themselves one from another, and not as if they were from His essence,[2] but in the sense that having come into existence from nothing, they are as the effect of a cause. As to the Maker, He has no cause, because He is infinite, which implies that He has neither beginning nor end; and since He has no beginning and no end, He has no cause. Everything is by necessity from Him. He is to them the generating cause through which they had their beginning. The first cause, which is the functional cause, is likewise from Him, and the same may be said of the second cause, but He placed the latter in the movement of the four simple elements, from which were derived the compound elements, which in their turn gave birth to the different bodies which emanate from them.

In the same way as a king bestows power on four of his servants as he pleases, in appointing one to an eastern region, another to a western region, a third to a northern region and a fourth to a southern region; and in the same way as these servants give [3] a similar order to others who are under them, and these to others, until we reach the low scale of functionaries; and in the same way as all these are bound together to the king, as with one chain, and he himself is not bound to any other king—in this same way God implanted movement in the elements, one towards another.

After having spoken of the cause of the composition of the compound elements, and of how they came into existence from

[1] Text "if you say."

[2] Lit. "from Him."

[3] Read the verb in plur.

the simple elements, and of other things, let us speak now—according to σύνταξις, that is to say, to the order of things, and according to the requirements of the first causes—of the causes of the second composition of the genera and species of bodies which emanate from the elements, that is to say, of the genera and species of sea, land and air.

CHAPTER IX.

On the cause of the coming into existence of the genera and species of sea, land and air, from the compound elements; on how something that does not resemble the compound elements emanated from them, in the same way as they emanated from the simple elements; and on why not one genus came into existence, 89
but three genera.

We will discuss at the beginning the grounds for the cause of the coming into existence of the three genera [1]: animals, animal-plants and plants, and the reason why neither more nor less than three of them came into existence. We will first seek the cause of this in the elements, which have three orders in the arrangement of this world: the high order, the middle order, and the low order. It is necessary that, according to the three-fold order in which their nature is placed, three genera should have emanated from them, in the shape of their arrangement. This order is that of animals in the high sphere, plants in the low sphere, and animal-plants in the middle sphere, because the latter possess affinity with the moving animals, the reason for which we will show later. The reason for the number of the three genera is the following:

So far as the functional cause [2] of their composition is concerned, a triad functions at the beginning of every composition, as there is at the beginning no composition of a functioning body consisting of two dimensions only. Indeed the beginning of the composition of the bodies took place in three dimensions, and

[1] This sentence is complicated, but its meaning seems to be clear.
[2] See Chapter IV.

no single composed body can consist of one or two dimensions only.[1] The arrangement of this world is also three-fold: past, present and future, or beginning, middle and end. It is therefore with good reason that at the beginning three genera came into existence from the elements.

Let us now speak of the elemental origin of their existence. Two elements are light, and two heavy. The light ones have more expansive and rarified parts, and the heavy ones have more contracted, condensed and thick parts. Both the heavy and the light elements possess a three-fold order. Indeed, in the light elements there are parts that are lighter than others, parts that are of medium lightness, and others that are less light. When these parts that are less light are compared with those that are lighter, those that are less light appear heavy. In the same way, in the heavy elements there are parts that are very heavy, others that are less heavy, and others of medium weight, and when the heavier parts are compared with those that are less heavy, those that are less heavy appear light.

In the first movement, the light parts found in the light elements—fire and air—increased more than the parts found in the heavy elements of water and earth; and then animals came into existence above the earth, near to the light elements. In this same way, when the parts of the heavy elements of earth and water predominated in quantity over the light parts of fire and air, trees came [2] into existence, which are more earthy, and stuck to the earth. And to establish equilibrium between the heavy and the light parts, there came into existence the animal-plant, which feels and shrinks from fear like animals, and contains also something resembling fleshy matter, but is fixed in the earth, and like trees does not move from place to place.

The cause of the coming into existence of the three genera from the compound elements is therefore as follows: trees, because of the excess of earthiness in them, received from the
90 earth what it possessed—a dense and compact body; but because of the small quantity which they possessed from the light and hot elements, they had also growth and increase. As to animals, because in the very movement of the first composition,

[1] See Aristotle, *De Caelo*, Book I, ch. 1.

[2] Read *hwāu* (in plural).

the light and hot elements grew in them in excess of the heavy and cold elements, they had good growth and increase; and because they obtained also expansion and rarefication from fire and air, and especially from fire, not only did they possess growth, but feeling and motion from place to place were also added to them. Indeed fire took them upwards and lifted them above the earth, and this is the reason why they possessed motion from place to place; but, because of the earthy part which they possessed, their feet leaned on the earth, and they were drawn by weight downwards, while the heat and lightness [1] which they possessed from fire and air drew them upwards, through the walking action of the feet [2] which tends upward. This was accomplished through the medium of the animal spirit that was composed of the four elements, and was kneaded with the body from the light parts in the light elements, and from the light [3] parts in the heavy elements, as if it were composed of them. It is lighter than the body, and in the same way as a light object moves a heavy one, it moves and directs the body. We have spoken at more length of the elemental cause of the existence of the animal spirit, and of how it directs the body, in the book which we wrote *On the Causes of all Fevers.*[4]

CHAPTER X.

On the functional cause of the four humours: red bile, black bile, blood and phlegm, and of their coming into existence from the elements; and on how the body, the bones, the veins, the flesh, et cetera, emanated from the humours.

The functional cause for which the four humours which are in the body came into existence, is that they should act as intermediaries between nutrition and the bodies. In one way they resemble nutrition, and in another way the bodies, so that through this mutual resemblance they may receive nutrition, and thus

[1] Read *ḳallīlūtha.*
[2] Read *deḥāya* for *reḥāya.*
[3] See above.
[4] This work of the author is lost.

impart increase to the bodies, and so that in this way these bodies may possess growth, increase and existence (itself). If the bodies had come into contact with nutrition suddenly and without any intermediaries, they would not have received it well, because of the great remoteness of their respective natures. It is for this reason that veins, arteries,[1] nerves, *et cetera*, came into existence, in the same way as trees acted as intermediaries between the earth and fruits. Some of them came into existence for the service of the bodies, and some of them for the sake of helping and building up [2] the bodies : the veins for [3] service, the bones for building up, and the other constituent parts for similar purposes, which it is not necessary to mention individually, and so lengthen unduly our discourse.[4]

The elemental origin of their existence is as follows : when in the first coming together (of the elements) the light and hot parts mixed [5] with the heavy and cold parts, through the medium of dryness and humidity, a seething took place, on account of the antipathy of heat and cold mixing through humidity and
91 dryness, similar to that produced by fire under a cauldron of brass, which, through its dryness, receives the heat of the fire, and through its heat [6] expands the humidity which is found in the water, and thus gives rise to boiling, and to the change in the cooking of the food ; and, through the boiling, the hard and heavy parts (in the food) dissolve and join with the humid part, through the hot parts, and in this way the change that takes place in the two antipathetic parts remains in the humidity, which itself dissolves the dryness (found in the food), in the same boiling, through the power of the heat, and from them one chyle is produced. This same process worked with the elements (in connection with the humours). Seething took place, which gave birth to one chyle (from the four of them), having four characteristics similar to those of the elements. One of these characteristics was thinner than earth and thicker than fire, and red bile was produced from the thinner, drier

[1] By translating the word *shiryāna* as " arteries " we do not imply that the author knew the difference between *veins* and *arteries*, as we know it to-day.

[2] Read *ḳūyāma*. [3] Read *lewāth*. [4] Lit. " merchandise."

[5] Remove the *Wāu*. [6] Text " cold."

and lighter parts that it contained; and from the part of the liquid[1] that was thicker, colder and drier, black bile was produced; and from the degree of cold and humidity which it contained, phlegm was produced; and from the hot and humid parts which it contained, blood was produced. Each one of these was produced separately. This process also takes place in the cooking of meat in a cauldron. The part in it that is hotter, that is to say, the fat found in the liquid, rises separately after the cooking, and the froth and the substance of the meat are also separated. If that (chylous) composition had not taken anything from the first elements, it would not have undergone any change. In the same way as that which it possessed was received through heat, cold, humidity and dryness, which formed its nature,[2] in this same way it also received its changes from them, and thus they gave it their characteristics. Any one part of this composition shows in itself all the elements, which it possesses.

Although all its humours received generally from the boiling the quality of being watery and liquid, yet in their power and in their colours they differed from one another. In the same way as its humours were composed of the elements through the changes of cooking and boiling, by the action of heat mixing with humidity, so also the bodies of animals were composed of them, by the warming action of heat. When there was heat, the parts that were thin and light departed from them, and thick parts were formed, and by their mutual composition a body that was harder and more earthy than they were came into being. In the same way as when we cook a body, such as barley, *et cetera*, the first thing that is produced is a liquid and watery chyle, and when we cook afresh this chyle, so that its humid and thin part rises up, it becomes itself a body thick, hard and non-liquid—in this same way the humours became bodies in the process of seething. Four different humours exist in the bodies; and these same humours which are in us are formed of the bodies that had dissolved and had reverted, after solidifying, to a body which contains the changes which they possessed from one another.

When the humours underwent a change from their humid state, so that they might become bodies, the heavier and drier

[1] Lit. "humidity." [2] Lit. "formed it as it is."

92 part found in them contracted to itself separately, and was solidified, and became bones; and the colder and drier part which they contained was solidified, and became nerves, veins, arteries and tendons; and the hot and humid part that was in them gave birth to flesh; and the part of them that was fatter gave birth to fat and adipose tissue. In this way the constitution of the body was composed through the general combination of like things with like things.

CHAPTER XI.

On the reason why, while the composition of the body is one, it has not one movement or one shape—say either round or long—but different shapes; and why it has an external skin.

There are seven movements in the elements: upward, downward, forward, backward, to the right, to the left, and circular. The upward movement occurs according to the nature of heat, the downward movement according to the nature of cold, the forward and the backward movements, and those to the right and to the left, according to the nature of humidity and dryness; and the circular movement is effected by the combined movements of all working together. Indeed fire, because of its heat, expansion and lightness, tends to go upwards, and it causes in the body an upward movement also. And water, because of its coldness, contracts, becomes heavy, and causes a downward movement, while air, through its moisture, which is rarefied and expansive, causes an expansion forward and backward (and to the right and to the left); and earth, through dryness, contracts that expansion forward and backward and to the right and to the left, until it reaches a level position.

As to the circular movement, it is effected by the elements in the following manner: earth is cold and dry, but when the heat of fire overcomes the cold of the earth, through the medium of dryness, the cold of the earth flees and moves towards water, that is to say, towards the cold which the latter contains and

93

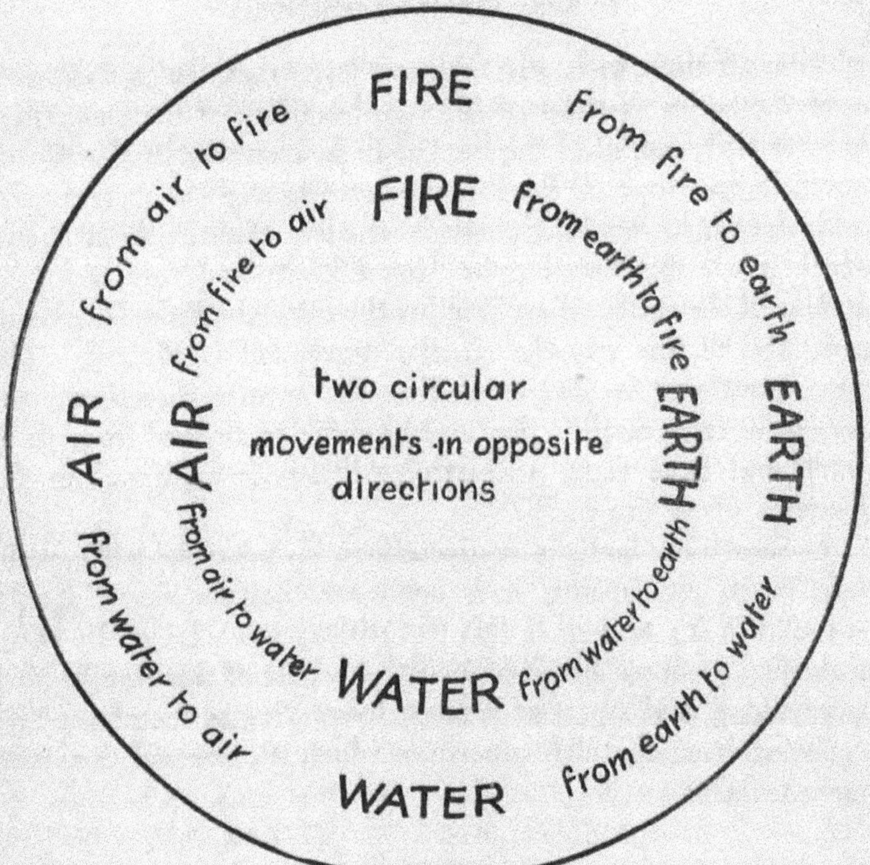

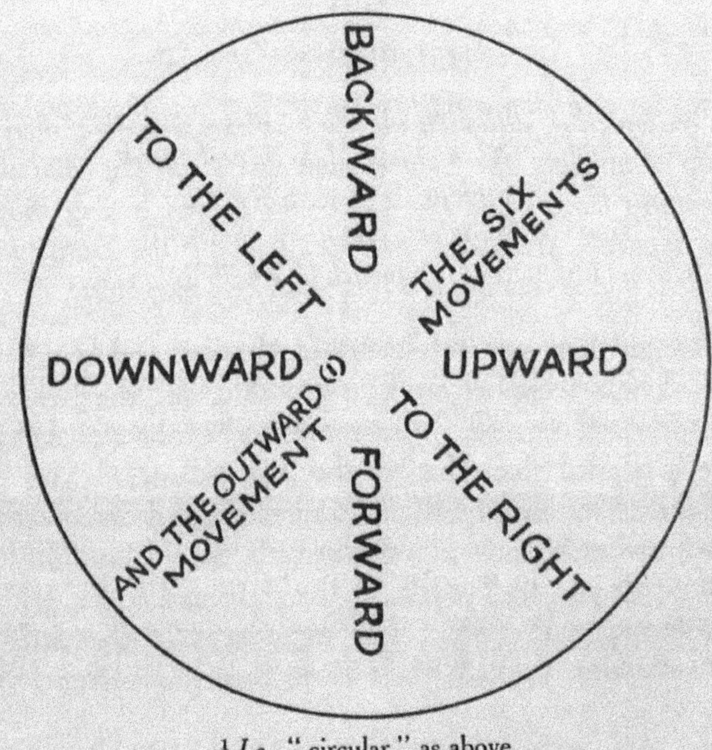

[1] *I.e.*, "circular," as above.

which is affinitive to it. In the same way, when the cold of water is overcome by the heat of fire, the humidity of the water flees towards air ; and when the heat of air is overcome by the cold of water, it flees towards fire ; and when fire itself is overcome by cold, since it has nothing above it which is affinitive to it, and to which it can flee, it contracts upon itself, and reverts to the dryness of the earth, where it fights the cold which the latter contains, and in this way the circular movement is effected. The same movement is also created in the reverse direction : the movement from earth to fire, and from fire to air, and from air to water, and from water to earth. All this is explained in the diagrams given (on p. 29).

Because our body is composed of elements which contain these seven movements, it is necessary that the seven should be found in it ; and since this (formation) was effected by them in the first composition, it follows that a part of this composition became long, and a part of it round, according to the power that emanated from the elements, from which it possessed the seven movements.

CHAPTER XII.

On the reason why, although all the movements were caused in the body of animals, the body of man and of all the quadrupeds is generally long, while the bones also possess a long shape and are separate from one another, but all the veins, muscles, tendons and nerves are one thick and long membrane.

We affirm that some elements are active, and some others passive. The active ones are heat and cold, and the passive ones are humidity and dryness. The movement of the active elements is more powerful than that of the passive ones. The movement of the active ones tends to go upwards and downwards, i.e. heat goes upwards and cold downwards, while humidity tends to go forwards and backwards, to the right and to the left ; and dryness contracts inwards. The movement of the active elements overcomes, through their more powerful natures, that of the passive ones. This is the reason why the length of animals overcame their width,[1] and this is the reason also why the bones, the veins, the arteries, the skin and the hide possessed a long shape, while, because of the weakness of the movement of the passive elements, they obtained their width from front to back, and the veins and other small membranes generally became round within that length.

CHAPTER XIII.

On the reason why bones, nerves, veins and arteries are one, from their root to their extremity, while the parts of the bones are separated.

In connection with the functional ground for their existence, the bones have been created to prop and build up the body. They act like foundations to the body. This is the reason why they have to be hard and dry. Indeed for the solidification of the building they must be placed with one another in such a way that they lean upon one another and have firm basis on the earth, and become as hard as stones which are placed one upon another, from bottom to top, and thus hold the edifice together and complete it, and make for the easy movement and the extension of the hands and the feet, *et cetera*. The veins and the arteries were created so that they might help towards the building of the
house, and act as aqueducts towards the feeding of its parts.[2] 94
The arteries help towards the ejection of vaporous matter, and the introduction of pure air to the movement of the heart, and the preservation of the animal spirit ; and the nerves towards the movements of the body and towards its feeling ; and the viscera towards the ejection of waste matter. This is the reason why they are not hard, but soft and humid, and form one whole, from their roots to their extremities. Indeed, if their parts had been divided from one another from their root to their

[1] The author is referring here to the length of an animal in an upright posture He explains later why animals lost their upright posture (see Chapter XXVIII).

[2] Cf. On nerves, arteries and veins, *Galen*, Vol. II, pp. 158-200.

extremity, they would not have been able to do their special work, together with other functions which they perform.

The elemental origin of their existence is the following: when the bones are compared with the other members of the body, they are found to be colder and drier, in the same manner as when the earth is compared with the other elements. The different parts of the substance[1] of the earth are also distinct from one another, on account of the excess of the dryness in them. This is the reason why, when different particles of earth are separated[2] from one another, each one of them stands by itself. As to water, on account of the excess of humidity found in it, it remains in the same place, in the same single mass, so that when we separate its parts in a level place, they run by necessity towards one another, until they become one again, because of their liquid state, through humidity. In this same way are the bones, veins, nerves, and muscles. The bones, because of the excess of dryness in them, have their different parts well separated from one another, but jointed to one another. Although they came into existence in round shapes, on account of the circular movement, as is the case with the vertebrae, the way in which they are placed one upon another is lengthwise. The veins, however, and the muscles, *et cetera*, because of the excess of the liquid humidity in them, became one whole, like water.

CHAPTER XIV.

On the reason why the bones are below, while the veins and the muscles are above, and why the bones, the veins and the arteries are hollow.

In connection with the functional cause, the bones became hollow so that they might contain fat marrow, which would moisten a little and soften their dryness, and so ward off injury caused by excess of dryness. The same thing happened to the veins, so that they might bring nourishment to all the body,

[1] Lit. "root." [2] Read *mithparshān*.

like aqueducts, and also to the arteries, so that through them the
animal spirit in the heart might be strengthened and invigorated. 95
The viscera also came into existence in the same way, to perform their function.[1]

The bones are placed below because they have no feeling. If they had been placed above, in an outer position, not being endowed with feeling they would not have fled from injuries, and the body would have suffered harm from dangerous accidents to it, and would not have profited by helpful occurrences. Further, if they had been in an outer position, they would have become still drier, and because of their thickness they would not have received air for the cooling of the body.

As to the elemental origin of their existence, it is the following: the bones came into existence in an inner position owing to the fact that their heaviness, dryness and coldness caused them to preponderate, as it happens to the four elements, any one of which that is colder and drier preponderates, as, for example, the earth, which sinks downwards. Such is the case with the bones of the bodies.[2]

The bones became hollow in the following manner: in the pressure which occurred in the first congealment[3] of the bodies from the humours, a substance came into existence from the four elements, for the purpose of feeding the different parts of the body. When these elements came together, each one of them received, in that very coming together, an impulse according to its nature. The air, when pressed by the (elements) that were heavy, cold and dry, jumped upwards, according to its nature, but the coming together of the other (elements) with it impeded it from rising unduly; and it jumped hither and thither, and in this way it created in them empty spaces, while they contracted upon themselves from the pressure exercised by the air. This is the reason why the members that are in an inner position in the body became hollow. The same thing happens when we fill skins with air. If we press it, it jumps hither and thither, in an attempt to escape, but because it is impeded on

[1] Lit.: "And the elements in the same way and the viscera came into existence."

[2] Read the genitive *Dalath* instead of *Wāu*.

[3] Read *baḳṭāra*

account of its being imprisoned in the skin, it inflates it. The same process took place in the bodies.

CHAPTER XV.

On the reason why the nerves are not hollow, like the veins, et cetera.

The veins take their root in the liver, and the arteries in the heart. Their root is in the centre, because there was in that place [1] a larger bulk of matter. There was much inflation by air, which exercised pressure on the membranes near it, and they became inflated. But when a small mass of the above matter reached the head which contained only air in the right measure of its capacity, that air which it contained inflated the cranium and made it spheric; but because there was a small quantity only of air, it was not able to return downwards and inflate the nerves which have their root in the brain of the head. Further, the nature of air is to ascend, and not to exercise pressure downwards.

As to the functional ground of the above, it is the following: if they were hollow, they would have been too weak to transmit the movement of the will; and the body is very much in need of this powerful movement.

CHAPTER XVI.

On the reason why there is marrow in the bones.

The great amount of humidity found in the bones came about in this way: when these contracted, it began,[2] according to its light nature, to rise upwards, but not being able to escape, because of their thickness, and seeing that they, like veins and arteries, *et cetera*, were hollow, it remained in them. Fatness, however, did not remain in the hollowness of the latter, because of the humidity of their nature, and on account of its lightness it leapt outside them, and there meeting with cold, it congealed. It

[1] Read *aika*.

[2] Read *sharyath*.

remained specially in a lower sphere, because it was deprived of heat. The congealment that occurred in this place is the fat which is harder and more compact than adipose tissue. Indeed, the adipose tissue is looser and softer. Both the fat and the adipose tissue emanated from the thick parts of the liquidity of the blood, and from excessive heat.

The reason why the marrow of the head, which is the brain, is larger than that of the bones, is, first that the place itself is 96
larger, and second, that a great quantity of humidity collected there through the rising vapours of the body, and remained there because of the width of the place. The same thing happens to an earthen vessel containing water. When it is heated by fire, the vapours rise upwards to its mouth, so that an alembic receives them, and they remain there and increase according to the width of the vessel. The same thing happened to the head, through the neck: when the humidity was moved by heat, it rose upwards and the cranium was filled with it and was inflated.

CHAPTER XVII.

On the reason why the apertures of ears, eyes, nose and mouth came into existence in the head.

With regard to the functional ground for their coming into existence, this happened so that we might receive through them from outside the knowledge of existing things: vision through the eyes, hearing through the ears, taste through the mouth, and smell through the nose. It happened also in order that they might be helpers to the body: the mouth in receiving nourishment, the nose breath, and each of the remaining ones in performing its respective function. This happened also so that they might drive away the waste matter that collects in the head, in order that no injury might come to it on this score: mucus through the nose, spittle through the mouth, mucoid secretion and tears through the eyes, and similar secretions through the ears, with other particulars which it is not necessary to describe.

The elemental origin of their existence is the following: when, in the first composition, the aerial matter rose to the head, its rising was impeded by the rigidity of the higher part of the cranium. This stoppage caused it to contract upon itself to all sides, in a circular way, according to its nature; and because by its nature it cannot go down, it underwent a great pressure, and fought against the bony matter. In this process it became fiery and hot, and leapt to escape; and it exerted pressure, and made for itself a way out through the four apertures: the nose, the eyes, the ears and the mouth. At the end of its escape, solid matter ran after it, according to the capacity of the mouth of the apertures, and stood outside them, but did not follow up the aerial matter, because of the expansive nature of the latter; and it made a kind of veil to those apertures: the pinnæ for the ears, the lids for the eyes, the lips for the mouth, and the alæ for the nose. This is the reason why, when vaporous matter becomes active, and collects in the head, it goes out through them; and if it were impeded there would be danger until it escaped, as this opening for escape has been cut for it from the time of the first composition.

CHAPTER XVIII.

On the reason why the apertures in the heads were made in front and at the sides, and not at the back.

The reason for this is that heat is more intense in the front than at the back. This happened in the following way: the heart, the liver, the gall-bladder, and the rest of the internal organs are placed nearer to the front than to the back, and, for this reason, when the vapour was heated and became active, it found for itself the parts that were in the front, which, because of their heat, were more expansive. Indeed the parts at the back are harder, because of their coldness. Since the organs of the front
97 resembled more its own nature, it made its way of escape through the front part.

CHAPTER XIX.

On the reason why the seat of the blood is near the seat of the red bile, the first being in the liver and the second in the gall-bladder ; and why the seat of the black bile is opposite the red bile, in the spleen, in a kind of diametrical position ; and why, with regard to its position in the chest, the phlegm is nearer to the black bile than to the seat of the blood, which is the liver.[1]

We affirm that the red bile is hot and dry, like fire, and the black bile is cold and dry, like earth. In the same way as fire and earth are in an outward position with regard to the other elements, so also are the organs which correspond with them.

As air is in the neighbourhood of fire, and water in the neighbourhood of earth, so the blood and the phlegm, the nature of which resembles theirs, had their place near the red bile and the black bile respectively.

The functional ground for their existence is the following: the black bile is on the left side, on account of the food going down to the stomach, so that the latter may digest [2] it through the fluid which comes to it and the acidity of which implants in it the desire for food. The function of (the gall-bladder) which is placed on the right is that when its fluid receives the food it may help its digestion by its heat. As to the blood, its function is to be sent from the liver through the veins to all the body, and in this way to nourish it and refresh it, like air. The function of the phlegm is to moisten the windpipe, so that it may not become consumed by the heat that rises towards the head; and also to moisten the vapours that rise to the brain, so that they may not suffocate it, and the animal consciousness thus be injured, and life itself perish.

[1] On this and the following chapter cf. *Galen*, Vol. I, pp. 278-378.
[2] Read *tishraiw* for *tikraiw*.

CHAPTER XX.

On the reason why the red bile was not placed in a higher position like fire, nor the black bile in a lower position, like earth, nor blood and phlegm in a middle position, like air and water.

We affirm that the elements differ from one another, and that when they united with one another, and became one mass, every one of them acted according to its nature: one of them rose upwards, another sank downwards, and the two others remained in the middle, because of their relative affinity with each other. In this way, one composition took place from all of them, but their union with each other was stronger in the middle; indeed, since the middle is greater than the two extremities—more bulk having occurred there—the humours obtained more help from each other there, and made their habitation in one place.

As to the functional ground for their existence, it is the following: the middle parts are large, and the outer parts are small; further, the outer parts are divided, while the inner part is united and compact, and it is there that the inner organs were fittingly placed, in order to feed all the body. Indeed, if the inner organs had been placed externally, the external part would not have con-
98 tained them, and in this way they would not have given food to all the body; and even if they[1] had done so, it would have been in a feeble way, and not[2] for long. If the liver had been placed with the stomach, and the bile above it, food would not have been digested; and if the spleen had been placed in a lower position, near the kidneys, there would have been no desire for food. It was with wisdom therefore that the Creator placed them in this middle position.

[1] Read the verb in plural.

[2] Insert the negative *lā*.

CHAPTER XXI.

On the reason why two movements were implanted in an animal, one of which is perpetual and involuntary—and this is the movement of the heart and of the arteries—and the other intermittent and voluntary.

We affirm that the functional cause for this is the following: the heart is the habitation of the animal spirit, and was placed in the middle of the body so that it might revive it. For this purpose it was in need of a perpetual movement, to direct the body in life, lest life should die out during rest. Lo, we see that when the heart is completely at rest, it is a sign of death. As to the voluntary movement, it takes place in the brain, in such a way that when the animal feels tired, it may take a rest, lest by perpetual movement its strength be weakened, and itself die. Indeed, we notice that when it is extremely tired, it is in great danger until it rests; and if the movement of the brain had been perpetual, together with the movement of the heart, the animal would not have gone on living even for an hour because of the two exertions which it would then have been called upon to make. It has therefore a perpetual movement so that it may live, and a voluntary movement so that it may be enabled to perform its natural and necessary work in this world.

Further, if the movement of the brain were perpetual, there would have been no beginning to those works which by necessity must come from rest to movement, nor would there have been an end to the works which by necessity must come from movement to rest. Indeed, if this were not so, food would not have been prepared, nor drinks, nor any other necessary act performed for the preservation of life, and the world would have been in vain, and its present purpose brought to naught. It was very fitting, therefore, that two movements were implanted in an animal.

As to the elemental origin of the existence of these two movements, it is the following: the arteries rise from the heart, and the nerves from the brain, and because the heart is endowed with a perpetual and essential[1] movement, the arteries that emanate

[1] Lit. "pertaining to its essence."

from it have also a perpetual movement ; the brain, however, having no perpetual movement, but only an intermittent one, the nerves which have their source in it and emanate from it have also no perpetual movement. This is the reason why animals have two movements, one voluntary and the other involuntary.

CHAPTER XXII.

On the reason why the heart is the only one of all the parts of the body to have a perpetual and essential movement.

In the first composition, in which the congealment of the body took place, after the four humours came into existence from the four elements, and took from those very elements a matter through which they thickened and gave rise to the body, a part of the fiery element stirred and predominated in that composition, and because of its strength in the middle it joined with the heart and endowed it with perpetual movement. No time intervened between the coming into existence of the humours and of the
99 bodies, but they came into existence together in that very same composition, in the same way as the simple and compound elements are simultaneous and do not precede one another in priority of time, but in that of order only. In our present state, our bodies come into existence from the humours in time, because the humours come from bodies, and bodies from bodies ; but at the beginning everything was still, as the first cause emanated at that time from the Maker, in whose nature there is neither increase nor decrease, neither beginning nor end. In the case (of bodies) He placed in them a cause, which is the elements, and this is the reason why they are naturally affected by increase and decrease. The heart, which has perpetual movement, came into existence from the fiery nature which predominated in it ; like fire, which takes effect in wood, and is seen in it with a perpetual movement. In the case of the wood, because the fire takes hold of it by accident only, it makes it move like the heart, to some extent only ; but because it was kneaded into the nature of the heart, the latter became endowed with a perpetual move-

ment. Indeed, fire decomposes and destroys the wood, but it does not give it its nature; but as to the heart, it constitutes it and gives it its nature. This is the reason why its shape is πυραμίς which means "fiery."

To follow up the illustration of the fire: when it mixes with an object, it gives it the semblance of a body with perpetual movement. The reason for this is that it kindles it strongly, and in this way it becomes endowed with perpetual movement, by the fire mixing with it and being kneaded into it, until it moves away from it. When, however, it mixes only slightly with it, it does not move it in the same way as it does when it becomes very hot.[1] In this very same way, because it mixed only slightly with the other organs (of the body) it did not move them; but as to the heart, because it mixed with it at the very beginning with an intense heat, it endowed it, like itself, with perpetual movement. Although the seat of the bile is in the gall-bladder, fire was not kneaded into its nature, on account of its dryness; but the high and thin part of the heart mixed with fire through its humidity.

CHAPTER XXIII.

On the reason why fire is endowed with a perpetual and essential[2] *movement, while the other elements have neither a perpetual nor an essential movement, but are moved by others.*

We affirm that fire is lighter than all the other elements and that its movement comes from its very essence. Its lightness, mixed with its great expansion, makes it rise above all the others. Indeed, every one of the elements obtains expansion, more or less, in the measure of its lightness, and is heavy in the measure of its compactness, and becomes all the less amenable to movement. Lo, the earth, which is more compact, and the parts of which are more dense than all the others, is by inference heavier than all of them, and in consequence is placed below all of them, and is more averse to movement. As to water,

[1] Add here *bāh*.

[2] I read *yāthānāi*, as above and below.

because it is more expansive than earth, it is lighter[1] than it, and is more amenable to movement, and this is the reason why it is placed above the earth. The air, in the same way, being more expansive and lighter, is more amenable to movement, and
100 is placed above both of them. Likewise, because fire is lighter than all the other elements, it is more expansive than all of them, and on account of the intensity of the expansion which it possessed in comparison with them, it obtained an essential movement, and was thus endowed with perpetual movement. Further, there was no other element above it, to impede and hinder it, in such a way that when that other element conquered it, it would be impeded in its movement, and when it conquered that other element, it would set it in motion.

What expansion gave in lightness and ease of movement to one element more than another, it gave with an increasing strength to fire, because of its high position; and it thus possessed, to the exclusion of the others, an essential movement. The other elements brook changes, while fire never changes. Earth is sometimes hot, sometimes cold, sometimes dry and sometimes humid, and the same thing happens to water and air; but fire is not sometimes cold and some other times humid. Because the other elements undergo contrary states, through change, they are precluded from being endowed with an essential movement, but because fire does not receive any change, contrary states are not found in it; and in this way nothing impedes its moving, and it is consequently endowed with an essential movement. It has the precedence over the other elements in movement, in such a way that it became endowed with an essential movement, for the reasons which we have stated. The other elements diminished in movement, inasmuch as they are moved by another; but fire became endowed with perpetual movement, because it does not exhibit contrary qualities, which would impede it, as they impede the other elements, which are sometimes in movement, and some other times not. This happens because there is no other element above it to impede it, as there is in the case of the other elements.

[1] Read *ḳallīlīn*.

CHAPTER XXIV.

On the reason why the brain is not[1] endowed with a perpetual movement, but experiences rest and stillness.

We affirm that the brain, when compared with the other organs, is soft and humid. The vapours that rise from the heart mix with its humidity, heat it, and give it a temperate constitution, in such a way that it becomes aerial and amenable to movement, like a musical instrument (ὄργανον). Since the soul is also contained in it, when the body wishes to accomplish the functions that belong to itself, and is in need of upward, downward and lateral movements, the soul conforms with the brain and, on account of its union with it, makes use of it as a kind of musical instrument. In this way the voluntary movement is strengthened by means of the nerves that emanate from the brain to the rest of the body. It has not been endowed with a perpetual movement, because it received the heat of its movement from another, and thus it does not pertain to its essence.[2]

We have spoken of these two movements at more length and in greater detail elsewhere. Here, because our intention is to write of things in a short way, we have refrained from amplifying.

CHAPTER XXV.

On the reason why all animals generally, such as man, ox, and horse, have feet and hands[3] in external positions, and why these are thinner and smaller than the trunk.

We affirm that the functional ground for the existence
of the above is the following: they are placed externally and 101
they are thin, so that they may fulfil the needs of the animal. The purpose of the feet is to walk and to bear the body above them, in order that the principal organs may not suffer injury

[1] Add negative *lā*. [2] Read *yāthānāi*.

[3] The words "foot" and "hand" may apply in Syriac to the whole of the leg and the arm respectively.

from their nearness to the earth; and also to carry them from place to place. If this had not been so, there would have been no movement from place to place. The purpose of the hands is to do all the acts that are indispensable to the needs of the body, such as sowing, reaping, building, tool-making, *et cetera*. If the hands had not been placed externally, they would not have caused movements in connection with the things that are useful for the preservation of the body.

The elemental origin of their existence is the following: in the very first composition of the body, a matter thickened in the middle for the purpose of the composition of the body[1]; and the two active elements, as we said above, caused a more powerful movement than that of the passive elements. In this way, length predominated over width. The matter which was in the middle, and was thick, was pulled upwards and downwards, and its thickness diminished little by little; and in the measure of its diminution, the different parts became thinner and smaller. This is the reason why the thighs are thicker than the shanks, *et cetera*, and the upper arms than the lower arms, *et cetera*; and at the end of all came the hand and its appendages, which are still smaller. The same thing happened to the feet. Because of the small quantity of matter in the hands and the feet, the bones which they contain became long, such as those of the shanks, thighs and arms; but since that matter weakened towards the end—because it diminished in quantity—it was divided, and fingers came out of the palm of the hand, and toes out of the instep of the foot; and when the outside air met that matter, it desiccated it. This is the reason for the division of thighs, shanks and toes. And because the final and external part became more desiccated, the appendages of fingers on the hands and toes on the feet came into existence in the final division. This happens also to the earth: when its external crust becomes dry, it gives rise to cracks. The palms and the insteps, however, were not disjoined, nor did they remain separate, but joined the main division, because of some humidity that stood there.

[1] See chapter XX.

CHAPTER XXVI.

On the reason why five fingers came into existence on every hand, and five toes on every foot, and why the thumb and the big toe are thicker and larger than the rest.

The functional ground for their existence is the following: they came about in this way in order that they might help one another. Indeed if there had been only one finger, it would have been very weak, and five were thus needed in order to assist one another. Some of them are long and some others short, and the thumb became thick and short, in order that it might be more powerful in comparison with the four others. This is the reason why it is also placed separately. Indeed when the four others get hold of an object their hold does not become strong until the thumb is stretched above them. The root of the thumb is also thicker than the root of each one of the others,
in order that it may have power and a strong grasp. The four 102
others are weaker than the thumb, because they are longer and thinner. When a long object contracts and becomes short and broad, it becomes steadier and stronger, and the contrary happens in the case of short objects becoming long. This applies to the four fingers, which are weaker than the thumb, because they are longer and thinner. This is especially the case with the little finger.

The middle fingers are longer, and the two which are placed on either side of them are shorter, and not so long, on account of the shape of the palm, which is indeed deeper in the middle, and its upper and lower sides are higher and above it; and because of their height and the depth of the middle, the middle fingers became long, so that through their length they might reach the end of the deep part of the palm. If this had not been so, when the hand was closed there would have been an empty space in the middle, and if we had taken with our hand an object that was either small or liquid, it would have escaped from our hand; and also, if we had pressed (the fingers) and they had gone down to the deep part, those of them that were on this side and that side would have lifted up, and an empty space would have been found between them, and thus they would not have accomplished their work well. Galen spoke at great length of the use of the fingers in his book *On the Function of the Organs.*[1] As to us, because it is not our intention to speak of the function of the organs, but of the elemental origin of their existence, we only speak of them in a cursory manner.

The big toe became also larger and thicker than the other toes, in order that when walking on the earth it might strengthen the others. This is especially the case when we walk in mud or in a slippery place. The big toe then presses down and becomes like a peg to the foot.

We may enquire now why hands, feet, fingers and toes bend inwards and not outwards, and can never be bent backwards. If they bent to both sides, their grasp would have become weak, and so the outward movement which they possess united with the inward movement, and they became more powerful by the combined strength of the two movements. Further, the back part, being devoid of fleshy matter, and not being hollow, is not useful for grasping things.

The elemental origin of their existence is the following: there are five elements in the composition of man and of the animals, the fifth element being the body, which is composed of them. When the middle part of the body was being completed, its matter was pulled upwards and downwards, and reached the extremities, and gave rise to the division of hands and feet; there each one of the elements struggled to separate itself from the others, and in this way the five-fold division of the five powers—that is to say of the four elements and of the compound that came out of them—took place. The body, which is composed of a quantity of matter greater than that possessed by any other single element, perfected the thumb, while the four other elements perfected the four fingers and the four toes.

The big toe is larger, taller and thicker than the other toes, while the thumb is shorter than the remaining fingers, for the
103 following reason: the body assumed more matter from the downward movement, on account of the heaviness of matter, and the

[1] Doubtless the *Περὶ χρείας τῶν ἐν ἀνθρώπου σώματος μορίων*. For the subject under consideration, see *Galen*, Vol. I, pp. 135-140, and 154-166.

ease of its movement downwards ; and in this way it made the big toe longer and thicker, in the place which is in the front of the foot in an inward direction (with regard to the trunk). This happened because there was a thick agglomeration of matter there. As to the other toes, because of their outward direction, and because of the diminution of matter there, the more they became remote from the big toe, the smaller and the thinner they became, on account of the thinness of the matter there.

As to the thumb, because the diminution of the matter that collected in the hand is even more accentuated than that of the matter collected in the foot, on account of the difficulty of an upward movement of heavy matter, and the ease of rising of thin matter, it did not receive as much matter as the big toe. The reason why the fingers became longer than the thumb and the remaining four toes, is that there was in the hand more heat than in the foot, and it elongated the fingers more than the toes. The little finger and the thumb became shorter because the heat in their position was smaller than the heat in the middle. The little finger is also thinner because the matter at its root is smaller.

CHAPTER XXVII.

On the reason why all the body bends forwards and not backwards.

The inner parts of the body are more humid than its outer parts, and this came about in the following way : the inner organs are at the front, and the humidity which they possess, and which passes through them, imparts to the front parts greater humidity, as it is near them ; and this humidity moistens the muscles, the nerves and the tendons which cause movement. As to the back part, because it is remote from the humidity of the inner organs, and is devoid of fleshy matter—with the exception of the thighs and the calves—it acquired a greater degree of dryness ; and because of the dryness which it possessed, it did not contract and bend backwards ; and because of the excess of dryness there, a tendinous matter was added, between muscle and muscle, which it made still harder, so that they might not add to the bulk of the back. The knees, however, can bend backwards, because humidity is greater in them in their back part than in their front part, as they received the humidity found in the back of the thighs and the back of the calves. The reason, therefore, why they move backwards and not forwards is because the front is devoid of humidity. Another reason is that the bone of the knee and the shin impede it from bending forward.

CHAPTER XXVIII.

On the reason why, at the end of the movement of elongation upwards, a round head came into existence, and at the end of the movement of elongation downwards our lower part became long, and the legs were made like a kind of triangle.

The functional cause of this is the following: the head became round because in it are found the five senses [1] which originate from it. This is the reason why this place is in need
104 of more power and help than the rest of the organs of the body; and a round and spherical shape is more powerful than the other shapes, because all of it constitutes a perfect unity, and is not, like the other shapes, composed of parts which lean upon one another, and each of which is in need of the help of its neighbour. This is the reason why the other shapes are weaker, and if the head were composed of them, it would have been in danger, arising from the movement of the senses in it, lest its composition should be dissolved. Therefore the head became round, in order to become the habitation of the animal spirit. It became also round so that it might avoid all injuries, through its circular shape, as it is less easily touched by things. A round sphere, when moving on the earth, does not rise easily when avoiding any shocks [2] (which it receives) from bodies. So far as a triangle or a tetragon are concerned, when they are moved on the earth, they are easily lifted up, and are easily pushed by bodies with which they come in contact. As to a sphere, because

[1] Read *reghshé.* [2] Read *deḥāyé.*

of the regularity of its surface, it suffers less shocks. This is the reason why nobody makes an axle in any other shape but spherical : because of its evenness it is not easily hurt.

Its elemental cause is the following : among all animals, man alone has an upright stature. We will discuss the cause of this later. When he extended upwards, he stood upright, and, in harmony with him, the matter which was in the middle also stood upwards ; and because he did not incline forwards, backwards or sideways, the head became round, not leaning to any special side. The rest of the animals, however, such as the horse and the ox, since their standing up was not upright, but leaned forward, their head also leaned lengthwise towards the earth, where its matter had leaned.

This may be explained in another way : when the body was composed and stretched upwards, the air that was contained in the middle was shaken and rose also upwards, in conformity with its nature, until it reached the extremity of the height (of the body) ; but when it was impeded there, and was thus unable to escape, it made a circular tour on all sides, because of the humidity of its parts and of the fiery nature that it contained, and in this way it inflated that place, and made the head in the shape of a sphere. The same thing happens to rainwater coming down to a place where there is water. The drop of rain goes down with force inside the water, and after it has sunk down, the air runs and penetrates after it ; and being pressed by the weight of the water, which wishes to keep it inside itself, it runs and struggles to go upwards, according to its nature, but being impeded by the water it creates round bubbles above the water. The air acted in this same way from the lower part of the body to its higher part, and thus made the head in a spherical shape. The same thing happens to the glass-blowers : when they blow a vessel, and the air reaches the outer extremity, it is checked there, and makes a round sphere.[1]

This may be explained in yet another way : the head being in a high position, the air serves it more than any other organ of

[1] Curiously enough the same illustrations to prove the cause of the roundness of the head, taken from the bubbles caused by rain, and the round spheres made by glass-blowers, are found in *Firdausu'l-Hikmat* of Ibn Rabban, p. 45.

105 the body, and surrounds it from all sides; and because the air, being humid, is one whole in itself, and since the particles of humidity make a circular tour upon one another, and have no special resting-place in any of their parts, the head became spheric.

The functional cause why the feet became long is the following: the legs did not become spherical in shape, because a spherical shape is not good for walking, as with such a shape a walking man might fall here and there, and the movement of life might then perish. (Man) became long, in the shape of a kind of triangle, so that he might stand up, and be fixed on the earth. The length of the foot was also made in a forward direction, because the movement of man is forward, and in this way the foot props him up when he walks, and even when he throws his weight forwards he does not fall. When he runs he leans towards his forward length, and moves the body and lifts it up above [1] the earth, while making a forward effort. The heel was made as it is so that the instep of the foot might become even, and so that the back part (of man) might lean on it, as his forward part leans on the ball of the foot and the toes. Indeed, if the heel of the foot did not hold and prop up the body, when man placed his instep on the earth, he would have leaned and fallen backwards. The four-footed animals did not need this, because, in the shock of the forward movement, the hind feet strengthen the fore feet, and the fore feet the hind feet, in the same way as in man the heel acts towards the ball of the foot, and the ball towards the heel. This is the reason why the quadrupeds are not in need of long feet. We shall speak later of the reason why their feet are generally round. The foot (of man) did not possess length on the right-hand side, because man has no sideways movement, and has thus been in no need of strengthening himself on one special side; and since the feet help each other in maintaining the right position, we can easily stand on one foot towards the left or towards the right.

The elemental cause of the legs is the following: the active movement that went downwards—because it was caused by cold—pulled down with it the thick and earthy matter more than the upward movement pulled it up, and according to the earthy

[1] Read *'al.*

nature which matter possessed, it made the legs in the shape of a kind of triangle, and long, because the parts (of that matter) contracted to one another, and man himself thus became a true triangle in relation to the earth. The matter of the body received this shape naturally, because its parts contracted towards one another, and created a division in the middle ; and the beginning and the end of the lines which it possesses form a triangle.

As to fire and water, because of their expansion, they did not form a triangle by their nature, because they are one whole. Indeed, the upward movement, because of its expansion and the evenness of its parts, did not contract towards itself. In the case of the downward movement, however, after its parts had run one after another, they divided themselves one from another, solidified into angles, and so completed a triangle or another shape.

The reason why the finger-tips are round, and the fingers themselves straight, arises from the fact that when an equilibrium was established between the pressure (of the elements) and the length (of the matter), the movement of the active (elements) predominated and made them long, but when the pressure reached the end (of the matter), it was checked, and caused a kind of hollowness at the end of the fingers, where it met with air ; in this way, according to the evenness of the parts of the matter, the finger-tips obtained a round form.

CHAPTER XXIX.

On the reason why nails came at the end of the fingers and toes, 106
and on their nature ; why hair and pores came into existence on all the body ; and why there was an outer skin.

The functional ground for their existence is the following : the nails came at the end of the fingers and toes, primarily in order to assist the hands and the feet. Indeed these two organs are external ones, and satisfy our daily requirements : the hands do all manual work, *et cetera,* and the feet perform all the movements of walking. In this way, when they come into contact with an outside body, the nails act for the tips of the fingers and the toes like shields which protect them from injuries that might arise from that body. They act also like coverings to the tips of the fingers and the toes, and through their hardness and rigidity they tie together those tips so that they may not crack in coming into contact with the injurious effect of the air. They serve also another purpose: they scratch the dirt from the bodies. When the pores are blocked, and a matter accumulates in the skin which causes itching, the nails scratch the mouths of the pores and open them, and through scratching they cause a heat-giving movement in that place, and in this way the accumulated matter is pushed outside. They serve also other purposes that are well-known and obvious.

The hair of the body is also useful for the following reason: any waste matter which collects under the skin, from the effect of heat, is driven outside through it. It serves also as a covering to the skin; in time of cold, it guards against cold, and in time of heat, it checks heat, so that these may not hurt the skin when they come suddenly into contact with it. This is especially the case with the head. The hair of the head serves also as an ornament and embellishment, *et cetera*.

The elemental origin of the existence of the skin is the following: two kinds of additional matter accumulate in the bodies from the digestion of food, one of which is natural, and the other accidental. The natural additional matter serves the sustenance of the body, which receives it for its growth and development. As to the accidental one, the body ejects it outside, because it is in no need of it. Now hear how both the natural and the accidental additional matter originate. When there is good digestion inside the body, food reaches all the organs, so that they may take nourishment from it. They take from it what is suitable and sufficient for themselves, and assimilate it. From this emanates the work that helps towards the movement of the increase of the body from childhood to manhood. The matter, however, of which they are not in need, and which is more than enough, they eject outside, as something superfluous. The same thing happens to water that flows on the earth. The earth first takes from it what it needs, and keeps it inside itself; but what it does

not need, it causes to overflow outside. In this same way, when the body receives the chyle, it takes from it what it needs, and
ejects outside what it does not need; and this is called 107
superfluity.

This quality is possessed by all trees and animals, and by everything that feeds, because when there is a superabundance of nourishment, there are superfluous things of which the body is in no need. Indeed trees exude a kind of moisture, such as frankincense and gum. Every organ also of animals ejects a superfluity peculiar to itself: the stomach, the stercoral matter that comes from the first digestion; the liver, those parts resulting from the nourishment of the second digestion; the small ducts,[1] the digested matter which is thicker than the above, and which shows itself in the urine. The same thing happens in the fourth digestion, which resembles that of the organs, in that in it also a superfluity occurs, which is ejected by the strength of the organs.[2] Indeed when these organs are weak, the contrary takes place, as that superfluity remains in them, and they become unhealthy from various kinds of diseases. As to the superfluity that results from the fourth digestion, it is more solid and firm, as it combines with its own strength that of the three other digestions; and since that superfluity occurs in an outside position, it becomes still more solid, and acquires firm congealment from that position.

So far as the hair is concerned, it has its origin in the humidity which was ejected towards the skin, and warmed by the heat which evaporates from the inner to the outer organs. That heat ejected from it the thin and dilative humidity, and then its different parts contracted and solidified into the skin. These same parts received then, little by little, appendages, which grew and elongated as they came out of the pores, and became thin in proportion to their bulk.

As to the pores, they occurred in the first composition and congealment of the body in the following manner: when the

[1] Lit. "veins," fibrous roots.

[2] This fourth digestion spoken of by the author relates to the superfluity ejected by every part of the body, after its nourishment through the first, second and third digestions. This superfluity, not being essential to the existence of life, is called "accidental".

parts of the body contracted to one another, the air was pressed by these parts which were coming together, and it leapt[1] from among them outside; and in its passage it made small apertures in the body of the skin. The inner pores are also caused in this same way.

As to the nails, they came from the superfluity in the bones, tendons and nerves, when this superfluity was ejected towards the finger-tips. Indeed, because of the nearness to that place of the end of the bones that are there, these nails possess a kind of solidity and humidity like bones, and insensitiveness like tendons, and thus resemble the nerves, the bones and the tendons. They are white in colour, like bones, but so that they may not be broken like them, they are malleable because of their humidity resembling that of the nerves and the tendons; and they do not possess any sensibility, on account of the bony nature which predominated in them, and also because they were superfluities, which had not obtained sufficient firmness in the congealment of the fourth[2] digestion. In the same way, hair does not possess sensibility, because of the superfluous character of the humidity that was exuded.

The skin came into existence in an outside position because, when the humidity of the outside position met the air, the latter destroyed the thinness which it possessed, and it thickened; and as a result its parts came together, solidified, and became
108 skin. The same thing happens when we cook grains of wheat, or other things: after they have dissolved and become chyle, if we leave them a short time exposed to air in a vessel or plate, the humidity rises above them and forms a skin in an outside position, in such a way that we can take it with our hands.

[1] Read the verb in sing.

[2] Add a *Dalath* at the beginning. About this digestion see above.

CHAPTER XXX.

On the reason why the body receives little by little quantitative growth and development, till about the age of thirty years, after which it remains in the same quantity without any development till the end, and none of its organs receives any development; while the hair and the nails receive development till the end.

We affirm that man undergoes four stages. The first is that of childhood, which is hot and humid.[1]

Why did this first stage become hot and humid?

When the first movement of the four elements occurred, in the first composition and congealment of the bodies, those among the elements which possessed expansion were lighter; and these are heat and humidity. They leapt to an upper position, and by their lightness overcame the dry and cold elements, because the parts of the latter contracted inwardly. This explains also why these became heavy. As it happens in the case of two runners, that the lighter of them outruns the heavier, according to the very nature and working of things, so it happened in the case of the elements, in their running through the bodies. The hot and the humid outran the cold and the dry, and obtained victory over them at the very beginning of the first composition. The first stage became therefore hot and humid, and consequently it frisks [2] and gambols hither and thither in a more accentuated manner; and on account of the dilative quality of humidity and heat which predominated, their parts expanded outwards, and consequently that stage became more amenable to nourishment and to digestion. Indeed, the same thing happened here as that which happens with matter, which, the more humid it is, the more readily it receives increase, imprints and shapes.

When dryness fought the humidity, and cold the heat that were in the body, it first happened that dryness overcame humidity a little before cold overcame heat.[3] Because the former are

[1] This seems to be irreconcilable with the author's statement below (Discourse II, ch. 5-6) that the age of childhood is cold and humid, and that of youth hot and humid.

[2] Read *daiṣa*.

[3] Text "heat overcame cold," but see below.

passive elements, the end of their fight must have taken place first, and in this way the stage of youth, which is hot and dry, came into existence. This stage received also quantitative increase, because of the excess of the expansion of heat. Indeed, this stage receives all the increases that take place through expansion, not, however, in the same way as the stage of childhood, since the latter possesses the expansion of humidity and heat combined.

Further, when cold fought heat and overcame it little by little, the stage of manhood, which is cold and dry, came into existence.[1] When cold and dryness predominate, the body does not receive increase and growth, because of the contraction in the parts of the former, and the body remains in a permanent state till the end.

As to the nails and hair, they did not come into existence for the purpose of the composition of the body, but they are (of the matter that was) ejected outside.[2] Indeed, from beginning to end (of a man's life) they are pressed outside, and flow
109 and emanate from that *accidental* superfluity which is not essential to any one of the stages, as the *natural* superfluity is. The former superfluity assumes its natural and final limit in the (different) stages, but does not impart any increase (to the body), while the latter possesses usefulness and purpose, and has no limit assigned to it.

The functional ground for their existence is the following: the body possesses a beginning and an end, increase and decrease. This is the reason why it grows until it reaches manhood, and diminishes until it reaches old age. If it had received increase continually, there would have been no end to its growth, and in this case it would not have grown old, but would have become immortal.

[1] *Cf.* Aristotle (*De Longitudine et Brevitate Vitae*, ch. 4) who states that old age is cold and dry.

[2] See the preceding chapter.

CHAPTER XXXI.

On the reason why in the first congealment and composition of the bodies of animals, a perfect composition took place consisting of a complete state of humours and organs, while now things do not happen in this way, but an animal comes little by little to formation, birth and defined stature.

We affirm that there are two causes in the composition of animals ; one of them emanates from the Creator, and the other from the elements and the parents. The one that emanates from the Creator does not brook change or decrease, on account of His strength. Indeed there is nothing antagonistic in His nature that would impede or check [1] Him when He wishes to make something ; nor does He do anything by degrees. As to the parents and the elements, because of their weakness and the antagonistic forces which they possess and which impede them, they complete their work by degrees. It is in the latter way that all the created movements, whether celestial or terrestrial, act. In the case of our adorable God, things come into existence concomitantly with His wish, and there is with Him no time which can be called first or last. He, therefore, completed the first man without any diminution, at the same time that He made him. In this way He made also all that He created. We accept by faith—and we do not doubt it—that He finished all His works in six days, but (we must assume) that He did so because of the weakness of our nature, wishing to lead us up gradually to the things which He made, by means of things which resemble our own works, which are accomplished little by little, and in well-defined days.

There are many things in the body which we have left unexplained, so that our discourse may not be too long, (but we shall speak shortly of the emotions), and if all rational natures accept as an axiom the principle which we have given concerning the effects of the elements, they will find in it the movements which give rise to these emotions.

We will write now on the composition of these different

[1] Read *wampakkar*.

emotions, such as anger, pain, fear, *et cetera*, which emanate from the bodies of animals through the movements of the elements. We will show that these (emotions) fall within the definitions (of the elements). Indeed anger consists in an outward movement of the body wishing to avenge itself on a thing that is harmful to it, through the movement of the hot and dry elements which are in the body. As to fear, it consists in an inward movement of the body dreading what is harmful to it and fleeing from it, so that it may protect itself from injury by means of the cold or
110 dry elements. Pain consists in an inward contraction of the body unable to fulfil its desire. Courage [1] consists in an outward movement of the body of the animal when not thwarted, and not hesitating in its desire to check what is harmful to it. Joy consists in the kindling of the natural heat and humidity of the body of the animal after it has fulfilled its wish or obtained some gain for itself [2] or escaped injuries. Desire consists in a movement of the body wishing to satisfy its wants, or to rid itself of unnecessary things, or to accomplish the things which it pleases. Weeping consists in an inward contraction of the body of the animal by means of cold and dryness, and in the union of these with the heat and humidity that are found inside, and their rising through evaporation towards the head, and then their flowing out through the eyes. Generosity consists in the desire of a man wishing to give his possessions to others with joy; while avarice or greed consists in the desire of a man wishing to take the things that belong to others, and withhold his own. Shame consists in an inward contraction of the body when a man is rebuked, or is seen doing things which are not lawful; and obstinacy consists in his scorning of rebuke, his recalcitrance to good things, and his transgression of duty. All these things take place when increase or decrease occurs in the movement of each one of the elements, according to the outside occurrences which move them. If, however, we were to speak of all the effects caused by them, our speech would be drawn out endlessly.

[1] Read *lebībūtha*. [2] Read *d-hāwain*.

CHAPTER XXXII.

On the reason why all animals when in pain experience an inward contraction only, while man alone weeps.

We affirm that this happens to man through two causes, one of which arises from his form, and the other from the superfluous matter.[1] So far as form is concerned, man is the only animal to possess an upright posture ; and when heat and humidity collect internally, heat flees, in conjunction with humidity, from all sides of the body, and rises upwards, according to its nature, until it reaches the head ; and since it can obtain no outlet in that place, on account of the cranium and of the contraction that takes place there with all the body, it becomes compressed, and finding no other outlet for the compression that it is experiencing than that of the path of the factors of sight, which are in the direction of the eyes, it follows them, and issues out of the eyes ; and it is then called tears.

As to the irrational animals, because their forms are not upright, but their length stretches along the earth, when the hot humidity which is in them is compressed, it does not rise to the head, but expands through all the body, and spreads in equal quantities throughout all the organs ; and thus, not having a thick accumulation in the head, it does not cause any tears, and consequently no weeping takes place.

So far as the second cause, that of the superfluous matter is concerned, it acts in this way : man makes use of numerous and different kinds of food, and also of many varieties of wines and of
the moist food of vegetables and fruits, of which other animals 111
do not make use. On account of the above superfluous matter a kind of heat-humidity that is dilative, thin and amenable to movement, is produced inside him, and when this is pressed by the contracting effect of the inward movement towards the body, which emanates from cold and dryness, it readily rises up towards the head.

Why does that (hot) humidity flow through the aperture of the eyes, and not through the aperture of the ears, mouth or

[1] Lit. "increase of superfluity". See above Chapters XXIX, XXX.

nose? We answer that from the first composition, humidity went out of the head through the nerves, to the eyes, because of the affinity which the same humidity which rose (from the body) had with the natural humidity (of the eyes). Further, it strove to flow towards the eyes, from the fact that the eyes are hotter than the ears, the mouth and the nose. Indeed in our classification we call heat a hot and bright element, and it is through its light that the eyes see. That humidity, therefore, flows only towards the eyes. The thick part of it flows towards the mouth and the nose, and comes out in the form of mucus and spittle, and the thin part of it comes towards the eyes, and never towards the ears, because of the cold and dry humour that the latter possess, since humidity flees from these antagonistic elements.

Tears became salty, because of the humidity that was parched by heat at the time when it was imprisoned[1] inside and compressed.

Why do some people, while weeping, sob, and some others make a loud laughing noise? We answer: when pain is accentuated by cold-dryness, this compresses humidity-heat inside, and some of the latter rises up, through evaporation, towards the head; but owing to its great quantity, all of it is unable to escape through the eye or through regular breathing, and people are obliged to eject it by violent breathing. This is the reason why they sob and make a noise. When, however, that heat is extremely powerful and sharp, and that humidity is thick and cannot easily escape, they experience a loud laughing noise and shaking of the head and of the feet, so that the humidity may escape through the mouth, the nose, the eyes, and the pores of all the body; sobbing and crying aloud occur together with the above laughing noise, especially among people with a hot temperament, because heat and humidity predominate in them, and they make a kind of seething noise, as does water when unable to escape from a well-constructed and well-closed pot.

[1] Read *ithḥabshath.*

CHAPTER XXXIII.

On the laughter[1] *and cry et cetera*[2] *of each one of the animal species, and how they come into existence from the elements.*

We must write at some length about the laughter of man, his sense of shame, his pride, his sleep, and his hatred; and about the neighing of the horse, the barking of the dog, and other distinctive characteristics that accompany each species of animals.

It has been said by the ancients that laughter is a characteristic exclusive to man, as neighing is to the horse, and that each 112
animal species possesses a distinctive characteristic of its own. Although these are exclusive characteristics, yet in the state in which they are found in each species, they necessarily emanated from the same elements. As witness to our assertion is laughter in man, which is accomplished through the seething of the body, and (causes) joy, happiness and suffusion of the cheeks. In the same way the neighing of the horse, with its concomitant heat and seething, denotes its provenance from the elements.

How do anger, passion and similar emotions belong generally to man, horse, dog, and other animals, while the elements which cause these emotions have no exclusive peculiarity of this kind?

We have shown above that the elements possessed seven movements in this world—upward, downward, to the right, to the left, forward, backward, and circular. Six of these movements—that is to say, the straight movements—predominated, some through heat, some through cold, *et cetera*, and caused anger, passion, pain, fear, together with other emotions. Here we shall speak of their predominance in different circumstances and of how the circular movement, even though its circumference may be increased or decreased, does not receive in its parts a change or deviation, but even when it is increased or decreased its movement is circular, and its parts do not undergo any change as those of the straight movement do. Indeed, when, according to this (straight movement), we bend[3] and incline to

[1] Read *guḥḵa.*
[2] Change the first *Dalath* into *Wāu.*
[3] Text adds here "forward."

the right side, the movement to the left side is brought about, and when we incline to the left, the movement to the right side is brought about, and when backwards, the forward one is brought about, and when forwards, the backward one is brought about. The circular movement, however, has none of this, and if any one element had predominated in it, it would have obtained a straight movement, and all its sections would have altered. This is the reason why through the balance existing in them these sections have a kind of peculiar working, which does not resemble that of any predominating single element. This working is called exclusive, not because of heat which predominates, nor is it accomplished through a predominating humidity, cold or dryness, but by the equal participation of their parts. This equality between the parts gave rise to distinct characteristics. In the same way as a circle possesses a movement which is distinguished from the straight movments, so it possesses a distinct and exclusive characteristic which is different from the characteristics of the other (movements).

What is the reason that man, horse, *et cetera*, have not the same characterisitcs, while the circular movement is equally balanced in them? We answer that each species possesses a greater or less proportion [1] of the elements. If this were not so, all the species would have had one form, one shape and one temperament; but since they are different from one another, this difference emanates from the fact that they have a greater or a less proportion (of the elements) in a way that is peculiar to them, and which cannot be separated from them.[2] The different species came into existence in accordance with the quantity of those proportions, and the circular movement came also in
113 this way to each species. Indeed we see that the more the parts of the bodies of a species are even, the more even also is the circular movement in it, and the less its parts are even, the less even is the circular movement in it, and when it is of a moderate evenness, the circular movement in it is also of a moderate evenness. Through this decrease and increase (of the elements),

[1] Lit. "some increase and decrease."

[2] Or "which is similar to that found in individua when compared with one another."

the species are differentiated from one another, and the variations in their circular movement are in the measure of the variations of the quantity which they take from the elements. Although, therefore, the circular movement is common to all—I think even to the first species—yet because of the degrees of the proportions of their borrowings from the elements, each one of them acquired a distinctive peculiarity which does not resemble that of another, in spite of the fact that they obtained this working from one balanced circular movement.

The above applies also to the medicinal herbs, and even to the stones, in their relative borrowings from the elements. Indeed some of the medicinal herbs cool, some of them heat, some of them moisten, and some of them desiccate. This characteristic is possessed by each one of them in a distinctive way, according to the measure of the power [1] (of the elements) which it possesses. The cold and dry ones among them astringe and bind bodies, and the hot and humid ones loosen and expand bodies. This is the reason for the medicinal mixtures. The cold, dry and contracting medicines astringe the stomach, and the hot and humid ones loosen it. Contrary to this rule, however, some medicinal herbs, although cold and dry, loosen the stomach, like chebules ; and some others, although hot and dry, also loosen the stomach, like spurge (εὐφορβίον) [2] and scammony (σκαμμωνία). Some others are cold and humid, and cause vomiting, such as the seed of *atriplex hortensis*, and some others are hot and dry, but are vomitive and astringent, such as cummin ; and yet others are hot and dry, but neither loosen the stomach nor bind it.

What is the reason for all these different effects while the moving power is one? We answer that chebules, through their coldness and dryness, astringe, press and move the stomach. They do this when they reach the dilative and humoral humidity which is found in the stomach,[3] which they compress

[1] The words "power" and "agent" as used in this chapter refer to the four simple elements. Further, the words "active power" and "passive power" refer to the active and passive elements respectively. *Cf.* Aristotle, *De Generatione et Corruptione*, Book I, ch. 6, etc.

[2] Change the *Kaph* into *Baith* in the text.

[3] Read *isṭumḳa.*

and force down. Pomegranate peels, gall-nuts, and the rest, do not act in this way, although their coldness and dryness is very strong, and they are powerful astringents. The reason for this is that chebules contain some active and thin power of cold, and through the thin parts which it contains, it presses downwards, according to its nature; a thing that is not found in gall-nuts, pomegranate peels and the rest. As to pomegranate peels, they contain more of the thick parts which come from the passive powers than do chebules, and possess also very few parts of the active powers. This is the reason why chebules loosen the stomach, while the others do not. As to scammony and spurge, and other herbs which resemble them, they are hot, and emit milky juices, but in their power they are dry, because the air has ejected from them the thin and humid parts. They do, however, contain a thin and cleansing power, due to[1] some thin parts of the hot agent (which they contain). This is the reason why, when they reach the body, they sink down through their thinness,
114 and do not press, but cleanse and purify the red bile, which resembles them by its thinness; and when they cleanse, because of the humidity that is imprisoned with the red bile, this humidity goes down, and is evacuated.[2] As to colocynth, it is hot and dry, and its parts are thicker than those of scammony. The latter has milky parts, while the former is hard and thick. This is the reason why, when it penetrates into the body, the bile, which possesses thin parts, opposes it on account of its thick parts, and does not give it opportunity for purgation. When, however, it meets with something the parts of which are similar to its own in thickness and in colour, that is to say, phlegm, it mixes with it through its thickness, and moves it through the sharpness that it possesses from its heat. It scrapes it, and then, after it has been separated from the organs, it sinks down, according to its nature, while accompanied by its power.

The reason why black hellebore, and herbs which resemble it, purge especially the black bile, is the following: this plant contains an active power with its dryness and thickness even more than colocynth; that this is more watery than black hellebore its very composition testifies; and when an object meets

[1] Lit. "and also."

[2] Read *mithnaphṣa.*

another object which is affinitive, it evacuates it. We have already said that there are in the active (elements) thinner, sharper and more mordant parts, which cause the movement of purgation. Peppercorns and ginger, however, in spite of being hot and dry, do not loosen, as scammony and the rest do. There are two reasons which account for this, the first of which is the one explained above, which deals with hard and thick parts. The second is that although they contain more active powers that are thin than passive ones, yet they do not grip the stomach, nor do they loosen it, because they have only a medium strength, both in their activity and passivity.

Why do some hot medicinal herbs cause the upward movement of vomiting, and why is the same thing also done by some cold medicinal herbs, while some hot and some cold ones cause a downward movement through the anus? We answer that both those which move upwards and those which move downwards do this according to the active powers and the thin parts which predominate in them, whether they are hot or cold. They do not move upwards except when the passive powers predominate in them more than the active ones. Indeed, some of them contain more active powers and some of them less; those of them that contain more active powers have a stronger movement, and those that contain less have a weaker movement. This is the reason why, when they are drunk, and go down to the stomach, they become its[1] enemies, and fight against it, and those of them that are stronger in the active power win a victory over the power of resistance of the stomach, which weakens; and then they cause a downward and cleansing movement; and those of them that are weak in active power, when they go down to the stomach fight against it, and the power of resistance of the stomach prevails, and it repels them backwards; and in the fight that ensues they get hold of the matter (found in the stomach) and seize it; and after they have been repelled, they drag that matter with them upwards. An illustration of this may be found
in a house which has two doors. When the enemy enters through 115
one door, if the owner of the house happens to meet him, he will repel him backwards, if he is stronger, but if he is weaker,

[1] Text "their."

the enemy will prevail against him and will take with his own hands from here and there any of the possessions of the owner of the house that he can find, and will then jump and go out of the next door, because of his strength. In this same way, when medicines enter through the mouth, which is the door of the body, into the stomach, if the latter is more powerful it will repel them backwards, but if it is weak they will cause a movement downwards.

There is sometimes a voluntary vomiting, and sometimes a compulsory one. The reason for the latter is that when food or drugs enter the stomach, they fight against one another, like enemies, and the stomach repels them, either because they are too heavy for it, on account of their great quantity, or because of their non-fitness; and in this way compulsory vomiting takes place. As to the voluntary vomiting, it takes place when we introduce our finger to the root of the tongue and to the head of the gullet, and tickle this head of the gullet, which communicates with the stomach through its lower part. When this titillation takes place, this lower part rises to help the upper part, and in doing so it drags with it, through a sudden upward spasm, the food that it had brought to the stomach, and drags it through the expelling power to where it emerges outside.

Through titillation there is a movement that passes through all the body. Not only one place is tickled, but other places as well. For instance, when the arm-pits or the throat are tickled, the body shakes and experiences spasms hither and thither, in such a way that it sheds from itself its coverings. This happens in the act of vomiting: the stomach is shaken, and expels the food.

Let us here put an end to the first discourse.

WE BEGIN TO WRITE THE SECOND DISCOURSE OF THE BOOK.

CHAPTER I.

On how we demonstrate that heat and cold are active powers, while humidity and dryness are passive ones.

WE have made mention several times of active and passive elements. We must now show the way whereby we know them. If someone asks: " How do we know that heat and cold are active, while the humid and dry elements are passive? " we will demonstrate this to him in the following way: first by the method of a blind man—through feeling:

When we put our hand on a very hot body, it will immediately cause in us a very painful sensation. The same thing will happen with a very cold body. But if a man puts his hand for the whole day in or on a dry or humid body, he will not feel any pain. From this we know that the former are active, and the latter passive.

We will prove the question in another way. If all the elements were active, how would the bodies have been composed in the way they are? And if all were passive, whence would have come the different shapes of animals, *et cetera*? In this same way, the shape of a house cannot come into existence without 116
an active and a passive agent, the active one being the workman, and the passive one the wood, the bricks, *et cetera*. If both the bricks and the workman were active, what would they have accomplished? Either nothing, or something in another thing; and there is no third thing, since there is no third thing apart from the elements. It follows that all the forms and compositions that emanated from the elements emanated from the active and the passive ones combined. Some of them are therefore active, and some others passive, and they flee from one another,

suffer from one another, and fight one another, in the following manner: when fire meets with water, it wages a well-known and severe fight against it, and sometimes it wins a victory, and some other times it suffers a defeat, while air and earth, although antagonistic to each other, do not engage in a known war and conflict.

Fire contains both dryness and heat, and water both humidity and cold, which are antagonistic to each other. The same may be said of earth and air, because the latter is humid and hot, and the former cold and dry. We affirm that the heat in fire is an active power, and so also is the cold in water; but the dryness found in fire is a passive power, and so also is the humidity found in water. When[1] fire meets with water, they fight against each other through their active powers. Fire, however, although containing an active power, which is heat, contains also a small proportion of a passive power. Nevertheless, because of the smallness of the proportion of the parts and also because of the nature (of this passive power) as found in it, its effect is not shown with that of the heat. The same may be said of the cold power of water, because if the latter did not contain a small proportion of a passive[2] element, (its parts) would not have combined with each other.

The passive elements also contain a small proportion of the active elements, and this is the reason why they receive them. Air, for instance, is composed of two parts: humidity, which is passive, and a small part of the heat which is found in fire. The same may be said of earth, which contains the passive power of dryness, and a small proportion of the cold found in water. And because passive powers[3] predominate in air and earth, they do not wage war against each other like fire and water, but wage only a small and insignificant war. It is because of the small proportions of the active elements which they contain that earth meets with air, and air with earth. When, however, the active power which they contain is stirred from outside—as may happen, for instance, by act of man or any other living being—they do wage war and cause disturbance. For example, when two portions of earth and air meet each other as a result

[1] Remove the first *Alaph*. [2] Read *ḥāshōshé*. [3] Remove the *Wāu*.

of a movement caused by man—as may happen if we shake the air by means of a piece of wood or bellows—it gives a sound when coming into conflict with the earth, and recoils from it because of the dilative quality (which it possesses) through humidity.

CHAPTER II.

On the ticklishness found in animals in such places as the sole of the feet, et cetera.

We have mentioned above the question of ticklishness, and we must now show its cause, and why it is not felt except in places the skin of which is thin and devoid of fleshy matter,
such as the arm-pits, the front of the neck, the sole of the feet, 117
and the part above the ribs. The reason for this is the following: when the hand reaches a part of the skin which is very sensitive, the thin and small nerves which it contains cause sensation when pressed [1] between the finger and the bones, as the skin acquires a greater sensitiveness when the nerves are moved hither and thither by the fingers. In this way a great and disproportionate sensation is caused to all the body. The places the titillation of which the body bears less easily are the hands and the feet.

Tickling may give rise to pleasure or displeasure: pleasure because of the superfluous matter which is found in those places, and which is ejected outside through the pores; and displeasure because of the excessive sensitiveness. Through pleasure, a man bursts out into laughter, and through excessive sensitiveness he flees from the tickler.

As to laughter, it is a peculiarity of man, and is effected by the balance of the parts of the circular movement, as the parts of the latter are balanced. The body experiences pleasure when (a man) laughs; and when the body experiences pleasure in an even way from titillation, the laughter, through its pleasure, mixes with the pleasure that emanates from titillation, and in this way the body is moved (pleasurably). When, however,

[1] Read *mizdarbīn*.

it is tickled disproportionately, although the humidity that is there becomes even, displeasure takes place, and man is moved to anger rather than laughter.

As to the places which are fleshy, like the calves and the thighs,[1] when the skin comes into contact with such places which contain much flesh, the nerves are not pressed [2] between the fingers and the bones, in such a way as to cause great sensitiveness. Only pleasure is caused by such titillation.

CHAPTER III.

On sleep : the reason for it, and how it takes place.[3]

The reason why (true) sleep comes to man alone of all the species of animals, and the mode in which it takes place in the other species of animals endowed with senses, are the following : man alone possesses thoughts, reasoning, memory and imagination, and this is the reason why man's [4] brain labours more than that of other animals. Because this labour causes displacement of the humidity that it contains, in a greater degree than in the other animals, (the brain of man) becomes drier, and thus is in need of rest to become humid. This is why sleep comes to it, which sets it at rest and moistens it. Indeed by sleep the brain rests with all [5] the body, and becomes moistened. Lo, when there is no sleep, we moisten the brain until it comes. It is through the brain that we generally feel the pain of sleeplessness and the happiness of sleep. This is why sleep came to the species of man.

How does sleep take place ? There is in the body a natural humidity scattered in the organs. This humidity rises to the brain and dissolves the dryness which took place in it. In dissolving it, it brings it to a state of equability, which is pleasurable to it. Indeed equability causes pleasure, and inequability

[1] Read *ḥerāwātha* (with a *Rēsh*). [2] Read *mizdarbīn.*

[3] *Cf.* this chapter with Aristotle's *De Somno et Vigilia*, ch. 1-3, where some difference in exposition may be noted.

[4] MS. repeats. [5] Read *ḳol.*

pain. The soul helps this humidity, because it impedes the
power of the senses from going outside and being lost until
it has obtained rest, and until the dryness has become moistened.
This change, effecting the pleasure of the head that takes place
in the brain, we call sleep. The brain lends some of this pleasure 118
to all the body, through the nerves, and in this way the muscles
which are tired rest also.

If someone says that all the other animals sleep, because they are seen to sleep, and close their eyes, we will answer that this is not true sleep, but a kind of rest which comes to the body; and this comes to an animal more especially when it is tired. Indeed some ten days or so may elapse before an animal suffers (from sleeplessness), while man cannot bear a single day without sleep, and cannot live without it. The sleep of animals is only a semblance of true sleep, in the same way as a painted man is a semblance of the living one; and these two things are not exactly the same. For the reasons which we have shown, sleep belongs to the species of man.

CHAPTER IV.

On the reason why, while animals have one single composition from the humours, in an equal way, they do not always beget males or females, but sometimes males and sometimes females.[1]

We affirm that, although there is an equilibrium of the humours in the composition, and although composition does not take place until this equilibrium is effected, yet this balance of the humours is not always of the same kind. If it were of one kind, it would have produced one variety, either males or females; but because it varies, sometimes a male is produced and some other times a female. A male is produced when there is not a destructive preponderance of cold. Lo, among males and females, some males are hotter (than others), and some females are colder (than others), but this relative predominance (of heat

[1] On this chapter, *cf.* Hippocrates, *Generation*, 6-7 (in Littré's edition, 7, 478).

or cold) not only does not destroy them, but rather establishes them. That males are hotter than females and females colder than males, their appearance itself testifies. Lo, males are more energetic, swifter, and more amenable to movement. They are also marked by their behaviour, and by the roughness of their skin. The reason why they have the membrum virile and testicles outside is due to their heat, that expands outward. As to females, their (genital organs) are inside, because of the particles of cold which contract inward.[1] And because of their humidity, their body is smoother and smaller, and their movement, together with their voice, is not so strong; and their brain is weaker.

CHAPTER V.

On the reason why the males and the females of mankind have at birth no hair on the whole of their body, except the head.

We affirm that males and females have hair on their head at birth on account of the heat and humidity that go up to their head at birth. The heat that goes up from all the body to the head, by means of vapours, moistens it, in the same way as the lid of a pot becomes greatly moistened through steam. This humidity does not collect in any other organ to the same degree as it does in the head. In the same way as when humidity increases in the earth together with heat, grass grows, so also it happens in the case of man. Humidity collects (in the head of man) because of the uprightness of his stature, which no other animal possesses; and since humidity collects there, hair grows.

The reason why no hair grows on the body of children till the age of puberty, is the following: the stage of childhood is more humid and less hot,[2] and because humidity is in excess
119 of heat, no hair grows. When, however, humidity diminishes little by little, and heat predominates, till these balance each other, hair grows on the body.

[1] Cf. *Galen*, Vol. II, p. 103.

[2] Than that of youth. This seems to be irreconcilable with the author's statement in Discourse I, ch. xxx (see p. 55 and note thereon).

In this same way, the earth does not produce much grass when rain falls on it and makes it too humid, till the sun shines on it and heats this humidity, and an equilibrium is established between them, and humidity rises up according to its nature. The same thing happens to the stage of childhood, in which no hair grows, until it reaches the stage of puberty, in which, through the addition of the strength of the heat, hair grows. It grows first on the thin and expanded parts, where there is more heat and humidity, as in the arm-pits and round the private parts. Then in the measure of the addition of heat, hair grows little by little on all the body.

CHAPTER VI.

On the beard that grows in men.

Men grow beards, while women, children and eunuchs do not, for the following reason: the stage of childhood, when compared with that of puberty[1] is cold and humid. The temperament of females is also cold, when compared with that of males; and the same is true in the case of eunuchs. Their temperaments resemble each other, although they are not identical in some things. Because children do not possess heat, which when joined with humidity causes hair to grow, they have no beard, or hair on their body. Indeed it is known that when the heat which was weak (in the time of childhood) becomes strong in the time of puberty, hair grows on their body; and they grow also a beard. Further, because the temperament of women resembles that of children, they do not grow a beard. Although hair grows on them at the time of puberty, it is by comparison less than that of men, as the heat that is added to them at the time of puberty is much less than that of young men; and because it is not intense, it is not sufficient to cover both the cold lower part and the higher part. (The heat) is also diminished in them because of the contraction which they experience

[1] Lit. "the sons of strength."

at this period, an experience which is not shared by men, and which drives away the humidity and the heat which have been added to them at this stage.[1]

The beard appears first in men round the mouth, because the natural heat which was added to the bodies at the time of puberty in rising and joining with the humidity found in the mouth, causes the beard to grow first on the lips of the mouth, as they are nearer to the humidity of the mouth. Heat increases then little by little, and gives rise to the growth of the beard round the mouth. Further, (the heat) that had gone up to the head, and there become superfluous, expands downwards, and reaching a lower part it remains there, as there is no room for it to go further down.

As to the functional ground of the existence of the beard in men and not in women, it is the following: it distinguishes males at first sight from females and children; and males and females were made so that the world might continue, according to the will of the Creator. Indeed if all males were females, or all females males, the world would have come to an end, and
120 the genera and species of animals would have ceased to exist long ago; and this would have been counted as a flaw (in the work) of the Creator, who created the animals.

CHAPTER VII.

On the reason why hair is found on the eyebrows and the eyelids at birth.

Hair is found at birth on the eyelids and eyebrows, because a small remnant of humidity bent the head downwards towards the front, and when a portion of (this humidity) reached the lower part of the depth and convexity (of the eyebrows), hair grew there; and when it reached near to the eyelids, a part of it remained there, and hair grew on their extremities. As to the ears they have no [2] hair at birth, because of the cold and dry nature which they possess, and humidity and heat do not

[1] This evidently refers to menstruation.

[2] Add the negative *lā*.

reach them, as they do in the case of the mouth and the eyes. The reason why they do not reach them is that a great part of the humidity goes up at the front, and wherever it goes upward, there also it inclines downward. Further, the front part is hotter, because the inner organs are found at the front.

The functional ground for the existence of the eyebrows and eyelashes is the following: the eyes are open, and in them is found the sense of vision. The eyebrows and eyelids came into existence with hair on them for the purpose of protection, in order that they might impede objects which might penetrate into the eyes and impair vision. Indeed we often notice that the eyelashes impede dust[1] from going into the eye. Further, hair came into existence on the eyelids for the purpose of ornament. Likewise the hair on the head is not only for ornament and beauty, but also to give an outlet, through its growth, to the noxious humidity that had collected there, lest by remaining in that place it should harm the brain.

CHAPTER VIII.

On the reason why hair does not grow on the forehead, or on the palms of the hands, or on the soles of the feet, or on the inner side of the curve of the muscles, or between fingers and toes.

We affirm that hair emanates from the superfluity that collects under the skin, and is parched and solidified there, and because the forehead is not protected from the air, like the rest of the body, the same air draws away the humidity that collects there before it is solidified through parching, and in this way it does not allow hair to grow there. As to the rest of the body, because it is covered with clothing, which impedes the air from coming into contact with the skin, and thus shuts in the vapours that come out through the pores, (these vapours) revert again to the skin, and help a great deal to keep the humidity that is in the skin in a parched state. Lo, the clothes with which we cover

[1] Read *ḥāla.*

ourselves in winter time impede the humidity of the vapours and cause it to revert to the skin of the body, which it warms and heats.

Further, hair does not grow on the forehead, and particularly not on the front of the cheek bones, because heat collects there to a considerable extent, and continually shakes the vapours, so that it thins them away. Indeed it happens always that when we get hot, or experience exhilaration, or drink a little wine, those places become red.

It is known that the humidity that is in the forehead vanishes by coming into contact with the air, from the fact that hair
121 grows inside the nostrils, because the vapours are imprisoned in them. On the outer side of them, however, no hair grows, because that side comes into contact with the air, which draws the humidity to itself. Damp clothes also, when protected from air, keep their humidity, but when they come into contact with it they dry up.

Further, the humidity found in these places is small, because it is nearer to the bones, and the flesh that is there is less thick.

If you say that the cranium of the head, and the place where the beard grows, come also into contact with the air, and in spite of that hair grows on them, we will answer that under the skull of the head there is much humidity; and although the air absorbs the humidity found in the outerl ayer, yet, because of its thickness, only a small part of it disappears, and the greater part of it remains; and thus much hair grows there. As to the place on which the beard grows, the humidity is small in quantity in the time of childhood; but when it becomes strong in the period of puberty,[1] it goes up to the jaws, and remains there because of the depth of the place; and although a little of it is absorbed in the air, much of it remains. When, however, it reaches the forehead and the cheek-bones, it does not remain there, as there is nothing to hold it in those places. Nature itself is proof that it is humidity and heat that give rise to growth, and that it is lack of this humidity that impedes it. It is indeed through lack of humidity that hair does not grow on the forehead, the palms of the hands and the

[1] Read *'ūza*.

soles of the feet. Because the hands are continually moving inward when they close and open, they compress and destroy the humidity that was there, by their continuous movement, their friction, and their handling of objects. Further, no hair grows between the fingers and the toes, because they also move continually, and rub against each other; and this movement and friction destroy humidity, and do not allow hair to grow there. The same thing happens between joints, because of their inward curve. Similarly, hair does not grow on the soles of the feet, because of this continuous movement, and because of the thickening and hardening which they experience through walking. In this way those places become dry, and no hair grows in them.

We have taken from nature all the demonstrations which we have brought forward, as these may be verified by all. They tend to prove the origin of the existence (of created things) from the elements, as set down by us in our present work. We have not deemed it advisable to bring testimonies which are not accepted by many people—a thing which we did in many places, and a thing which some people are in the habit of doing. We have brought forward, therefore, demonstrations that are accepted by men of all shades of opinion: Indians, Persians, Greeks, Syrians, and the rest of the peoples who wrote books of all kinds. We have not read this (elemental explanation of the origin of the genera and species) in any ancient book. Possibly such a book, treating of the origin of (things found in) the universe, that is to say, of the origin of the different genera and species from the elements, exists, although we have not had the good fortune to delight in it. Indeed (if these things had already been discussed) our testimony would have been either useless [1] or self-evident,[2] having, that is to say, the self-evidence of the elements which fall under the senses, and which no one 122
denies. As to the functional grounds for their existence, many have written about them, and this is the reason why we have treated of them briefly.

[1] Lit. "thin air."

[2] Before this word the author uses the word *tortoros*, which he explains by "testifying to itself." (Possibly *Gr.* ταὐτολογός.)

The functional ground for the non-existence of hair (on the forehead and cheek-bones) is to prevent dust, sand and other things which happen to fall on those places, from getting a footing there, and thus impeding vision ; and also to distinguish man from other animals ; and again for the sake of beauty, because if hair had grown there, it would not have been seemly. Indeed, when a beard appears on the cheek-bones of some men, they always shave it, because of shame ; and if they let it grow, the place looks ugly.

If you ask why the air does not destroy the hair that grows on the face of other animals, we will answer that it is because of their thick humour, due to the homogeneity of their food, and to the fact that they do not add to their food and drink when they are satisfied, and do not abstain from food when they are hungry. As to man, because the multiplicity of his diet and the addition to his necessary food and drink cause in him thinness of humour, his humidity is thin, rarefied and less hot, so that when the air comes into contact with it in the face, it evaporates it. So far as the face and body of other animals are concerned, because (their humidity) is thick, the air only draws it out. The reason, therefore, why from the beginning the animals had hair on all their body, is that their humidity was thick from the beginning, and in the generality of cases, no addition was made to it ; while that of man was from the beginning much thinner, more volatile and rarefied, and thus, until it became thick, it did not give birth to hair.

CHAPTER IX.

On the reason why eunuchs have no beard, or hair on their body, and their voice is thin and unbroken ; while the voice of men is deep and broken.

We affirm that heat collects in large quantity underneath the place where the testicles are, and from there it evaporates and goes up to the higher places of the main body, for three reasons : firstly and principally because the semen assumes its vigour from all the body, and comes and collects there in great strength, as all man is found in it potentially ; secondly, because the urine, which is very sharp and hot, collects there, together with the fæces ; thirdly because that place is protected by the buttocks at the back, and by the thighs on both sides and these impede the heat that collects there from evaporating.

As to the semen, it comes there at the time of puberty, as the testicles possess a power of drawing to themselves the semen, through canals and passages, in the same way as the stomach does with food, which it makes into chyme, and as the kidneys [1] do with the watery matter found in the food in the liver. The testicles perfect the semen, and give it as it were their final stamp towards reproduction. It is for this reason that if a constriction or a concussion, or any other affection occurs, the semen is not secreted for the purpose of reproduction ; and similarly when the testicles are completely removed, the path of the tubules through which the semen comes is cut off, and the secretion and the drawing of the semen from the body, which 123
they effect, come to an end ; and in this case the semen, which used to heat and warm that place, together with the humidity that was in it, ceases to collect there, and the place becomes cold, as in the time of childhood, and its nature resembles that of children. This is the reason why (in the case of eunuchs) heat and humidity are not sent in the time of puberty to the head to cause a beard to grow, as is the case with those whose testicles have not been removed. This is the reason also why their skin resembles that of children and women, and the hair that grows on them is less thick, Their voice also becomes thin and high, while that of men is deep.

The voice of men is deep for two reasons : first because heat predominates in the throat, and thus expands its contracted parts and loosens them little by little, according to its own nature. In this way it differentiates the throat of men from that of children, and expands (the vocal chords) outward, according to its nature, thus widening the throat.[2] When the

[1] Before the word " kidneys " the author makes use of another word *bīshātha*, for which I can find no adequate meaning. Is it a corruption of *bi'ātha*?

[2] Read *gaggarta*.

throat is expanded, the voices of men become deep, while those of children, of women and of eunuchs are the contrary. In the same way as in a stringed instrument the thick cords give a deep sound, and the thin cords a high sound, in this same way, here also, the voices of men become deep on account of the heat that is added to them. As to the voices of children, women and eunuchs, they are unbroken because the parts which come into contact (with the air) are tighter, on account of cold; and when the voice comes out, it harps on those tight parts, and they give a high note. As when we strike two bodies, such as brass and wood, the brass gives a high sound and the wood a low sound, the former because of the contraction of its parts, and the latter because of the expansion of its parts, so also are the voices of men, women and eunuchs.

CHAPTER X.

On the reason why women have a womb, but not men; and why women menstruate, while men and the females of other animals do not.

We have affirmed that the temperament of men is hot, while that of women is cold, when compared with each other. In the very first composition of bodies from humours, the preponderance of heat in the temperament of men caused them to have genital organs and testicles outside;[1] while the cold found in the temperament of women contracted inward, and caused the womb. The male casts the semen outside and the female inside.

Take a demonstration from the trees, which in winter time send their heat inward, and matter is imprisoned inside with it, and thus the hardness of the wood is strengthened and increased. In the summer (this matter) is driven out, because of the heat that preponderates, and this causes fruits to grow in the outward parts. In this way menstruation occurs in women,

[1] See above, p. 72.

and not in men, because the heat which is in men causes more expansion of the body, and dilates its pores, and thus more discharge takes place than in women. The cold, however, which is in females, because it contracts the body and the pores, does not cause such a discharge ; and so the humidity and the heat which in men are discharged outward, are in them kept inside. 124
Humidity finds there a congenial place, as the superfluous matter in women is of the same nature as humidity. It collects there, and then it is driven downwards in one direction only, because it is not able to escape through the pores by different paths.

The females of other animals do not menstruate, because their food [1] is thick and dry, and they do not eat beyond their requirements. Further, because they labour more, their temperament is drier and harder than that of women, and because of this they have no superfluity to be driven out, beyond what their own needs require. In the case of women, however, because they labour less, and make use of different and numerous kinds of food and drink, their nature is more humid, and so a superfluity collects in them which exceeds their own requirements, and which has to be driven out. If you ask why, when animals conceive, they have blood for the nourishment of the embryo, like women, and why, after they have brought forth, it changes to milk, we will answer that there is no superfluity of which the animal is not in need, and which has consequently to be driven out, like that of women ; it is found in them only in sufficient quantity to satisfy their needs. When, however, an embryo is conceived, the flow takes place for its nourishment, coming, as it were, from the natural dwelling-place of the embryo, and from its very own nature. Just as with the increase of every organ there is an increase of matter for its sustenance, and as when that organ is cut off the matter itself vanishes, because it grows out of the very nature of the body—so also, when the embryo is formed, it is from the very nature of the animal, and resembles it also in the increase of matter for its nourishment. When, however, the young is brought forth, the flow comes also with it in the form of milk for its

[1] Read *saibarthain*.

sustenance; and when it has been weaned from milk, that matter does not flow any more and become a superfluity to be driven out.

The functional ground for the existence of the above things is the following: our Creator, who is all-wise in Himself, created males and females for the propagation of the world, so that this procreation might continue according to His will. It is He who made one active agent, that is to say the male, and one passive agent, which is the female. In this way there are females and males among all animals, as all of them are in need of these two agents—the active and the passive. He made woman the recipient, and man the agent from whom she conceives and begets. He made for the woman a womb as a receptacle and dwelling-place for what the active agent—which is the male—has worked in her, in order that she may receive it and keep it towards gradual completion. He made its food from that addition and superfluity which is left over from the blood, and which is the menses. Because the male is the active agent, a genital organ was prepared for him outside. He implanted in (the sexes) passion for each other, in order that they might by necessity unite, and thus enable propagation to be accomplished. He divided procreation and labour between them: the man working, gathering, and feeding the woman; and the latter conceiving, bringing forth, nursing and rearing.

CHAPTER XI.

*On the reason why the incisors and molars came into existence in
the mouth, and not in another place; why the number of the
125 teeth is thirty-two, and why they are different from one another;
why the molars, which are larger, are inside, and the incisors,
which are smaller, outside; and why the incisors fall and
then grow again, while the molars do not.*

We have said that appendages came into existence at the top of the organs, and at their outward extremities, from the superfluities (of the body) that were driven out; and because the upper and lower jaws have an extremity and are round, the superfluity which had remained there, from the matter of which that part of the jaws had no need for its nourishment, was driven out by its natural driving power; and the part consisting of the upper and lower gums of the jaws emerged[1] out of it, and the incisors and molars were made; and since that place was hard, what remained from it had to be bony. This is the reason why the teeth are bony.

Why are all the teeth not one, like the bone of the gums? We answer that if that bone had drawn the superfluity to itself for the purposes of building up, (the teeth) would have been formed of one bone, like itself; but because that superfluity was an extraneous addition, it was driven outside and became as a kind of appendage to the bone. It did not give to the superfluity its uniform structure, by drawing it to itself and making it a part of itself; but it drove it out, and it was scattered. Anything that is driven out from inside expands, and anything that has been driven out and expanded, its parts become divided. This is the process of the coming into existence of the teeth.

Why did (the teeth) come into existence in the mouth and not in another place?—Because there is no place in the body that resembles the mouth. If any such place were found, it is possible that incisors and molars would have grown in it.

Why were they divided into the number of thirty-two, and not into another number? We answer that the evolution of the number of the years of the development (of the body) is completed at about this number thirty-two. Indeed until this time there is a development of all organs of the body, after which they remain stationary without increase or decrease. The reason for this is to be sought in the fact that (at this age) dryness preponderates over the humidity of childhood, and because of its contracting power impedes the adhesion of the parts of the body and thus precludes any increase in it. Further, the evolution of the number thirty-two is complete, because it is divisible without change till it reaches unity; and similarly,

[1] Read the verb with a *Wāu* at the beginning.

when it is multiplied, it keeps the same number till thirty-two; its quotient and its divisor are also the same. It thus receives reduplications and divisions which other dividends do not. Indeed, other numbers vary, both in their division and in their multiplication.

Because this evolution is natural to the growth of the body, and sets a limit to it, after which no increase is effected in its growth, this same evolution gave the same number, thirty-two, to the molars and the incisors. It would have been possible for the matter to have reached in its division the number of forty
126 at its highest, or of twenty at its lowest; but because the nature of the movement of the development of the body has its limits in the number thirty-two, it gave this very same number to the molars and incisors.

The reason why the molars are round and placed towards the inside (of the mouth), and are compound, while the incisors are small and placed towards the outside, is to be found in the fact that the place of the gums of the incisors is thin and narrow, because, the front part being hotter, its matter diminished in volume. Indeed, when the place of the molars is compared with that of the incisors, the former will be found to be cold; and because of its coldness, its matter contracted, and gathered in that place more superfluity than in the place of the incisors. This is the reason why the molars are larger than the incisors.

Further, both males and females lose their incisors, but not their molars. The reason for this is to be sought in the fact that in the time of childhood the body is small and humid, and thus its bones are soft and weak; but when the stage of youth is reached, heat and dryness predominate, and the body, together with the bones, becomes hard. Because the teeth of children are a congealment of the humid matter of childhood, when this matter hardens at the stage of youth, the bone of the gums of the incisors hardens also. Then the superfluity which is a remnant of it, and which forms the incisors, acquires also great bulk; and this matter, because of its hardness and heat, pushes the front teeth outside, since its nature does not allow it to join with them or be added to them. Thus they are pushed out, and fall. It is known that this is so from the nature of the teeth that grow after them. Indeed no other matter is joined to these second teeth, and the first teeth do not fall until the head of the other teeth appears and pushes them out. This is also confirmed by the hair that grows on the body of children, which is thin, weak and not black. When, however, the heat of the time of puberty begins to warm, the black hair appears and pushes out that of childhood.

Further, the second teeth, because of the heat and dryness which predominate in them, have sharp points, and are not as white as the first teeth, and are broader.

The incisors appear first, and then the molars, because of the heat of the place of the incisors and the coldness of the place of the molars ; and the molars do not fall like the incisors, because the incisors grow some time during the first year and are thus made out of a matter that is amenable to change, transformation and flexibility, and fall with the growth and change of stage, and others appear in their place. As to the molars, they attain their completion about the twentieth year, at a stage which is the driest and hardest ; and they are stronger than the (comparative) humidity of youth, and do not possess the weak power of childhood. Because, therefore, the place (where the molars are) is cold, and does not receive heat easily, as does the place where the incisors are, the molars do not fall ; and because they do not fall, no others grow in their place.

The functional ground for their existence is the following : the Creator so ordered the mouth as to receive food, masticate it and knead it, so that it may be more easily digested in the
stomach. This process requires a hard matter with which to 127
grind. Indeed if flesh only were found in that place, it would not have been in a position to crush things, and no man would have been able to eat dry and hard food, because that flesh would have been torn and would have become sore, and man would have been prevented from eating. A bony matter was, therefore, given to him for the purpose of crushing the hard food and masticating it with his saliva. This is the reason for the existence of molars and incisors. The incisors are at the front for the purpose of cutting the food into small portions, and handing it to the molars, which grind them like a mill. When we eat

meat, or some other kinds of food, we sometimes bite it a little in the front of the mouth, and tear it, as very often the mouth cannot contain the portion which is taken, on account of its great bulk, but when it is cut little by little by the incisors it is placed on the tongue. The latter then receives it, and hands it to the molars, and after a part of it has been ground by the molars, the tongue takes it back to itself, and hands another part, which is kneaded in its turn. The reason why it has been called in Greek "the vaulted tongue"[1] is that when it carries the food, it holds it until it is ground, and sent by the swallowing power from itself to the gullet.

The molars are round so that food may stand on them and be kneaded. If they were sharp, some food might have fallen away from them and not been ground. They are in the shape of a mill, which grips the grains of wheat and grinds them; and they become the flour which is kneaded, leavened and baked in the fire. The molars grind dry food in the same way, and then it is kneaded by the saliva and goes down to the stomach, where it is kept, fermented and baked in its heat as in fire.

Teeth serve also for speech. When air penetrates into them it is compressed by the tongue, which affects the sound of some letters between them, such as ṣ, p, s, ṭ, t, *et cetera.* This is the reason why, if one of them is broken, speech is impaired.

Galen wrote on the function of the incisors, molars, mouth, lips, *et cetera*, in his book on *The Function of the Organs*, and he wrote also at some length on other things; but he did not touch at all the subject of their coming into existence.

[1] The text gives here the incomprehensible Greek word *ora log losmos*, which has been misread by the copyists. It may possibly be a corruption of the two words οὐλός and γλῶσσα.

CHAPTER XII.

On the reason why the bodies of men are tall, short, black, white, light, heavy, straight-haired, or curly-haired, according to the countries; why those dwelling in northern countries are white, fat and straight-haired, while those who dwell in the confines of the eastern countries, or in the western and southern countries, are black, thin, and curly-haired.[1]

These things happen in the following way: some men are 128
tall because at the beginning of their composition at the time of the movement of the (first) elements towards one another for the purpose of the building up of the body, the movement of the active ones, which are the hot and cold elements, prevailed over that of the humid and dry elements, and caused length, in the way we described above, when we treated of the subject of the length of animals. The hot element went up, and the cold down, according to their nature; and thus an opportunity was given for extension. This is the reason why the shape of their members became long, and (the body) received additional matter according to its length. Some men are short because at the beginning of their composition phlegm predominated in them, its characteristics also predominating in the body, and its whiteness in the skin.

Some men are swarthy because the black bile predominated in them over the other humours, and showed its colour in them. Some are red, because the movement of the blood predominated in them, and left something resembling its colour in the body. The same may be said with regard to the yellow bile (in connection with the yellow races). Indeed the colours of the humours affect the bodies, and when one of them predominates it shows its power and its colour in those bodies.

Some men became heavy and some others light in their movements in the following way: those in whom the earthy element predominated acquired, through its weight, heaviness; and those in whom the fiery element predominated, became light like fire.

[1] *Cf.* Hippocrates, *De aere, aquis et locis*, ch. iv, xii, xvi, xxiii-xxiv; and *Galen*, Vol. I, pp. 74-78.

Some are curly-haired because heat and dryness increased in the matter from which hair comes out; while in those who are straight-haired, humidity predominated. Lo, when we wish to stretch the contraction of leather, we moisten it, and when we wish to contract it, we dry it.

The reason why those who live in a northern country are white, generally fat, red-haired or fair-haired, and straight-haired, is the following: the northern countries are cold, and the air, being also colder, absorbs to a lesser degree the vapours from the body. This causes the heat to keep inside, and in this way the outer surface of the body becomes cold. When (bodies) are cold, they generally become white. The fact that bodies become white is illustrated by water which, when very cold and frozen, becomes white; and even after thawing, some whiteness still remains in it. The above men are fat owing to the fact that, when the vapours are kept inside, they accumulate and congeal, through the predominating cold; and after they have congealed, they thicken and moisten the body all the more. That those men become fat through the moisture that has accumulated is illustrated by the fact that we see all bodies which have been moistened become inflated and swollen; and when moisture is taken from them they become thin. This happens to wood, vegetables, *et cetera*.

The reason why those who live in the countries of the east, west and south are dark and thin, is to be sought in the fact that the sun, being nearer to those countries, makes them hotter;
129 and because of its heat, it scorches the bodies, and after it has scorched them, it darkens the outer surface of the body, which is nearer to the ambient hot air. Indeed the bodies of those who stand in the sun in winter time do not become so black, on account of the remoteness of the sun, which also causes the air to be cold in that season; but when they stand in the sun in summer time, their bodies become black, because of the proximity of the sun, and the heat of the air. Those who have less humidity on the outer surface of their body, owe this to the fact that their humidity vanishes through evaporation; the hair of their skin is less thick, and what little remains of it from the scorching of the sun is curled, stunted and short, on account of the small

quantity of humidity. This is known to be so by the fact that when we bring skins near fire they contract.

As regards those who, in their colour, hold a middle position between black and white, the blackness of their skin is in proportion to their nearness to the sun.

Ethiopians have thick lips and noses, and large teeth, because, when the humidity of the composition of the building-up of the body reached the mouth, it caused a certain recurvation, on account of the heat and dryness which predominated in the air; and having increased in volume, it accumulated, and became fixed there, and bent the lips and the nostrils upwards, according to its nature, and thus they became large. When we bring damp skins near the fire, that part of their outer surface which comes into contact with fire doubles against itself.

The above men became thin because of the excessive evaporation which takes place in their body in those countries.

We have written these lines about the above things in brief terms.

CHAPTER XIII.

On the three kinds of divisions in the differentiation of the genera of animals.

After having spoken of the composition of bodies in general terms, let us now speak of the division of the three genera: animals, animal-plants (and plants). First, about the division of animals: animals are divided into terrestrial, aerial and aquatic animals; and each one of them is sub-divided into other genera. The terrestrial genus is divided into different species, such as man, ox, horse, *et cetera*. The aerial is divided into eagle, vulture, sparrow, *et cetera*. The aquatic is divided into fish, siluridæ and crustaceans. Animal-plants are divided into sponges, molluscs and shell-fish.[1] Plants are divided into trees, shrubs and herbs.

[1] Or, oysters.

Let us now speak about their origin from the elements. In the terrestrial animals, at the time of their very composition from earth, fire, water and air, the earthy element predominated over the others, when compared with the aerial and hot animals. This is the reason why, according to the predominance of the earthy matter that occurred in them, they dwell on the earth, and neither in the water nor in the air. This is also the reason why they are heavier in their movement, and possess a thicker temperament than either the aerial and the aquatic animals. We have spoken above of the causes of this predominance.

130 In that very first composition, liquidity [1] predominated in the aquatic animals, more than in the aerial and terrestrial animals, and their habitat and dwelling-place were in the water. This is the reason why their temperament is loose, expansive, humid and liquid. In this same way the aerial element predominated in the aerial animals over the elements of water and earth, and for this reason their habitat was in the air, in which they fluttered and flew.

We will speak now of the origin of all this. In their first encounter in the general movement of the composition of animals, the elements acquired their own respective movements; and each one of them, according to its special nature and movement, drew to itself what was affinitive to it, and thus gave rise to a distinct division. (In this process) the earth predominated, through its particular movement which was concomitant with the general movement of the composition of the body, and gave birth to plants. It is for this reason that the latter are more earthy. By the general movement, I mean the one which affected all the elements for the purpose of forming the genera and species, as none of the latter is deprived of any one of the four of them; and by the particular movement, I mean the predominance of one of them in the above genera and species. Thus, as we have just now said, the plants are fixed in the earth because they are more earthy, the birds are in the air because they are more airy, and the fish in the water because they are more watery. The terrestrial animals hold the middle position between the two extremes.

[1] Remove the *Alaph* from the beginning.

Further, in the very particular movement of the composition of the earth, the part which was drier gave birth to trees, the part which was less dry gave rise to herbs and vegetables, and the part which was of medium dryness gave rise to shrubs, which hold a middle position between herbs and trees. As such may be counted the mustard plant, *et cetera*, which are of the category of herbs, but possess some woodiness. In the same way, the animal-plants also have three divisions.

After this let us speak of the difference in the species of the above genera, and first of all, of the differentiation in the species of the terrestrial animals, such as man, horse, ox, *et cetera*. We will at the outset speak of the reason why man alone, of all the terrestrial animals, is the only one to have an upright posture, and to walk on the earth not on all-fours, like other animals; and of the reason why he does not possess all his qualities from birth, but acquires them little by little, through teaching. Indeed the quadrupeds and other animals perform all their actions from childhood, and not through teaching. For example, some swim from their childhood, or know their mother, or distinguish between different kinds and qualities of food, or flee from enemies, or possess from the very beginning the full power of movement with others of their own kind.

CHAPTER XIV.

On the reason why man alone, of all animals, is upright in his posture, and why he acquires everything by teaching, while other animals acquire what they possess from their nature.

We affirm that man is composed of two natures: one of them from the body, which is from the elements,[1] and the other from 131
a rational, immaterial and unelemental soul. The nature which is composed of the elements is a body which falls under the senses, and the existence of which is unquestionable; while the other nature is immaterial and does not fall under the senses,

[1] Read *isṭūkhsé*.

and its existence is thus questionable. We must therefore demonstrate its existence, and then add its attributes. Because it is not perceived through the senses, it is necessary to demonstrate its existence by reasoning. We will demonstrate this through its workings, over which sense has no power, and it is from them that we will deduce our reasoning.

On the existence of an immaterial soul.[1]

The eyes of the body, when open, perceive external actions ; but when these eyes are closed and unopened, there are inside the man other eyes which perceive ; and since there is no room for eyes inside the man, with which he can see, as there is no place for such eyes, either inside the head of man or in any of his other organs, these internal eyes are only a semblance of the external eyes.[2] Because they are only a semblance, and are not a composition from the body, like the external eyes, they have not even a place. Where would such a place be? They are also something different from the body : lo, the eyes of the body are limited, and their vision may be impaired ; while those others are imperishable, as they are not composed of the antagonistic elements from which the eyes of the body are composed. The eyes of the soul are therefore not material like the external eyes. Since man has two sets of eyes, one external and the other internal, the external one being composed of the elements, perishable, perceptible through the senses, and consequently not denoting a body—man, therefore, is composed of two natures : an immaterial soul and a material body.

Animals do not possess this, because they do not walk when their eyes are closed, as they have only the material eyes ; while man walks even when his eyes are closed. Lo, men blind from birth walk ahead. They walk, therefore, through the eyes of the soul.[3] Further, when we plan to do something, for instance

[1] The author's views on the soul differ considerably from those of Aristotle. See Aristotle, *De Anima*, Book II, ch. 2-3, and Book III, ch. 4-8.

[2] The text is here corrupt, and the copyist has put in red the heading "quality of the soul," which is incorrect, and which may have been taken from a marginal note.

[3] There is after this an incomprehensible sentence which literally means "that thou mayest forget the fact that it has been given." These words may have belonged to a longer sentence omitted by the copyist.

to build a house or make a door, we first see, with the internal eyes, all the work completed, before it is made; and after we have completed it, it is perceived by the eyes of the body in the same form as it was perceived by the internal eyes; but because the animals have no internal eyes, they cannot look at a thing before it comes into existence. This is the reason why they do not make any work of art, but only look, with the external eyes, at a thing which is made. Indeed, they possess nothing which would give them an internal sense, because they have no immaterial soul.

Further, we see that a man performs two contrary actions, one of which is for the building up of his body, and the other for its destruction. The action which builds up his body is
the strong desire, formed from the elements, for food, drink, 132
sleep and marriage, which constitute him, and without which it is impossible for him to live. This desire is effected through the harmony of the elements, when they do not fight and combat one another. This is known to be so from the fact that when they fight one another in any illness, no desire takes place. Desire, therefore, takes place when harmony is established between them. We see, however, that in spite of the fact that they are in harmony with one another, and their power [1] is unimpaired, and they do not contradict one another, but are at one to do a certain thing, yet the strong desire for food, sleep and marriage is checked in them forcibly, and is not fulfilled, while the material desire persists. This cannot come from the elements, because of their mutual harmony; but it comes from the soul, to which the performance of the action is repugnant; and in this way the desire is checked by another thing, which is the immaterial soul. Animals do not possess this, because they cannot by themselves check themselves when they are animated by a strong desire towards food, sleep and marriage; while man can do so. This happens to animals because they have no other nature that is antagonistic to the elements; but such a nature is found in man. It follows that, since man possesses two antagonistic workings, the one perfecting his body, and the other destroying it, the

[1] Lit. "feeling."

elements in him encounter another thing which checks them, and that thing is the immaterial soul.

Further, we see that all animals in general fear man, work for him, and are subject to him, although they are equal with man in the matter of their provenance from the elements. There is no species of animal which works for, is subject to, fears, or satisfies the wants of, any other animal, but man alone, for whom all work, and to whom all are subject. The reason for this is to be sought in the fact that he possesses another nature, which is stronger, higher and more sublime than that which emanates from the elements, and which is immaterial. If, therefore, animals fear man generally, work for him, and are subject to him, in spite of the fact that he is in everything equal to them in his provenance from the elements, there is in him an immaterial nature, which is not from the elements, but which unites with the known elements, and dominates the animals. How it is united, and how it is possible that a material body should be united with an immaterial one, we have discussed as fully as possible in our *Book on the Soul*, which we wrote in twenty chapters, and in which we treated, by means of natural and cogent proofs, of the existence of the soul, its nature, and all the necessary questions relevant to it.

The reason why man, of all quadrupeds, is the only one with an upright posture, and the only one that walks on two feet, is the following: we have shown that man is composed of two natures—a material nature and an immaterial one, which is the rational soul; and because the nature of the soul is light—since it is immaterial—it rises upwards, and through its union with the body stretches it upwards by means of its lightness.
133 As to the body, it weighted man's feet downwards, according to its nature. This is the reason why man was lengthened towards his upper and lower limits, and in this way the uprightness of the stature of man was accomplished. This explains why he, alone, of all animals, moves slowly in an upright position towards the heaven, making use of the eyes of both the soul and the body. So far as other quadrupeds are concerned, because their nature is one — that nature emanating from the elements — although they possess more length than width, as we have shown and

demonstrated above, yet they walk on all fours. This happens because there is in them no other nature to unite with that which emanates from the elements, struggle with it, and lift their feet upwards. This explains why their head inclines downwards and is straight, while that of man is round, in accordance with the uprightness of his stature.

If you say that an immaterial thing does not possess in its nature the faculty of going upwards or downwards ; and that if it were able to go upwards and downwards, it would also go to the right, to the left, forwards and backwards, and thus it would be material, and not immaterial—we will answer that the immaterial soul does not possess in its nature any one of the movements of the body, but only possesses them because of its union with the body (through the animal soul).

How could the rational soul unite with the body through the animal soul ? The animal soul possessed very thin parts, which united with the rational soul ; and this same animal soul united with the body through its thick parts, but because it possessed a predominance of the elements which tend to go upwards, over the elements which tend to go downwards, its nature tended to go upwards ; and having become more volatile, when united with the immaterial soul, it took a greater *élan* upwards. And because of this predominance of the upward drive which it received, it pulled man upwards, and made him upright in his stature, while it formed a connecting link between his two opposite ends, and caused an upward movement like a perpendicular line. The irrational animals do not possess an upright stature, because this predominance is not found in them. It is in this way that the immaterial soul may be said to have straightened the stature of man.

The immaterial soul, being circumscribed, possesses a particular movement, but this movement is not localised, like that of a body whose movement is effected by increase and decrease through heaviness and lightness caused by the antagonism (of the elements). The soul reaches where it wishes, through the movement of the mind, immediately it wishes it. If the light of the sun, immediately after it has shone, reaches the ends (of the earth), and in the same way, immediately a body stands

before it, it is checked and makes a reflected movement, how much more will the soul do so? Indeed it flies instantaneously through the mind[1] wherever it wishes, without taking a straight line, as such a line is made by a thing which possesses thickness, coldness, and density. Because the soul does not possess any of these things, it is not animated by any straight movement forwards or backwards, but, through its swift nature, moves
134 swiftly wherever it wishes, and its nature itself testifies that at the very moment when it wishes to be somewhere, it is there. As to the body, it does not possess, like the soul, the faculty of being quickly everywhere, but it takes years, months or days to do it, because it is imprisoned in a dense and thick matter. Since, however, the soul is immaterial, there is nothing to check it.

It follows from this that it is not strange for an immaterial soul to extend the body upwards, in accordance with its nature, even when it does not possess by nature this upward extension, but only a circumscribed movement. It is through this circumscribed movement, which is neither straight nor curved, that it is joined to the body, and moves away from it.

The union of the body with the soul is in the following way: it is inside the body, but not in the way in which a body is inside another body, for instance water or oil in a skin; but it is in it as an integral yet composite entity; and it is not mixed with it, because, like the body which keeps its entity intact, it also keeps its entity without change. It is in the body according to its size, whether large or small, while itself is neither large nor small. This is illustrated by a piece of iron or other metal, in which fire is seen. The greatness or smallness of the fire is seen according to the greatness and shape of this object, although fire itself has neither greatness nor smallness. This is known from the fact that when from a lamp we set fire to two pieces of wood, one of which is eleven cubits long, and the other five, although the lamp is one, with a definite size, yet when the fire moves to the piece of wood, it possesses in itself neither measure nor shape, certainly not those it had before. If fire, which is a material thing, is like this, how much less would the immaterial

[1] Read *hauna*.

soul, when united with the body, have in itself shape, greatness or smallness ?

Why are all animals endowed with all their modes of behaviour from birth, and not from teaching, while man is in need of teaching ? We answer that animals possess all these things from nature, because, as we have shown, they have only one nature ; and since they have one nature, and that from the elements, nothing impedes those elements from putting into action the natural impetus, together with the natural knowledge. As to man, because he is composed of two natures, united with each other, when his nature that emanates from the elements takes its drive to accomplish the things that belong to it, his immaterial nature impedes it, and allows it to accomplish them only little by little, while joining some of its rational movement with the irrational impetus of the former, so that all its actions may be accomplished rationally, through teaching. This takes place in two ways : either inwardly, through natural thought, which gives rise to all arts, inventions and sciences, or outwardly, through a teacher, who has received them through tradition from his elders. The animals do not possess any of these things, as they possess all (their modes of behaviour) without increase or decrease from their birth. Lo, they run, cry, swim in water, and distinguish one another, their suitable food, their masters and strangers. To the animals was given from the beginning, by the elements, through the animal soul,
a natural knowledge and a movement which is natural, simple, 135
and unchecked by any antagonistic thing.

As to man, because he has in him something which is not elemental, when the elements strive, through the animal soul, to do a thing, the immaterial and rational soul inhibits them ; and after it has checked them, it allows the action to be accomplished gradually. It is well established that an action that emanates abruptly from the elemental impetus, ends also abruptly ; while all the works of man, because performed rationally and gradually, have duration. Lo, the four rational sciences, and all the arts, have continuity, because they came gradually into complete existence ; and they point, as with a finger, to the rational soul, their maker, and testify to its being lasting and immortal. On the other hand, there is not a single action of animals that is preserved after their death, to testify to us that they have an immortal and lasting nature. They have only their composition, which perishes and reverts to the elements.

CHAPTER XV.

On the reason why man alone has white hair in his old age.

The reason for this is the following: man makes use of numerous and different kinds of food, and also after his hunger has been satisfied, he adds to this satisfaction more food and drink. This gives rise in him to many superfluities, which join with the skin and render its external part soft, tender and liable to changes, like all the characteristics found in the stage of childhood. After the age of puberty, heat joins with his humidity, and kindles and parches this excessive humidity, and gives birth to black hair on all the body. When, however, man reaches the last stage of old age, the heat found in the humidity diminishes, and quickly evaporates, because of the width of his pores and the displacement of his humidity. The cold then predominates, and his hair becomes white, through cold and humidity. Lo, when cold predominates in water, to the point of freezing, it becomes white. The head and the beard become white first, because it is there that a great quantity of the superfluity which causes the growth of hair accumulates.

As to the other animals, their hair does not become white in their old age, because they do not make use of numerous and different kinds of food, nor do they add to it after they are satisfied. This is the reason why the superfluity which accumulates near their skin is less in quantity, and is hard and dry, and presses and squeezes itself against the pores, and in this way it becomes not very liable to change, and the evaporating matter that accumulates there becomes small. Because the nature (of animals) is thick, and the superfluity which collects near their skin is dry, their pores are narrow. This explains why in their

old age they do not experience much evaporation, and since
their heat is kept inside together with their superfluity, and
does not receive further increase, their hair remains black, and
does not whiten in their old age. This explains also why their
hair does not grow much, but keeps the same length from
childhood to old age, with the exception of goats, sheep and
other animals which, because of the humidity and heat which
their nature possesses more than others, have an everyday 136
growth of hair.

CHAPTER XVI.

The reason why all animals, such as horse, ox, dog, eagle, pigeon, et cetera, are not one species, although all of them emanate from the elements.

Increases and decreases occurred in the very first composition that took place from the elements, on account of their mutual antagonism. It was indeed impossible that one species only should emanate from them, because all the species did not come from one element.[1] If there had been only one element, or if the (four) elements had no mutual antagonism, and consequently no increase or decrease had occurred, it is just possible that one species might have resulted. Many species, however, of terrestrial, aquatic and aerial animals, of animal-plants and of trees, came into existence, because different and innumerable parts were formed from the increases and decreases that took place in the first composition from the elements.

First the general division took place, which gave rise to the three main genera of animals, animal-plants (and plants), as we explained above. Then the more particular division took place from the general division, and gave rise to three subdivisions from each one of the genera. And so took place the division of the animals into terrestrial, aquatic and aerial; of the plants into trees, shrubs and herbs; and of the animal-plants into shell-fish,[2] sponges and molluscs. Further, the

[1] Cod. erroneously *iskema* (Gr. σχῆμα) instead of *isṭukhsa*.
[2] Or, oysters.

terrestrial animals gave rise to innumerable species : man, ox, horse, *et cetera ;* and the aquatic animals gave rise also to innumerable species : fish, siluridæ, crustaceans, *et cetera ;* while the aerial animals gave rise to eagle, pigeon, vulture, *et cetera.*

Further, each one of the species (was sub-divided) and gave rise : in man, to individuals,[1] such as Peter, Paul and John ; in the ox, to a white ox, a black ox, *et cetera ;* in the horse, to a powerful horse, a swift horse and a beautiful horse ; in the fish, to a megacephalous fish, a microcephalous fish, a broad fish, *et cetera ;* in the eagle, to a beautiful eagle and a large-winged eagle. The same thing applies to the other animals, as found in the diagram below :—

Fat	Large	White	WHEAT	*Herbs*	PLANTS
Dark	White	Black	BARLEY		
Thin	Large	Round	LENTILS		
Dark	Fat	Sharp	MUSTARD	*Shrubs*	
Yellow	Light	White	RUE		
White	Thin	Large	ALMOND	*Trees*	
Sweet	White	Black	FIG		
Sour	Sweet	Red	APPLE		
Sponges	*Molluscs*	*Shell-fish*			ANIMAL-PLANTS
Broad	Thin	Long	FISH	*Aquatic*	ANIMALS
Fat	Microcephalous	Megacephalous	SILURIDÆ		
Large-winged	Swift	Beautiful	EAGLE	*Aerial*	
Black	Beautiful	Megacephalous	PIGEON		
Deep-voiced	Small	Large	VULTURE		
	Bold Paul	Old Peter	MAN	*Terrestrial*	
Black	White	Horned	OX		
Piebald	Black	Red	HORSE		

This diagram is found in the MS. on the following page, owing to lack of space.

[1] The same word as above, translated " individua."

In the same way as changes took place in the main genera from increases and decreases so also they took place in the particular genera that followed them, and so on in all the genera, species[1] and individua. If there could have been any division below that of individua, these would have given rise to yet another division.

All animals alike possess bones, flesh, veins, arteries, *et cetera*, as we pointed out above ; trees generally possess woodiness and fruit ; herbs possess a stramineous and volant nature ; while shrubs are endowed with the characteristics of both the above. The same (division) applies to all animal-plants. The elements imparted through their movements to each of the genera some characteristics that are common to all, and some others that differentiate them. They imparted also to the species characteristic differentiations whereby each of them is distinguished from the other, as the horse and the ox are distinguished from the ass. This was affected in the first composition of the elements, through the proportionate quantities of heat, cold, humidity and dryness, the antagonism of which gave rise in each species to
an increase or decrease different from that of its neighbour, 137
and gave it distinctive marks. This is the reason why they acquired through this movement, different forms : long, short, or round ; and also colour and other different characteristics, which it is not necessary to enumerate individually, and thus prolong our discourse endlessly ; and those characteristics have been handed down to the present day in each species.

Accidental variations also occurred in individuals of each species, not, however, through that variation which perfected this species in its essential existence and constituted all the genera and species, but through the accidental variation that occurred to each individual at the beginning of its composition from the semen. This resulted either from addition to food, or from climate, or from any other external accident that affected individuals of the species, and not all the species. In this way, one became red-haired, another snub-nosed, and another tall. In this way also this one may be said to be the son of that one,

[1] In the text, " species " is placed before " genera."

and both of them the children of another ; but all (these characteristics) as combined in any one individual will not apply to another. The first variation of the species is effected by the movement of the union of the elements with one another, and emanates from God, who implanted this movement in them, through their mutual antagonism ; while the second variation comes from the same antagonism of the movements of the elements, in accordance with what the Creator placed in them, but through the medium of the parents.

This may be illustrated from speech, which is composed of substantives, verbs, and μετοχαί, that is to say, participles, *et cetera*. Although these substantives and verbs denote the same genera and species in different countries, yet the (individual) names of these genera and species are different. For instance, one Alaph is not distinguished in its genus from another Alaph, or one Baith from another Baith, but in their species they differ, as one Alaph may be long, another short, this one beautiful, and another one not ; and so also with Baith, Gamal, *et cetera ;* while all of them have been formed through the process of genera, species and individua, from the various movements : long, short, forward, backward, round and tall.

Genera and species are always and invariably the very same, while individua may or may not be alike. It is in this way that I understand the classification of the genera, species, persons and individua, which I said emanated from the causative increases and decreases.

138 The functional and utilitarian ground for the division of the above not into one genus or one species alone is the following : if all the world were composed of one genus or one species of things, the species and the individua would have been identical with one another, and this would have given rise to confusion between the species of man and of different species of animals, and in this way it would have been impossible to distinguish the species of man from that of the ox, of the bear, or of any other species, and to know which was man and which was not ; and in the species of man, who was a king and who was a slave. If this were the case, the (order of the) world would have come to an end through the mistakes that would have arisen ; and a

worse result would have been that weakness would have been imputed to the Creator, as it might have been believed that He created the universe in one species haphazardly, or that something prevented Him from creating different genera and species. Immeasurable benefits therefore accrued from the differentiations of the genera, species and individua. Indeed the number of these differentiations of genera and species bespeak the infinite wisdom and omnipotence of the Creator.

Further (these differentiations) teach us justice in our social intercourse with one another, in our (agreements and) disagreements with one another when buying and selling, and in other things. Indeed (without them) one individual would have taken the thing which we were wishing to give to another, and we should not have recognised our sons and daughters and their mothers, and this would have given rise to innumerable injustices. It follows that the (Creator) acted very wisely in not ordaining the genera, species and individua to be identical with one another.

CHAPTER XVII.

The reason why in the first composition the genera and species of man, horse, ox, et cetera, came into existence from the elements, and did not vary from the beginning up till now ; while other species come into existence in our days, such as flies, midges, tape-worms, et cetera.

We affirm that the first composition of the genera and species, such as man, ox, and others, which took place at the beginning, came into existence from the general movement of the simple elements, whereby the composition of the four compound elements took place from the simple ones. That movement did not occur again up till now, and cannot occur again. Since there are no simple elements that come together for the purpose of (general) composition, there are no genera that emanate from them. Indeed the genera and species came into existence once, in the same way as the coming together (of the elements) occurred once ;

and as there is no time intervening between the simple and the compound elements, but only priority of order, as we have shown above, so there is no time intervening between the compound elements and the genera and species which emanated from them, in such a way that we may assert that this one was made at this time, before that one at that time. On account of our weakness, this composition was extended with us to years, but so far as the Creator is concerned, He (made the world) all at once.

139 The unimpeachable testimony of the prophet shows us this: "In the beginning God made the heaven and the earth."[1] This implies that both of them were made in the beginning, and not the heaven in the beginning, and after this beginning, the earth. It is on account of our weakness that the division of first day and second day, *et cetera*, was made. Everything emanated from Him, and was completed by Him as with a sign, because there is no delay[2] in His infinite nature. In the same way as in our days compound elements do not emanate from the simple elements, in spite of the fact that these are as simple now in their nature as they were, so also species do not emanate to-day from the (compound) elements, as they did in the beginning. Indeed when they acquired at that time their separate existence they no longer possessed the same qualities that they possessed when they were united in the beginning, and therefore no species emanate from them now.

Why then have some species come into existence from the elements from the beginning up till now, for example flies, midges, tape-worms and the like? We answer that the elements possess two increases: that of generation and that of growth. One of these concerns their substance and does not change; while the other concerns generation, and takes place after the species of the first composition, emanating as they do from the elements, and not changing in their substance, have undergone changes (in their composition), deteriorated and rotted. This is the reason why (the species of the first composition) have come down, through generation, from this being to that being, up to the present time, the composition of the true substance which is in them remaining, like leaven, and taking in

[1] Gen. i. 1. [2] Read *tuhhāya*.

some of them the form of semen, and in others the form of eggs. The essence which is in the semen draws to itself, through food and drink, the substance which emanates from the elements, and which has not changed (from the beginning); and in this way the growth of the bodies of animals takes place.

As to the species that come into existence to-day from the elements, they emanate from the increase that takes place after (the first composition) has rotted, deteriorated and undergone changes: examples of these are tape-worms, midges, flies and lice. The reason why these have no generation is that they come into existence by accident, from the corruption that takes place; and in this way they come into existence and cease to exist accidentally. This is known to be so from the fact that until meat has rotted, deteriorated and decayed, no worms emanate from it; and until dung has rotted, it does not breed [1] fleas.

Let us now take an example of how things in the cosmos grow, from man, who is a microcosm, and who comes into existence from semen. That semen is a small particle of matter, but it possesses great potentiality to come into activity. It takes from father and mother all things belonging to their nature: head, hands, feet, eyes, nose, liver, and all the other organs of the body. It afterwards receives from outside, through the movement emanating from nutrition, the things that resemble it—that is to say, the true substance (which emanates from) the elements; and, subsequently, little by little, it attains the state of resembling its parents. Similar things happen to all the species that have offspring resembling them.

After they have received their food, however, something remains of it, which deteriorates and decays, and which the first matter [2] does not receive for the purpose of growth. From this residue, when found inside, worms emanate, and when driven outside, lice and nits are generated. And as there is
change in the decay, which is continually making a fresh be- 140
ginning, and accidentally coming into existence and ceasing to exist, genera and species continually emanate from the elements. Because, however, the first composition took place once in the

[1] Read *'ābedh*, with a *Dalath*.

[2] *I.e.*, the first composition, as above.

beginning, and will not take place again, no more genera and species will emanate from it.

CHAPTER XVIII.

On the reason why all quadrupeds—with possible small exceptions —and all birds have a tail, while man has not.

We affirm that man possessed an upright stature because in his first composition a continuous and straight upward movement occurred in his body. This is the reason why no superfluity could collect in the middle of it, through the cessation of the movement, and thus cause a kind of inclination outwardly ; although, as we stated above, he had his genital organ and testicles outside, on account of the heat which predominated in males over that of females. As to all the other animals, such as the horse, the ox, the ass and the rest, no straight upward movement occurred in them, but their length inclined towards the earth ; and when the impetuosity of the movement was checked from the direction of the head and the middle, it went to the buttocks, and there matter, becoming heavy and preponderating downwards, made the hind legs, and thus caused a kind of angle. Then matter expanded there, and caused an accumulation outside, which made the tail. As those who set off to run along a certain path, when they reach their goal and turn to their right into another path, delay their movement in turning round, and then begin to run in the other path : in this same way delay was caused in the movement of the composition of animals near the buttocks, where the matter that had been delayed caused a superfluity which leapt outward, and gave rise to the tail.

The functional ground for the existence of the above is the following : animals have tails to act as a protection and a covering to their anal orifice, so that it may not receive injury from an outside blow or any other accident, and thus suffer, or be hurt or damaged, with the consequent cessation of the animal function, and death. As to man, he is in no need of a tail, because, on account of the uprightness of his stature, his buttocks cover his anal orifice. The same result is also obtained by his clothes. Further, tails came into existence for animals, to act as hands in that place, and thus drive away flies[1] and other insects which settle there. Thus they serve as a hand-fan in the direction of the testicles and thighs, and all round the buttocks.

Animals walk also on all fours because God created them in order that they might serve and assist man generally, and that he might ride on them and load them with all his movables. Indeed if their posture were upright, they could not be of such service, as they would have been too weak to carry things, since their hind legs alone would not have been adequate to carry both their body and the burden of the carried things. When, however, the burden is placed on the middle part of an animal, it carries it equally with its fore-legs and its hind-legs.

CHAPTER XIX. 141

On the reason why some animals have horns, and some not; and why of those that have horns, some have one only, in the middle; why some of them have hoofs, and some not, and of those that have hoofs, some have cloven hoofs and some round; and why some others have claws.[2]

We stated previously that in the movement of the general composition, changes took place that affected the particular compositions of the species and their differentiations. And owing to the antagonistic parts of the elements, which caused their increase and decrease through the balance or lack of balance of the movement, as we wrote at the beginning of this our work, more than one species came into existence. Owing to the above increases and decreases, there emanated from them species of animals, some of them with, and some without, horns. The change that took place from this increase and decrease, and gave some animals horns, came about as follows: the bony,

[1] Read *dabbābé*.

[2] For a similar statement concerning the origin of hoofs, horns, etc., see *Firdausu'l-Ḥikmat* of Ibn Rabban, pp. 28-29.

venous and cartilaginous matter increased in them in comparison with other animals, and thus an excessive matter of this kind came into existence in them; and owing to the fact that this matter did not become thin and evaporate, because of the thickness of the animals—such as the ox and the goat—heat accumulated inside: both the heat found naturally in the composition that had accumulated, and the one found in the second addition; and this heat lifted that matter upwards, as happens inside a pot. When it reached the head, it collected there near that part of the back that is denser, firmer and colder. Then it was driven outwards towards the front, because of the excessive heat that is found there; and through the driving power it was driven outside, like a superfluity, and in this way it made the horns. These came first small and soft, and then dried up, through their contact with the air. What was thick and earthy in that matter, and consequently did not go upwards through heat, preponderated downward, and made the hoofs.

The animals that possessed not much, but a little, of the above cartilaginous and bony matter, had neither horns nor hoofs, but only claws. This is the reason why those of them that have claws have no horns. In general horns, hoofs and claws are of one substance, although differing in some respects, and participate, like tendons, in the nature of bones, sinews and nerves. When we see them and feel them, we know that they are hard and unbreakable, like bones, and do not bend, as tendons do on account of their cartilaginous and sinewy humidity. Their composition, however, is weaker than that of tendons, because when bent they are seen to be on the point of breaking. It is known that horns are of the nature of bones, from the fact that those animals that have horns have no incisors in the upper jaw.[1] This arises from the fact that when the matter of the body was driven up to the head, it mixed with the other matter that had come to the upper gums, and drew it to itself, on account of its affinity with it; and it became bone.

The reason why some animals have one horn in the middle,

[1] Lit. "high teeth." It is interesting to note that this statement is found in *Firdausu'l-Ḥikmat* of Ibn Rabban, p. 28, where we read that the animals that have horns have no incisors in their upper jaw.

while others have two horns, one on each side, is the following : 142
a sufficient quantity of matter to give birth to one horn only collected in the head ;[1] and because it was not sufficient to give birth to two horns, it stopped in the middle, and did not divide itself to this side and that. The animals, however, that have two horns, possessed more of that matter than was necessary for one horn, and consequently it divided itself to this side and that, and gave rise to two horns. This is illustrated by the practice of those who mould bricks : when the mould holds the matter of one brick only, they mould one, but when the matter is above the requirements of one brick, they push aside the residue, with which they make another brick. It is in this way that two horns came into existence in some animals.

Why are the hoofs of some animals round and not cloven, and those of some others cloven ? We answer that in the case of round hoofs, a more humid matter collected in the feet of those animals, and this humid matter gave rise to round hoofs ; while in the case of the cloven hoofs, a drier matter collected. Indeed we see that humidity joins separate parts into one, as when it makes dust into mud ; and dryness separates these parts, as happens in the case of a humid soil, which cracks after it has become dry. That dry parts contract inwardly, and humid parts expand outwardly, the very nature of hoofs shows. Indeed the hoofs of oxen and goats are harder and drier, and are not easily broken, when compared with those of a horse or an ass.

As to the animals that have claws only, they possessed only a small and liquid portion of matter, which gave birth to these claws at the extremity of the toes. The small quantity of this matter is explained by the proportion of the increase and decrease that arose from the antagonism (of the elements) in the temperament of these animals ; and their toes became bony on account of the great quantity of bony matter that increased there. And because of the dryness of these toes—since they are found in an external position—division occurred in them in the fore and hind feet. It is in this way that toes came into being. Further, this place is devoid of flesh ; and cartilaginous and bony matter predominate there, on account of its remoteness from the

[1] Read *brīsha* "in the head," instead of *barnāsha* "man."

middle (of the body). Because of this remoteness, the more liquid and thin part of the matter that came to the toes evaporated before it reached them ; but the part that was thick and dense reached that place. This explains why the fore and hind feet are drier (than the rest of the body).

CHAPTER XX.

On the reason why particular species of animals have only one food, which in the case of horse, ox and the like, is grass, in the case of lion, wolf and the like, meat, and in the case of the birds, some meat and others grain ; and the drink of all of them is water ; while man eats meat, bread and innumerable fruits, and drinks wine, water and various other kinds of beverages ; and this in spite of the fact that all of them emanated from the elements.

We affirm that man has by nature the power to abstain from
143 food when desiring it, and to add to it after being satisfied. This characteristic none of the animals possess, as we demonstrated above, since it does not pertain to the nature of the elements, but to the rational and spiritual soul, which brought about changes in man through its union with his body, and gave him a desire to add to his food after satisfaction. This desire the nature of animals does not possess. This very power enabled him to take all varieties of food. It is through teaching that we take from one another the different kinds of food, and it is through it also that we make additions to our food every day, till the end (of our life). As to animals, they have from the very beginning, and without any teaching—from the nature of the elements, and through the animal spirit which they possess—one and the same kind of food, without addition or diminution. Further, as the rational soul gave to man, through the power which it has possessed from its creation, the prerogative whereby, also without addition or diminution, birds, trees, corn, fish, *et cetera*, are subject to him—in this same way it made their food to become

his, and made them to become both the carriers of his food, and his actual food. It imparted to him also the power to prepare all the varieties of his food, such as bread, meat, vegetables, *et cetera*, with rational thought. The same thing applies to his drinks. As to animals, they do not act in this way, for they do not take their food after rational preparation ; and because they have one nature only, they eat their food in the simple state in which it came from the elements. Indeed it generally happens that when their food is prepared with thought, they do not take it. This arises from the fact that they do not possess in themselves a power akin to that which has changed the food : like things are acceptable to like. This is the reason why man makes use of many varieties of food and drink, while animals do not.

The reason why some animals eat vegetable food, and some others meat, and likewise birds, is the following : we stated and demonstrated above that two of the elements are active, and two passive. The two active elements predominated in the carnivorous animals in their very first composition from the elements. This explains why, on account of their strength,[1] these animals were in need of a stronger [2] food, which is meat. In this way they became carnivorous, and generally more active and savage, and not placid. The passive elements predominated in the herbivorous animals, which found themselves in need of a weaker food, which is grass, barley and straw ; and they are consequently more placid. This is known to be so from the fact that the active elements are more animated ; and they wage an open war, and become tumultuous, when they meet each other ; while the passive ones are not so.

The functional ground for the existence of the above is the following : man is higher than all animals, birds, creeping things, trees and grass, and God made him their lord, and created them for him, and not him for them. Having thus made them
for his service, the Creator imparted to them the power of 144
supplying him with his different kinds of food, just as they were subjected to him (in other ways). In the same way as a king makes use of different varieties of food, clothes and other [3]

[1] Lit. "activity." [2] Lit. "more active."
[3] The copyist puts here the word *haiwātha*, "animals."

luxuries, more than his servants, so also man is richer in varieties of food than all other animals.

After having completed our discussion of the species of the terrestrial animals, let us now discuss the species of the winged birds.

CHAPTER XXI.

On the reason why winged birds are oviparous, and do not conceive in the womb, like the quadrupeds, and have no womb, no renal bladder, no outward place for the private parts, and no hands.

We spoke above of the coming into existence of birds, and of the fact that their habitat was the air, because in the first composition of the species, they were made in the place in which the nature of air predominated over that of earth and water. Because earthiness diminished in them more than in the terrestrial animals, the quantity of the thick, earthy matter in them also became smaller. Owing to this diminution, there were also in them fewer of those organs that are more earthy and cartilaginous, such as the womb and the bladder. Indeed, when earthiness accumulated, that part of it that was most earthy and dry brought the bones into being in the impetus of the movement of the first composition; and after this, the part that was smaller in quantity, less earthy and less dry, made the nerves, the ligaments, the tendons, and such like. On account of the smallness of the matter, the two above organs that were not necessary to the life and growth of the body were left out. The natural movement made the necessary organs first, and left out the other organs, which are only for the reception of superfluity, and without which animals can live—that is to say, the bladder, which is for the purpose of urine, and the womb, which is for the purpose of conception. Indeed the animal may live without the reception of the semen and the conception which take place in the womb; and (in the case of birds) their urine flows through the path of the food digestion. The animal, however, cannot live without bones, nerves, veins, arteries and the like. This is the reason why the above two organs were left out in them.

Birds have eggs for the purpose of procreation, because their flesh is more humid than that of the terrestrial animals; and owing to their humidity, and to their lack of the earthiness possessed by the other animals, when the semen falls in them it does not thicken inside them, so as to give rise to bones, nerves, head, feet and the rest of the body, until the eggs are cast out and come into contact with the earth, and with the dry air. Then the latter desiccates that humidity which had predominated in them; and through the intimate contact with the mother, who imparts from herself a kind of natural and penetrating power, that humidity which is in the egg becomes hot; and through this power the embryo is fed from the matter that is inside the egg, in the same way as an embryo is fed [1] in the womb by the heat and humidity of the blood of menstruation. After it has formed itself into a body, and reached perfection, it comes out of the egg, like the child which has finished its course in the
womb of its mother. The outer shell of the egg prevents the 145
body from being injured, as the womb does in the case of the child.

The reason why the egg is soft and watery in its composition, and possesses an outer shell that is hard and earthy, is the following: we affirm that in the semen which gives birth to the egg are the four elements, and the watery power vastly predominates over the earthy power; and because of this predominance of the watery power, and the lack of the earthy power, when the first movement towards the composition of the body occurred, the earthy power was overcome, on account of its small quantity, and was driven out by the powers that overcame it, towards an external position. Lo, those who escape take refuge in a foreign country! That (the egg) has much humidity can be seen in the quantity that is found in it. Indeed humidity predominates over the outer nature (or the shell) many times over.

The egg became round [2] because when movement occurred, and earthiness was driven out from the middle towards the outside, it was driven out from all directions: from the front,

[1] Read the verb in masc. [2] Read *galilta*.

from the back, from the upward direction, from the downward direction, from the right and from the left, and so the egg became round. The fact, however, that it is not spheric, but slightly elongated, was brought about by the active powers which all animals possess, as we demonstrated above, where we discussed the φυσιολογία, that is to say, the science of nature, and spoke of the length of animals.

Why have birds no hands? The reason for this is to be sought in the fact that in the very first composition the matter which made the hands was lifted upwards, because of the predominance of air which birds possessed. This matter gave birth to wings, in the same way as in the terrestrial animals earthiness predominated and gave rise to hands, which preponderated downwards. The (matter) which was most earthy preponderated downwards, and gave rise to feet, so that (the creature) might walk on the earth, as (terrestrial) animals do. Further, birds became possessed of feathers, from the superfluity which was produced in them, was parched, and was driven outside, like hair on the bodies of terrestrial animals. This process became more pronounced in the feathers of the tail, because it is there that the matter ran, on account of the perpetual movement which birds possess there, in their flying, more than in all the rest of the body.

The functional ground for the existence of the above is the following: birds have neither womb nor bladder nor outward private parts [1] as the quadruped animals have, because if they had all these they would have possessed too much width, and would have become heavy; and thus they would have flown weakly, would not have been able to fly high, and would have experienced fatigue more quickly. When the wise Maker created them, and wished them to dwell in the air, He deprived them of the above parts, so that they might not become heavy. Further, He made their chest pointed, in order that when it came into contact with the air, it might cleave it without hindrance. It is in this way that the builders of ships make the prow of a ship pointed, so that in moving she may rend the air and the water. If she had been built in another way, she would have been

[1] Cod. strangely "stomach protruding outside."

impeded in her course by the water. Our God, the wise Artisan, made birds after this fashion, with pointed chests.

Further, they are oviparous, and not viviparous, because they perform the function (of procreation) frequently; and if the chicks had been inside them, the principle of this procrea-
tion would have ceased, on account of the exiguity (of the 146
place). Further, they are not viviparous because they have no womb.

Birds have also thin legs, on which there is no flesh, in order that they may not become heavy in flying, and also in order that they may be able to bend them and insert them under their feathers, so as not to be impeded by air. Instead of hands, they have also wings, on which feathers grow, so that through the latter the wings may obtain more extension, and so that when the wings shake the air which accumulates under them, they may not pull down the body of the bird in its *élan* upwards or forwards —like swimmers who push the water with their hands and feet backwards, and drive the body either forwards, or upwards or downwards, and thus proceed in their course in the water.

CHAPTER XXII.

On the reason why the winged birds have no incisors, no molars, no ribs and no spinal vertebræ.

We stated above that in the terrestrial animals, earthiness predominated over airiness, and because of this earthiness which predominated in them, they possess, more than the aerial animals, bones which are cold and dry like earth, and very hard and rigid. They were thus endowed with many ribs and spinal vertebræ in the lower part of their body, and with jawbones, together with incisors and molars, in the upper part of their body. Since, therefore, there was lack of earthiness in their tail portion, birds possess neither ribs nor spinal vertebræ (in that part of their body); nor do they possess molars and incisors in the upper part of their body, but have a part only of the veins, nerves, bones and tendons (possessed by the terrestrial animals).

Some of the matter went up[1] immediately towards the head, and made for them beaks, the nature of which resembles that of horn, and which in them takes the place of molars, incisors and lips.

The upper part of the beak of carnivorous birds, such as the eagle, the hawk, *et cetera*, became bent downwards, while both parts of the beak of herbivorous birds, like pigeons and sparrows, were bent equally, for the following reason: the upper part of the beak became bent, in the carnivorous birds, because of the heat and dryness that more than in the other (birds) predominated in them, in the nature of their upper part. Indeed, dryness bent the beak downwards, and heat elongated it; and its bending took a downward and not an upward direction on account of the impetuosity of the movement of composition which affected the upper part of the beak more than[2] the lower part. Further, this class of birds have longer, sharper and more bent claws than the herbivorous birds, because of the preponderance (of heat and dryness) of which we have just spoken.

The functional ground for the existence of the above is the following: birds do not possess incisors, molars, ribs or spinal vertebræ, lest they should be too heavy for flying. Further, they are in no need of incisors and molars, because they do not require to masticate their food, and thus render it more easily digestible; but their food goes down as it is. Because the nature of birds is hotter and more humid than that of the terrestrial animals, when food goes down to their stomach, they
147 experience a more powerful digestion than the terrestrial animals. Lo, grains of wheat and barley, and seeds of millet[3] are never digested in the stomach of terrestrial animals without being first ground (by the teeth); while they are quickly dissolved in the stomach of birds, and nothing of their original shape can be recognised.

Birds have also, for their convenience, a beak of the nature of horn, in the place of lips and teeth. It became non-fleshy in order that it might not easily receive injury, owing to softness and humidity, when coming into contact with the hardness of

[1] See p. 108, on the origin of teeth, nails, etc.

[2] Read *min* for *'am*.

[3] So I translate the word *bashshūḳé*.

the earth, and when pecking seeds. It serves also as a weapon for them in their fight with their enemies. Carnivorous birds have the upper part of their beak long and bent, in order that they may tear meat and cut it to pieces little by little; while the others, because [1] they eat seeds, do not need to have the upper part of their beak long and bent. The former have also long and sharp claws, in order that by means of them they may grasp pieces of meat, and tear, cut and eat them; and also for the purpose of fighting their enemies.

Birds have a long neck in order that, when extended in their flight, it may penetrate the air, and thus rend it with a thin head, and bring after it all the rest of the body.

Terrestrial animals have molars and incisors to help digestion, because they possess less heat and humidity, and the parts of the latter which they do possess are thick; while the parts of them possessed by birds are thin. Further, terrestrial animals have spinal vertebræ, in order that the weight of the body may lean on them, and the body itself be attached to them, so that when standing or walking on the earth, it may not bend downwards. This holds good especially in the case of man. They have vertebræ also in order that the ribs may be attached to them, and hold inside [2] them the inner organs. As to birds, because they fly in the air, and are not constantly in need of being supported on the earth—since the air carries them—they do not require spinal vertebræ. Further, on account of the paucity of their internal organs, they do not require ribs.

After having spoken of terrestrial and aerial animals, let us speak now, in the measure of our capacity, of aquatic animals, such as fish, *et cetera*.

[1] Read *meṯṯol.*

[2] Read *b-ghawhain.*

CHAPTER XXIII.

On the reason why fish are longer than all the terrestrial and aerial animals, have no hands and feet, and are not drowned in water ; and on the reason why some of them have white scales, some of them black scales, and some of them no scales at all ; and why some of them have shells.

We said above that in the first general composition (of bodies) the genus of fish came into being where[1] humidity and cold predominated. This is the reason why their habitat was in the water, an element that is affinitive to them, in the same way as birds have their habitat in the air, and terrestrial animals and plants have theirs on the earth.

The reason why fish became long and without hands or feet is the following : humidity predominated in them more than in the rest of the animals, and because of this, when the movement of the active elements occurred in the first composition, and expanded their length, as in the case of the other animals, it
148 found in them a matter that was more humid and liquid, and thus it had a better opportunity for expanding them lengthwise. The reason, therefore, why length was added to them more than to the other animals is that their nature was more amenable to expansion, and so length predominated in them. Lo, humid plants give more, and dry ones less, scope for expansion. Indeed in the same way as the matter of dough is more easily shaped and elongated, but when it becomes dry it is elongated with difficulty, or not at all—in this same way the matter of fish acted.

Fish have no hands or feet because the matter that was fit for hands and feet was elongated in the same process, on account of its softness, and thus added to the length. The small remnant that was left of it became to them something like wings, to help their movement.

The reason why they are not drowned in water, in spite of the fact that it penetrates into their mouth, is the following : they open their mouth, and the water goes into it when they wish to eat food ; but they press that water, and it comes out

[1] Read *aiḳa*.

of their ears (or gills), while they seize the food with their mouth and send it into their stomach. As to the rest of the animals, they have no openings from the mouth to the ears, and when they receive water from the mouth, it goes into the stomach, and they are drowned, because they cannot endure in water without breathing, except for a short time. When they wish to cool hot vapours, through the breathing of air, they cannot do so when they are in the water ; since instead of air, water goes into them, which not only does not cause any cooling, but extinguishes the natural heat, and thus causes the death of the animal. When fire is kindled, and through air fumes emanate from it hither and thither, and it flares up, if we throw a great quantity of water over it, not only will it not flare up, but it will be extinguished. The same happens to the animal fieriness, which exists in the heart of animals.

As to fish, because the heat that is in them is very small, it is not kindled, and does not give rise to vapours ; and because it does not give rise to vapours, it is in no need of cooling or breathing. This is the reason why fish do not possess lungs for the in-drawing of air. The reason, therefore, why they are not drowned in water is that they do not breathe. Even when they are in need of cooling, their cold and humid temperament cools them instead of air ; and when they come out of water they immediately die, because air penetrates into them, and heats and evaporates the natural humidity that is in them, and thus causes their death.[1]

The nature of fish is intermediate between that of the terrestrial and the aerial animals. They differ from the[2] terrestrial ones in the fact that they have fins on both sides, and in this they are more akin to birds. They do not, however, require powerful and large wings, because water takes the place of these for them, on account of the contraction which it possesses more than air. In this way they cleave the water, and penetrate through it, as a bird cleaves the air and penetrates through it.

Fish do not walk on the earth, and in this way they resemble aerial animals more than terrestrial animals ; but they do not

[1] *Cf.* Aristotle, *De Sensu*, ch. 5, and *De Respiratione*, ch. 15, etc.
[2] Add here *min*.

fly in the air, because when compared with birds, they are colder, more watery, and heavier.[1]

149 The reason why some of them have scales and some not, is to be sought in the fact that when dryness and humidity mixed with heat and cold, in the very first composition, there was a fight between these two antagonistic forces, and because dryness was overcome by humidity in the temperament of fish, it was driven by it inside, where it gave birth to small and weak bones; and some part of it was driven outside, and gave birth to scales.

The reason why, when water moves on and around bodies, it makes them round, is the following: all of it encircles at once the whole of the body, and does not remain in any one spot, on account of its liquidity. This explains why the scaly body of the fish is round. Lo, pebbles in all rivers are generally round and smooth, from the movement of the water, and their parts are even all round.

The reason why such fish as the siluridæ have no scales or thin bones, but only a backbone, is to be sought in the fact that in their composition the dry part found in them was smaller than that of the other fish; and on account of this smallness it was not able, in the fight which it made in the first composition, to give birth to scales outside and small bones inside, but gave birth only to a backbone, so that the animal might be supported by it. What was necessary (to the animal) came into existence, and since there was no superfluity, nothing else came into existence.

Why are scaly fish white, while the siluridæ and some other fish are black? We answer that some parts of the superfluity are more contracted, and some others more expanded. The bodies which have no scales emanated from the humidity that was more contracted. The latter fish, therefore, became black, because of this contraction, while the former became white, because of the expanded nature of their humidity. Lo, we see that when the bodies of animals contract and become cold, their external surface becomes livid; and when they expand, that lividness departs from them. That from contraction blackness is produced is seen in clothes which keep off from us the white-

[1] Read *yaḳḳîré*.

ness of light in proportion to the closeness of their texture, and show us its whiteness in the measure of the looseness of their texture. Indeed their very appearance [1] indicates the closeness or otherwise of their texture.

The reason why some sea-creatures have shells, such as crabs, turtles and other shell-fish, is the following: the earthy part that is very dry predominated in them, and hardened their external parts. This is the reason why they walk on the earth, and their composition and temperament are hard. If anyone controverts this, let him resort to the principle which we stated at the beginning, examine the elements, and refer that which predominates in animals to the predominance of these elements: if dryness, to earth; if humidity, to water, *et cetera.*

We have written briefly these things on the coming into existence of the aquatic animals. On the subject of their history, on how they bring forth their young and rear them, on their genera and all other similar things, many naturalists have spoken.

The functional ground for which God made fish long is the 150
following: they became long so that they might be fit for swimming and for movement in the water. Indeed, if they had been round or square, they would have moved less easily, and would not have been able to rend water except with difficulty. When the front part of a long body rends the water, the whole body is impelled after it easily. Lo, seamen, noticing this shape, made all ships long. I believe that they took this shape from that of fish. They made also for their ships oars on both sides, in imitation of the fins of fish.

Fish have also no feet, so that they may not become heavy, and be weighted down by them, and their impetus be checked in its course.

Fish are not drowned in water because their temperament is similar to it; and having less of the aerial nature in them they are in no need of air, and have consequently no lungs.

Let us here put an end to this second discourse.

Here ends the second discourse.

[1] Lit. "touch."

THROUGH GOD WHO PERFECTS, WE BEGIN THE THIRD DISCOURSE.

AFTER having spoken of the differences of the genera and species of terrestrial, aquatic and aerial animals, of their coming into existence from the elements, and of their functions, we will begin now, by the assistance of God, to speak of the differences of the genera and species of tastes, colours, smells, sounds and feelings, and of how the perceiving power receives them through each one of the senses.[1] We will begin first with tastes.

CHAPTER I.

On the genera of tastes.[2]

The single name " taste " indicates a kind of *summum genus*, which is divided into seven genera, namely : sweetness, bitterness, sharpness, saltness, sourness, acerbity, and ἄποιον—that is to say, something tasteless—or tastelessness. Each one of these genera is sub-divided into species, and these species themselves into individua. Sweetness, for instance, is sub-divided into the tastes of honey, sugar, dates, pomegranates, *et cetera ;* while honey itself is sub-divided into sweeter (honey) and less sweet (honey), and pomegranates into less sweet

[1] The reader may advantageously compare all this discourse of the author with Aristotle's *De Anima*, Book II, ch. 1-12, and *De Sensu*, esp. ch. 3 and 4. Our author, in contrast to Aristotle, gives the elemental reasons for the working of each sense, and divides the different sensibles into different sections in greater detail.

[2] *Cf.* on these tastes Aristotle, *De Anima*, Book II, ch. 10, where the classification is not similar.

(pomegranates), *et cetera.* Each of the remaining tastes is similarly sub-divided, as is shown in the following diagram :—

Degree	Example	Taste
Less sweet Sweeter Medium sweetness	Pomegranate Honey Sugar	8[1] Sweetness 200
Very bitter Less bitter	Aloe Wormwood	8 Bitterness 12
Very sharp Biting	Pepper Ginger	8 Sharpness 12
		8 Saltness 12
Very sour Powerful	Common vinegar Citron	100 Sourness 200
Very astringent Less active	Gall-nut Sour-dock	100 Acerbity 12
		Tastelessness

SEVENFOLD DIVISION OF TASTES

There are seven tastes, because each of the active elements is divided into three orders : a high order, a middle order and a low order. Each of the three is found in both heat and cold. The last taste is composed of all of them.

The functional ground for their existence is the following : 151
God made them for the utility of the animals and their feeding, especially man. Indeed the food of (herbivorous) animals, which consists of barley and straw, has also sweetness, and is agreeable to their palates. This is the reason why they take their food readily. If it were tasteless, they would not have enjoyed it, and would have received less nourishment from it. This is borne out by men who, in the case of food that is agreeable to the palate, eat more of it, while in the case of food that is less agreeable, they eat less of it, and very often leave the table hungry and unsatisfied. This holds good also for the carnivorous

[1] I cannot understand the meaning of these numbers, unless they refer to the numbers of the individual tastes contained in each species.

animals. The food of men is salted, because no taste is agreeable without salt. Sourness increases the desire for food. Acerbity is useful in different ways, such as strengthening fruits at their beginning, so that they may not fall on account of the humidity which they always contain, and thus perish before they are ripe. It helps also to strengthen soft bodies. As to sharpness, it flavours tastes, and by its stinging and heating, it facilitates digestion. The other varieties of tastes serve also the requirements of all animals. It is not [1] necessary, however, to speak of each one of them in particular, as they are known and patent to everyone ; but we will show the cause of their coming into existence from the elements, in accordance with what the Creator placed in their movement.

CHAPTER II.

On sweetness.

Sweetness comes about when an even quantity of heat permeates the humidity (of a body), and through its predominance softens its contracted and hard parts, or impedes them from forming. This is shown to us in the species which undergo changes ; for instance, when the heat of the sun joins with the humidity found in grapes or apples or pears *et cetera*, it softens little by little their hardness and the contraction of their parts, and expands these parts that had contracted on account of cold. That heat does this naturally, when mixing with humidity, and causes sweetness in them, is known by the fact that those of them which come into contact with the heat of the sun ripen more quickly. This occurs when well-balanced proportions of heat join with well-balanced proportions of humidity ; and after this humidity has been warmed by heat, the result is a balance of parts agreeable to the tasting palate. This is called sweetness. Indeed all things that have well-balanced parts in all species are agreeable, and are tasted with pleasure. Lo, the pleasantness and sweetness which occur in cooked food are brought about by the balance of the quantities of the constituent parts, which

[1] Read *lā*.

causes pleasantness; and their lack of balance causes unpleasantness. Why should we say more about these things, while life itself is sweet and pleasant through the harmonious balance between the elements?

Sweetness causes pleasantness to the palate in the following way: sweetness comes into contact and joins with the humidity 152
of the mouth, which itself consists naturally of a well-balanced and even humidity mixed with heat. Because the humidity and heat found in that sweetness are affinitive to the humidity and heat of the mouth, the latter does not show aversion to the sweetness, but receives it and presents it to the palate, which takes pleasure from it on account of the general balance of proportions, which is not impaired by any increase or decrease of other tastes that might cause repulsion to the palate.

The reason why the mouth, and not the ears, the nose or the eyes, became the tasting organ, in spite of the fact that they were composed, like it, from the four principles, is the following: it acquired this function from its composition and its position. Indeed the shape of its composition does not resemble that of the ears or the nose, nor do the objects received by the latter from outside remain in the middle of them, as food remains in the mouth after being received by it. Unless the mouth closes and opens, and thus introduces and ejects air, and this air joins with the palate above and with the tongue below, taste is not perceived. As to the other openings, they do not possess this function, for they neither receive food nor possess a tongue or a palate. This is the reason why we do not taste with the above organs. Indeed if the mouth were like the ears, it would have accomplished the things which the latter perform, and the same would have happened if the ears were in the place of the mouth, or of the eyes, or of the nostrils. It is obvious that the mouth possessed all these functions from its first composition, to the exclusion of the other organs of the body. Lo, the liver, the heart and the kidneys acquired a particular function of their own, and their respective functions do not resemble one another. They acquired all this from their composition, and from the position in which they were placed. This same thing happened to the mouth.

We have spoken of the five functions of the five senses at great length in a special book which we wrote on the subject.

On the bitter taste.

The bitter taste occurs when heat is in excess of a well-balanced quantity, and when a well-balanced humidity is lacking, and dryness predominates. This is known by the changes which tastes undergo. When, for instance, honey or any other sweet taste is heated in the fire, and its well-balanced humidity vanishes, and heat is added, it immediately acquires bitterness. This explains why, when the palate receives it, it imparts unpleasantness to it on account of the lack of balance which has come to it from the addition of dryness and heat, as compared with the balance found in the component parts of the palate. This is the reason why men generally have an aversion to bitterness, and an affection for sweetness. This is especially true of children, because the humid and hot temperament of their palate is more suited to sweetness.

On sharpness.

A sharp taste occurs when heat and dryness, especially the former, predominate in an object.[1] It is known that (sharpness) comes to objects through the heat and dryness which predominate in them, from the fact that when we bring any object near to fire, it renders that object sharp-tasting by imparting to it a great
153 amount of heat, as, for instance, when a stone is thrown into the fire and is well-baked, it becomes lime, which is very sharp to the taste. This is explained by the heat and dryness which have come to it. Before it was thrown into the fire, it did not possess sharpness.

Sharpness not only causes a feeling of aversion, like bitterness, but also destroys the balanced parts of the humidity found in the mouth, through the predominance of heat and dryness, as fire would. Indeed pepper, ginger, the root of horse-radish [2] and the like, which are strongly impregnated with heat and dryness, cause this sensation to the mouth.

[1] Lit. "species," according to the author's terminology, as above.

[2] The word used means generally "hemp."

Bitterness holds a middle position between sweetness and sharpness. This is the reason why anything that is lacking in sharpness is near to bitterness, such as cummin and cyperus roots ; and anything in which sharpness strongly predominates has no (bitterness), such as ginger ; and anything that is lacking in bitterness is near to sweetness, such as the root of liquorice. These three varieties of tastes came into being from the predominance of one [1] element, heat, which united to itself the humid and the dry (elements).

Three other varieties of tastes emanate from the other active element, cold, which works in the two passive elements, the humid and the dry, and brings into being the remaining three tastes, which are saltness, sourness and acerbity. As to the remaining ἄποιον taste, it comes into existence from the combination of all of them, when none of them predominates in a special way, so that it may be singled out, recognised, and distinguished by the sense (of taste).

On saltness.

Saltness occurs when a balance of elemental heat, cold and dryness is found in an object. This gives birth to an even balance, pleasurable to the palate. Lo, salt is not unpleasant to the palate, and even renders pleasant all varieties of food. It is not, however, so pleasant to the palate as sweetness, because sweetness possesses in a higher degree an even balance of parts and also of expansion. The marked taste of saltness emanates from a medium balance of cold, heat and dryness, as found in sea water. Indeed water is cold and humid ; when, however, heat evaporates the thin humidity found in it, it thickens it by intensifying the earthy matter in it. This happens when the water of the sea has been heated by the sun ; and when its thin humidity has evaporated, it has become thick and assumed a small amount of matter from heat, it becomes salty, and acquires a balance (between its elemental parts) through the diminution of its coldness and humidity. This is the reason why, in comparison with its first nature, it is hot ; and this explains why some people call bitter, hot. Lo, when we throw sweet

[1] Read *ḥadh* in masc.

water on ashes, it becomes bitter, and hotter than its first nature, because it has drawn to itself the parching effect of the fire. It has, however, received heat accidentally only, since by nature it is cold, and its aquatic nature has not changed. When sea water contracts, thickens and becomes salt, it dries up; but its dryness is less than that of the earth, and of the earthy
154 species, such as trees *et cetera.* This is known from the fact that when salt is thrown into water, it dissolves quickly, while earth does not.

If you say that, on the contrary, salt is drier, since it desiccates bodies and keeps them from deteriorating—a thing which earth does not do—(we will answer that) this happens not because it is drier than earth, but because it is more humid than it; and when it comes into contact with the humidity of fleshy bodies, because its own humidity dissolves and penetrates deeply into the bodies, it joins with them and imparts to them the dryness which it possesses in a higher degree than these bodies, and in conjunction with dryness it keeps them from deteriorating, through the balance of parts which it possesses by nature. Indeed a balance of parts preserves and does not injure, while a non-balance of parts injures and causes deterioration.

On acerbity.

Acerbity occurs when there is a great predominance of cold and dryness. Because cold and dryness contract the parts (of a body), everything that is acerb contracts them also, as gall-nuts and pomegranate peels do with damp and soft skins and other bodies [1] which resemble them. This is the reason why, when it reaches the palate, it contracts the humidity that is found in it, and draws it together, while drawing also its parts to its own composition; and by bringing these parts together it causes a non-balance of parts, by an excess of cold and dryness. This explains why, when we taste something acerb, and lift our tongue to our palate, we find the tongue acerb, on account of the lack of balance of its parts with one another. Because (the acerb body) has contracted the (balanced) parts (of the tongue), it imparts a

[1] Lit "species," according to the author's terminology.

very unpleasant sensation, as a result of which we express aversion to it, and shake our head, as if we were shrinking from it.

Lo, all fruits are generally acerb at the beginning of their formation, because earthy matter has predominated in them. This explains why they are hard and inedible. When, however, humidity increases gradually in them, and heat conquers their coldness, they lose their acerbity. That they are cold and dry is evident from the effect which they produce, because when they are brought near humid and hot bodies, they chill and desiccate them.

On tastelessness.

Tastelessness occurs in bodies when there is no predominance of either cold or heat, humidity or dryness, and so no marked trace of any taste is shown in them. When, therefore, it comes into contact with the palate, it does not cause contraction, balance of parts, increase, or decrease to the humidity that is found in it, or in its own body. This explains why there is neither pleasure nor displeasure in this taste.

On sourness.

Sourness occurs when cold is less in quantity than in the case of acerbity, and more than in that of saltness, that is to say, when it is found in a medium quantity between the two. The same also applies to humidity. This is the reason why sourness causes both pleasure and displeasure to the palate. It is pleasurable to it when the balance of its parts brings it nearer to a salty or a sweet thing. This is seen in the fruits of trees, such as
grapes or apples, which are first acerb when they possess a 155
woody quality, but when water has been added to them, and that water has been heated by the sun, and the latter has imparted a balance to its coldness, they reach that state of balance which causes sourness.

We have written these things ἐν συντομίᾳ, that is to say briefly, about the seven tastes. Let us now proceed to write, according to our promise, on colours.

CHAPTER III.

On colours.[1]

The name " colour " also denotes a kind of *summum genus*, as it is divided into six genera: whiteness and blackness, which are the universal, true and first genera, of which the remaining ones—called " genera " in a relative sense only—are composed, namely redness, saffron-yellowness, greenness and gold-yellowness. These are sub-divided into species, and the species into individual colours. As to whiteness, it is sub-divided into the whiteness of the swan (κύκνος), that of snow, *et cetera*; and blackness, into the blackness of an Abyssinian and that of a raven; and the other genera are sub-divided in the same way.

Why are there six genera? We answer: for the same reason as that which we stated above with regard to tastes, which emanated from the active and the passive elements.

On their functional ground.

Colours are very useful in the world, because they distinguish for us objects from one another. Indeed if there were only one colour—whatever it might be, white or black *et cetera*—first of all we should not have been able to distinguish between night and day, nor between evening and morning; and great confusion would have arisen, because times, years and days would have ceased to exist, and on this score, weakness would have been attributed to the Providence of the Creator. Different colours were, therefore, very necessary for the working of knowledge.

Further, if there were only one colour, there would have been by necessity one element only, and not many; and if there were one element only there would have been no generation, since there could have been neither increase to, nor decrease from, that element. It is, therefore, with just reason that there were many colours, so that the elements might demonstrate the truth of the facts (of the universe). Indeed if there were one colour only,

[1] *Cf.* Aristotle, *De Sensu*, ch. 3-4, where some similarities and dissimilarities may be found.

confusion and lack of knowledge would have resulted in connection with all species which undergo change and modification: firstly in the species of fruits, such as grapes, apples, and others, which undergo modification, for these would not have shown to the mere sight the modification which affects them; secondly the humours would not have been known, as the black bile would not have been distinguished from the red bile, nor phlegm from blood; and the science of medicine would have come to an end, since (physicians) would not have known the cause of an illness, nor would they have prescribed its remedy, and cures would not have taken place; and many other things in the world, of which it is not necessary to speak here, would have been impaired.

As to their elemental origin, it is the following:[1] whiteness occurs sometimes in bodies when heat predominates in them[2] and is joined by dryness more than humidity.[3] Redness occurs when bodies are affected by heat, and when dryness and humidity are found equally balanced in them. This is known to be so from the fact that when fire takes hold of woody matter, this matter exhibits three colours: red, white and black; white when dryness has conquered humidity, and after the latter has been affected by heat, has been conquered, has evaporated and become ashes. When, however, the humidity predominates at
the same time as heat, it causes blackness. This is known by 156
the fact that when we throw water on burning charcoal, and when the humidity of the water conquers the dryness of the charcoal, because this humidity sticks to the parts of the wood, and does not evaporate, since the wood has received its humidity from the water, it becomes black. This shows that blackness may occur sometimes from the predominance of humidity with heat. When, however, fire takes hold of wood that still keeps the humidity which brought together the parts of the wood, and after dryness has predominated and joined to this wood through heat, and when that woody matter itself holds the middle position between woodiness and ashes, it becomes red, through the heat which

[1] The same explanation, coupled with the same illustration of a piece of wood and charcoal, as to the cause of the black, white, red and green colours, is found in *Firdausu'l-Ḥikmat* of Ibn Rabban, p. 363, where the author introduces his sentence by: "Some others have said."

[2] Read pronoun in masc.

[3] Cod. "by humidity more than dryness."

kindles and heats. It is known that redness is an intermediate state between whiteness and blackness, from the fact that when redness predominates in an object, it inclines to blackness, and when it is less red, it inclines to whiteness.

The black, white and green colours came into existence from the other active element, which is cold. Indeed when it predominates with dryness, a white colour is produced, but when it predominates with humidity, a black colour is produced; and when it is in an intermediate state, a green colour is produced. This is known by their working: when snow and frost occur, and cold and dryness increase in objects, and their parts contract to such an extent that they become like stones, they are white; but when that dryness is softened in them, and cold diminishes and they become humid and their parts expand, whiteness is removed from them, and they incline in colour to blackness, as compared with their former state. When humidity [1] and dryness together with heat are found in an intermediate state, green is produced. We can test the truth of this by the grass of the earth. Indeed at the beginning of its growth and at its end it is white, but in its intermediate stage it is green. This is accounted for by the fact that at this stage humidity and dryness together with heat are in an intermediate state and in balanced quantities, since (the grass) has received humidity from rain and heat from the sun; but at the beginning and at the end, all the above are not found in it in a balanced proportion, but dryness predominates in it, heat is diminished, and it is no longer green.

The saffron-yellow colour occurs in species by the mixture of the power which causes red with that which causes white. This is known from the fact that when we mix red colour with white, it becomes saffron-yellow; as when saffron, which is red, is mixed with a white colour, it renders it saffron-yellow. This is established also by the fact that when saffron-yellow is intense, it inclines to redness; and when it is not so pronounced, it inclines to whiteness. In this same way, a livid colour comes into being from the mixture of white and black; and this is the reason why it possesses in itself both colours.

[1] Delete the *Dalath*.

There are other colours which are composed not of the two above principal (colours), but of intermediate colours that emanate from them; but because the difference between them is not so pronounced, we have not classified them (below) under the name of colours. The principal and universal colours are two, because the active powers are two, and their antagonistic qualities are also two; and from additions to and diminutions from (these two colours) innumerable other colours emanate, of which there is no need to speak.

White	Black	Red
Snow, Cotton	Raven, Pitch	Vermilion, Minium

Saffron-Yellow	Green	Yellow
Wax, Orpiment	Verdigris, Leek	Citron, Gold

CHAPTER IV. 157

On how the eye perceives colours.[1]

After having spoken of the coming into being of the differences of colours, let us now speak of how the eye perceives them and their differences from a distance, and how it draws them to itself. Lo, the mouth does not taste from a distance. (We will discuss later) how the nose smells and the ear hears.

We affirm that this quality came to the eye from its composition and its position. Indeed the composition of the eye differs from that of the mouth, and even from that of the nose and the ears; it differs from them also in its position. The eye

[1] The structure of the eye is well discussed by *Galen*, Vol. I, pp. 607–651, but there is nothing in him about the process of vision.

is clearer, more transparent and more brilliant, and on account of its clearness, transparency and brilliance it is able to perceive from a distance colours and forms, through its affinity to the light of the sun in clearness and transparency. Indeed, the light of the sun shines in the clearness of the eye and its sapphire blue, and causes therein a reflection (ἀνάκλασις) of the things which are affinitive to itself. But because the parts of the mouth are more contracted and earthy—and so also are those of the nose and of the ears—they do not receive any reflection from the light of the sun. In the same way as when a polished and bright vessel of brass or silver comes into contact with the light of the sun, it causes a reflection of the sun on a near wall; and in the same way as water, when the light of the sun shines in a house, reflects the light of the sun on a high wall—in this same way, when the light of the sun shines on the eye, it causes a reflection in it of the outside objects and forms. The eye sees these from a distance, and draws their image to itself, while the other senses do not act in this way.

How does (the eye) perceive colours and distinguish them from one another, as the mouth distinguishes between tastes? We answer that the eye, on account of its brightness and clearness receives the bright and perceiving power of the soul from inside, and in addition it receives the light of the sun from outside. Thus in it the perceiving and bright power of the soul is united with the light of the sun, and this light of the sun is carried by it till it reaches the reflected bodies; and it (the soul) then perceives their shapes and colours, and draws them first to the eyes from outside, and then to the imagination from inside. This explains why, when we close our eyes immediately after having perceived colours, we see them in our imagination as if with the eyes.

The organs of the other senses do not act in this way, because of their thickness and lack of balance. In the same way as when we throw water into a glass vase, it receives the light of the sun, which penetrates both the glass and the water which is in it, and shines in a house—while an earthen vase does not do this because of its opacity—in this same way the eye receives the perceiving and enlightening power of the soul, and brings it to the light of the sun. That the eye is the recipient of this power of vision[1] which is in it, is established by the fact that when we close our eye in a dark place, and place our finger on the pupil of the eye, and shake it hither and thither while closed, we seem to perceive a kind of circular light, on account of its shape; and the same thing happens if we place our finger in one of its corners.

CHAPTER V.

On how[2] the perceiving power of the soul receives bodies and colours that are thicker than itself; and how the eye receives them from it.

We affirm that the eye is composed of the four elements, 158
and that the bodies, images and colours are also composed of the four elements. This causes an affinity which empowers the eye to receive from outside that which is affinitive to it. The eye possesses also in itself the different colours, because it has the white and black colours—which are the principal colours, of which the other colours are composed—and in addition it has the grey colour, or any other colour. Thus when the perceiving and bright power of the soul, which is joined with the eye from inside, receives—because of the affinity of which we spoke before—the bodily factor which is in the eye and which is from the elements, it receives also the colours, and carries them to the outside bodies and to the outside colours. The powers which are in it draw from the bodies that which is affinitive to themselves, and also the power of the colours from the colours, and bring them to the eye. The eye first receives them through the affinity and relationship which it has with them, and after the perceiving power (of the soul) has also received them, it presents them to the memory and stamps them in the back part of the brain. Lo, what we see outside we always see inside in exactly the same way.

How are colours, forms and memories stamped in the back part of the brain, and how is it, if they are stamped in the body

[1] Substitute a *Dalath* for the *Wāu* at the beginning. [2] Lit. "why."

of the brain, that the latter is not [1] strained and does not perish? And how can that small space hold the images of towns and countries, and, which is more wonderful, of all the world? We answer that the animal spirit [2] is composed of the four elements, especially of their thin, light and volatile parts. We have already spoken at length of these things in the book which we wrote on *The Causes of Fevers.* The earthy matter which the brain possesses is joined with the vision of the eye, and expands in conjunction with it towards the bodies, and draws from them, as we stated above, the earthiness which they possess and which is affinitive to it, namely colours, forms and images, and returns with the vision of the eye towards the animal spirit. This spirit receives them, and through its union with the rational soul—of which we spoke above, in our discussion on the soul—and through the senses, it offers them to the rational soul, which takes them through its union with the animal spirit, and keeps them by means of the uniting link which is found between them, and which is endowed with both material and immaterial qualities. Through this process, it imparts its unlimitedness to the limited body of the brain, so that it may keep the images which are in it, and hold them, and not be strained or suffer injury, and so that they may become its own, through its union with them.

I mean by the unlimitedness of the soul, that by means of which it is in an instant everywhere, through the swift movement of the imagination, immediately it wills it, without any intervention of time, which bodies have to endure. (The soul) possesses this quality in its nature. Why? Because it is simple, and has in it no antagonistic forces, as we have shown clearly in another place. This is the reason why it has nothing in it which affects its movement or delays it. If the light of the sun, immediately it shines, has it in its power to reach the confines of the earth, when there is no intervening obstacle; and when there is an intervening obstacle, there is no time required for its reflection (from that obstacle)—how much more will it be possible for (the soul), which is swifter than the light of the sun, to move immediately to any place and return?

159 Bodies have it also in their power to reach everywhere, but

[1] Insert the negative *lā*. [2] Lit. "soul" in all this passage.

on account of the obstacle of heaviness and lightness, caused by heat and cold, time intervenes between them. This explains why it is (the soul) that takes over impressions and images and, in the unlimitedness which it possesses, holds and keeps them, in such a way that immediately it wishes to be in a place it is there, through the shining of the rays of its imagination. In spite of the fact that (the soul) is in the body through its union with it, it is everywhere without delay; and it never contracts or expands in the work of memorising in the back part of the brain, which is better suited (than the front part) for this, on account of its solidity; it is also better fitted than the front part to hold the impressions which are in the soul, because it is relatively colder. This is the reason why we say that the memory, being in the back part of the brain, receives impressions through its union with the soul, and its body is not strained, nor does it suffer and perish. This is illustrated by the fact that when we wish to look at a given town, it is as if we were travelling in it—in spite of the fact that there is no movement which extends to it—through the vision of the mind. That this act belongs to the rational soul is known from the irrational animals, which have no memory, because they perform no arts and crafts.

The particular work of memory is to take forms and images from bodies, keep them in itself, and form them into other bodies, as happens in building houses, making instruments, *et cetera*, and in all other human acts. Since animals do not possess these faculties, they have no memory. They have nothing but a kind of material knowledge, gained through the senses. If you assume that they do not do the above things because they have not, like men, hands or fingers suitable for such things, we will answer that some animals have them, such as bears and monkeys. That all the above things are not done by the hands is established by the fact that the work of the hands is impaired when the thought, which causes the brain [1] to be moved by the imagination-memory, is not linked with them.

Further, if you assume that there are some animals which build houses, like swallows, we will answer that these animals possess this faculty—as they possess also all their other natural actions

[1] Lit. "head."

—from nature, and not from learning; and for this reason they do one thing only, without adding to or diminishing from it, since that which emanates from nature is invariable. As to man, he does not possess the innumerable works of art from nature, nor does he simply receive from outside the impression of something he did not possess, and then make something identical with it. If any animal possessed memory, it would act (like man; but since none of them acts like man) they have no memory.

CHAPTER VI.

On smell.

Smell also is a name which denotes a kind of *summum genus*, and is divided into other genera, species and individua. Smell, for instance, is divided into a good smell and a bad smell; the good into a pleasant smell and a sweet small, *et cetera;* and the bad into a putrid smell, a rancid smell, a decaying smell, *et cetera;* and so on, till the individua are reached.

BAD SMELL: DUNGY, STINKING

GOOD SMELL: PLEASANT, SWEET

A good smell occurs when the parts of heat or cold are
160 balanced with humidity and a dryness that has not deteriorated but has been preserved in its original state. Indeed such a balance is pleasing to the sense of smell, on account of the lack of excess among the parts. The smell becomes more pleasant when the balanced parts of dryness predominate greatly over the balanced parts of humidity. This occurs in the following manner. The balanced parts are contracted by dryness towards one another, and thicken greatly, and thus they are better able to keep the balance and also the pleasant smell. When, however, humidity is greatly in excess of dryness in the same composition, it expands the parts and softens them; and these parts becoming remote from one another on account of this expansion, their

balance diminishes, and consequently the pleasant smell also diminishes. We see that when fruits are eaten in their right season their smell is more pleasant, but when they become soft [1] it is less pleasant. This is accounted for by the fact that the parts of those that are eaten in the right season are more contracted towards one another, while the parts of those which are soft are expanded, the balance of the humidity in them having been destroyed, together with that of the dryness.

The unpleasant smell that emanates from the deterioration caused by heat is explained by the manner in which compositions leave their natural state, and become rotten.[2] This is illustrated by the bodies of animals and by fruits. Indeed when humidity and balance (of parts) begin to deteriorate through heat, the lack of balance [3] which appears then in animals and in fruits causes their smell to be unpleasant. It is through a balanced humidity and heat that the growth of animals and fruits take place, and it is through them that they become pleasant-smelling. But when the contrary—that is to say, a lack of balance—takes place, bad smell, decomposition and deterioration [4] set in. Even the preservation of the life of animals is effected by balance (of parts), and illnesses and pestilences by lack of balance.[5] The same thing happens to inanimate objects, such as earth or water. Indeed when water is thrown on the earth, it gives at the beginning a pleasant smell. When, however, it remains on it for a time, some heat joins with it ; and when there is no evaporation, that heat increases, and causes putrefaction in the humidity. This happens because its natural parts, which were balanced, become divided against one another, and when this division sets in, the parts are disturbed ; and when these are disturbed and have relinquished their former and natural position, they cease to possess their former natural smell and colour.

The stinking, decaying and putrid smells, *et cetera*, arise from increases and decreases which cause a rotten and deteriorated humidity, and of them there is no need to speak.

[1] Read *wadhsharya*.
[2] Read *methmasyān*.
[3] Read *shawyūtha*.
[4] Read *ḥubbāla*.
[5] This idea is repeated below, Fifth Discourse, c. XI, p. 217.

CHAPTER VII.

On the sense of smell.

Why the nose receives smells through inhalation, while the eyes, the mouth and the ears are not in need of this inhalation of air.

We affirm that smell is thicker than sight; and sound is thinner than taste. Taste and smell are earthy and watery respectively, while sound and sight are aerial and fiery respectively. Because sight [1] is fiery, it is in need of a thing that is affinitive to it, in order to receive the colours and shapes of bodies, and that thing is the light of the sun. And because light has thin parts and reaches everything instantaneously, it does not require to be drawn in. Indeed it reaches everywhere instantaneously and without any effort, because of its lightness; and although air intervenes, light pierces through it because it
161 is thinner than it. Further, hearing is aerial; and because air extends to everything by its lightness, and fills everything that is empty, when a beating movement occurs it brings the parts of this air [2] to the ear, because they are expanded and light; and this is the reason why the ear does not need to draw it in.

As to taste, because of its earthy nature, it makes use neither of light nor of air, but is effected by the coming into contact of two earthy bodies, namely the outside bodies and the sense of taste. Indeed unless the former penetrate inside the mouth, the latter does not taste them. As to smell, it is watery, as we stated, and as water is thinner than earth, smell also is thinner than taste. This explains why smell does not receive, as the sense of taste does, a body, but receives only parts which are very thin and which rise upwards in the air, like water which evaporates from bodies that are humid and thick. When these are brought to it, it does not receive them directly, as the mouth does, without any drawing upward. (The nose) requires parts that are very thin in liquidity which, in rising upwards in the air through evaporation, become light, and, through their lightness, are brought to it by the air. If this were not so, the natural thickness which is in the

[1] Read *ḥezātha.*

[2] Text, "its parts bring this air."

(smells) would not have allowed them to rise upwards. Consequently the nose makes use of the lightness of the nature of air, to inhale them upwards, and through a powerful indrawing they reach the brain, which distinguishes between them by means of the powers which it possesses, and which are affinitive to the nature of smells. Indeed smells are not [1] brought near the nose, unless they are raised up. This is due to the earthy and liquid weight which they possess. The lightness of the air draws them upwards, and by means of it the earthiness of the smell is raised and reaches the brain. Further, the shape of the nose being like a chimney,[2] this indrawing of the air is upwards, and since the nature of the indrawing of the air is in the category of vaporous things, and smells are also vapours, the nose alone, to the exclusion of mouth, eye and ear, is able to become the organ of smell.

CHAPTER VIII.

On the sense of hearing.

Hearing also is a *summum genus*, which is divided into genera, species and individua.[3] It is divided into four genera: sound, speech, wind [4] and breath. Sound is sub-divided into other genera: thin, thick, clear, harsh, melodious, *et cetera*, and these themselves into others; speech is sub-divided into eight parts: noun, verb, particles, *et cetera*, and so on down to the individua; wind [5] is sub-divided into animal breath and aerial (wind), and these into other sub-divisions; breath is sub-divided into that made by rational and by irrational beings, and these into different species and individua.

SPEECH		SOUND		BREATH		WIND	
NOUN	VERB	HARSH	CLEAR	RATIONAL	IRRATIONAL	ANIMAL (Breath)	AERIAL (Wind)

[1] Add negative *lā*. [2] Read *kawtha*.

[3] Add negative *lā* before the word.

[4] Lit. "spirit." In Syriac, as in Greek, the word "spirit" means also "wind." Early man believed that the spirit was a kind of wind which moved the body.

[5] Here also the word means both "wind" and "spirit."

Sound.[1]

Sound is the movement of air which, having come into contact with bodies, struck them, and been driven by them, brings that striking to the ears, through the instrumentality of which it reaches the brain. Sound is either breathy or non-breathy.

The non-breathy sound occurs, for instance, when two stones come into contact with each other and strike each other,
162 and from the concussion thus caused between them, the air leaps out and carries their striking and brings [2] it to the ears. The breathy sound occurs when an animal inhales the air by means of the indrawing power inside the lungs, and exhales it through the expelling power, and in its exit it strikes the larynx and the τραχεῖα ἀρτηρία (trachea), and then goes out and reaches the ear.

Sound is much more general than the other divisions (of hearing). Indeed even verb and noun cannot have an outlet without it, and the same is true of the other divisions. A sound that is thin or thick, *et cetera*, is like a genus divided into species. A thin voice [3] is said of man, of ox, of horse, *et cetera*, and so also is the case with a thick voice, *et cetera*. The same applies to each species, when its separate individua are compared, each one having its own characteristics. Thus it may be said that one individual has more, and another less, of a certain quality.

We have spoken of the origin of the thin and thick sounds. A thin voice occurs when the part of the organ of the voice does not possess any superfluity, but carries it all out with it in its exit. It becomes harsh when some humid and thick matter collects in the path of its exit, which, becoming swollen, renders exiguous the space of the width, and in this way the voice does not carry out all (of the superfluity) at the same time with itself. This impedes the voice from reaching far. That this is so is known by the fact that when we wish to speak in a harsh voice, we act as if we were contracting the throat, and in this way we check the voice.

[1] *Cf.* Aristotle, *De Anima,* Book II, ch. 7.

[2] Read the verb with a *Yodh* instead of a *Ḥeth.*

[3] In Syriac the same word *ḳāla* means both "sound" and "voice."

As to μελωδός,[1] that is to say melody, it occurs when we sing or perform Church Services, and divide the voice in the throat into notes, producing some of them near its lower part, some others near its middle part, some others near its top part, some others in its left side, and some others in its right side; and in this way notes are formed and brought together in sequence. A melody takes place when thin and thick notes are mutually balanced, and none of them is in great excess over its neighbour, but all of them are formed in harmony with one another. In this way the voice becomes beautiful and delightful. This is illustrated by the strings of a guitar, in which, when the vibration of the thin part is made to balance with that of the thick part, and the high note with the low note, they are moved in harmony with one another. A discordant note occurs when the contrary happens.

There are four ἦχοι, that is to say, general notes, of which all melodies are formed, and which are called πρῶτος, that is to say first (note), (δευτερός, that is to say) second (note), τρίτος, that is to say third (note), and τέταρτος,[2] that is to say fourth (note). There are also four παραγεννητοί,[3] that is to say, notes which are beside them, or derived from them, out of which again four others are derived, which are called παραπλησίοι.[4] From these twelve (notes) is derived the number of all melodies, in which are included all the myriads of innumerable tunes found among all the peoples. As all the countless numbers are included in the ten units, and are formed of them; as all the innumerable speeches are included in the eight parts of speech; as all the days of the world are included in the seven days of the week; and as all the innumerable movements of the seven planets are included in the twelve signs of the Zodiac—so also all the melodies are included in the above twelve notes.

Why should there be four genera of musical notes? Because

[1] Read *mehlodos*. It is curious that the lexicographer Bar 'Ali, as quoted in the *Thesaurus Syr.*, column 2026, also writes this word *mehloros*, with a *Resh*, instead of with a *Dalath*, and translates it into Syriac by *ni'mthānaya*, all exactly as in our MS. Is it possible that Bar 'Ali took his information from our author?

[2] Correct the word accordingly.

[3] Read the word accordingly.

[4] So I read the word *parapathoro* of the text.

163 of the four elements, of which the body which produces these sounds is composed. The process whereby four genera of musical notes are composed and formed, is through the expelling power, which drives the voice (in the case of the first, or high note) towards the head of the throat, while gently contracting it. This is the reason why the voice becomes thin. The expelling power causes it to penetrate hither and thither in the throat, forming sounds which harmonise with one another, and out of which a delightful melody is produced.

The second note is formed when the expelling power drives the voice downwards towards the lungs into the lower part of the throat, and opens it considerably. This is the reason why this note is thicker than the first, and preponderates downwards, as if it were going down into a pit.

The third note is formed below the first one, is not so clear as it, and is thicker than it, since it produces its vibrations from all sides and mixes them with one another, and thus produces a musical note.

The fourth note occurs in a position immediately above the second one, between it and the third. This explains why it is less thick than it, but thicker than the third.[1]

The notes are formed at both extremities of the throat, the first note being at the top of the throat, and the second note at the bottom of the throat, while the two others are in the middle, according to the position of the four elements. Indeed two of the latter are active and two passive, the two active ones being found in this world at the two extremities—the hot at the extreme height, and the cold at the extreme depth—while the two passive ones are between them.

The first note has been called "the musical note," because it is more piercing than the others on account of its position in relation to the high element (of heat) which is more powerful than the other elements. The second note became thick [2] in

[1] As explained below, the author does not count the notes successively from top to bottom, but counts as the first note the highest, and as the second note the lowest; the two intermediary notes being counted as the third and fourth, in accordance with their nearness to the first and second notes respectively.

[2] Cod. "became thin."

accordance with the nature of the second element (of cold); and the other two, which are in the middle, are in an intermediary state between thickness and thinness, in accordance with the nature of the passive elements, which are in the middle.

The remaining eight (notes) were formed in the intermediate space between the above four, as species from genera. Indeed every species of notes gives rise to individua, by endless and innumerable additions and diminutions, among all peoples and tongues; although people do not always form them with premeditated art, or know how to distinguish which notes are derived from the first, which from the second and which from the rest. All people possess naturally in themselves the faculty of composing tunes, in the same way as every one of them possesses, from natural thought, the power of building a house; but the question of an *artistic* production belongs to the artist. Anyone can compose tunes and believe that he is composing them in a masterly way, while only those of a musician are so.

When the voice of animals and of man comes out in a natural way, it is not delightful to the hearing, but is, on the contrary, jarring to it; only the one which is divided into melodies is delightful, although the natural voice, being one and evenly-balanced, should have delighted the hearing, on account of this balance. Since, however, like is pleasing to like, and since the body is composed of varieties of different parts, and its pleasure is caused by the harmony between them, which, in creating an 164
even balance and an even temperament causes a happy life, it follows that the body is delighted only by sounds which are formed with one another, through the diversity of their parts, in a balanced harmony. When no balanced harmony of the elements exists in the body, it does not derive any happiness out of life. In this same way, when there is no even harmony of sounds no pleasure is produced. When the voice comes out, naturally, and reaches the hearing, it does not delight it, because it is not composed, like the body, of many parts, but penetrates abruptly and does not enter the hearing in parts which are harmoniously composed, and so does not cause it any pleasure. This explains why, when the sense of hearing draws to itself an unpleasant sound, if there is a balance found in the composition

of the harmony that sound becomes[1] melodious, however long and however hard and high it may be said to be, and pleases the hearing, as we have said.

CHAPTER IX.

On speech.

Speech does not give a pleasure similar to that of melody, in spite of the fact that speech is composed of noun, verb, *et cetera*, and that noun and verb are included in the division of sound, and are a breaking-up of it. First of all speech strikes the root of the tongue, and from there it reaches the palate and the lips, where it is formed, in the same way as the voice is divided into melodies in the throat, by striking first its root, then from its root coming to its middle, and thence to its head. We affirm that the movement of melody belongs to the body, and is accomplished through the body, and for this reason it is pleasing to the body; while speech is not a movement of the body, but a movement of the rational soul, and for this reason, while it is being uttered, it pleases the soul only and not the body. When, however, melodies are formed together with speech, then pleasure is derived for both soul and body. This is known from the fact that when a breaking-up of a melody takes place in the throat, and does not reach the mouth in order to join with speech, it pleases only the body; but when speech is joined with a melody, and both of them are in motion together, they please the intelligence of the soul together with the body.

As melody does not give rise to pleasure until many notes come together and are formed with one another, so also is the case with speech. Indeed a single word, or one syllable only of a noun or a verb, gives no pleasure to the soul, because it shows no meaning; as, for instance, when we say "made," and then stop; but when we add noun to verb and say, for example, "God made," or "man walks," and noun and verb have thus been joined together, then the soul is pleased. This is all the

[1] Read *hāwé*

more so when speech is long and composed of many nouns and verbs and the rest of the parts of speech ; as, for instance : "God is the Creator of heaven and earth, infinite, invisible." When the soul has received this, it gathers from it great profit, just as great pleasure accrues to the body from different melodies.

CHAPTER X.

On wind.[1]

(The word) wind is used in a two-fold way : in animate beings it is called "breath," and in inanimate things "wind." Breath 165
takes place when air is drawn in by the inhaling power, penetrates the mouth, reaches the lungs, and is then driven out through the exhaling power. When it reaches the mouth, that air is checked by the mouth, and being pressed by the exhaling power, which would drive it out, it causes a slight opening of the lips, and through this exiguous exit it gives rise to sound. In the case of inanimate objects, however, sound is produced when bodies are driven and shake the air in the midst of the antagonism of the elements. When this sound is heard by the ears, it is called "wind."

We shall speak of the nature of wind, when discussing the causes of the movements that occur in the air. When hot air is shaken by cold air, through the antagonistic movement that exists between heat and cold, the air, being thus shaken, gives a sound when coming into contact with bodies. This sound is known as wind. This is illustrated by damp pieces of wood. When we set fire to them, and the cold which is in them fights with the heat of the fire, and their humidity with its dryness, the air which is in them is shaken by the contact of the two antagonistic forces, and reaches the extremity of the pieces of wood. When pressed at this extremity of the wood, it comes out together with the humidity, and gives a sound. The same thing happens in a pot full of water, the mouth of which is covered. When fire is kindled under it, it immediately causes an opening, and gives the same sound as that which is made by wind.

[1] Remove the *Lamadh*. See further, p. 191.

CHAPTER XI.

On breath.

Breath takes place when we draw up the air through inhaling, and exhale it little by little and gently through the mouth and the nose, the air giving only a slight and imperceptible sound.

CHAPTER XII.

On the ear.

Why does the ear receive sound, while the mouth and the other (organs of sense) do not? We affirm that its position and composition [1] are the cause of its reception of sound. How? The aperture of the ear is connected directly with the brain, without any obstacle in the middle. When, therefore, the air is shaken and carries sounds, it is not checked by anything, until it penetrates and strikes the brain, which perceives it. As to the rest of the organs of sense—that is say, the mouth, the eyes and the nose—they are not apertures connected directly (with the brain) like the ear, so as to receive sounds and bring them to the brain and strike it.

Further, the nature of the ear is very cartilaginous, and this makes it more suitable for receiving sound, on account of the contraction of its parts. Indeed we see that when the air is shaken and comes into contact with expanded bodies, such as wool and the like, it gives no sound, or very little. This is explained by the fact that the sound given by air is due to the force with which it is reverberated, while when it reaches the loose parts of wool, it penetrates into them and joins with the air which is already in them, where it remains quiescent, or penetrates hither and thither because it is not checked. In contracted bodies, however, such as stones, brass and iron, it gives a sound, because their parts are contracted and are thus contrary to the expanded nature of air; and when it comes into contact

[1] Read the word with a *Wāu* instead of a *Dalath*.

with them, not finding in their parts anything affinitive to its
nature, so as to join with it, it is driven away from them; and 166
when it is thus driven away, it returns backwards, and gives a sound. Lo, when we strike with a stick on a garment of woven wool, it does not give much sound. This is especially the case with a garment of very loose texture. But when we strike two pieces of brass or iron together, they give a loud sound. In this same way, because the ear is very cartilaginous and sinewy, sounds strike it, and it receives them. The parts of its nerves also are contracted.

As to the eye, its nature is soft, and also it has no aperture with a direct connection (with the brain). If the eye were in the place of the ear, it would receive sound, and if the ear were in the place of the eye in its composition and position, it would receive colours. As to the mouth, it is not suitable for receiving sounds, because it is generally closed, and when it is open it is not sinewy like the ear. Further, what it receives it does not carry to the brain, but to the lungs, an organ which possesses a very expanded nature. As to the nose, it has no direct passage. It also works through inhalation and indrawing; and further, it is very humid, while the ear is very dry.

CHAPTER XIII.

On the sense of feeling.

The sense of feeling is also a *summum genus*, divided into other genera, species and individua. It is divided into soft feeling, hard feeling, rough feeling and smooth feeling, and into others with a lesser or greater degree of these qualities.

Smooth		Rough		Hard		Soft	
Gold	Marble	Brick	File	Iron	Stone	Byssus	Wool

Objects are called soft when the parts of their body are humid (and expanded) under the touch, such as mud or a garment of wool. Because the parts of the humidity of such a body are, through their expansion, not pressing against one another, they are, therefore, affinitive to the humidity (of the skin), and consequently the body is called soft. Objects are called hard when their contracted and dense parts come into contact with the skin, and press it from outside, endeavouring to contract its expansion and desiccate its humidity. As such are stone, wood and the like. An object is rough when the parts of its outer surface are not even, or are remote from one another, one being high and another low, such as a brick and the like. A smooth object is one in which the parts are even—one not being higher than another—and polished, glossy and close to one another, such as marble, gold and the like.

The species and individua of the above possess likewise differentiations, on which there is no need to dilate, lest our discourse should become unduly long. It is indeed our aim in this book, here as elsewhere, to avoid as much as possible lengthening our discourse, and to choose only the most important causes (of created things), and the coming into existence of the species from the elements, and to set aside all other questions which are self-evident and obvious, so that we may be able to bring utility, without much trouble on their part, to our brethren the φιλοπόνοι and φυσιολόγοι,[1] that is to say, lovers of labour and investigators of the hidden laws (of nature).

CHAPTER XIV.

On the reason why the sense of feeling is spread over the whole of the body, while the other senses are not.

167 We affirm that the sense of feeling is produced [2] by the thin and small nerves which come from the brain to the skin, and which are spread over all the body. This is the reason why the sense of feeling extends to all the body, while the other

[1] So I read the word *pusiaharhono* of the MS.

[2] Remove the *Dalath*.

organs of sense, not being spread over all the body, but being in one place only, perform individual functions, in accordance with the composition of the body, as we have shown.

On the fact that feeling emanates from the elements.

The body is composed of contrary parts : hot, cold, humid, dry, even, uneven, rough and smooth ; and the skin possesses these qualities from the body, and takes from it also its superfluities. As it draws the impressions of the above qualities to itself from inside and drives them outside, it draws them back from outside and brings them to the body, through the affinity which it possesses with all the parts of the latter.[1] When antagonistic things come into contact with antagonistic things, conflict occurs, as when heat comes into contact with cold, and cold with heat, *et cetera ;* and because the skin possesses all these qualities, when the heat which is in it comes into contact with the cold which is outside, it suffers from it. The same happens in the case of any other opposites. This suffering is called "feeling." In the same way as the skin suffers from things that are antagonistic to itself, so also it is pleased with things that are affinitive to itself and with things that are of similar temperament.

Feeling does not take place outside only, but the inner parts[2] also feel a cold, hot, heavy or light object, though not to the same extent as the skin does, because the latter feels more intensely, on account of the greatness of the surface found outside, and the more powerful movement of bodies which may strike it from a distance.

CHAPTER XV.

On the reason why there are only five senses, and not one or six or any other number.

We affirm that there are four senses in the head, two of which are above and two below : the ears and eyes above and the nose

[1] Read *dīleh.* [2] Remove the *Baith.*

and mouth below. In the same way as there are two elements in the high sphere of this world (and two below), so also there are two (senses) in the head—the eyes and the ears—which are above like fire and air, and (two others)—the nose and the mouth—which are below like earth and water.

Further, in the same way as all the genera and species of this world form a fifth composition from the elements, and this composition, although of the same nature as theirs, is distinct from them in number—so also a fifth sense, distinct in number from the above four, came into being. Although they have a preponderance of one element over another, they are one in their nature, which comes from the elements; and just as there are, in all this immense universe, four elements, neither more nor less, together with the fifth composition which emanates from them and constitutes the fifth in number—so also in man, who is a microcosm, there are only five senses, neither more nor less. In the same way as this fifth composition which emanates from the elements is much larger than any one of them, so also the fifth sense of the body—namely the sense of feeling—is larger than any of the other senses.

The functional ground for the existence of the (five senses) is the following: God created them so that through them the requirements indispensable to our existence might be satisfied. That every living being is in need of the five senses is known from the fact that if any of them is impaired, the course of our life
168 is hampered.[1] Indeed if we suffered from blindness, the function of walking straight, of performing any art, of choosing between beautiful and ugly things, and of distinguishing between different colours, together with many other things, would come to an end in us. If the sense of taste were missing, with what should we evince a desire for a sweet thing, a sour thing, bread, meat, or any other variety of food? And if desire were absent, there would be no reception of food; and if reception of food were absent, there would be neither growth nor development, and our life would come to an end. If the sense of feeling were absent, the same thing would happen, because we should not feel the heat of fire, and should be burned, and our life would

[1] Remove the *Wāu*.

perish. The same thing would happen to us with regard to our handling of objects. Further, if we did not feel the pain of a cut or of a thrust in our body, we should be in constant fear and anxiety in our life, and our life would perish. Nor should we be able to avoid the injuries that come upon us, since we should not feel them; and we should perish, or should not even have come into existence, and our life would become a mockery.

Further, if we had no hearing sense, we should not receive knowledge, nor should we learn to speak, for it is through hearing and from teaching that this speech is learned; and we could not be distinguished[1] from irrational animals; and what is worse, the great benefit of the knowledge of, and faith in, God would perish in us, as "faith cometh by hearing,"[2] as well as other calamities which would befall us. Further, if we did not perceive a delightful smell, and did not enjoy it and take profit from it for the benefit of the body, that delightful smell would be to no purpose; and if we did not flee from an evil smell, it would penetrate frequently to the brain, and cause some injury to it.

Since the senses are the constituents of our existence, it is with great wisdom that God created them in us, wishing to vouchsafe happiness to us and bring us into existence from non-existence, which is a great gift. Because it is in His nature to do good, He created for us these five senses in order to maintain our life.

CHAPTER XVI.

On the fact that colours, sounds, tastes and smells are not essences, as some people believe.

After having finished our discourse on the origins of the five senses and their differences (we will discuss the same theme), although digressing slightly from our subject, on account of the utility that will accrue from it to the public.

It is erroneous to assert that colours, smells, taste, sound and

[1] Add the aux. verb *hwain*.

[2] Rom. x. 17.

feeling are essences and not accidents,[1] as some new philosophers of our days have tried to show. We met the head of their heresy and its innovator at the time when we were writing this book, and refuted his false reasonings (παραλογισμοί). He dared to assert that the colours of objects, sounds, smells and tastes are essences and not accidents. Further (his followers) have endeavoured to refute the great opinion of Hippocrates and Galen, and of all the physicians and philosophers, by saying that bodies
169 do not emanate from the four elements only, but from many elements: the light, the heavy, the contracted, and the expanded elements, and others which they assert[2] to be similar to the hot and the dry elements and the rest. They have also written a book on this subject. If our Lord[3] and God vouchsafes life unto us, we shall refute their opinion at the end of our words in this book, provided that the waves of the rough sea have subsided from us, together with the murders and conflagrations in different countries, the tribulations in all towns, and the wars between nations, which are occurring at the time in which we are working at this book, such as have not been heard of since the beginning of the world.

We shall first define what is an essence and what is an accident, according to our and their statement, together with its implications. We shall set forth also its premises (πρότασις), and show, by methods of analogy, through self-evident reasonings (ἀποδεικτικοί συλλογισμοί), that their arguments (συλλογισμοί) are false.

An essence is that which exists and does not cease to exist, while an accident is that which happens and ceases to exist. (The adversaries) believe in this. Let us see now to which of the above definitions colours, tastes, smells, sound and feeling conform—to that of essence or to that of accident. If they exist and do not cease to exist, they are essences.[4] Let us first speak of colours. We wrote a book against those who do not believe at all in accidents, but assert that everything is essence, and also against those who assert that everything is accident. We spoke there at great length about colours, *et cetera*.

[1] The text of this sentence is complicated, and the copyist has written "essences" in the place of "accidents," and vice versa.

[2] Read in plur. [3] Remove the *Wāu*. [4] Cod. "accidents."

We notice a colour in a body, say a white body, and we notice that this white colour moves away, and (the body) receives a red colour, and that the red colour also moves away, and the body receives a black colour. As an illustration, let us take pieces of wood, kindled by fire. They receive these three colours—white, red and black. At the beginning we see a red colour in them, but when we throw water on them we notice a black colour; and when the fire has consumed them completely they become white ashes. (The adversaries) claim that the redness was inside the (pieces of) wood, and came outside them, and so also was the the case with the whiteness. They assert with reference to all bodies that receive colours, that their colours are essences which come outside them. But if the three above colours were in the wood at the same time, why were they not seen [1] when the wood was cut, before it was thrown into the fire? They were not in the wood, neither inside it nor outside it. Indeed they did not come out of the wood, because they had not been inside it, but they were accidents which ceased to exist.

Further, if you say that the black and the red colours were hidden on account of the excessive whiteness found in the wood and of their own small quantity, how is it possible that that which is quantitatively small can conquer suddenly that which is quantitatively large, in such a way that that which is large becomes completely invisible, perishes and never reverts to what it was before? Since those things which did not exist conquered, and since essence does not become a non-existent thing, (those colours) were, therefore, accidents which came into
existence from non-existence. Indeed essence does not move 170
away from inside to outside, nor from outside to inside, being always that which it is, without any change.

Further, if you say that they were found in the wood, although they were not visible,[2] and that when they came near the fire they thickened and fell under vision—this also proves that they were accidents which came into existence from non-existence. How did they become thicker in their union with fire? And (how was it that) the white colour and the other colours which were in the wood not only did not thicken, but ceased to exist?

[1] Read *methḥzain.*

[2] Again read *methḥzain.*

By their ceasing to exist, they show that they are not essences but accidents. Lo, the red colour of fire and the black colour of charcoal (καρβώνιον), when the wood becomes ashes, cease to be red or black; and if they undergo such a complete change, they necessarily cease to exist; and if they cease to exist, they are accidents. They are known to be accidents by the fact that they were not in the wood, and are not seen in the ashes, but came accidentally into existence from the union of fire with wood. Indeed if we cut the piece of wood in two, and throw one piece into the fire and not the other, the piece which is in the fire receives the red and black colours, while the other piece remains as it was. It follows, therefore, that it is the fire which brought colour to it, through its heat and dryness, and that something came into existence in the wood which was not in it before.

If they assert that both the red and the black colours are found in the ashes, let them prove this to us, either through the eye or mentally; for if, through any process, that redness or blackness is perceptible in the ashes, even if it is not visible to the eye, it should be so to the mind. This may be illustrated by the fact that when an ounce of water is thrown into an ass-load of flour, and becomes invisible in it, when people tell us that it contains water, we do not believe them as long as we do not see it; but if they extract that water out of it into a pot through an alembic, by the process of heating, we become convinced that there was water in it. This may be also illustrated by another example: if a person says that in this house there is a man, but he is hidden by the walls, we shall not be assured of the truth of this assertion unless he comes out to us, or unless we go in and see him there; but if it is not proved to us that there is a man in the house, that person will be found to be a liar. It is easy for anyone to say a thing without proving it. It is proof that distinguishes between truth and falsehood.

If you further assert that the colours evaporated away into the air that absorbed them, lo, they are not seen in the air with the eye, and how can we know that they are in it? Your assertion can only be verified if they are found in the air some time after, but since no colours that have left the bodies have ever been

seen in the air, it may be inferred that they are not in it, and that they have ceased to exist. (The adversaries) bring forth demonstrations to prove that they are in the air, and that they are absorbed in it, but are invisible, just as water, which is an essence, when thrown into a pot rises up through the heat of the fire, thins away, is absorbed by the air, and becomes invisible. It is in this way, they assert, that colours are in an object but are not seen in it because of the smallness of their parts. This demonstration, however, is unsuited to the nature of things, and is known to be so by the fact that water has a separate existence,
and does not come into existence accidentally, like colours, 171
which have never been seen to exist in themselves. Water, because it is an essence, only undergoes displacement. Indeed it is thrown into earthenware, and thence into glassware, and thence into brassware, *et cetera*, and always remains the same without any change; and when its parts are expanded, they move away into the air, and remain in it of the same nature as when they were in the vessels.

A further illustration: a man takes ten grains of wheat, counts them, and separates them from one another, but when they are milled and have become flour, he can no more count them, because they have been divided into small parts; and when he blows away these parts or casts them into the air not only can they not be numbered, but they are not even visible. It is, however, known that they have not ceased to exist, but that they are in their own essence, and have only moved away from one place to another, while their essence exists in itself. In this same way, when the water moved away, it was not visible, but nevertheless it was found in the air. Colours differ from this, and having no separate existence in themselves do not move away. Is it indeed possible to take a black colour or a red colour from a vessel, and introduce it into another vessel, and from that into another, as we take earth, water, fire or air from one vessel to another, and again from that to another, while they themselves exist without change? Never.

If you say that in the same way as fire—which is an essence—is visible to us only when it ignites pieces of wood or any other object, and that in the same way as when it moves from them, it thins away, penetrates into the air, and becomes invisible, in this same way colours are also essences, and are seen only while they are in bodies, but when they move away from matter they thin away and become invisible—we will answer :—

Fire is an essence because we see it move from place to place without any change. Indeed it is in a piece of wood, and leaps out of it to another ; and can be taken from a piece of wood and brought to another, while remaining in both. And when we strike fire from steel, it goes out of it, and penetrates into the air, to the length [1] of a cubit, until it reaches a body (such as wood) which it ignites, and when fire flares up in it, we see it outside the body of the wood. Indeed it rises many cubits above the wood, when it is well kindled, and acquires a pyramidal shape by itself.

Further, it is seen to be an essence, from the fact that when it comes into contact with water it wages a severe war against it, and sometimes conquers it ; and anything that conquers an essence is an essence. As to colours, none of them has ever been seen by itself separately, as an essence is seen, neither did they set up for themselves a separate position in the world.[2] Moreover, we do not see the red colour in a garment or in anything else move away from one object to another, nor do we see it separately by itself. That fire,[3] however, is seen [4] separately in a garment is established by the fact that the air receives the fire and contains it. The following particular case may be generalised to apply to all cases : when we bring fire into a house, and it remains a short time in it, its heat is kept in the air of the house after we have extinguished it. This shows us that the
172 air has received it, and that in spite of the fact that it has been deprived of its working and of its visibility, yet it remains and is perceptible in the air to the sense of feeling and touch. Further, when a spark comes out from between two stones which strike against each other, it penetrates the air. It is in this way that from the particular we proceed to the general.

As to colour, when it ceases to be in wood or a garment, what sign does it give us that it has moved away to the air ?

[1] Read *rewīḥā'ith*. [2] Lit. "made the heaven from a position."
[3] A short space in the MS. The word omitted was evidently *nūra* (fire).
[4] Read the verb in fem.

Indeed it is not seen that it has moved,[1] nor is it felt, nor does it fall under any sense. This demonstrates to us that it is not found in the air, and that it has ceased to exist.

Further, we often notice that when fire is burning in a lamp, in which the oil has diminished and is about to come to an end, the fire leaves the lamp and rises in the air to a distance of three inches,[2] and then is absorbed in it. This proves to us that fire is able to exist by itself, and to rise in the air.

Further, we see that fire which is ignited in a lamp causes by itself, when increased or diminished, a downward or a forward or a backward movement; and this is a sign of an essence. What kind of movement has a colour ever had, either by itself or otherwise, such as that we see possessed by the elements—fire, water, earth and air?

Further (if colours had been essences), it would have been necessary for them to be separated, and positions assigned to them, according to their number, like the positions assigned to the four elements. Because the red colour and the other colours do not exist by themselves and in themselves, but are sometimes in fire, sometimes in earth, and sometimes in water, in the same way as contraction, expansion, heaviness, lightness, length, *et cetera*—which they have also called elements—come into existence sometimes in fire and sometimes in the other elements, all are, therefore, accidents and not essences. Indeed, do we ever see that earth in is need of water for its existence, or to be known through water, in the same way as the colour which is in it? Or that water is in the same need in regard to earth, or air in regard to fire, or fire in regard to earth? Never. This is because they are essences, and each one of them exists by itself. We do not hesitate, therefore, to say that colours are accidents, and when, without any proof, the adversaries pretend that colours are in the air and in the ashes, they resemble the man who says that he has pieces of gold, and when asked where they are, answers: "In the air, or in this house, or in such and such a place," and does not give any proof for his statement.

[1] Read *Dalath* instead of *Wāu*.

[2] Text, *ṣeb'a*: an ancient measure of length, equalling six barley grains placed side by side.

CHAPTER XVII.

On the sense of taste.

(The adversaries) say also that the seven varieties of tastes: sweetness, acerbity, *et cetera*, are seven essences. These, however, come into existence also by accident and cease to exist, as is demonstrated by the following example. Let us take our illustration from a pomegranate or a bunch of grapes. We see, for instance, that in a bunch of grapes there is at the beginning an acerb taste, then sourness, and after that sweetness. Is it not the case that the acerb taste came into existence (by accident)? If it was an essence, to what (particular) place did it move away? When the sour taste came, the bunch of grapes did not modify its shape of grapes; it only received size and other additions. Indeed, if the three tastes have not moved away, and are still there, we should find them together in the grapes when they ripen.

If you say that sweetness predominated over the other two tastes, how can we know that this is so, while they are inside the
173 grapes and are not perceived by the sense of taste? It is indeed the sense of taste that demonstrates their existence, and distinguishes between them; and since their existence is not perceived by the taste, they have no essential existence. Your pretension is, indeed, similar to that of a man who says that there are three men in a house, but when we enter it we see only one, in spite of the fact that he,[1] by the word of his mouth, says that there are three men.

Further, since there was first acerbity (which you say is an essence) in the grapes, how did sourness, which you say is also an essence, but which was invisible and intangible, conquer this acerbity? And how did sweetness conquer in its turn the latter? If you say, according to your habit, that all these tastes were there, how is it that they were not manifest at the beginning? Indeed it is known that they were not there, but that they came accidentally into existence, by the fact that grapes did not receive the three tastes from the water or from the vine,

[1] Text, "you."

nor did they receive them from the nature of the air, or from the heat of the sun, as none of these in itself contains tastes. Tell us, O new philosophers, whence did the grapes receive their acerbity, sourness and sweetness? They received their size and heaviness from the water and from the nature of the woodiness of the vine in which they were; and the latter were visible before the grapes received them.[1] From what nature then did they receive their tastes? Since there was no other thing (from which they could have received them) apart from those elements which we have enumerated, did they not receive them from the different combinations of these elements?[2]

Shout, therefore, with us in a loud voice, that tastes are accidents, and that they accidentally come into existence[3] and cease to exist.

CHAPTER XVIII.

On the sense of smell.

(The adversaries) say also the same thing about smells, namely that they are essences, because they enter a body and come out of it without ceasing to exist. We shall prove to them, however, that this is not the case, and illustrate our argument by an apple, which at the beginning, before it ripens, has no smell, but after it has ripened acquires smell. Whence did it obtain that smell? It came from the apple, but not as an essence, and adhered to it. (If it was an essence) where was it (before it appeared in the apple)? Further, if we throw the apple into fire, and it is burnt and scorched by it, an unpleasant smell comes out of it. Whither did the first pleasant smell move, and whence did the unpleasant smell come? And why, if that smell was an essence, did it not endure in the fire, like the earthiness of the apple? And if it was thinner than the earth, why did

[1] Remove the *Alaph*. The copyist has made many grammatical mistakes in this paragraph, which it is not necessary to mention.

[2] On how tastes resulted from the combination of the elements see above, pp. 124-129.

[3] Read *ithaihain*.

it not separate itself from the apple, like the water?[1] Indeed, one can grasp the water of the apple while the latter is burning, when the water goes up into an alembic. If you say that the smell was thinner than the water, show me by all means in which way it moved away; and since you cannot possibly show me its moving away, it did not move away, and it is therefore an accident.

CHAPTER XIX.

On sounds.

That sounds are neither bodies nor[2] essences, but come into existence and cease to exist, may be established as follows: when we bring two bodies near each other, and look at them, we do not perceive any sound; but when they strike against each other we hear their sound. Now, where was that sound before they struck against each other? Was it inside the stones or inside the air? If inside the stones, they were in no need of striking against each other (to produce it); if inside the air, we should have heard their sound before they struck each other, and even before they came near each other! Further, when the sound is quiescent, where is it? Inside the stones or inside the air? If inside the stones (since, as you say) it is an essence,[3] and since an essence does not cease to exist, the stones would have
174 given sound at all times. If inside the air, it would have been likewise necessary that its essence should not cease to exist.

If you say that the movement of striking causes the sound to come out from inside the stones, it will follow that when it caused it to come out, it imparted to it a kind of movement which it did not possess before it came out. If it imparted such a movement to it, how can an essence receive something that it did not possess, in the above sense? And if it received something that it did not possess, since (in your opinion) the movement of striking imparted to it the power to go out and to be heard,

[1] *I.e.*, juice. [2] Read *w-lā*. [3] Lit. "nature."

it will follow that the faculty according to which a sound is heard does not belong to the sound, but to the movement, and it is the movement that causes the sound to come into existence from non-existence, and the sound ceases to exist after[1] the movement has ceased to exist.

If you say that the sound was inside the stones, and that the contraction and the thickness of their parts, combined with its thinness, caused it to be imprisoned in such a way that it could not be heard outside, but that when the stones struck against each other it leapt and jumped outside, because of its lightness :—

We answer that when those stones struck against each other, or a piece of iron struck against another piece of iron, they did not undergo any change, nor were they pierced or broken, nor did any other untoward accident happen to them. How, therefore, did the sound which was imprisoned in them escape? Lo, when air, water or earth are imprisoned in a body, such as a potter's vessel, until it is pierced or broken or opened they do not come out of it, even if it is struck several times by other bodies; how then can a sound come out of bodies? If you say that being thinner than a body it penetrates into it and comes out again, how is it that when the ear is plugged it does not penetrate into it? If the sound was inside the air before and after the stones struck against each other, although not heard there, how was it that the air, which is very tenuous, impeded that sound from reaching you before the stones struck against each other? Indeed when sound occurs, it is the air which carries it and brings it to the hearing organ. O new investigators, since the sound is neither inside the bodies nor inside the air, it is, therefore, an accident.

We affirm that when, owing to the distance that separates them, two bodies strike against each other violently, the air is unable to penetrate between them on account of the contraction and density of their parts; and when it is pressed by them it leaps out from between them, escapes, takes over the striking and carries it to the ear. This is called "sound." It is known that the sound consists in their striking, from the fact that a particular object emits a particular sound. Indeed when iron

[1] Remove the *Wāu.*

strikes iron, or stone stone, or wood wood, *et cetera*, we distinguish their sound, even if we do not see them.

All this proves that sounds are accidents, adhering to bodies.

CHAPTER XX.

On the sense of feeling.

The adversaries pretend also that the sense of feeling which embraces rough, smooth, soft and hard feelings, *et cetera*, is an essence. O men diligent in the vision of recondite thoughts, suitable to night and to evening gatherings (σωματεῖα), the roughness of stones or of any other object, does it not come from the non-evenness of the parts of the body ? When its parts are smooth, where is roughness found ? Inside the body ? But how can the inner parts be uneven ? Indeed both the unevenness and the smoothness of stones are outside, while the inner parts are not said to be either smooth or rough ; and because both roughness and smoothness are outside, and are sometimes
175 found and sometimes not, they are therefore accidents, which come into existence and in the same way cease to exist. The same may be said about other varieties of the sense of feeling.

Contraction and expansion, *et cetera*, are also accidents, which come into existence and cease to exist, and sometimes exist and sometimes not. Shortly, all things that fall under the five senses are accidents, with the exception of a thing which exists at all times, does not undergo any change, and does not possess its existence through another thing, namely an essence, as for instance when we point to a man or a horse. A man, such as Socrates, is living, rational, cold, humid, hot, dry, and these qualities constitute a living being in such a way that if any of them decreases or increases he becomes ill, and if any of them disappears completely he dies, because they are constituent qualities of his composition. When we say, however, that Socrates is white, short, soft, rough, thin-voiced, *et cetera*, these qualities are not constituents of his nature ; and this is the reason why, when they increase, decrease or disappear, they do not destroy the *substratum* (namely,

Socrates). Indeed when he is blackened by the sun, his composition is not dissolved, and when he becomes tall at the stage of youth, he does not deteriorate, nor does this happen when, in his old age, his skin becomes hard and rough, or when it becomes soft and thin at the stage of childhood and youth, or thick at the time of puberty.

Further, although we notice that the body of Socrates is impaired or suffers from cold or from humidity, we[1] never say that fever came to him from colour or from any other accident, such as tallness, shortness, *et cetera*. This is the reason why, to cure him, we offer him something hot or cold, dry or humid, and by it he is made whole ; but we never offer him colour and such things ; and if we do offer them to him, he will not obtain any benefit from them, and we shall become an object of derision. Even the sick man himself suffers from, and complains of, heat or cold, *et cetera*. The fight or suffering brought about by a thing which exists in itself is known, but how can a thing which is not self-existing, and the very existence of which is in another, fight or bring about suffering? It has first to constitute itself, and then fight with another thing[2] and bring about suffering. It is imperative that the one who does not wish to do injury to truth and to himself should confess that all the above things are accidents, which come into existence in the elements, through the movements of the latter.

As an illustration let us take a body which is liable to contraction, expansion, heaviness, and lightness, such as wax. When we bring it near fire, it melts and expands, through its heat, but when we bring it near cold water or any other cold object, or air, it contracts and thickens ; but neither when it is contracting nor when it is expanding does it change from being wax ; and when both contraction and expansion combine evenly in it, it acquires its natural equilibrium.

Further, we notice that bodies which are very contracted are very heavy, and those which are very expanded are very light. Let us take as an illustration two stones of equal size, one of which is black and the other white. When we weigh them, the black one is found to be heavier and the white one lighter, when 176

[1] Lit. "you."

[2] Read *ḥeraina*.

compared with each other. It cannot be said that the heaviness and lightness of bodies, which are effected by the predominance of contraction and expansion respectively, are essences.

Further, we notice that wool is endowed with expanded parts and great size, but when we have made it into a garment and have contracted it, if we take a quantity of it equal in bulk to that of the wool, it is much heavier than the wool. This may be applied to other bodies.

As to us, we are able to show that what we call essences are self-existing, and are not in another object. We are also able to show their position and their movement from place to place. Earth is situated below, water above it, air above water, and fire above air. We can take every one of them separately from its position, and return it to its place. As to colours, heaviness, lightness, *et cetera*, they have no form, defined position, or separate existence. (If they had) they would necessarily have been liable to separation from one another, in a way that would enable us to take one of them from its place and return it to it. This is the reason why they have no other power but to be now in this or that element, and now in this or that body which emanates from the elements. They are therefore accidents. If they were essences and adhered to one another, why did they not give birth to one composition and collect in one place, as did water, earth, *et cetera*, and the bodies which are composed of them? It would have been also necessary for their characteristics to be according to their number, and for their collecting into one place to give rise to a distinct body. It follows from this that anyone who does not wish to be obdurate, to be known as an arch-heretic, and to court the empty notoriety of being honoured by men, should relinquish his opinion, hold to ours, and say that colours and such like are accidents, which sometimes exist and sometimes cease to exist.

When (the adversaries) are pressed from all sides, they say: If colours, smells, sounds, contraction, heaviness, *et cetera*, are accidents, remove them in your thought from a *substratum* (ὑποκείμενον), and show us [1] then its essence, in the same way as we [2] show you the above things, (demonstrating) how each

[1] Text in singular. [2] Text again in singular.

one of them has an individual existence separate from that of its neighbour.

This objection has been prepared and uttered by the head of their heresy, and we answer it as follows : It is impossible to remove in our thought all accidents from a *substratum*, such as earth, or the body of an animal, *et cetera*, as long as that *substratum* remains.

Every statement must be included in the category (ἕξις) of the impossible, of the necessary, or of the possible—which may or may not be. What is in the category of the impossible cannot be ; what is in the category of the necessary is by necessity ; and what is in the category of the possible sometimes is and sometimes is not. An example of the impossible would be to say that a man is composed of an ox, an ass, a horse and a lion, or that he flies, as it would not be possible to imagine this in the thought. Even if it were possible to imagine it in the thought, it would be impossible to bring it into actual fact, and it is consequently untrue. An example of the necessary would be (to say that) fire is αὐτοκίνητος [1] and καυστικός, that is to say, self-moving [2] and burning, and cannot help being so, nor can it at one time be so, and at another time not be so. An 177
example of the possible would be the following : Sophronius is sitting and speaking. This may or may not be so. The category of the possible embraces all accidents at all times. This is also true of the category of the necessary. But the possible is not by necessity in such a way that it is impossible that it should not be. It is known that it is possible that it might be, from the fact that sometimes it is and sometimes not. For instance when (a man) is said to be sitting or speaking, it is possible that this may be the case or not ; and consequently the statement may be true or false.

The only thing that falls under discussion is the possible, as it sometimes is and sometimes not. Indeed when we say that a thing is possible, it is because sometimes it is and sometimes not. Since this has not been denied by any man, there is no question about it. When we say, however, that we should take in our thought all accidents from a *substratum*, we wish that

[1] Read the word accordingly. [2] Read *metzīʿath.*

which is impossible. This we do when we think, for instance, of man as being composed of four species of animals, or as flying, because when we imagine this we imagine the impossible, and so imagine an absurd thing. A question like this should not be asked by those who are seeking after truth.

To separate in our thought from a *substratum* one or two [1] accidents, and then to point to this same *substratum*, is feasible, because it is possible that this should actually be the case. It is indeed possible to think of a man as white or black, tall or short, and then, in pointing to him as the *substratum*, to verify whether he is black or white or rough. It is impossible, however, that all accidents should be removed from a *substratum*, as there would be no *substratum* left in which to verify them. This can happen only at the dissolution and death of a composition. When all accidents have been removed from the *substratum* which received them, the constituent essence of its composition vanishes; and as the accidents have no receptacle in which to exist, they cease to exist. Now what is a *substratum*? It is a true and essential substance: hot, cold, humid and dry, which was composed when, say, Sophronius became alive; and it receives accidents, remains the same after the dissolution of the composition, and does not perish.

How can we prove that a man's (essential substance) came into being when he was born, and that it was not in him an extraneous composition, like the accidents? We answer that man is always man, from childhood to old age, and that he possesses by necessity heat, cold, humidity and dryness, sometimes in balanced, and sometimes in unbalanced, proportions. But it is impossible that any of them should be completely removed from him from childhood to old age, and that he should remain alive. Persons suffer from four kinds of sickness: they cry and complain that they are hot, or cold, or dry, or humid, and physicians offer them medicines which create contrary effects.[2] To the existence of these four elements all rational beings testify, because when any of them is completely removed from a composition, or conquers the others completely, the composition [3] perishes.

[1] Read *train*.
[2] Cf. *Galen*, Vol. II, pp. 271-273.
[3] Remove the *Dalath*.

The fact that the elements are four in number is known from the seasons of the year, which are four. Indeed if they were more than four, the seasons would of necessity have been more 178
numerous. In the spring, the hot and humid elements predominate; in the summer, the hot and dry; in the autumn, the cold and dry; and in the winter, the cold and humid. If any of the above accidents were an essence, it should have given rise to another season.[1]

Lo, from the childhood of a man to his old age, accidents come into existence in him, which never cease to exist, and which do not destroy his composition, because this composition is not constituted of them. A Greek, for instance, is white from his childhood to his old age, and never becomes black; but when browned by the sun, in spite of the fact that black is contrary to white, his composition does not cease to exist. An Indian is black from his childhood to his old age, and his composition does not suffer. This applies also to a man whose body is contracted or white, or to one whose body is expanded or fat, or to one who is short, or to one who is tall, thin-voiced, thick-voiced, voiceless, or to one who is blind and for whom colours have ceased to exist, or to one who is deaf and dumb and for whom sounds have ceased to exist. These and all other accidents which come to him from childhood to old age do not modify him, in spite of the fact that they are contrary to one another. Indeed a Greek will never be said not to be hot, as he is said not to be black; nor will an Abyssinian ever be said not to be cold, or dry, or humid, as he is said not to be white. Nor will a man in his old age be deprived of the four elements, as he has been deprived of innumerable accidents. In the matter of heat, cold, humidity and dryness, Greeks, Romans, Indians, all men, all species of animals, birds and insects[2] are one from their birth to their death, and it is impossible that they should be dispossessed of them, except in so far as an essence is liable to (accidental) increase or decrease.

If you say, according to your wont, that (accidents) sometimes go inside (a body) and sometimes come outside it we will answer: When did the blackness of a Greek go inside him or

[1] Read *zabna* for *gauna*. [2] Read *raḥsa*.

come outside him? This we have discussed[1] above.[2] It follows from this that the essential substance of Sophronius remains in a composed state all his life, without any change, and after his dissolution it remains also the same, in a separate existence. But not so the quantity which was added to it from childhood to old age, nor the weight, for instance, of forty litres of food or drink.[3] It receives food from earth, water, air and fire. But fire has never changed from its fiery nature, and has a separate and individual existence, and the same may be said of air, water and earth. Their very position bears witness to our words that none of them moves away from its place, or undergoes any modification. When the quantity of Sophronius' body is dissolved, it is dissolved into the first principles from which it emanated, and because it is from them and returns to them, like them it remains and does not cease to exist.

In the same way as the quantity of the composition of a house,[4] with its tetragonal shape, its figure, its beauty or ugli-
179 ness, and its other qualities, emanated from the composition of such things as earth and nails, and did not exist before the composition of the house, but when the composition of the house dissolved, its tetragonal shape, its beauty or ugliness, and all other accidents which accompanied it from the time of its composition are also dissolved, but the earth, the stones, *et cetera*, remain, because they are essences—in this same way I imagine the case of Sophronius.[5] His esssential substance remains during his composition,[6] before his composition and after his composition, and heat and the other elements do not undergo any change after all the accidents which adhered to him from their coming together and kept him, have ceased to exist.

It has been shown that when by dissolution or some extraneous action, the accidents have been removed, the essence remains self-existing, both at the time of the composition and after it, and both in life and after death.

O wise men, do not rejoice at this priceless invention of your thought, which you have only invented because it is so common-

[1] Read *imarnān*.
[2] When speaking of colours.
[3] Read *shiḳya*.
[4] Read *baita*.
[5] So correct the text.
[6] *I.e.*, in his lifetime.

place! There were some arithmeticians who believed that numbers, colours, tastes, *et cetera*, were essences, and when Agathinus, the head of the philosophers, heard that a new knowledge had appeared, he desired to hear it; and when the arithmeticians were summoned to the assembly of the philosophers, and the latter heard their words to the effect that everything that changes and falls under number is an essence, they laughed and said: "They are like children whose only care is counting numbers, instead of (forming) opinions. All the thought of a child is for his play, and it is a player who has babbled this new kind of play. Leave them alone to enjoy their play like children." And (the arithmeticians) not having been found worthy of an answer, went out amid booing and laughter. And their opinion was despised by everybody, and ceased to exist, until now it has sprung up from you.

The adversaries further object to us as follows: if colours, tastes, *et cetera*, are accidents, and come into existence from non-existence, and again cease to exist, from what then did they emanate, from an existing or from an non-existing thing? And after they ceased to exist, did they move to nothing? Show us how a thing can emanate from nothing and pass into nothing.

We answer: They (the accidents) came to be [1] from a thing that was; and they came to be through a thing that was, while they themselves were not before they came to be; and they did not come to be from nothing. Indeed if they were already, they were in no need of coming to be. That which is, is not said to come to be. If it were said to come to be, then the expressions "it came to be" and "it is" would mean one and the same thing. If this were the case, it would not be necessary that the figure of a man should come to be from earth, nor a chair [2] from nails and wood; and the same applies to any artistic production. To speak in short terms, the movement of our life in this world would be of no use, since everything that came to be would be already; and since this is absurd, there is, therefore, a difference between a thing that is, and a thing that comes to be.

[1] Or "became," and so throughout this passage. See Aristotle, *Physica*, Book I, ch. 7, and *De Generatione et Corruptione*, Book I, ch. 3.

[2] Remove the *Wāu*

So far as essence is concerned, it cannot be said that it comes to be, but that it is. A man also is a living and rational being, and cannot be said to have become living and rational. He can, however, be said to have become a child, dark, white, tall, heavy, old, large-headed, or small-eyed. The attributes "living" and "rational" belong always in exactly the same way to all men at all times, and are not found in one man and lacking in another, like the attributes "dark," "white," *et cetera*.

We do not say that earth, water, air or fire came to be, but that they are. From earth a tower, a brick, or a manger, which were not before come to be; and all of them are earth. A brick or a manger is not self-existing, but came to be from a thing
180 that was, and not from nothing; and it exists now in a state in which it did not previously exist. A house also was not a house before it came to be, because if it was, how could it come to be? And if it was not, how could it cease to be? Lo, earth is, and we do not say that earth comes to be; and because it does not come to be, we do not say that it ceases to be, like a house, earthenware, or any other object. Further, since they came to be from a thing that was, and are not self-existing because they came to be, they will cease to be, and when they cease to be, they are no longer. Lo, a gold figure comes to be from gold, which was before the figure came to be; and the figure could not be said to have been before; and when the gold is melted and the figure ceases to be and becomes gold, we do not say that the gold came to be.

To express ourselves in short terms: every accident which comes to be from an essence through an active movement and then ceases to be, does not take the fact of its being from the nature of the essence, nor does it revert, when it ceases to be, to the nature of the essence, but it comes to be and is formed through the movement, and perishes and ceases to be[1] when both the causative movement and its effect cease to be. It neither comes to be from a thing, nor does it revert to a thing.

After these things let us discuss, according to the order of our discourse, the μεταλλικά, that is to say, the metals.

Here ends the third discourse.

[1] Text, "does not perish and cease to be" (*sic*).

AGAIN, BY THE ASSISTANCE OF THE BENEFICENT CAUSE, WE BEGIN THE FOURTH DISCOURSE.

CHAPTER I.

On metals.

AFTER having finished our discussion of the genera and species of animals, animal-plants and plants, of their divisions, and of everything found above the earth, which was brought to our notice, we will now speak of the metallic natures, and study first the question of how the earth, which is cold and dry, gives birth to species which do not resemble it, such as gold, silver, brass, iron, lead, tin, mercury, yellow orpiment, alum, sulphur, vitriol and other species.

We affirm that gold, silver, and the other genera and species found in this world, were composed of the elements by the power of the Creator, through the antagonistic movements which He implanted in them. The metals resulted from the increase and decrease of the four elements ; and, the earthy part together with the watery part having predominated in them more than the other elements, they are found in the earth. Indeed the genus or nature of gold, silver, brass, lead, tin and mercury is one, but their species are different ; and although their nature was watery and expanded, the cold of the earth predominated in them, and they became contracted. That this is the case is established by the fact that when we bring them near fire, it removes from them the earthy cold, and they melt like water and expand. As water congeals and becomes ice in winter time when there is severe cold, but melts and reverts to its first nature when brought near fire, so also is the case with gold, silver, *et cetera*. Their species, however, differ, in accordance with the proportion of the elements found in each one of them,
when compared with one another. Constituent accidents adhered 181

to them from the predominance (of this or that element in them), and distinguished them from one another.

CHAPTER II.

On gold.

The nature of gold is more humid than that of silver, and for this reason it is more malleable ; it is also heavier than silver, because its parts are more closely joined together. The colour of gold is yellow, while that of silver is white because the latter is colder and the former hotter.[1]

Brass is drier than gold or silver, and its colour is more red, because it is hotter. Tin is more humid than gold or silver, and so also is the case with lead ; and mercury is more so than all of them. This explains why they melt in the fire more quickly. As to iron, it is more earthy than all of them, and drier, and on account of the contraction of its parts, it receives fire with difficulty, and does not melt like the others, except when the melting power is brought into intimate contact with it.

All metals melt, are moulded, and stand a certain time in fire without perishing, with the exception of mercury, which flees immediately from fire, evaporates, and cannot be moulded. The reason is to be found in the fact that the other metals possess more earthiness than mercury, and so endure in fire a certain time. Mercury, however, is nearer to water in its humidity, and is liquid like it, and moves hither and thither ; and because of this it is like water in no need of melting, but is evaporated quickly by fire, in the same way as water, when placed on fire, rises in the air, on account of its humidity, and is thus in no need of undergoing the process of melting. Ice, however, is in need of the process of melting before rising into the air, because of the earthiness which predominates in it. This applies also to the other metals in which earthiness predominates, in contrast with mercury.

[1] Almost identical statements on metals with those found in these chapters are also found in *Firdausu'l-Ḥikmat* of Ibn Rabban, p. 369.

CHAPTER III.

On the reason why gold does not rust.

It is necessary to enquire into the reason why, while all metals have one nature, gold alone does not deteriorate, nor does it suffer from earth, water, fire and air ; and next to it comes silver. The other metals deteriorate, emit an unpleasant smell, taste badly and rust, in spite of the fact that the nature of all of them is watery, but contracted because of their earthiness. How did they acquire this great difference ? We affirm that although their nature is generally watery, but contracted because of their earthiness, yet their species are different from one another, through the individual changes which they underwent in the following way : in all bodies are found active and passive elements, and each one contains a different proportion of active parts and of passive parts. This is due to the fact that when, in their first composition, their active parts met their passive parts, like joined with like, and (the active parts) were drawn towards the activity of the parts which were affinitive to them. This applied also to the passive parts. When they joined together through humidity, thus forming the species of gold, silver, *et cetera*, they gave rise, in the place where the active parts predominated, to gold and silver, and in the place where [1] the passive parts predominated, to brass, tin, *et cetera*.

The hot parts gave rise to gold, and the active cold parts to silver. I mean by these parts, the high parts of the active elements ; for there are three kinds of parts in both active and
passive elements, and they form a high order, a middle order 182
and a low order. When the high part of an active element is compared with the middle part, the high part is more active than the middle part ; and the middle part is less active, and is passive in comparison with the high part. The low part of an active element is very passive in comparison with the high and middle parts. The high part causes increase and decrease in the middle part, and likewise the middle part in the low part. The

[1] Read *aiḳa*, with a *Yodh*.

middle part both acts and suffers, while the high part only acts, and the low part only suffers. When the low parts of the active elements are compared with the passive elements, they are by comparison active ; and because the high parts of the active elements act and do not suffer—since there are no other parts above them—and because gold came out of them, gold, therefore, never suffers. In the same way, silver comes out of the high parts of the active power of cold ; but because the active power of heat is more powerful than (that of) cold, gold is more powerful than [1] silver. Although everything is composed of active and passive elements, there are some species of bodies in which the active elements predominate over the passive elements, and *vice versa ;* and gold contains more parts of the active elements than any other species.

Just as artisans are active agents, and the stones and pieces of wood of a house are passive agents, the former acting and the latter suffering ; and just as there is one high artisan who is more powerful than all others in activity, and all the other active agents are under him and suffer his command, and they in their turn make others who are under them suffer their command, in such a way that those who are in the middle are both active and passive—so is the condition of the middle parts of the active agents.

It is in the above way that gold acquired an active agency higher than that of all other active agents, and does not submit to any other agent. As to brass, it is also hot, but it emanates from the lower [2] part of the active agents, and from the hot part found in the passive agents, through which an active agent joins with a passive one. As to lead, tin, iron and the rest, the passive part predominated in them over the active part, and this is the reason why they constantly suffer from earth and fire, as we stated above.

[1] Read *min* instead of *mā*.

[2] Read *taḥtaita*.

CHAPTER IV.

On the reason why brass, iron, tin (et cetera) emit an unpleasant smell and rust, while gold and silver do not.

Gold and silver possess balance of parts in their composition, while brass, lead and the other metals do not possess balance of parts. This is the reason why the smell of gold and silver is pleasant, together with their taste. This is established by other known species which, when they possess balance of parts and no modifying addition predominates in them, emit a pleasant smell. Let us take the example of an apple. Until a predominance of extraneous things destroys the balance of its composition, it does not lose its pleasant smell, but when heat predominates, and it has become humid, it decays and emits an unpleasant smell. In this same way the other fruits and meat are very liable to deterioration ; but in the case of a thing which possesses balance of parts, its movement, like that of a circle, is not checked, and this is the reason why it does not suffer division of parts, and is completely balanced
and one, and remains intact always. In this same way, gold and 183
silver which possess balanced parts, remain intact, and are not affected by any division of parts, nor do they suffer any change and deterioration, nor does their taste become unpleasant. Lo, illness causes suffering to the body, through lack of balance of parts, while good health causes happiness, through balance of parts ; and deterioration, death and decomposition are also caused by lack of balance of parts. It follows from this that the deterioration, unpleasantness of smell and change of colour which affect brass and other metals, are caused by lack of balance (of parts).

Let us take an illustration from a composed object, like a tripod (τρίποδα), which stands on three balanced feet. When one of them is missing, the shape of the tripod is impaired, and it falls. Not to lengthen our speech unduly, everything that is balanced remains intact.

Let us now discuss how sulphur, yellow orpiment, bitumen, vitriol, *et cetera*, come also out of the earth.

CHAPTER V.

On sulphur, yellow orpiment, bitumen, alum, and the rest of the metallic species.

Sulphur, yellow orpiment and bitumen are composed of the four elements, but the fiery and airy parts predominate in them, while the earthy and watery parts are found in them in a very small quantity ; and because of the fiery parts which predominate in them, when fire is brought into contact with them, it flares up quickly. They are contracted like stones, on account of the earthy part which they contain and which they have assimilated through their nearness to the earth, from the parts of it that were less heavy. This is the reason why they are easily kindled by fire, which dissolves their contracted state until it receives from them that part of them which was from itself. The air also takes back their airy part, because that earthy part in them which was comparatively lacking in heaviness is conquered and subtilised by the numerous fiery and airy parts which they contain, and which draw that earthy part upwards, in the same way as air draws, when moving, the small and light particles, like dust and chaff, and raises them upwards. This explains why, when these (metals) are destroyed by fire, nothing remains of them. In the case, however, of pieces of wood, *et cetera,* which contain a thick, earthy, heavy and compact part, when fire is kindled in them, it takes from them that which is affinitive to it—as does also air—and what is left of them remains in their place.

As to alum and vitriol and the metals which resemble them, the watery and earthy parts predominate in them over the airy and fiery parts, and when fire is brought near them from a lamp, it does not kindle them, because when those watery and earthy parts come into contact with fire, they drive it away. They contain an earthy part greater than that found in gold, silver, *et cetera,* in which the watery part predominates. This is also the reason why they cannot be moulded, like gold and silver. Further, their taste is acerb and sharp, on account of the lack of balance of parts which they possess from the contraction which they underwent, and from the dryness and cold of the earth which predominated in them. They contract the parts of the palate and of the mouth towards one another, together with the humidity which is in them, and thus cause acerbity and sharpness of taste.

The functional ground for the existence of metals is the following: they were made by the Creator for the fulfilment of the needs of the world, and for the glorification of His wisdom, showing as they do that He did not create it haphazardly. For the benefit [1] of men He made differences in them, so that some of them are valued more highly than others. He made gold more valuable than all the others, for if it had been of the same
value as the rest of them, the advantage of its value and its good- 184
ness would not have been known. He thus imparted to it the distinctive virtue of non-deterioration, a quality not possessed by other metals. In this respect silver comes after it, as it possesses also non-deterioration, though in a less marked degree than gold, so that many weights and pieces of silver may be counted as one of gold; likewise brass, many pieces of which are counted as one piece of silver; likewise lead which deteriorates more easily in comparison with brass, but endures more than some others. The Creator did this for the fulfilment of the needs of men, so that they might possess some things more highly valued and some others less highly valued, the former to be bought with gold and the latter with silver. In the first category may be counted fields, wells, precious objects, *et cetera*, and in the latter—that is to say the less highly valued—category, may be counted vegetables, cereals, fruits, *et cetera*. If there were only one species (of metal), men would have encountered great difficulty when buying [2] precious things, as they would have been obliged to use many pieces of that single species. Lo, there are many small countries which do not make use of gold and silver, nor of buying and selling, and, their transactions being in disorder, they live irrationally, like those who do not know how to read and write.

Further, gold and silver are also useful for their own beauty, and for the adornment and embellishment of the trinkets of women; while brass, lead, tin and iron are useful for domestic requirements. The Creator made iron hard, contracted, dry

[1] Read *yūthrāna*. [2] Read *nizbūn*.

and generally not liable to injury from anything, so that it might cut and cleave objects, and be useful for the performance of crafts. Indeed without it no craft can be performed. It is also useful to soldiers in fighting against their enemies.

Further, bitumen [1] was made so that it might check and impede humidity from objects of use, such as ships. As to the use of alum, yellow orpiment and sulphur, it is so obvious that we do not need to speak of it.

CHAPTER VI.

On the reason why different colours are found in the earth, of which some is white, some red, some black, et cetera.

We affirm that earth is generally cold and dry, but that it contains partial predominances (of one element over another). Indeed, when compared with one another, some parts of it are drier, others less cold, and others hotter ; and it is on account of these predominances which are partially found in it, and which emanate in great or small proportions from the antagonistic forces of the elements, that different colours are found in it. Indeed when we prepare clay (for pottery), it is generally black at the beginning, but when we have moulded it into a certain shape and placed it on the fire, it first becomes red and then white. This happens to it from the greater or smaller quantity of the elements.[2] We must conceive that the same thing happened in the first composition of the earth.

The question of how colours are produced in different species we have discussed above in our discourse on colour.

CHAPTER VII.

On the reason why some parts of the earth became soil, some others mountains and rocks, and some others plains.

185 We affirm that in the first composition of the earth, parts of it became drier and parts of it less dry. Those which were less

[1] Read *ziphta*. [2] The first letter should be *Dalath* instead of *Wāu*.

dry joined together on account of humidity, adhered to each other and condensed, and from them rocks were made. As a potter's vessel becomes hard through the excessive dryness which it receives from fire, in this way I picture to myself that they contracted and became hard. Those of them which possessed expanded parts consisting of soil did not adhere to each other through humidity, and on account of the expansion of their parts they are lighter than the rocky parts. This may be established by a quantitative comparison of their parts. When a certain measure of soil and an equal measure of stone are weighed,[1] we find that the measure of stone outweighs that of soil. This explains why soil is above and rocks below.

If rocks are below, how is it that mountains are above the earth, although they are not soil, and how were they lifted up? We affirm that in the first movement of the coming together of the simple elements, when they joined with one another and became compound, every one of them strove, in accordance with its nature, to separate itself from its neighbour: fire and air went up, and water and earth went down. When the fire and air which were imprisoned in the earth were moved by the straight upward movement, and came into contact with those parts of the earth that were more contracted, dense and stony, they pushed them with a powerful movement and lifted their parts upwards, until their force was spent on account of the weight of those parts, which became then mountains above (the earth). But when (fire and air) met with parts that were less dense, or rather expanded, because they were not compressed by them they escaped between their interstices and thus did not lift them up, because they were not compressed inside them; and in this way plains came into being.

This is known to be so from the fact that northern countries, which are more contracted (by cold), have more hills than southern countries, which are more expanded. Let us take an illustration from a potter's vessel: when we hermetically close its mouth, so that no air can escape from it, and we immerse it forcibly downward in water and leave it; since the air is compressed and cannot escape, on account of the hardness, dryness

[1] Read *teḳīl.*

and contraction of the parts of the vessel, it will leap upwards above the water, in accordance with its nature, and lift the potter's vessel above the evenness of the level of the water. This is a well-known movement; and this is how mountains came into being.

CHAPTER VIII.

On the reason why the northern countries are higher than the southern countries; and on how we know this.

The sun passes over the southern countries, and there has its rising and setting. This is the reason why the southern countries are hotter and more expanded than the northern countries, and thus imprison less quantities of vapours in the earth; while the northern countries, being colder and more contracted, imprison in the earth a greater quantity of vapours. Wherever great quantities of vapours rise, they cause decrease of matter, and wherever they are imprisoned [1] they cause increase of matter; and wherever there is increase there is bulk, and wherever there is decrease there is smallness of bulk. Because the northern countries are more contracted, matter increased in them, and by its increase it acquired bulk, and on
186 account of its bulk it was pressed hither and thither and rose upwards. The contrary took place in the southern countries.

Let us take an illustration of this from the bodies of animals. In those of them that have evaporation through their expansion, the fleshy matter is small. Lo, when bodies are expanded through heat in a bath, and evaporation takes place through perspiration, weight is taken away from us, and our body becomes light and diminishes in bulk. This is the reason why the northern countries are higher than the southern countries.

That northern countries are higher than southern countries is established by the fact that all the rivers that come out of them immediately flow southwards, until they reach the sea; and although there are rivers in southern countries, we seldom

[1] Read *methḥabshīn.*

find them flowing towards northern countries; and if they happen to flow northwards, it takes them longer, except occasionally from one specified place to another, owing to its particular configuration.

The functional ground for the existence of the above is the following: mountains came into existence for us, so that they might contain forests of wood for the use of men. Some trees are large and tall, as cedars, cypresses and such like, which are used for the roofs of dwelling-houses, and for fortresses, bridges, *et cetera*; while some others are not used for building, such as the oak, the service-tree, *et cetera*, but are used for heating.

Further, mountains came into being so that out of them large rivers might come, and might water all regions, flow in the expansive and numerous regions of the plain, and give sweet water to men—since the water of southern countries is generally brackish—and also so that they might water the fields. Indeed southern countries have less rivers and springs, and even rain falls less frequently in them.

If the earth were all flat and even, and if the northern countries were not high, rivers would not have flowed, nor would springs have been found in it, except to a moderate extent, and even so, water would not have flowed from place to place, but would have collected and given rise from time to time to floods,[1] as the Euphrates and the Tigris and other rivers do, and the dry land would have deteriorated. If there were no sea (to serve as a drain for water), the countries of the earth would have become too damp, the roads would have been cut off, there would have been no corn or trees, towns and villages would have floated, and complete destruction would have resulted. Indeed after ten days' rain on the earth, the water cuts off roads, and by collecting in one place, strangles the soil and ruins it, together with its corn. How much more would not large rivers have done this! The Creator, therefore, acted wisely in making mountains for the help of mankind, and in making them high so that their water might flow. They serve some other purposes which are obvious, and of which there is no need to speak.

[1] Lit. "a kind of sea."

CHAPTER IX.

On the reason why springs generally come out of mountains, and their water is cold ;[1] *and on the reason why mountains have trees.*

We affirm that in the upheaval which took place through fire and air going upwards from the inside of the earth, and through which mountains came into existence, some[2] places were found to be weak and soft, and through them air leapt upwards, and in doing so gave rise to slits and clefts formed in an upward direction inside the mountains, at their top or at their side. In this way it opened up fissures and escaped. This very air drew
187 also the water after it, through the affinity which its humidity possessed with water, and springs came into being at the extremities of the mountains, from which water flowed. As it happens when we draw air by suction from a hollow tube, and the water is drawn up after it, which we cast aside, so it happened in the mountains. This happens only seldom in the plains, owing to their expansion ; and this explains why springs, rivers and trees are fewer in them.

(Mountains also) have trees more than plains because of the contraction of their parts, which imprison[3] within themselves heat and humidity, which give rise to many trees and plants. Their water also is cold and sweet, because it is removed from the sun. Indeed when the sun is removed from us in the winter to a southern country, our water becomes very cold, and when it comes near us in the summer, our water becomes warm.

In the southern countries there are only hills, and springs are few in them, if any ; and if trees grow in them they are small in size. Their water also is not cold, and it is thick because heated by the sun, with the result that its thin and light parts evaporate ; while in northern countries the water is sweet.[4]

[1] On springs and rivers see Aristotle, *Meteorologica*, Book I, ch. 13.

[2] Read *meddaim.*

[3] Read *methhabshīn.*

[4] It should be noted that in the author's terminology the word "northern" is often synonymous with "high" and the word "southern" with "low."

CHAPTER X.

On the reason why snow falls frequently in the mountains, but very seldom in the plains.

In the winter high countries are cold, and low [1] countries warm. The reason for this is to be sought in the fact that the sun shines directly on low (southern) countries, and is far from northern countries. The heat which was in the air in the summer on account of the nearness of the sun, penetrates (in the winter) to the depth, and conquers the cold which in the summer [2] was in the earth. (The heat) then rises upwards, and the more it rises upwards and becomes remote from the heat which is in the earth, the colder it becomes. This is the reason why the countries of the high mountains are cold, and their air also is very cold. This causes the humidity that is above them to condense, on account of the extreme cold that is found there, and it becomes snow, which falls on the mountains and remains on them. Even if it happens that sometimes it falls in the plains, it melts quickly on account of the heat that is inside the earth. In the mountains, however, it remains, on account of their remoteness from the heat which is inside the earth.

CHAPTER XI.

On the reason why the seasons of the year are four, neither more nor less : winter, summer, spring and autumn.

There are four seasons in the year, in accordance with the number of the four elements of which all the universe is composed. Two of these seasons act, and two suffer, like the elements. Summer acts through heat, and winter through cold. Spring does not act, because of the predominance of humidity, and autumn does not act, because of the predominance of dryness, except when there is in them an accidental predominance of

[1] Remove the *Lamadh*. [2] See below, Fifth Discourse, ch. I.

heat or of cold. If there were ten elements, there would have been ten seasons, and if there were only two elements, there would have been only two seasons ; but because there are four elements, it is right that there should be four seasons.

How can seasons emanate from the elements? We affirm that the sun rises towards the zenith and sets in the southern region every year at a certain time, but when it rises towards the zenith in the northern region the hot and dry element becomes
188 stronger with us, on account of the heat and dryness of the sun, which in this way gives rise to summer. This occurs because, the position [1] of this element being higher than that of the others, it predominates any time the sun rises. It may be known that the sun is in the zenith, from the shadow of a body which becomes small. This may be illustrated by a burning lamp under which is a body: the shadow of that body becomes very small, and the farther that lamp is from the body, the longer becomes its shadow. In this same way the shadow of a body becomes small in the summer, on account of the nearness of the sun, and long in the winter, on account of its remoteness. In this way cold holds sway, and we have winter with us. Indeed as the nearer a furnace of fire is, the hotter becomes the ambient air and the more it heats the bodies which are in it, so also is the case with the air which is with us, and which becomes heated by the sun, and then itself heats the bodies, and gives rise to summer, while the contrary gives rise to winter.

The functional ground for the existence of the seasons is the following: They were constituted in this way for the benefit of men, animals, plants and corn. Indeed if there were only one season, there would have been neither generation nor corruption, neither growth and increase, nor decrease; and if there were only the two passive [2] seasons, there would have been no composition, but only decomposition, and there would have been a complete cessation of composition. It is with great wisdom, therefore, that the Creator made the four seasons.

The winter was made so that through cold it might contract, collect and condense the humidity which the species receive from rain, from air, and from other causes. If this did not take place

[1] Text, "nature." [2] Read *ḥāshōshé*.

there would be no budding and no increase in the trees (and plants), since the humidity, before being contracted and condensed, would evaporate outside, and no fruit or corn would grow. This happens also to the seed in the womb, because if it is not caught at a propitious moment, it will not increase little by little, and perfect the fruit which is the embryo. Since, however, the humidity is imprisoned,[1] it increases and condenses, and being invigorated from outside by the earthiness, airiness and fieriness in which it is imprisoned, it gives rise to budding.

As to the summer, it expands, through its heat, the humidity which the bodies of animals, plants and corn have received, draws it out, and causes it to evaporate outside. (This humidity) first causes in plants green leaves, and then blossoms ; but when dryness overcomes humidity little by little, it hardens the blossoms and produces fruits ; and after they have ripened through heat and humidity—and this is even more true (of the semen) of animals—humidity diminishes, and dryness predominates and greatly contracts this humidity inwards.

In the autumn season, the moisture of the humidity is completely evaporated into the air, with the result that the leaves fall, and the trees cease to have fruit.

CHAPTER XII.

On the reason why the water which comes out of the springs and wells of the mountains is sweet, while that which comes out of the wells of the south is either saline or bitter.

The northern countries are, as we stated, cold. This is especially the case with the mountains which are in them. Because of this coldness which they possess, the surface of their earth and their air have some contraction. Owing to this, the thin part of the water being imprisoned with its thick part, and both being mixed together, an equilibrium is established in the

[1] Read *methḥabsha.*

water, and sweetness is born of this equilibrium. This is the reason why the water of the springs and wells found there is sweet. Further, the water found there is cold, because the inside of the earth is cold.

189 As to the southern countries, because they are hot, their air and the surface of their earth are expanded. This causes the thin part (of the water) to rise away from it, and the thick part to remain, thus establishing a lack of equilibrium in it; and from this lack of equilibrium which has predominated in it, it becomes saline. Indeed when the thin and airy part (of the water) rises away from it, and the earthy part remains, it joins more closely with the earth, on account of their mutual affinity, and in this way it thickens all the more. It is known that it is the airy part found in it that sweetens it, from the fact that when we take sea water or any saline water, and put it through an alembic—or if we take [1] it in a sponge and press it out—it will become sweet, and what remains below (the alembic) will possess a greater salinity, since its thin parts will have risen, and its thick parts which are very salty and heavy, will have remained. Indeed when we fill vessels with sweet water, and weigh them, and fill them again with salt or sea water, and weigh them, it will be found that the salt water is much heavier than the sweet water. Lo, salt water even dries up a body, while sweet water moistens it.

(Water) becomes bitter when its thin part evaporates from it and is scorched by heat. This may be illustrated by cooked food: when humidity evaporates from it and heat is intensified in it, it becomes bitter. This happens sometimes with meat or with a dish of lentils or rice. If they are not stirred, the water dries up in them and they adhere to the body of the utensil. We affirm that air sweetens foods in expanding them and rendering them thin; or if you like to put it so, it is like a passive agent acted upon by an active one.[2] This is the reason why, when the parts of the food that is being cooked adhere to the body of the pot, and air does not intervene between them, the heat of the fire is intensified and scorches the food, and the latter becomes bitter.

[1] Read *meḳabblīnan*.

[2] I translate this sentence by removing the *Lamadh*.

CHAPTER XIII.

On the reason why earthquakes take place.[1]

An earthquake takes place in the following way : Dryness and cold give rise to contraction in the earth, as all dryness does. That they give rise also to inertia and heaviness, is known from the fact that (earth) does not move by itself ; and since it does not move by itself, and there is no other thing which is as large as itself which can move it except an element which is antagonistic [2] to it, what is, then, the element which is capable of moving it? It is known that the element that is contrary to it is the hot and humid element, which is air. This is because the expansion which the air contains opposes the contraction of the earth. Air is hot and humid, and earth is cold and dry. When heat comes into contact with cold, and humidity with dryness, they contend with one another and flee from one another, and in this way an antagonistic movement is caused. This (antagonism) comes about not only through heat and cold, humidity and dryness, but also through their respective contraction and expansion.

After having shown that an earthquake is caused in the earth naturally by air, let us discuss now how and why it sometimes takes place and sometimes not. There are in the earth cracks and fissures, in some of which there is water and in others air. When air moves from place to place in these fissures, and has no opening through which to escape, it drives away before it the contraction of the earth through its own expansion, and drives away also the cold of the earth through its own heat, and its dryness
through its own humidity, and finding no way of escape in the 190
fissures, it returns and causes commotion and tremor, which is the earthquake. The air moves when a certain heat comes into contact with the earth and fights its cold, and thus the air is shaken. This is known to be so from objects that are above the earth, in which, when heat fights with cold, the air shakes the intervening object. When, for instance, we place grains of maize [3] on the

[1] About earthquakes see Aristotle, *Meteorologica*, Book II, ch. 7-8, where the explanation is not identical with that of our author.

[2] Read the verb with the negative *lā*.

[3] So I translate the word *ḳadhmānāye*.

fire, and the fire begins to get hold of them, some parts of them leap up towards the outside, and drive away with them to a distance the other parts that are above. This is a well-known result which, however, is not obtained until the fire begins to get hold of the grains, and the air, becoming hot inside them and not being able to escape because of their contraction, shakes the body violently from inside, and drives it away. The same process goes on inside the earth when it becomes hot and the air which is in it cannot escape.

We may illustrate this by a pot in which food is cooked. When we cover it with a powerful lid, after having poured into the pot water and set fire under it, the air drives away the lid and throws it to a distance, and gives a violent sound at its exit. This same thing occurs in the earth.

The same thing happens also in the bodies of animals. When air moves sometimes in order to escape from a certain place, and finds contraction in the skin, it causes several times at that spot a quivering and shaking from the inside towards the outside, until the place expands and dilates.

Why is there not always earthquake, but only now and then? Because heat does not always predominate in the earth to cause this. Although (cold and heat) are antagonistic to each other, they have times of inactivity in their coming together, and do not always fight and cause movement. As heat and cold do not always fight in the air and shake it, but do so only now and then, so also is the case inside the earth. Indeed when heat predominates in the air, and fights with the cold which is in it, a movement is caused by the antagonistic power of heat and cold, the former upwards and sideways, and the latter downwards and sideways. Through this antagonistic movement of heat and cold, the air is shaken between them from place to place, and being imprisoned at the extremities, is driven back, and in being driven back causes not only movement but violent agitation.

CHAPTER XIV.

On wind.

After this let us speak of the nature of wind: what it is and how it comes about. Some Indians have said that it is a fifth element, but we refuted this opinion in our book on *Fevers and their Causes*, where we discussed the animal spirit, the natural spirit and the rational soul. Although Aristotle asserted, in his work *Meteorologica*, that wind emanates from the dry vapour that rises from the earth, in saying: "Dry vapour is the origin of all winds, and rises constantly from the earth "[1]—yet (we may answer) that in this case it would have been necessary that winds should be constant and unceasing. Since there are sometimes winds and sometimes not, while that vapour constantly rises from the earth, in a great or small quantity, it does not seem likely that it constitutes the origin of the winds or the measure (of their strength).

Further, how can that dry vapour cause movement and expand? Indeed all dry parts contract inwardly and give rise to 191
inertia, while wind is an outward movement. How is it then possible that the origin of its nature should be from a dry vapour? If (you say) simply that (the movement of the latter) is also outward, (we answer) that it would have been necessary for such a movement to take place through a humid or hot vapour, which would cause an expansive outward movement; and if you say that it does take place through a hot and dry (vapour), (we answer that) when heat is mixed with dryness, contraction is effected. Indeed when we bring to the fire meat or a skin, or any other humid object, it causes contraction in them. As to humidity and heat, not only do they not cause contraction in them, but they add further humidity to them and expand them. How therefore can expansion originate from a nature that is hot and dry, and give rise to wind which moves from place to place and expands? Because we see that wind is sometimes at rest and

[1] See Aristotle, *Meteorologica*, Book II, ch. 4: "The dry vapour is the source and substance of all winds. . . . The vapour takes place constantly." As is seen in this chapter our author combats the idea of the Aristotelian dry vapour. For winds see also Aristotle, *Meteorologica*, Book I, ch. 13, etc.

sometimes moving, since sometimes there is wind and sometimes not, its origin is therefore not from a dry nature, but from something else ; and because there are two elements that are active and two others that are passive, and the dry element is passive, the story (of the origin of wind) is different (from that given by Aristotle).

Let us now set forth our own opinion about winds, and let anyone choose what he wishes :

Wind is a sound-giving movement of the air from place to place, and is brought about in different ways. The air is shaken either by hard bodies which cause it to move, or by reason of the fight between heat and cold, humidity and dryness.

The former method may be illustrated in the following manner : when we inhale air through the mouth, and exhale it violently, it causes wind in its exit through the lips, in such a way that when it comes out with great force it drives away any (small) object that is placed before it.

Since air is sometimes hot, sometimes cold, sometimes humid and sometimes dry, wind cannot emanate from a dry vapour that rises from the earth. This may be illustrated by the bellows of goldsmiths, which when they are filled with air and are pressed produce a sound-giving movement, and shake (the air) so that the sound reaches far ; and whether the air is hot and dry, as in summer, or cold and humid as in winter, the air contracts in the measure of the violence of the movement, and gives rise to wind ; and it is only when the hand is lifted from the bellows that the wind subsides.

The second method by which wind arises may be illustrated as follows : when, owing to the conflict [1] which takes place between the elements, heat sometimes predominates in the air in a certain place, and fights against the cold which is there ; or when the cold which is in another place fights against the heat which is in it ; or when its humidity fights against its dryness, the air is driven out violently between them, and moves from place to place. This is known to be so by the fact that when we bring damp pieces of wood near fire, we perceive the air escaping through their extremities, blowing and giving sound.

[1] Read the word with a *Dalath* instead of a *Resh*.

This occurs when the heat of the pieces of wood fights with their cold, and their dryness with their humidity, and the air, being pressed in the hollowness of the pieces of wood, moves from inside to outside and gives rise to wind.

It is known also that wind is caused in the air by the im-
pulsion of bodies, from the fact that when we blow it with a 192
fan or with another object, we move it [1] for the purpose of cooling the heat that is in our bodies. Such a process, which takes place at all times, summer and winter, evening and morning, night and day, shows that wind is the movement of the air from place to place.

The functional ground (for the existence of moving air, or wind) is the following: it is very useful for everything, and is pre-eminently the life of every living creature. Indeed if air did not move from place to place, as for instance from the inside to the outside—through which process we exhale hot fumes and steamy vapours,[2] and inhale pure and cool air for the purifying and cooling of the natural heat—no flesh could have lived. Further, when it passes over bodies, it soothes and expands them; and it ripens fruits when, through its movement, it causes their humidity to evaporate, and thus strengthens them.

Again, if (air) did not move, clouds would not have watered the earth, because it is the wind that carries them from east to west and from place to place, collects them and condenses them by bringing them closely together, and also dispels them and gives rise to fine weather. It is also the wind that by its movement causes the springs and rivers to flow; and it is the wind that kindles fire, for if there were no air to move the fire, it would be extinguished. It is the wind also that carries seafarers to their destination without any trouble on their part. In short, it brings innumerable benefits to all the world.

Because it is not our intention to make this our book long, but rather to write it in a compendious way, we will abstain from quoting the opinions of the ancients about wind, either to refute or to confirm them, in spite of the fact that we have referred briefly to the opinions of some people.

[1] Remove the *Wāu*.

[2] Read *lahgué*.

CHAPTER XV.

On the coming into existence of the sea, and on how it came to be above the earth and under the air.

The sea is an agglomeration of great quantities of water, larger in volume than all other agglomerations of water which constitutes one of the four elements. It came into existence in the following way: when the first elements came together and gave birth to the compound elements, one of the compositions which they formed was water, which is cold and humid. On account of its coldness, its position was below the hot air, the position of which is above that of water. Because the parts of heat are expanded and those of cold are contracted; and because that which is expanded is light and rises upwards, while that which is contracted is heavy and preponderates downwards; and because air is hot and water cold, it was necessary that air should be above and water below.

Further, its position is above the earth on account of its humidity, and the earth, which is dry, is below it because that which is dry has contracted parts, while that which is humid has expanded parts; and since that which is contracted is heavy and that which is expanded is light; and since earth is dry and water is humid, it is right that water should be above earth, and earth below it. Even the appearance and touch of (the above elements) demonstrate to us the fact that when water and earth are together, water rises up, earth sinks down, and air rises above water. If there were another element possessing the same relation towards earth as water possesses towards it, and as air possesses towards water, one might have supposed earth to be placed above it, and in such a supposition even fire might have
193 been below another element. This explains why, when the elements separated from one another in the way we stated above, and affinitives mixed with affinitives, water was above earth, in the places occupied by the seas.

CHAPTER XVI.

On the reason why, in accordance with the natural law of water, the sea did not cover all the earth, but only parts of it.

We affirm that in the first union of the elements with one another, some water remained in the hollows of the earth, and joining with the soil of the earth, on account of the affinity of coldness which the two of them possess, remained in it ; and because the earth is not smooth and even, but possesses some places which are high and others which are low, for the reason which we discussed above,[1] water collected in certain places. As when water irrigates the earth, after the latter has taken from the former the part which it needs and which joins with it according to the measure of its need, the part that remains collects in one place and rises above the earth, and the places upon which there is no water remain dry, so it happened then, when the watery element mixed with the earthy element.

CHAPTER XVII.

On the reason why the water of the sea became saline.[2]

The water of the sea has been heated by the sun from the beginning of the world up till now, and in this way the heat of the sun acts upon it and draws from it, through the air, the thin, volatile and expanded part which it contains and which is affinitive to itself. The water that remains consists of the thick parts which collect together and bear in themselves the result of the scorching of heat which changed them to saltness. They further receive from the nature of the upper surface of the earth a certain dryness and thickness. It is known that this happened to the sea from the scorching of the sun and from heat, by the

[1] See ch. VII, *on Mountains.*

[2] On this subject *cf.* Aristotle, *Meteorologica*, Book II, ch. 3, where the explanation is somewhat different. Ibn Rabban, in his *Firdausu'l-Ḥikmat* (p. 506) treats also of the subject, in terms similar to those of our author.

fact that when we throw water on ashes which have been consumed by fire, and then collect the water, it is saline because of the scorching which it has received from the fire.

Further, water found in a southern country is generally brackish, on account of the nearness of the sun. This happens because the sun sucks from it the expanded and thin part. As to the water of northern countries, it contains the above part, and so it is sweet, clear [1] and thin.

The ancients had innumerable opinions about the sea, some saying one thing and some another. As to us, we have found it convenient to set forth our own opinion about it in short terms. Let every reader choose what he wishes.

CHAPTER XVIII.

On the reason why there are in the earth hot springs, the water of which is sulphuric and hot.

We affirm that the springs which jet forth from the earth take their nature from that of the earth in which they are. Different parts of the earth have different tastes in them. If its taste is saline, the water will be saline ; if astringent, it will be astringent ; if bituminous, it will be bituminous ; if sulphuric, it will be sulphuric, *et cetera*. The emanation of (these different parts of the earth) from the elements we have discussed above.[2]

We will discuss now the reason why sulphuric springs are hot, while astringent ones are cold ; and why of the remaining kinds of springs, some are hot and some are cold. Sulphuric springs are hot because the nature of sulphur is hot ; and when their (water) moves inside the earth, in order to escape, it moves with it the fieriness which is in (that part of the earth), and thus
194 becomes hot, being pressed in the narrow and exiguous fissures of the earth. So far as astringent water is concerned, because it is astringent it is cold, and it is cold because when it moves in

[1] Read *raḳḳīḳé*.

[2] See ch. V of the present discourse, on the Metallic Species.

the fissures of the earth to come upwards, there is nothing in that part of the earth to heat it. Bituminous springs are also hot, but not like sulphuric ones. Bitumen melts, while sulphur burns, because the nature of sulphur is thinner, lighter and hotter than that of bitumen, and consequently more inflammable. This is illustrated by the fact that when we bring sulphur near a small piece of burning charcoal, it catches fire and flares up, while bitumen does not act in this way, on account of the thickness of its nature, but only melts. It does, however, burn when it finds an inflammable matter, such as wood, *et cetera.* Saline water is not hot, because the earth does not contain (in that place) a heating power to heat it.

Let us here put an end to this discourse.

Here ends the fourth discourse.

AGAIN THE FIFTH DISCOURSE.

CHAPTER I.

We will discuss first the formation of clouds and rain.

CLOUDS are formed in the following way: when the sun moves to the lower countries of the south, and the air is deprived of its heat, cold predominates in the air and drives away the heat which is in it; this heat then goes down to the earth where it remains. In remaining in the earth, it heats the water which is inside it. In this way the season called winter sets in. This is known to be so from the fact that at this season the water of springs and wells is warm, and the air which is above is cold, while in the summer the water [1] of the earth is cold and the air is hot.

The heat which in the winter is inside the earth heats the water which is in it, and this water rises upwards in the form of vapours, and mounts up in the air, in the same way as the water which is in a pot, when heated by fire, leaves it and rises upwards in the air. When, however, the vapour becomes cold it ceases to rise. When the vapour rises upwards to the surface of the earth and above it, the heat which caused it to evaporate flees as it reaches the coldness of the upper air, and leaves it. The (cold) air then receives these vapours through the affinity of cold with humidity, and condenses the parts which heat had expanded; and after they have condensed and become cold, they [2] collect together. The part of them which is thicker and which possesses a more earthy [3] quality, condenses first in the upper air, contracts in accordance with its nature, and gives rise to clouds. Then that thick part imprisons through its contraction the remaining parts of the vapours, which are thinner, more expanded and lighter than itself, and from them rain emanates.

[1] Remove the *Baith*.

[2] Place the first *Wāu* before the preceding verb.

[3] *I.e.*, solid.

To explain the above process in greater detail: the coldness of the air contracts the outer parts of the vapours, since it is with these parts that it first comes into contact. After these parts have been contracted and have thickened, they imprison within themselves the remaining parts, like a sponge which has become swollen and thick through its thick, watery parts which are more closely akin to earthiness, and these (thick) parts imprison the thin parts of wateriness, although both these parts emanate from water. It is in this manner that clouds and rain are formed from the vapours which rise from the earth, rain from the thin parts of the vapours, and clouds from their thick parts which imprison water.[1] 195

When cold increases and compresses the outer parts, and these in their turn compress the inner parts, and the outer parts are compressed also by the inner parts, these inner parts, being compressed by the coldness of the air, and having no room to penetrate inside any further, preponderate downwards, and their natural weight rends the clouds and they come down on the earth in a kind of downpour, which in quantity is larger than some great rivers. When that downpour meets with the air, the latter breaks it up into parts, and these parts in their turn break up into other parts, until small drops are formed which come down on the earth. This is rain.

(The downpour) is broken up in some such way as the following: air is light and hot, and rises upwards, while water is cold, humid and heavy, and sinks downwards. When, therefore, the water which has come down from the clouds joins with the air, and tries, in accordance with its nature, to incline downwards and thus to drive the air with it, the air resists coming down with it, because its own nature tends to rise and to leap upwards, on account of its expansion and thinness. When, therefore, the air meets with the water, it eats into the body of the water, which is thicker than itself, in order to rise upwards, and in this way it breaks up the parts of the water on all sides, until it forms it into small drops.

Aristotle said, in his work *Meteorologica*, that clouds, rain, snow, winds and all other accidents that happen in the air are

[1] The text of all the above sentence is corrupt.

formed of two vapours that rise from the earth.[1] He did not, however, explain distinctly their respective formation. Indeed in the summer there is no rain, except accidentally and occasionally, as the inside of the earth is cold and the water also is cold, in consequence of which coldness its parts are contracted towards one another, and do not give rise to vapours.

It is in the above way also that rain is formed both from the sea and from places where a great quantity of water is collected. From (these expanses of water) rise vapours which condense in the way we have explained. Indeed the inhabitants of sea-shores notice clouds rising from the direction of the sea, which pour heavy rain on them.

Why, while vapours continually rise up into the air, do they not always condense and give rise to clouds and rain?

We affirm that the vapours which rise from the earth are expanded, but when they reach the upper layers they collect together. Such collecting together, however, does not take place immediately and every day, but once in many days. When this collecting takes place, and the vapours condense and give rise to clouds and to rain which comes down on the earth, the air becomes empty of them until many days have passed and new vapours have collected, to give rise again to clouds and rain.

How is it that this process does not happen regularly on specified days?

The reason is that the vapours do not always collect and rise in the air in a defined quantity, but their quantity is sometimes large and sometimes small, and just as their increase or decrease is not well defined, so their collecting together is not well defined. As it happens that water in a pot sometimes gives rise to a small quantity of vapours, and some other times to a large quantity; and that when the vapours that rise are few, the drops take a long time to come down from its mouth, but when the vapours are numerous and thick, the drops come down thickly—so also it happens with the vapours that rise from the earth into the air. If they are very thick and numerous, they collect together in large quantity and continuously, while the contrary takes place if they are small.

[1] See Aristotle, *Meteorologica*, Book II, ch. 4, Book I, ch. 10, *et passim*.

CHAPTER II.

On the reason why clouds are sometimes white, sometimes black and sometimes red. 196

We affirm that clouds are white when they are empty of water, which has not yet collected in them on account of their expanded nature. Because of this, the light of the sun penetrates through them, and they appear to us white, like its light. When, however, the parts of the clouds contract together, and wateriness collects in them, the light of the sun is impeded from penetrating through them from one side to another. This happens on account of the quantitative thickness which is inside them, and which renders them very contracted, and collects their parts towards one another, in such a way that the light of the sun does not penetrate through them, and so they appear black.

Let us take an illustration from a sponge. When it is full of water, we see it inclining towards blackness. Not only a sponge, but other similar objects, when damp, incline to blackness, and when dry, to whiteness. In the same way the sun penetrates but little through all contracted objects, and their colour is black. When, however, they are expanded, it penetrates through them and they become white.

Clouds become red when the sun is either rising or setting, but when it is in the middle they are never red. This arises from the reflection (ἀνάκλασις) of the sun. Indeed when it rises it comes into contact with them in a directly opposite way, and so also it does when it sets ; and the sun which is red causes a reflection, and is seen to give them its own colour. When, however, the sun is in the middle, it is not directly opposite them, but is above them, and so does not give rise to reflection towards us, and therefore they are not seen to be red. This is illustrated by a red glass which we place in a window. When the sun rises, it comes into contact with it from the east in a directly opposite way, and penetrates through it, and its ray reaches the wall of the house, where it forms the reflection of the colour of the glass on the wall. This does not happen at any other time but at the rising of the sun. It is in this same way that the sun shines on the clouds. It comes into contact with the air and imparts colour to it; and the air then takes this colour to the clouds, and they become red.

Why does the above process not take place every time the sun rises or sets [1] and there are clouds opposite it?

We affirm that this happens every time there is a watery moisture in the air. These red clouds do not occur in all the sky thickly, but only in a scattered way.

Why does the humidity of the air give rise to red clouds?

The sun is by its nature hot and dry, and the air of the winter time is cold and humid. When the dryness of the sun comes into contact with the humidity of the air, and the heat of the former with the cold of the latter, if that humidity (which is in the air) is smoothly and evenly spread out, it receives the colour of the sun, which causes a reflection on the body of the clouds, when coming into opposite contact with them.

Further, if humidity and coldness have predominated in the air (thus destroying its even balance), when the sun comes into contact with it and fights with it, as a result of this fight the air does not receive anything from the sun; for it is its even balance that causes reflection on the clouds.

This may be illustrated by the fact that when we pour water into a white glass goblet, and place it opposite the sun, the ray of the latter penetrates through it; [2] and if we place an object on the side towards which the ray penetrates, we see that that
197 spot upon which the ray shines acquires a red colour. Further, if we put, on that spot where the ray shines, a piece of cotton and some black scrapings from a pot, and leave it there for a short time, the cotton catches fire. This happens because the sun comes into contact with the humidity and coldness of the water, and the latter, on account of the smooth and even distribution of its humidity, causes, by means of the ray of the sun, a reflection on the body on which that ray of the sun shines, and this body receives the redness of the sun. Because of the fight which takes place through the fieriness of the sun and the humidity and coldness of the water, the heat shines, through the

[1] Add *au* before the verb.

[2] Read *beghāwāh*.

rays, on the piece of cotton, and sets fire to it. It is known that this takes place on account of wateriness, from the fact that if we do not put water in the goblet, colour will not appear nor will fire break out.

CHAPTER III.

On snow and hail.[1]

Snow and hail occur when there is severe cold accompanied by dryness, in the upper region, which contracts the parts of humidity inwardly, and thickens them. The formation of snow is different from that of hail. When wind blows with cold and dryness, it penetrates humidity and expands it, and thus snow is formed in the same way as, when wind mixes with humidity through the movement of water, the water [2] undergoes a change, and scum is formed.

As to hail, it occurs when there is no wind in the contraction of cold and dryness to expand the parts of humidity. These parts thus come closer together, harden and come down through the air like stones. They are generally round, like spheres. They come down in the air in this round shape on account of the weight which they possess. When the air comes into contact with them from all sides, it makes them even, in accordance with its nature, as air is even in all its parts. It does this through the affinity which its own humidity has with that hail.

So far as snow is concerned, because of the expansion that it possesses, it does not fall in a round shape, but is elongated during its passage through the air. The air penetrates through its parts, while it does not penetrate through hail. This is the reason why the latter is rounder on all sides.

That snow and hail occur from the severe cold found in the upper region, our lower region testifies. Indeed when we have with us a severe and dry cold on the surface of the earth, ice is formed.

Further, hail takes place either at the end or at the beginning of winter. In the vernal season the sun begins to rise towards a

[1] *Cf.* Aristotle, *Meteorologica*, Book I, ch. 11-12.

[2] Read *mayya.*

northern region, and when its heat begins [1] to reach the cold found in the air, that cold contracts itself inwardly, and thus imparts extra coldness to the air. Then the parts of humidity collect together and condense, because of their cold, and give rise to hail. Hail also occurs in the autumn, when the sun begins to move away, and the air becomes cold. At this time a fight takes place between cold and heat, and in this fight humidity becomes condensed through the cold which is contained (in the air), and hail occurs.

Dew occurs generally either in the summer or in the winter, in the following way: a humidity that is thin, expanded and small in quantity, rises in the form of vapours from the earth towards the air which is nearer to the earth; and because it is small in quantity and thin, it does not rise much upwards; and when
198 the cold air of the night comes into contact with it, it contracts it to some extent, and after it has been contracted [2] it preponderates and comes down on the earth; but because it is endowed only with a weak contraction and a weak coldness, it does not form itself into drops which can be seen, as rain does, nor does it give rise to the contracted formation of clouds, but only to something that can hardly be noticed by the senses. This falls down on the earth and moistens it a little, but does not irrigate it like rain.

CHAPTER IV.

On thunder and lightning.[3]

Thunder occurs when the humidity which is imprisoned in a cloud collects, joins with wind, and is pressed by other clouds. When clouds press against each other on account of the wind which drives them from one side, they become closely compressed; when they become closely compressed, they are squeezed; and

[1] Read *damsharya*. [2] Read *ethraspath*.

[3] On thunder and lightning see Aristotle, *Meteorologica*, Book II, ch. 9, where the explanation is not always similar.

when they are squeezed, the air leaps to escape from them, in accordance with its light nature. In doing so it demolishes and rends them, and it is this rending that gives sound, in a way similar to that of a skin in which there is water and air, and which when closely compressed by a hard object, bursts and gives sound, because of the air which was in it and which escapes violently. It is in this way that we often hear the sound of thunder like something that is rent or bursting or being demolished.

The sound of thunder occurs when (two) clouds, one of which is fast in its course and the other slow—on account of the heaviness of one and the lightness of the other—come into contact with each other. When an impact takes place between them, and they impinge upon one other, they give rise to sound, in the same way as when two bodies meet with each other and impinge upon each other. When one of these bodies is swifter in its course than the other, and comes into direct contact with it, it gives rise to a sound like thunder. Indeed the sound of thunder resembles that brought to us by hand-mills when they are grinding.

As to lightning, it also takes place when the air is compressed, in its exit from the clouds, by the impact which takes place between them when moving at a great velocity. The air is then heated, is kindled, catches fire and gives light. Its nature itself is amenable to this, because it is hot. Bodies which are with us do this also when coming into contact with one another. When, for instance, flint comes into contact with iron, fire leaps out between them, through their movement. They first give sound, and then fire comes from them, resembling lightning which illuminates all the earth. This occurs even in damp bodies. When, for instance, a man strikes a hard blow on the cheek of his companion, it gives rise to a sound, and the one who is struck sees something like lightning coming out. Indeed this is no strange thing in the movement of dry and humid bodies, when they come into contact with one another.

CHAPTER V.

On the reason why lightning is seen with the eye before the ear hears the sound of thunder.

The sound of thunder reaches the ears through the movement of air, while the light of lightning reaches the eyes through the movement of fire; and because fire is swifter than air, its movement is also swifter than that of air. This is the reason why, while both (thunder and lightning) are in action, the lightning comes first and reaches us before the thunder. It is well known that what the eye receives comes to it first, and what the ear receives comes to it afterwards, from the fact that when we see from far a tree [1] being chopped, the eye receives (the impression of) the movement, when looking at it, before the ear receives the sound.

The functional ground for the existence of the above [2] is the following: clouds are formed so that they may contain water within themselves, and spread out from place to place, and thus give rise little by little to streams of water which come down broken up (into parts) on the earth, which they irrigate.
199 If they came down too swiftly on the surface of the earth, its inhabitants would not derive any benefit from them, and the earth would be less inhabited. Indeed when rain comes down slowly, the earth receives it better and it sinks into it thoroughly. Further (if there were no moving clouds) only those places from which many vapours rose up would have had rain, and it would not have fallen on other countries which were lacking in vapours. It is, therefore, with great wisdom that our God willed that clouds should be formed, so that water might be imprisoned in them and come down gradually. Indeed anything that is done gradually has a good ending, and anything that is done hurriedly and quickly gives rise to confusion and disorder. God made clouds also in an upper region and not in the vicinity of the earth, in order that when they shed their water it might become scattered, on account of the distance which it has to travel, and thus give rise to drops which evenly irrigate the

[1] Read *īlāna*.

[2] *I.e.*, clouds, rain, hail, snow.

soil on which they come down. Indeed if the clouds were in the vicinity of the earth, when they shed their water this water would not be scattered in the air, but would collect in one place, and would flood the towns and houses on which it fell. Part of the earth they would deluge, and part of it they would not irrigate at all, and men would perish[1] by drowning (in the plains), while the mountains and high places would not receive any water.

Further, God caused rain to come down from above primarily to save man from the labour of irrigation. Even as He Himself is Lord, so He honoured man in this and made him a lord, in the same way as He relieved him in other ways, as for instance in making the animals work for him, serve as his mounture, and supply all his innumerable needs, such as his different garments and the varieties of his food, *et cetera*. Indeed if man had wished to irrigate all the plains and flat pasture lands, he would have experienced great labour. Further, he would not have been able to irrigate all the land, but only that which was level, and he would never have been able to irrigate the mountains and hills. He should remember the grace of God who took away from him the hardship of universal irrigation, by granting him the favour of having to irrigate his lands only a little, and thus saving him from experiencing great labour.

He saved him also from severe conflicts, buffetings and murders which occur about water, as we often see happening. For this He ordered that the irrigation of rain should profit equally the rich and the poor, the strong and the weak, and that in this matter no man should be able to inflict or suffer injustice.

Further, because rain-water is very sweet, and contains no foreign and harmful properties, like that which flows on the earth and takes from it its properties, the fruits that grow through rain-water, which is pure, are more useful for food than those that are (artificially) irrigated. This is known to be so from the fact that all fruits and other kinds of food which are watered by rain have a better taste, a better smell, and are more digestible.

Snow also is very useful. When it falls down on the soil,

[1] Read *gmūrya*.

it kills harmful insects which destroy crops and trees, and strengthens the outer surface of the earth, so that the plants and roots may be invigorated. Further, it remains in the mountain tops, so that in summer, when the sun rises towards the northern regions, it may melt and increase the water of the rivers which
200 irrigate the countries which are deprived of rain.

God made also hail for chastisement, in such a way that when rain comes and the fruits benefit,[1] and in spite of that we do not desist from our evil ways, He sends hail and through it chastises us so that we may be punished and desist from our evil ways and do good works. A two-fold benefit accrues to us from hail: on the one hand it acts as a chastisement, by the destruction of our crops, and on the other hand it causes us to turn back to God and to desist from our evil ways.

CHAPTER VI.

On the reason why a rainbow is formed, and why different colours are formed in it: green, date-red, and yellow; while the white and black colours, which are the principal colours, are not found in it.

(The rainbow) appears to us in the shape of a bow because the heaven, in the place that is opposite the rainbow, appears to us arched. Indeed the heaven appears to us like a vault, and we imagine to ourselves the colours of the rainbow in accordance with this vaulted shape. Because the side of the sun that is in our direction comes into contact with the colours; because the sun is seen by us in the shape of a sphere, and a sphere has a vaulted shape on any one of its sides; and because it is the sun that is the cause of the rainbow—we see its colours in the shape of a bow.

The explanation of the different colours in[2] the rainbow is the following: after rain has come down and has ceased, and a clear sky has begun, a certain thick, in addition to a thin,

[1] Read *yathrīn*.

[2] Read *d-hī*.

even and smooth humidity, remains in the air. When the sun shines on the clear and smooth part of that humidity, it causes a reflection on its thick part, and creates in it to our vision the different colours found in the rainbow.[1] Indeed when the sun shines on a polished and smooth utensil of brass or gold, its smoothness and brightness cause a reflection on the bodies which are near to it, and they are seen to be of different colours ; but when the sun shines on an object that is not bright and smooth, it does not cause this effect. In the air also, when some of its parts are clear and smooth and the sun shines on them, it causes a reflection on its other parts.

The sun never causes a rainbow in the middle of the heaven, because it is not then in opposition : for reflection takes place in an opposite direction.

When the eye, which is humid, bright and clear, looks straight at the sun for a short time, and then turns its vision and looks at the earth, it sees the image of the sun before it ; but when its brightness is impaired, it cannot do this. The same thing happens with regard to the rainbow, when the air is humid, smooth and clear.

A rainbow is not formed every time there is rain and the air receives humidity, but only now and then, because it is only now and then that the humidity that remains in the air is smooth, clear and thin, and opposite it is a thick and contracted humidity, while the sun is in a place that is fit for reflection, in opposition to the humidity, and the parts of the latter in which colours are seen are of the right kind. In the same way as a mirror does not always cause reflection, so also is the case with water. If the former is not clear, pure and opposite, though the air is bright and not misty, or if it is dark, in spite of the fact that the eye is bright, it does not cause a reflection. The same thing applies to the humidity found in the air.

Why are three colours seen in the rainbow : date-red, green 201
and yellow ? We affirm that the sun is hot and fiery, and that when its fieriness comes into contact in an opposite direction with a matter which is contrary to it, namely the smooth and

[1] That the rainbow is caused by reflection see Aristotle, *Meteorologica*, Book III, ch. 4.

watery coldness of the air, it fights it, and in the first instance imparts to it its red colour, and by means of this colour establishes a slight union with it, in the same way as fire does with wood and other objects. When fieriness is not united with the matter of a living body, but is only in opposition to it, the air intervenes [1] between the fire and this body, and reddens it, and just as the skin of those who stand in the sun for a short time, or near fire for a whole day, becomes in the first instance red, in this same way the sun acts when it comes into contact with the humid and thick parts of the air and heats them, and through this heating a kind of disposition is formed in them to receive redness, which appears to our vision, through reflection, as a red colour, or various other colours, as the case may be ; and because we see that redness from a long distance, if any of its parts are thick and contracted, they incline to our vision slightly towards blackness. This explains why we imagine that colour to be date-red.

The green colour is formed thus : when the humidity which is in the air together with cold comes into contact with heat and draws this heat to itself and is conquered by it, it receives a red colour ; but the part of it which has not been conquered, and has hardened and drawn completely together, gives us a green colour in the vicinity of the above date-red colour. Indeed humidity has it in its power to give rise to this green colour. When, for instance, there is a great bulk of water, parts of it give rise to green, like the green which is seen in rivers and which is called by some people "the dwelling-place of frogs." Further, all grass is green when its humidity thickens.

The saffron colour is formed through a composition of red and white, found in wateriness. Indeed the deep saffron dye is formed of red and white.

The black colour is not found in the rainbow, because there is no seething of the humid matter by the fieriness of the sun. Indeed the black colour is formed by the seething of matter by fire, and since no seething takes place in that humidity, neither is its evaporation effected. This is the reason why neither the white nor the black colours are seen in the rainbow.[2]

[1] Remove the *Wāu*. [2] See Third Discourse, ch. III, *on Colours*.

CHAPTER VII.

On hurricanes and whirlwinds.[1]

Hurricanes occur when a wind is stirred in one place, comes into contact with another wind, and is checked by it. Going back, it comes into contact with yet another wind, and is thus driven by one wind forwards, and by another wind from another side. When it has been driven from all sides, it turns round and whirls, and when it has been pressed from the upper side, it comes down on a region, with a violence measurable by the velocity with which it has been driven in a complete movement. This explains why that wind blows in a disorderly way, and turns round with violence, as if it were wishing to shake mountains and houses. This occurs because the heat and cold which are found in the air fight against each other from different sides, and the air which is between them is pressed from all sides.

As to whirlwinds, they occur on the earth when winds turn and rise upwards, carrying with them dust and chaff. They 202
are caused as follows : on summer days air rises in small quantity from between the fissures and cracks found in the earth on account of the dryness of the summer, and rises above the earth (in the form of vapours that are) humid and cold because the (inside of the) earth at that time is cold and humid. The hot and dry air comes into contact with (these vapours), and imprisons them in itself, because it is larger in quantity than they are, and comes from all sides. Wishing to flee hither and thither in the fight which takes place on account of their antagonism with the air, and not being able to escape, they cause a circular movement inside the hot air, in their desire to flee on all sides. Then the light and airy part found in them leaps[2] upwards, and through the circular movement which was implanted in it because it was not able to escape in any direction, it draws to itself through this movement sand,[3] dust, straw, *et cetera*, and raises them upwards. That which happens to a man who is surrounded on all sides by enemies, and who, being impeded in his desire

[1] On hurricanes and whirlwinds see Aristotle, *Meteorologica*, Book III, ch. 1, where the explanation is somewhat different.

[2] Remove the *Wāu*.

[3] Read *ḥāla*.

to escape eastwards turns westwards, and is impeded there also, and then turns northwards and southwards, where he is also impeded, and in this way makes a circular movement—this same thing happens to whirlwinds.

It is only seldom that whirlwinds occur in the winter, and they are not so severe in their upward violence, because the fissures of the earth are smaller, with the consequence that although vapours rise in the winter from the earth more than in the summer, they do not bring about a circular movement, the reason being that they rise from the thin, small and insignificant fissures of the earth, and from all the earth in general, and not from one place only which is rich in vapours. The vapours which give rise to whirlwinds rise out of the fissures abruptly, and not from every place. Lo, mountains have no whirlwinds, because they are rocks which have no fissures, while whirlwinds occur often in the plains because of the fissures which they contain.

CHAPTER VIII.

On κεραυνοί, that is to say, thunderbolts, and on shooting stars.[1]

A thunderbolt is fire that comes down suddenly from the heaven to the earth. It occurs in the following way: air is by nature hot and humid, but it does happen that cold, dryness, and a heat which is in excess of what is natural to it, may be found in it. Then heat and cold fight against each other in it, with the result that sometimes heat is conquered by cold, and dryness by humidity. In this case the light and expanded parts (of the air) join with heat, and its heavy and contracted parts join with cold. When the cold parts of the air become more numerous than the hot parts, and the hot parts strive according to their nature to escape, and draw the air upwards, they drive in their fight the cold parts in an upward direction in accordance with their nature. When that heat has been conquered in the upper region by the parts of cold and contraction, because they are more numerous there, that heat comes down, against its

[1] Join both words in the text, where they are wrongly separated by two dots.

nature, and meets there with cold and contraction, and conquers them with the great violence by which it was blown, and comes down on the earth; and in its violent passage through the air it heats and kindles it, and in coming into intimate contact with it, causes a flame. Indeed fire has this quality of coming downwards in the air against its nature, when pressed by a body. When, for instance, fire comes out of the striking of a steel and
flint, it moves violently downwards until it reaches [1] the earth. 203
When the same thing happens through the air found in the upper region, and the fire comes down on the earth, it is called a thunderbolt.

When the above eastern, western, northern and southern parts (of the air) have weakened in their coldness, and fire comes into violent and direct contact with the side that is weakest, the phenomenon of shooting stars occurs; but if heat is able to escape from all sides, it gives rise to a kind of agglomeration of fire which is often seen. It is known that this movement is caused by an antagonistic force, from the fact that a thing decreases in volume when it meets with a thing not affinitive [2] to itself and increases when it meets with a thing affinitive to itself. Indeed it flees from that which is antagonistic to itself. Since fire cannot flee from that which is affinitive [3] to itself, it must, therefore, flee from that which is antagonistic to itself, namely cold. If it were heat that caused this movement in the fire, it would have empowered it to rise upwards. Indeed we notice that when a burning fire comes near another burning fire, there is an increase of volume. It follows from this that the above movement is caused by an antagonistic power.[4]

[1] Read *d-maṭya*.

[2] Add the negative *lā*, and read *dāmé*, with a *Dalath*.

[3] Read *d-dhāmé*.

[4] On thunderbolts and shooting stars, *cf.* Aristotle, *Meteorologica*, Book I, ch. 4-5.

CHAPTER IX.

On γαλακτίας (the galaxy).[1]

What is the reason for the formation of the galaxy—called by some people "the milky way"[2]—in the heavens, in which a kind of whiteness is seen, which possesses a defined shape extending a certain distance?

The stars found in that place are closer to one another than in other places in the heavens, and it is because of their vicinity and close proximity to one another, and their remoteness from our vision, that (the galaxy) appears to us as a kind of white patch resembling a path. Indeed the sense of vision perceives parts which are remote from it indistinctly, as if they were one, when these parts are in close proximity to one another. For instance, two lamps in close proximity to each other are seen as one when placed at a great distance from us, but when we come nearer to them we distinguish between them. We should think in the same way about the milky way which is in the heavens.

The eye is deceived not only by lamps, but also by bodies. When we look from a distance at a big town, it appears to us a thick mass, without any separate objects, but when we approach it little by little, the distinctive objects which it possesses become visible to us. The eye suffers this (disadvantage) because it is through the light of vision that it distinguishes parts from one another, and since the light of vision is expansive, when it reaches some parts it shows their expansion, but when these parts are remote it does not do so. This is due to the fact that the expansion and diffusion of the light of vision are due to the heat which is found in the eye; and because the sharpness of this heat is impaired by being spread out to a distance, so that only a small part remains in it, when it reaches the bodies which are separated from one another, cold predominates there over the heat which is found in the light of the eyes. When cold has predominated, contraction shows the parts to the eyes as if they were joined together in one. Since cold contracts and heat

[1] On the Galaxy, *cf.* Aristotle, *Meteorologica*, Book I, ch. 8, where the explanation is different.

[2] The Syriac expression is "the strawy way."

expands, it is when we go nearer little by little, and heat conquers, that the eye begins to distinguish the parts. 204

Let us take an illustration from old men and young men. The former, because of the predominance of cold in them, do not see far, while the latter, because of the predominance of heat, see far and distinguish between small parts.

CHAPTER X.

On the halo of the sun and of the moon.[1]

What is the cause of the halo round the sun and the moon?

There is a halo round the moon when the air which is before it, in our direction, is very thick, contracted and misty. When the light of the moon penetrates through it, in our direction, we see this light as a circle. This light penetrates through the air as if it were driving away in a circle that thick humidity before our eyes. When this humidity has been pushed away, its parts become contracted in a circular way, and it is seen by us as a halo round the moon. That such a thing does happen may be illustrated from a high room [2] which is full of smoke. When we place a lamp at one end, and we stand at the other end, we perceive against the light of the lamp the air of the room [3] forming something like a ring. If there is no smoke the above phenomenon does not occur. It is known that this phenomenon occurs because of the thickness of the air, from the fact that (in other seasons) there is no thick humidity of the air as in the winter. When the air becomes thick and dense, to be followed generally by fine weather, that which we call fog in the air gives the appearance of a ring (round the moon), until the condensation of humidity is driven away.

We see also the halo like a ring round the sun, because the light of the sun is seen by us against it in a round and circular way, and for this reason, immediately [4] it penetrates the

[1] *Cf.* Aristotle, *Meteorologica*, Book III, ch. 3.
[2] Lit. "house." Read *b-baita*. [3] Read *b-baita*. [4] Read *meḥdha*.

thickness of the humidity it illuminates it, and thus we see it as a circular ring. In this way the halo of the sun appears to us in this shape on account of the thickness found in the air.

CHAPTER XI.

On the coming into existence of the heaven and the stars from the elements, and on their nature.

We said at the beginning of our speech [1] that the elements were endowed with two antagonistic movements, antipathetic to each other, and with two other movements affinitive to each other. The antagonistic ones are those of heat with regard to cold, and of humidity with regard to dryness. The affinitive ones are those of heat with regard to humidity and dryness, and of cold with regard to humidity and dryness. We showed there also how all this world of ours emanated from the elements through the predominance of the antagonistic movements in them, with the result that generation, corruption, increase, decrease, *et cetera*, came into being in it. When the simple elements moved to join together, and the two antagonistic and affinitive movements emanated from them, the antagonistic elements were drawn together, because of a relation to each other brought about by the affinitive movements, and in this way increase was effected through them. The affinitive elements were also drawn together through their (natural) relation to each other, and gave rise to increase. A small decrease, however, occurred in the latter's composition, due to the movement of the antagonistic elements, and likewise a decrease occurred in the composition of the antagonistic movements, due to the movement of the affinitive elements. In this way a particular composition came into being from the antagonistic movements, and also a particular composi-
205 tion from the affinitive movements.[2] The antagonistic movements gave rise to our universe, and the affinitive and harmonious movements gave rise (in it) to life and health for a short time. The circular movement emanated also from these harmonious movements, which, however, look lame when compared with the degree of continuity and equilibrium found in the higher universe.

[1] See First Discourse, ch. I.

[2] The copyist repeats this sentence.

This equilibrium and this balance remain but a short time (in this world) because they are not absolute in it, and because of the multiplicity of the antagonistic movements which predominate (in it) they deteriorate quickly.

In this same way there are antagonistic movements emanating from heat and cold, humidity and dryness, in the higher universe, but they are overcome by the multiplicity of the affinitive movements, and are not absolute, and in this way they do not cause any corruption and changes of increase, decrease, *et cetera*. It appears that this is so, from the fact that some of (the higher bodies) are hot, like the sun and Mars (*Ἄρης*), and some of them cold, like Saturn (*Κρόνος*), while some others are humid, and others dry. The bodies that are with us are also like this, as some of them are hot and some others cold, *et cetera*.

The bodies that are above consist of elements, like those that are below, but while (we know how) the bodies that are with us receive generation, corruption, increase and decrease, we do not know how the harmony came about in the bodies that are above. We do not know whether an accident happened which brought this upon them, or what introduced the change between life and death, or what is the nature of the perfect harmony and balance that exists there between the elements. (We do know, however, that) life is the preservation of a body, so that it may remain as it is without change, while death is the dissolution, decomposition, modification and destruction of an existing thing. If, therefore, balance and harmony give rise to the preservation of an existing thing, so that it may suffer no modification, and if death is the dissolution and destruction of an existing thing, and if the heaven, the sun, the moon and the stars are preserved in the condition in which they are, and do not suffer any modification, they must possess this quality through balance and harmony between the elements.

As in this world the preservation of an existing thing consists in its not suffering any changes—although this preservation is not perfect, because it dissolves and comes to an end, through the predominance of the straight movements, and because it comes into existence after much labour—so also is the case with the higher world,[1] where, although an unknown antagonism

[1] Lit. "there."

of some kind exists which causes movement, yet, being overcome by the equilibrium and harmony that exist there, it does not produce destruction. If with one hundred parts of gold one single part of brass is mixed, the effect of the deterioration of the brass is not visible, on account of the large quantity of gold, but (the brass) is kept in the gold and preserved in it. In this same way the balanced and harmonious movement covers up the antagonistic movement which in this world causes destruction. On the other hand, if with one hundred parts of brass one part of gold is mixed, the parts of brass overcome that single part of gold, which then ceases to have any practical effect, and is known only mentally. The same thing applies to our world, in which, because the harmonious parts are few and the antagonistic parts numerous, changes, modifications, increases, decreases and corruptions take place.

CHAPTER XII.

206 *On the reason why this world of ours is endowed with a straight movement, and the higher world with a circular movement.*

We have shown above that in the composition of this world of ours there were predominances of the antagonistic movements over the balanced and harmonious movements. Because the antagonistic movements—that is to say, that of heat with regard to cold, and that of humidity with regard to dryness—are effected in a straight contact, the movement of heat being from down upwards, and contrary to that of cold, and the movement of cold being from up downwards, and contrary to that of heat; the movement of dryness being from outside inwards, and that of humidity from inside outwards—the straight movement was brought about in this world as follows: upwards, downwards, forwards, backwards, to the right and to the left.

The above process does not apply to the higher world, because the balanced and harmonious movement predominated in it, and became circular, like the movement of heat with regard to humidity and dryness, and that of cold with regard to humidity and dryness. In this process the movement of heat may go from up downwards, and cold from down upwards, because this same movement of heat does not go in the same way towards humidity as it goes towards dryness, but goes towards the directions of the straight movements; and since it goes towards the straight movements, it must go in a circular way, from this side and from that side. The above process applies also to the movement of cold. This explains why the movement of heat is linked with that of cold on both sides, and the movement of cold with that of heat, also on both sides: heat from up downwards, and cold from down upwards, and they thus make one circle. Through this linking, the movement of heat did not flee from the movement of cold, nor the movement of cold from that of heat, and so, although antagonistic and antipathetic to each other through the straight movement, they became harmonious through the circular movement, as the diagram placed below shows.[1]

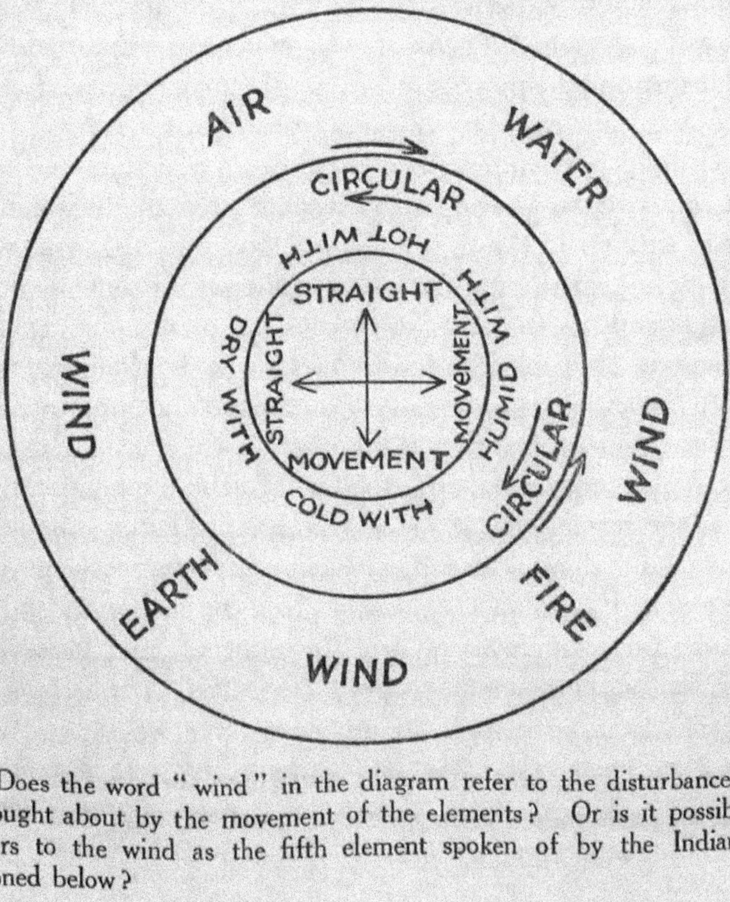

[1] Does the word "wind" in the diagram refer to the disturbance of the air brought about by the movement of the elements? Or is it possible that it refers to the wind as the fifth element spoken of by the Indians and mentioned below?

If you say: "How can the circular movement arise while the position of earth is below, and that of water above it, while air is above water, and fire is above air? Lo, there is no empty space left for a circular movement!"

We will answer that the elements wage war with one another, and in this war sometimes dryness predominates in the air, and sometimes humidity, while sometimes there is an equal quantity of both. In this way the air is able to receive them both, and since they are in it, it receives in itself the two antagonistic principles which through it are linked with each other. Lo, we find the air drier in one region and more humid in another. Indeed the region of the east and that of the west do not possess an air having the same temperature, nor does the region of the south in comparison with that of the north, as the air of one is hot and that of the other cold; that of one humid, and that of the other dry. It is in this way that it became possible for the circular composition to be formed, through humidity and dryness, and through heat and cold; while the straight movement was effected from up downwards and from down upwards, from south to north and from north to south, from east to west and from west to east.

207 The great Aristotle and some others, as if they had forgotten that this circular movement emanates from the elements, stated that in this world the four elements have only the straight movement, while in the higher world the circular movement only is found. Indeed Aristotle shows, in his book *De Caelo*,[1] that it is because this world of ours possesses a straight movement that it is affected by generation and corruption, increase and decrease, and proves that the higher world, because it possesses the circular movement, is not subject to these (changes), and that this is the reason why it did not come into being, but is eternal, rational and divine, and does not receive increase or decrease. He reasoned as follows: since a circle possesses one movement and one line, which is one and the same without beginning and end, and without increase or decrease, it is therefore eternal and the dwelling-place of God; and since we do not see that any of its parts suffer change, we know from this that all of it

[1] See Aristotle, *De Caelo*, Book I, ch. 3, etc.

is without change and without accidental existence ; and from this it is known that it is a fifth element.

(Aristotle) said this because his mind was not seasoned by the condiment of the salt of divine knowledge, which has enlightened our intelligible eyes and lifted them from the darkness of the ignorance of the nature of created things, and increased our knowledge of God. Moses said : " In the beginning God made heaven and earth." [1] He did not, then, make them without a beginning. They came, therefore, into existence at a beginning, being made by God and not resulting from nothing,[2] as we showed at the beginning of our speech to have happened through natural processes.

(Aristotle) said that the heaven constitutes a fifth element, and is also a body.[3] He named as four the elements which are in our world, and (he named) that of the heaven as a general element. He equalised the elements which are in our world and in the higher world, by saying that they are (all) elements and bodies. He separated, however, the elements which are with us from one another, in calling one (hot), another (cold), another humid and another dry. He also named and described their coming together. As to the element of the higher world, he only named it, but did not describe it or state whether it is composed or simple, but merely said that it is a fifth element, without any further explanation or definition. He resembles in this some Indians who believe in the existence of five elements in our world, four of which are those in which we ourselves believe, while the fifth is the wind ; but when we press them to distinguish it for us by a special name apart from the names of the four elements—the hot, the cold, the humid and the dry—because they have no name for it, they fall back on the four elements, and say [4] that it is a composition that emanates from them and is thinner than they are.

I believe that (Aristotle) did the same thing with his fifth element. He did not show how the heaven and the stars are the fifth element distinct from the four elements which are with us, nor did he distinguish it as he had distinguished the

[1] *Cf.* Gen. i. 1.
[2] Lit. " from God and not from nothing."
[3] *Cf.* Aristotle, *De Mundo*, ch. 3.
[4] Read *āmrīn*.

hot element from the cold, and the humid from the dry, in our world. He merely asserted that the fifth element acquired the circular movement, and our (four elements) the straight movement. He defined the heaven [1] by saying that the heaven was the essence of the circular movement, and the end of all other movements.[2] We see, however, that the essence of the circular movement is not only the heaven, but also all the bodies which
208 we have with us. Animate bodies possess the circular movement voluntarily, and inanimate bodies forcibly. Indeed all animate bodies possess the circular movement, although not in an absolute way, for in this case, in what would the heaven be distinguished from them? In nothing! As animate bodies emanate from the elements, so also does the heaven, through the process of the circular composition.

(What he said) about "the end of all the movements" presupposes a beginning, as an end is the counterpart of a beginning, and because beginnings are the counterparts of ends, and ends of beginnings, it follows that this world of ours belongs to the movement [3] of a beginning and the higher world to that of an end. It follows also that both of them are one and the same essence, whether that essence be near or far. It follows further that, as the whole connotes the nature of the part, there is no distinction between (the higher world) and our world, apart from the fact that (the higher world) is general, and (our world) is particular.

As to his (opinion) that the (higher world) does not receive increase or decrease, and consequently did not come into being, will not perish, and is eternal—you have only looked with one eye at the straight movement of the elements, O Aristotle, and have not looked with the other eye at their circular movement which we have shown in our above diagram!

A straight movement truly receives increase, decrease, and the rest (of the imperfections), on account of the antagonistic composition of the elements which gives rise to this increase, and also to generation and corruption. When the hot (element)

[1] Read *lashmayya*.

[2] For a right understanding of Aristotle on the circular movement in relation to the other movements, see his *De Caelo*, Book I, ch. 8, and Book II, ch. 1–2.

[3] Read *mettzī'ānūtha*.

meets in a direct line with the cold (element) there is flight, and victory or defeat. It is in this way that increase, decrease, generation and corruption take place. When, for instance, any composition in this world [1] completes from beginning to end the distance of a hundred cubits in a straight line, first it begins to move, then it covers one cubit, and then ten cubits, and thus increase and decrease occur in the straight movement; and when the movement reaches its limit, it goes back again in the same straight movement, and in doing so experiences a change, since the left-hand side takes the place of the right-hand side, and the right-hand side that of the left-hand side, the front that of the back and the back that of the front. In this way change is brought about, and the very movement of the line deteriorates and is thrown into disorder. When, however, increase, decrease or change take place in the circular movement, they do so in a uniform way, as it is the very same movement which circulates throughout the whole of its course, and it reaches its limit without hindrance. No beginning is seen in it, and it is not broken in the middle, and in going forwards it is always the very same movement as when it goes backwards, or to the right-hand side or to the left-hand side. With us this happens not constantly, but with the heavenly bodies it happens constantly.

The straight movement had a beginning, on account of the predominance of the hot and light power, and had an end on account of the predominance of the cold and heavy power. This happened in this way: the hot power took a small *élan*, and expanded [2] in accordance with its nature, but on account of the fight of the cold and dry power with it, the air checked the *élan* of its expansion little by little. It then came to a standstill, and the movement came to an end. As to the circular movement of the heavenly bodies, because it has no predominance of the hot and cold powers, and no beginning or end, no increase or decrease are seen in it, but the whole of it is the self-same movement in its beginning, middle and end.

With us [3] the equilibrium and balance of parts through which

[1] Lit. "which is with us."

[2] Read the verb in masc.

[3] Read *lewāthan*.

life, health, preservation and the circular movement take place are not complete, but in the higher world there is great equilibrium and balance of parts, and this is the reason why its com-
209 position is preserved without dissolution, increase or decrease.

As with us the circle of an axle continually moves and is not checked when it begins, and we cannot see in it either beginning or end, although we know that it did begin to move, having been previously not in motion, and while the movement is in the axle we are not in a position to see its beginning or its end, or to perceive any increase or decrease in it; whereas the movement of the triangle, because it possesses straight lines, when it reaches the end of each line is checked and stops, and afterwards begins again with another straight line; while both (the circle and the triangle) emanate from the same matter, whether that matter be iron or wood, and are only distinguished from each other by the fact that they are endowed with balance and proportion or with lack of balance and proportion—so is the case with the higher nature of the heaven and of the stars, and the nature of this world of ours. The nature of both is one and the same, namely a nature that emanates from the four elements, the only difference being that the movement of the higher world began and moves circularly of its own accord, on account of the equilibrium and balance of parts which it possesses, while the movement (of this world of ours) is straight, on account of the antagonism which predominates either in increase or decrease.

The heaven is a body, fixed and very large, having by nature a self-generating and circular movement, and composed of the four elements in a state of balance; and on account of this balance between the four elements it does not receive in its nature dissolution or change. The heaven is a body which moves circularly of its own accord, and which is fixed and large, and in our world there is no body that possesses all these qualities together.

Some men have seen fit to say that the heavens are three,[1] but this does not contradict the truth of our words, in which we showed that the heaven is composed of the same elements as

[1] *Cf.* II Cor. xii. 2.

those we have with us, and that it does not constitute a fifth element.

We will now make another demonstration: if gold and all the other species of metals, and all bodies, are composed of the four elements, and gold is the only one of which the nature does not deteriorate and decay, but remains always the same in earth, fire, air and water, when compared with other species; and if it acquired this quality from the balance of parts found in its nature, although this balance is not absolute, how much less will the nature of the higher world, which possesses an absolute balance of parts and equilibrium, deteriorate or change in spite of the fact that it is not eternal on account of the antagonistic powers which it possesses?

An eternal being must be one, simple, infinite and uncreated. An infinite being must be one, but the heaven and the stars fall under plurality of number; and number falls under division (of parts), and division connotes a beginning; therefore the heaven and the stars had a beginning, and are not eternal but created, according to the words of the prophet Moses: "In the beginning God created the heaven and the earth."

Further, if some of the stars are cold and dry, such as Saturn (Κρόνος), some others hot and dry, such as Mars (Ἄρης), some others cold and humid, such as the moon, and some others possess these qualities in a medium degree, such as Mercury (Ἑρμῆς)—it follows that they possessed in themselves, at the very beginning, increase and decrease, when compared with one another, and are, therefore, not eternal.

Further, since we notice that the stars have two movements, one of which is westwards and the other eastwards—the sun and the moon and all the moving stars eastwards, and Aries, Taurus, *et cetera*, westwards—how is it possible that they should possess two antagonistic movements, if they are from one and the same element which does not change? Lo, the elements which we have
with us each possess one movement: fire and air upwards, 210
and earth and water downwards, in spite of the fact that they are compound elements. It follows that because the stars have two antagonistic movements—whether their nature be subtile [1]

[1] Lit. "hidden." Read *kasya*, in the sense of "thin, spiritual, volatile," as below. The text of the sentence is corrupt.

or thick—they are the outcome of the antagonistic movements of the four elements; and so, being composed of the four elements, they are not eternal, but are created from nothing, as we showed above.

Further, if the sun completes its orbit in twelve months, the moon in thirty days, and Saturn in thirty years, how can these different movements—a slow movement, a swift movement and a medium movement—be found in one single element?[1] To what kind of category does this hidden nature found in the stars—or as you call it, this spiritual nature—belong? Does (the movement of the stars) belong to their outward and visible nature, or to the one that is hidden inside? If it belongs to the one that is outside, the one that is inside will have, therefore, no movement (of itself), but the one that is outside will direct the one that is inside; and if it belongs to the one that is inside, since the one that is outside is changeable, on account of the changes undergone by its movement, (the inside nature must also be changeable) and not one element; and because it possesses antagonism, it must be constituted of the elements which possess antagonism.

Further, if both slow and swift movements pertain to the spiritual nature found in the stars, and (similarly) both heavy and light qualities pertain to the different elements which are with us, (the element of which you speak) is, therefore, of the same nature as the elements which are with us.

Further, even if this (difference in movement of the stars) happened on account of the nearness or remoteness of places, why did they not possess one temperament, hot or cold? Lo, Saturn, remotest of them all, has a slow movement and is cold. The reason for this slowness is its coldness, and not an alien nature that is subtile and spiritual.

Further, a hot and swift body like the sun is not one single and spiritual nature, for if it were the self-same one, having in it no increase, decrease or change, different and antagonistic movements would not take place in its nature. (Since different movements do take place in it) it is, therefore, not one element, and is not eternal.

[1] See Aristotle, *De Caelo*, Book II, ch. 10, where the explanation of this point is different.

You say that heat, cold, humidity and dryness do not belong to the nature of the body of the fifth element, but that it is the bodies in which these elements dwell that are called (the fifth element), and (that this element) possesses a nature not possessed by the four other elements.

(We answer :) If we take away from this fifth body the qualities of the four elements, what name shall we give to it? Has it any distinctive nature which might designate its identity? If it has, it would by necessity belong to a fifth body. Lo, earth is a body and an element, and possesses a name which designates by itself its particular identity, which is not like that of the other three (elements). The same may be said of the remaining elements.[1] Since your element has not (a name which designates its identity) we cannot, therefore, say that it is a fifth element. If (you say that) it is like the body of animals which is composed of the elements, while itself is not those elements, (we will answer that) its composition must necessarily possess the four qualities like the body of animals ; and if you say it is not so, you will be calling this body and element ὁμοιομερής,[2] without showing that its essence is different from that of the (other) elements.

If you pretend thoughtlessly that it is a nature which is subtile, rational, spiritual, endowed with a soul, and what is
more disgraceful, also divine and eternal, and that as the soul 211
is in the body, so also it is in the stars, and that as the soul directs the body rationally, so also it truly directs the stars :—

(We will answer that) the nature of the soul is subtile, rational, spiritual, endowed with free-will, immaterial and unelemental. As to the stars, they are bodies and elements, and are not like a soul. Let us see now whether they are rational and endowed with free-will like the soul which directs the body. Lo, the rationality of the soul has given birth by itself to many inventions.

[1] *I.e.*, air, fire and water.

[2] So I read the corrupt word *onomomos* of the MS. For this word "homoeomerous" (ὁμοιομερής) which means "having parts like one another and like the whole of which they are parts," see Aristotle, *De Caelo*, Book III, ch. 4, and the note thereon by J. L. Stocks in his edition of the *Works of Aristotle* (Clarendon Press, 1922). The subject is also discussed at some length by *Galen*, Vol. I, pp. 64-66.

Indeed it has invented the four natural sciences, and given birth to all the arts, which did not previously exist, and works continually on them, every day adding to them and taking from them. That it is also endowed with free-will, all its hidden ways testify. Sometimes it wills and completes its actions; sometimes it wills to complete them, but checks itself from doing so; and some other times it begins them and ceases to perform them in the middle of its work. Its works and its inventions bear witness, therefore, that it is a nature different from that of the four elements, and that it is spiritual and subtile. There is no other nature in creation which does the things which it does.

Now by what kind of action did the stars show that they are endowed with free-will, or that they are rational? Have they been able to perform any living action apart from the action which they possess in their nature? Indeed from the time they were created up till now, they have performed no new action of any kind which did not previously exist and which came into existence through them, to which they added and from which they took something every day. They perform one self-same natural action, which they performed from the time they came into existence, and from their birth. All their actions have been accomplished, and they are not able to add anything to them or to subtract anything from them, and they make no new invention which denotes another nature in them. Everything that they perform is by necessity, and not as in the case of men, who when born perform no actions, and in spite of that make every day new inventions through their rational soul. What inventions did the stars make which denote their rationality? Are not all the things which they accomplish done by necessity? Where is the free-will which they possess and which may be shown in its working? Are they able to modify the path of their course by one single hour? Has the sun or any other star when reaching the middle of the heaven ever stood still or checked its course or speeded up its running or delayed it, as the rationality and the free-will of human nature is able to do, in standing still in the middle of a movement or speeding it up, or delaying it? They are like the elements which possess everything they have by necessity and compulsion. Indeed fire rises upwards by necessity, and never goes downwards by nature; likewise earth goes downwards and never upwards, and is not able to give birth either to an increase or to a decrease in its movement, because the elements are endowed neither with rationality nor with free-will. Since, therefore, the stars, and also animals, are like the four elements in possessing all their qualities by necessity, they are composed of them, and have no other nature. Man, however, not being like them, has another nature which is rational, endowed with free-will, immaterial and spiritual.

These things will suffice as a *résumé*.

CHAPTER XIII.

On the coming into existence of the heaven and of the stars from the elements.

The heaven and the stars were composed, as we showed
above, of the balance and equilibrium of the elements with 212
one another. This balance and this equilibrium, however, possess different parts. One part of them is better balanced, and another part less balanced, while another part possesses a medium balance. This shows that these compositions possess increase and decrease with antagonism. Indeed if there were no[1] increase or decrease, how could modification and number be found in them? Their balance and equilibrium which is at the top and the highest, has parts more contracted to one another. Indeed even the parts of bodies which are with us, the more they are contracted and the nearer they are to one another, the more they possess balance and equilibrium. And since the heaven has been formed and composed of parts that are very well balanced and the highest in equilibrium, its parts are more contracted to[2] one another. In this the prophecy of Moses conforms to the nature of things, when he says: "God said, let there be a firmament in the middle of the heaven,[3] and let it divide, and God

[1] Add *lā* to *ellū*. [2] Remove the *Wāu*.

[3] *Cf.* Gen. i. 6-7, where "in the midst of the waters." The author has omitted all reference to "waters" in this passage, apparently as not being in harmony with what he stated above on the origin of rain (Fifth Discourse, ch. I).

made the firmament and called this firmament Heaven." Indeed the word "firmament" denotes parts which are close to one another, contracted and endowed with balance and equilibrium, and which have no greater or lesser balance, as an expanded body has. Because the heaven is a firmament, it is not similar to the stars, the parts of which are not balanced, and this is the reason why its movement is wholly one, is well-balanced, and possesses the same characteristics in all its parts. As to the stars, on account of the lack of balance with which they are affected, one of them is lighter and hotter than another, and one is colder and heavier than another. Further, (the firmament) because of its well-balanced parts, has only one colour, and is in this way dissimilar to the stars, one of which is redder, one whiter, and one more lustrous than another.

If you say that (your supposed fifth element) is a body composed of the four elements, tell me what kind of body it is, in the same way as you name to me the bodies composed of the elements which are with us, such as man and horse. We affirm that the name "body" is a high genus, divided into other genera, and also into high and low species. As the bodies on this earth which are composed of the four elements are divided into animate and inanimate beings, and this main division of animate and inanimate beings is divided into genera and species of different varieties, so also is the case with the higher world, which is a body composed of the four elements, fire, air, water and earth, and which is sub-divided into the heaven and the stars, some of which are fixed and others moving. The fixed ones are Aries, Taurus, *et cetera*, Aries itself being sub-divided into separate stars; and the moving ones are the sun, the moon, Mars, *et cetera*.

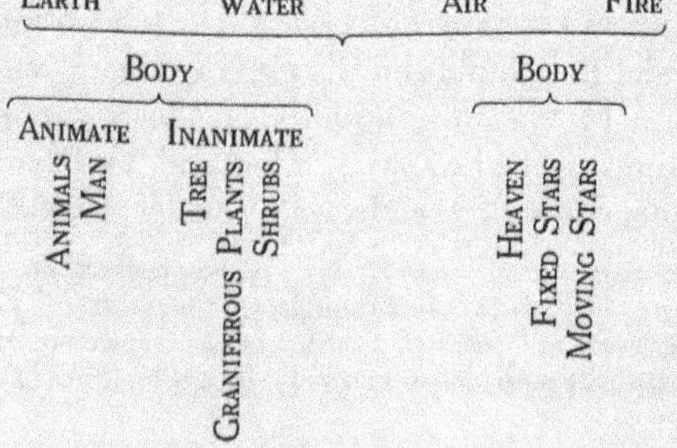

In the same way as we give names to the bodies which are with us, and when we are asked, "What is this body?" we answer, "That of a man, or of a winged creature"—so also we give names to the higher bodies, and when we are asked, "Of what is this body?"—we answer, "Of the heaven, or of the sun, or of Saturn."

The moving stars, such as the sun, the moon, *et cetera*, and the fixed ones, such as Aries, Taurus, *et cetera*, have the very same nature as the heaven, the latter being composed of the best-balanced and best-proportioned parts, while the stars are composed of the intermediate and last degree of proportion and balance: the moving and expanded stars of the middle degree, 213
and the fixed stars of the last degree; and because the moving stars have on account of their intermediate degree of balance a greater resemblance to the nature of the heaven than the fixed stars, they possess a movement of their own accord, like the movement of the heaven, which moves of its own accord; but they are inferior to the nature of the heaven in one respect, namely that they are not one in number, on account of their lack of the first degree of balance. As to the fixed stars, they are less perfect than the nature of the heaven in two respects: first in their lack of movement of their own accord by reason of their extreme remoteness, and second, in the fact that they are not one in number, on account of their lack of the first and medium degree of balance.[1]

The balance of each one of the stars is accompanied by a certain increase of heat, or cold, or humidity or dryness, but they are more especially affected by an increase and decrease of fieriness. Lo, the sun is hot and is larger than all, the moon is humid, Saturn is cold, and Mars is dry. They have received this increase and decrease although their general nature is one, from the fact that they possess a certain small degree of lack of balance,[2] and this is the reason why this (increase and decrease) occurred in them.

The seven moving stars possess also the straight movement in their rising to, and setting from, the middle of the heaven, but do not experience any change or corruption because of the

[1] Lit. "medium remoteness." [2] Lit. "part."

predominance of the balanced part which increased in them. This explains why each one of the stars has a different temperament, exactly in the same way as it happens with our world, where every species possesses a particular temperament from the elements. Indeed a lion is very hot, a fish very humid, a scorpion very cold, and a dog very dry, and a lamb holds a middle position between them. Further, one lion may be hotter than another, one fish more humid than another, and one dog drier than another, and this increase which is found in a species not only does not destroy the species, but rather constitutes it. This applies in the same way to its individua.

If one wishes to change the lion[1] into a moderate temperament, he will destroy it; and so also is the case with fish. It is in this way that I imagine that the nature of the stars is from the elements. Each one of them acquired a certain increase through which it is distinguished from its neighbour, and this increase keeps its temperament from destruction. Thus the sun has a particularly hot temperament in the first degree of heat, and that heat is not opposed to its temperament. The same principle may be applied to Saturn, and also, similarly, to the fixed stars.

As to the reason for the variation in the temperament of the stars, some of which are hotter, and some drier, than others, it is the following: they came into existence from the four simple elements, which acquired changes from one another. When they met one another for the purpose of becoming composed, they completed, from the whole of themselves, one general body of nature and of the stars; but a certain increase from each one of them was left over from the composition. Then the parts which belonged to that increase ran towards one another, the affinitive to the affinitive: the cold part towards the cold part, the hot part towards the hot part, the humid part towards the humid part, the dry part towards the dry part, and the temperate part towards the temperate part. When the hot part collected in great quantity
214 in one place, it was imprisoned in a certain part, and formed a starry matter, which was composed of the four elements, and which was near to that great quantity of heat. In this way the sun came

[1] Read *arya*.

into being, and separated itself[1] of its own accord. In the same way the cold part ran towards the cold part, and gave rise to Saturn. In this same way acted the humid and dry (parts). Thus the stars and the heaven came into being.

Increases (and decreases) occur necessarily in a composition from antagonism. This is known from the composition of the animals which we have with us, and of which, although they emanate from the four elements, one species is hotter than another, and one colder than another. As to the heaven, it was not divided into parts like the stars, for the reason which we stated above. In this way the sun is hot and dry in its composition, and Saturn is cold and dry, while the moon is cold and humid. The temperament of the fixed stars is likewise. If the nature of the sun were not composed of the four elements generally, how could it receive the composition[2] of the two elements (of heat and dryness), and how could Saturn also in this case receive (the composition of) the two elements (of cold and dryness), and similarly in the case of Aries, which is temperate, and also of Mercury?

As with us a lion is composed of the four elements, but heat and dryness predominate in him, so also is the case with all the remaining species, as well as the higher bodies. An affinitive thing receives what is affinitive to itself; so fire does not join with water when it comes into contact with it, nor water with fire, because there is no affinity of any kind between them; but because bodies contain heat and dryness, cold and humidity, they sometimes receive heat and sometimes cold, *et cetera*. This applies also to the stars; because they are bodies composed of the four elements, each one of them receives heat or cold or humidity or dryness, as the case may be. They do not constitute, therefore, a fifth element,[3] but are a creation composed of the four elements, as we learned from the blessed Moses, when he spoke of the coming into being of the Creation.

[1] Text, "they separated themselves."
[2] Text repeats "composition."
[3] Lit. "nature."

CHAPTER XIV.

On the reason why the heaven and the stars came into being[1] *outside our world, and not inside it.*

Our world came into being from the antagonistic principles[2] through the straight movement, and straightness comes into being in the middle. The[3] higher world came into being from the equilibrium and balance between (the elements), which emanates from the circular movement. It is impossible that the circular and straight movements should emanate from the same power, as we showed above, because in its nature the straight movement occurs in the middle, and the balanced and circular movement outside. It is, therefore, rightly and even necessarily that our world, because it came into being from straight movements, should have been in the middle, and that the higher world, because it came into being through the circular movement, should have been established outside our world. Indeed in all circles which we have generally with us, the straight line extends always within the circular line from the middle towards the extremities; and although the straight line may overstep its mark outside the circle, because the circle which we have with us is not compact, this does not[4] happen in the higher world, because (there) the circle is compact, and the line does not[5] overstep or pass through.

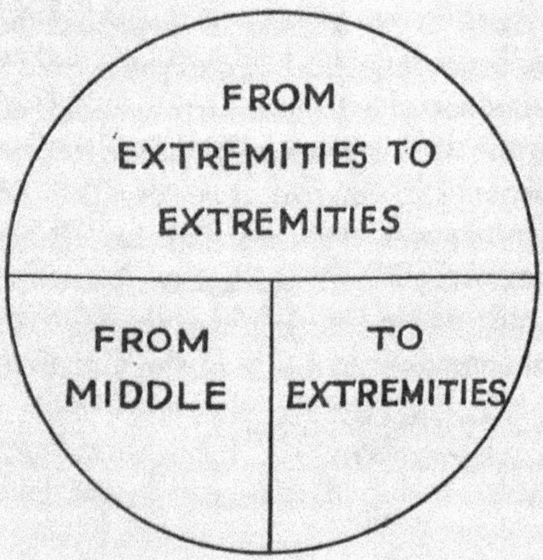

[1] Remove the first *Wāu*.

[2] Elements.

[3] Read the pronoun in masc. *hāu*.

[4] Add here the negative *lā*.

[5] Read *w-lā*, instead of *ella*.

CHAPTER XV.

On the reason why twelve fixed Signs of the Zodiac came into being.

The nature of the heaven [1] is composed of an even balance of the four elements, as we showed above. Each element acquired 215
a beginning, an end and a middle, and thus gave rise to three localities. This explains why round the pole of the heaven, which is composed of these elements, twelve localities, and consequently twelve Signs of the Zodiac, came into being, as four times three make twelve. (The elements) caused this from the balance of parts from which the nature of the heaven is made. If (the Signs of the Zodiac) were not from it they would not have possessed a distinctive quantity, but only a bulky quantity (without distinction between parts). This may be proved from the temperaments and natures of the Signs of the Zodiac, which do not possess one temperament only. Indeed some of them are hotter than others, some of them less hot, and some of them of medium heat. Leo is hotter and drier, Virgo [2] moderately so, and Cancer less so; Aries is hotter and more humid, Taurus less and Gemini moderately so; Scorpio is colder and drier, Libra less so, and Sagittarius moderately so. We know this to be the case from the order of their division in connection with the four seasons of the year. Lo, Leo, Cancer and Virgo [3] belong to the summer, which is hot and dry; Aquarius and Pisces to the winter, which is cold and humid; Scorpio, Libra and Sagittarius to the autumn, which is cold and dry; and Aries, Taurus and Gemini to the spring, which is warm and humid.

As in the cycle of the year there are, from the increase and decrease of the four elements, twelve fixed changes which never move, so also in the cycle of the heaven there are twelve changes which never move, and which emanate from the increase and decrease of the four elements, and from the compact and even movement (of the heaven).

[1] Read *shemayya.* [2] Read *bathūlta.* [3] Again read *bathūlta.*

CHAPTER XVI.

On the reason why seven moving stars came into being.

We showed also above that the heaven and the stars came into being by a natural process from the balance and equilibrium of the elements, but that they contained a part emanating from the antagonistic movements which are not balanced and which are six in number. Indeed the antagonistic movements have six variations with the addition of a seventh, which is the circular, and which accompanies them at their extremities, from one end of the line to the other end of the line on both sides, at a point where it rises and passes over all of them. This is the reason why seven movements and variations came into being to differentiate the movements of the elements.[1] The seven movements affect the seven moving stars as follows: forwards, backwards, upwards, downwards, to the right, to the left and circularly, as illustrated in the following diagram.[2]

All bodies that are with us [3] possess these seven movements from the elements. How they emanate from the elements we have shown above.

Because the elements gave rise to seven varieties of movements in the straight movement, and because a small proportion of this straight movement is found in the higher world, the seven moving stars come down from the middle of the heaven towards
216 the south, and rise up through the straight movement in a circular way. This explains why (the moving stars) possess seven movements. Further, their movement varied, sometimes going forwards and sometimes backwards—because the straight movement possesses increase, decrease and variation—and in this way they possessed variation from the circular movement of one line. This movement is one and the same without any variation, although in the circular diagram we placed a line in it as an illustration, according to the laws of the bodies which we have with us. We did this in order to show the seven movements which emanate from the elements. In the circle

[1] See above, First Discourse, ch. XI.

[2] The line joining the words "forwards" and "backwards" has been inadvertently omitted by the copyist.

[3] Read *dalwāthan*.

of the higher world, however, we do not look at the movements in this way, because there the straight lines are swallowed up by the predominance of balance and equilibrium, with the consequent predominance of the circular movement.[1] We only wished to demonstrate by this that the variation of their seven movements took place through the power of the elements, and not from anything else, at the time of their creation by God.

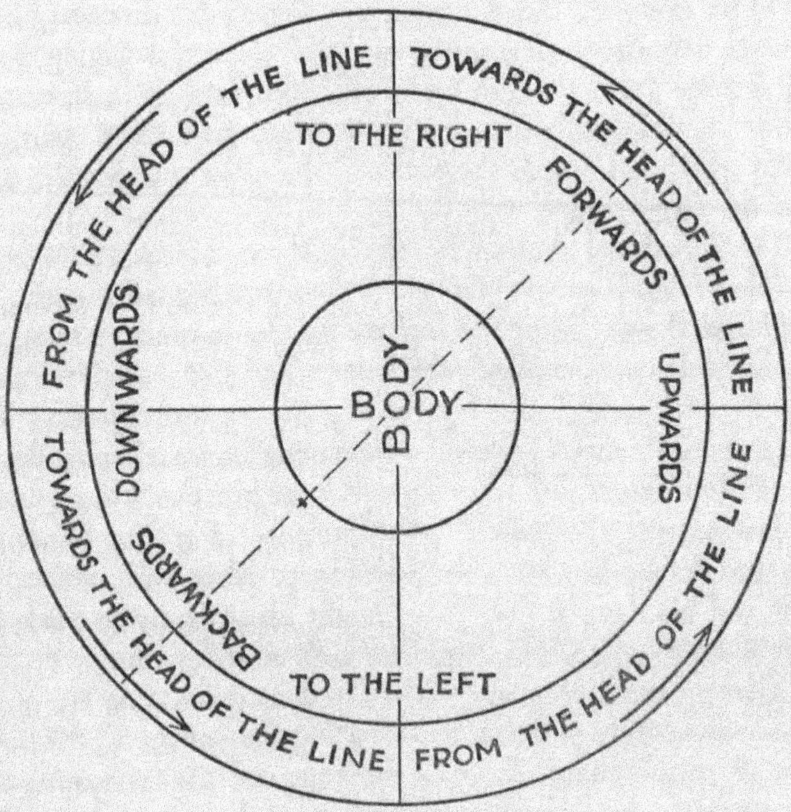

CHAPTER XVII.

On the reason why two antagonistic movements occurred in the higher world, one of which is eastwards and the other westwards.

We affirm that there are in the circular impulse two movements which emanate from the elements, and which are opposite

[1] Read *zaw'a* for *shab'a*, and read the preceding verb in sing.

to each other. These we have shown in our above diagram,[1] by the circular line of which it is explained that the hot element made a movement towards the humid and the dry elements, and that contrariwise the cold element made a movement towards the humid and the dry elements. These two movements are contrary to each other, for it is impossible that the hot movement should be directed towards the humid and the dry elements in the same impulse, but (it must necessarily be directed) in an opposite impulse and movement; and since the higher world was created from the hot and cold (elements) it is imperative that it should be endowed with two antagonistic movements, one of which is towards the east and the other towards the west, from the same pole.

The functional ground for the existence of the above is the following: God—may His glory be worshipped—made our world and the world of the heaven for the honour of man, and gave all to his service, both the heaven and the earth: the earth to grow and yield fruits for his sustenance and preservation,[2] the terrestrial animals, some for his sustenance and some for his use; and birds and fish likewise. It is for him that He made also the heaven and the stars for the purpose of giving him light, and completing for him, through their rising and setting, his night and his day, so that these might cause fruits to grow and bring seeds to completion, together with all other things of which the human race is in need. It is for this reason that He placed them in an outside position, in order that they might all the better serve all man's requirements. As some of the servants of a king surround him in an outside place, and every one of them serves him from outside, and prepares for him the things which are useful to him, in this same way God ordered the heaven and the stars to be, in an outside place, the servants of man, the king of the earth.

It is written that God made two great luminaries, a greater luminary and a lesser luminary, and set them and let them be for signs.[3] God willed that the cycle of the year should be in twelve months, and the cycle of the week [4] in seven days, so that the life of men might be ordered in this world in a rational

[1] Lit. "theory." The diagram referred to here is that on p. 29.
[2] Read *ḳuyāmeh*. [3] *Cf.* Gen. i. 14-16. [4] Read *shabtha*.

way, and so that (man) might know the number of months, days and years, and delight in the changes of times. If this were 217
not so, men would have lived irrationally like animals. Because God separated us from them and created for us a spiritual soul, possessing the power of discernment, He made also for us the separation of days, months and years, so that we might live rationally. With this end in view He created the twelve Signs of the Zodiac, so that they might make the twelve months of the year through the annual cycle, and also the seven days of the week.[1] In ordering the cycle of seven days in every (week of the) month, and twelve months in every year, He made the seven moving stars to circle in conjunction with the four (groups of) Signs of the Zodiac, and thus perform the one cycle of the year. He created also two movements in them, because if they had only one movement in one direction, the moving stars would not have made a distinct movement when passing through the twelve Signs of the Zodiac, nor would the seven days of the week, which circle in the month, have been distinguished, nor would the year have been completed by months which come into being by passing through the (Signs of the Zodiac) in their rising and setting, each one of them finishing its course in them in thirty days, and the sun completing the year in twelve months.

If there had not been two antagonistic movements, this would not have taken place, and the rational composition of this world would have been thrown into disorder. It is, therefore, with great wisdom that God made two antagonistic movements.

CHAPTER XVIII.

On the reason why, while the nature of the heaven and of the stars is one, the latter are bright and shining, while the heaven is neither bright nor shining.[2]

The heaven is a thick and firm body, the parts of which are close to one another. As to the stars, they are separated from

[1] Read *shabtha* as above.

[2] On this point see Aristotle, *De Caelo*, Book II, ch. 7, where the explanation is similar.

one another. It follows that because the movement of the heaven occurs in a circle, the stars come into contact with the ether which is in the upper regions, and which passes between them, heats them and kindles them, so that they become ignited, bright and shining. In the same way as when we move a piece of wood which is on fire, from east to west with a quick movement, it becomes kindled, bright and shining, in this same way the stars when moving come into contact with the ether, which, in passing round them, kindles and ignites the fieriness which is in them, and they appear to us from far to be bright. As to the heaven, because it is thick and very firm, and does not contain separate parts which the ether can penetrate and kindle, when the ether comes into contact with it, it turns back and does not penetrate between its parts. This is the reason why (the heaven) is not bright, fiery or shining.

The moving stars are brighter, and appear [1] larger, because, as they are never joined (to the firmament), the ether surrounds them from all sides in a circular way. As to the fixed stars, because the air does not penetrate between (them and) the firm parts of the heaven (to which they are fixed), they appear to us less bright.

CHAPTER XIX.

On how [2] the bright day and the dark night come from the elements.

The sun performs its course in a circular way. When it is above the earth, it expands the particles of air by its light, which is then shed on bodies, and in dissolving the contraction of the night gives cognisance of them to the power of vision which
218 is in the eye. This happens after (the sun) has penetrated through the air which has been thus expanded, and shone on the bodies. This is called day. When, however, the light (of the sun) is checked by the intervention of the earth, the earth impedes by its contracted parts the expanded parts of light from penetrating through the air which is with us. When this happens, and the particles of air contract because of the (intervention of)

[1] Change the first *Dalath* of the verb into *Wāu.* [2] Lit. "why."

the earth, and when this contraction reaches the bodies, it impedes the power of vision from reaching them and from acquiring cognisance of them ; and darkness is set over the earth. This is called night.

That this is so is illustrated by the bodies which we have with us. When we completely impede the light of the sun from penetrating into a house, the colour of the house becomes dark, like night. The reason for this is to be found in the earthy body of the walls of the house, which by their contraction preclude the light of the sun from passing through them and penetrating into the house. The air which is in the middle is then contracted, and impedes the power of vision of the eye of those who are in the house from reaching the bodies and distinguishing one from another.

Further, it is known that the earth is the cause of darkness—which is nothing but absence of light—from the fact that the air is cold in the night. The earth gives it something of its cold nature, as it gives it something of its contracted nature. That darkness comes to the earth by accident, and not by nature, and that it is a mere absence of light, while the light of the sun [1] is something pertaining to an essence,[2] is known by the fact that when the light of the sun reaches (the earth), darkness disappears from the air. If darkness were something pertaining to an essence,[3] it would have received by itself, like light, a part of the air, and would have fought against light, sometimes winning a victory and sometimes suffering a defeat, as fire and water, which are by nature antagonistic (fight against each other). It is in this way that light and darkness would have acted. But because we see that light never ceases to exist, and there is nothing to change it, while darkness comes into being and ceases to exist, light is, therefore, an essence, and darkness is of the nature of an accident.

The functional ground for the existence of the above is the following : God made day and night for the benefit of man and the rest (of the Creation). Indeed [4] He divided the life of animals into labour and rest. Labour takes place in the day, and

[1] Read *shemsha.*
[2] Lit. "natural."
[3] Again lit. "natural."
[4] Delete the *Lamadh* before *gair.*

rest in the night, lest by labouring both night and day their power should be impaired and should perish. He made the night to set at rest that which the day fatigued. It is for this reason that He made the day bright and the night dark. He made the day bright so that (men) might distinguish between objects and bring them together, and perform all the arts and crafts of their daily life; and so that when darkness followed and they did not distinguish between objects any more, they might go to rest. Indeed He foresaw their ἀπληστία,[1] that is to say, their insatiability, and this is the reason why He ordered this, so that in spite of the fact that the lights of art expand before them, they might not labour in them to such an extent as to exhaust themselves to death.

Further, He caused the night to be cold, so that it might cool the heat of summer days. Indeed if men and animals did not rest and refresh themselves they might possibly have perished.

Further, if there were no night, there would have been great disorder, as there would have been no clear distinction between day, month and year.

CHAPTER XX.

On the reason why, while the nature of the moving stars is one, the sun is in the middle, Saturn above, and the moon below.

219 We affirm that the largest of the (moving) stars is the sun. We shall speak below of the reason for its size. The active and first element predominated in it more than in the other stars, and because it is the largest and most powerful, and because the powerful first and active (element) predominated in it, it occupied a certain place, and the other stars stood on this side and on that side of it. This place became the middle, and each of the other (stars) stood remote from the sun in proportion to the antagonism which it possessed with it. Lo, Saturn, which is cold and dry, is in a place above (the sun), and the moon, which is

[1] Read the corrupt word of the text accordingly.

cold and humid, is in a place below[1] (the sun). As to the remaining four (stars), two of them, on account of the medium (temperament) which they possess as compared with that of the sun and that of Saturn, they occupied a place between them. In the same way, the two remaining ones are between the sun and the moon, on account of the medium (temperament) which they possess. On account, therefore, of the size and greatness of the sun, it occupies a middle position, and the other (moving) stars stand on either side of it.

If you say, "Why is the sun not above, in accordance with its fiery and light nature, and the other stars below, on account of their coldness and heaviness"?

We will answer that it is on account of the antagonism which predominates in the straight movement. This happened also because the sun is in the middle of the circular (movement), and this happened to it naturally, in spite of the fact that a light and hot thing should be above and a cold and heavy thing below. How this process takes place from the elements we showed above (when describing) the circular movement. In this process, when complete balance and medium balance predominate, that which is more powerful, hotter and lighter occupies a middle position, on account of this complete balance and medium balance, and that which is colder and heavier occupies an outer position on either side. In this way the powerful and thick part of the fieriness of the sun drew together through the circular movement, and contracted its parts in an inner position, and (drove away) that which was antagonistic to an outer position. Lo, we see in the circular movement which is with us that when the wind has a circular movement it collects within itself the scattered particles of dust, and gives volume to objects which were very small.[2] In this same way, when water makes a circular movement, it collects the dreggy part in the middle. When, however, wind and water move straight, they scatter[3] and drive away collected objects.

The functional ground for the existence of the above is the

[1] Lit. "inner place."

[2] Lit. "as small as nothing." The text of the last part of this sentence is corrupt.

[3] Read the verb in plural.

following: God wisely set the sun in the middle of the other luminaries for the benefit of this world. If it were set in a lower circle, it would have burned the earth through its proximity to it, and if He had placed it in a higher sphere, its heat would have reached us but imperfectly, and all flesh would have perished from cold; fruits would not have developed, nor would plants have grown. He placed it in the middle so that when its heat came into contact with the moon and with the other (stars) it might cool, and might give us through its medium position a medium temperament, and also so that through the increases and decreases of its rising and setting a preservative equilibrium might be established.

CHAPTER XXI.

On whether the circle of the sun and of the other stars passes under the earth, or round it in the northern countries.

220 The ancients held different opinions on this subject. Some of them said that they pass under the earth, and some of them (that they pass) across the northern countries. In order not to lengthen our speech, it is not necessary to refer to their testimonies, nor to the earth which they said is in the shape of a sphere in the middle of them, like an axle (κέντρον).[1]

Those who held that the sun and the stars pass under the earth demonstrated their point as follows: in the time of the equinox (ἰσημερία), that is to say, equality between night and day, we see the sun rising straight (overhead), its circle being like a wheel.

Further, to whichever country we go, whether it be towards the east or towards the west, and whether we travel for one day or for a thousand days, we see the heaven curving downwards, both in plains and on mountain tops. The earth is also seen in this way, together with the sun, the moon and the stars.

Further, some stars are seen to be over a northern country by the people who dwell in it, while the same part of the heaven is

[1] *Cf.* Aristotle, *De Mundo*, ch. I, 391*b*.

known also to those who dwell in a southern country. (If the stars did not pass under the earth) these people would not have seen these very same stars from their two different places.[1]

Those who held that (the stars) pass across northern countries, and not under the earth, (affirmed this because) when they looked at some fixed stars (it seemed to them) that they moved in a circular way from south to north, from north to east, and from east to south, as the Great Bear and the Milky Way. They asserted that the sun and the stars do not pass under the earth, but that they circle above the earth, and that the light of the sun is obscured from our vision (when it is) in northern countries, on account of the high mountains which are found there.

Whether they pass under the earth or round the earth, their movement is circular.

When some men considered the composition of the heaven and the stars, that they are from the four elements like this world of ours, and that they possess the two movements of the active elements, one of which is in an upper position and the other in a lower position, and (the movements) of the two passive elements which are in the middle, and that the impulse of their circular movement is from down upwards and from up downwards—they concluded that the circle of all the stars passed under the earth.

Further, if the sun passed across northern countries, these would not have been so cold that some of them are uninhabitable on account of the severity of that cold. Lo, in southern countries there are places which are uninhabitable on account of the excessive heat[2] of the sun, and so also is the case with eastern countries, where it rises, and with western countries, where it sets. The black colour of the men who live in those countries testifies to this. If, therefore, the sun passed across the northern region, even if it did not heat it to the above extent, it would for a time (τέως) have rendered their cold temperate. It is said that this shows that the sun passes under the earth.

[1] The passage is complicated.

[2] Read *ḥūma*.

CHAPTER XXII.

On the reason why the movement of the sun, of the moon, and of the (other) moving stars does not occur in one place, but sometimes above, sometimes below, and sometimes in the middle.

The bodies of the sun and the stars are composed of the four elements, two of which, fire and water, are active, and the two others, earth and air, passive. These are found in them in balanced and even proportions. In accordance with the nature of the two active elements, (the sun and the stars) sometimes move upwards and sometimes downwards, and this without any antagonism or fight. Indeed if these elements were fighting against each other, all of the stars would not have moved upwards
221 together, nor would they have moved downwards (together), but a certain hot part in them would have remained above, and a certain cold part below. This, however, does not take place because there is no antagonism in the parts which compose them. Indeed when the sun goes down from the middle of the heaven towards the southern region, when it stands in the middle, and also when it returns to rise again, there is a balance of heat with humidity.

It is therefore rightly that the moving stars rise above because of the hot element, and go down because of the cold element which are in them.

The sun does not rise over the northern region as it sets over the southern region. (This may be explained by the fact that) all parts of this world were composed of the four elements, and in their composition the affinitive was drawn to the affinitive: cold towards cold, and heat towards heat; and because heat accumulated in the southern region, cold was drawn to northern countries, where it increased as in diametrical opposition to the heat which increased in the southern region. And because heat predominates in the sun, the latter makes its circle in the country that is more affinitive to it, that is to say, in the southern region. It rises until it reaches the middle of the heaven, and after standing there, it does not pass across the northern region, on account of the cold which has increased there, but returns and makes its circle towards the southern region. This applies also to the moving stars, because, although they are said to be relatively cold and hot in comparison [1] with one another, yet when their nature is compared with the coldness and hardness of the northern region they are hot. Further, because they are bodies composed of the four elements, and the cold found in the north is only one of these elements, compound bodies (like the stars) flee from its excessive cold.

CHAPTER XXIII.

On the reason why the sun is larger than all the other stars, and next in size to it is the moon.

There are two active elements which always conquer the passive ones, and of these active elements one is more powerful than the other, that is to say, heat is more powerful than cold. This is the reason why the sun assumed more power than the active element of cold, and also than the passive elements, to all of which it is superior in size. That anything that is expanded acquires more size and becomes larger on account of this expansion, may be established by the bodies which we have with us. Dust has a small size when collected in one place, but when it is scattered in the air and its parts are expanded, we see that it assumes a size that is several times larger than its former size. This principle may be extended to other bodies. It is known that the fiery essence is more expanded (than the other elements) from the fact that when water becomes expanded by the action of heat which moves it, and becomes vapour, it occupies several times more space. Because in its nature the sun is the most powerful part of the hot element, it predominated (over the other stars). As to the other stars, because the hot parts did not predominate in them to the same extent,[2] but in some of them there was cold and in some of them a medium temperament, the sun became larger than all of them.

The moon became the second largest (star) after the sun, because the other active element [3] predominated in it, and it

[1] Read *b-peḥma.* [2] This sentence is corrupt. [3] *I.e.*, cold.

thus became superior to the others in size. Further it appears to us large because it is the nearest to us of all the stars.

The functional ground for the existence of the above is the following: God made the sun larger than all the other stars, and of a fiery nature, so that its heat might reach us and heat the lower region, and thus cause fruits and plants to grow and
222 ripen, and develop by its heat all the bodies of animals. If it were small in size like one of the other stars, its heat would not have reached us, nor would it have given us light, and there would have been always darkness and night, and all this beautiful order of Creation would have been disorganised. There would have been also innumerable mistakes,[1] as our vision would have been too weak to discern the identity of objects, and the years, months and days of the world would not have been distinguished, and from this great confusion would have ensued. Further, God made the moon large and near to us so that it might shine in the night, lest our vision should have to endure deep darkness when we wished to perform the necessities of our life.

It is with great wisdom, therefore, that God made the two luminaries larger than the other stars.

CHAPTER XXIV.

On the reason why the moon receives light from the sun, to the exclusion of the other stars.

The orbit of the moon is smaller than that of all the other stars, and because of the smallness of its circling, and the lightness of its course, it does not acquire a sharp impulse and movement. This is the reason why it does not ignite and become hot like the other stars, but only to a small extent which is not perceived by our vision. That it does not become hot is known from the bodies which we have with us. Those of them which have a sharp movement and a swift course are greatly heated and ignite, while those that move slowly either are not heated at

[1] Read *paudé*.

all, or are heated only moderately; and anything that is not heated does not ignite and give any light.

Further, because the moon is near to the earth, it receives more than all other stars a certain thick and dense humidity from the vapours which rise from the earth and obscure its bright nature. Because, therefore, of the shortness of its circuit and the weakness of its movement, it is not heated by the ether;[1] and also, because its nature is obscured by thick vapours, it does not give us light. When, however, it comes into contact with the sun and receives its rays, the sun heats it and removes from it the thick humidity which adhered to it, and gives light to it. In this way the heat of the sun takes in it the place of the heat which its movement lacked, and which did not ignite it in the middle of the ether. When it becomes little by little deprived of the light of the sun, its nature weakens, and the vapours which rise from the earth become stronger and obscure it, until all its light disappears; but when the light of the sun shines on it again, it gives light.

As when a mirror, the nature of which has deteriorated owing to a thick humidity that collects on it, is cleared by the movement of heat coming into contact with it, and its bright nature again becomes visible to us and reflects for us our own image, so also is the case with the moon. The hot light of the sun comes into contact with the moon, and its active nature removes the thick humidity from it, and brightens its nature; and by means of this brightening and clearing it shines; and through reflection it gives us back its light through which we see shapes and images; it possesses, however, a certain dark hollow which denotes thick humidity, which came to it from an unknown source, and which remained in it as an indelible accident.

[1] The author uses in this chapter, and in chapter XVIII. of this same discourse, the word "ether," in the sense of "the air in the higher regions," or "the upper air." This view does not seem to be in harmony with that of Aristotle (*De Caelo*, Bk. I, ch. 3, and *Meteorologica*, Bk. I, ch. 3), who holds that the ether constitutes a fifth element in the celestial region.

CHAPTER XXV.

On the colour of the heaven.

Why does the colour of the heaven not appear to us white or black, or like the colour of the stars?

We affirm that the nature of the heaven has contracted and firm parts, and owing to this contraction it does not ignite,
223 since the ether does not pass between these parts, heat them and ignite them, so that they may appear white through expansion, or red through burning, like the stars. It does not possess the black colour because it does not possess a cold contraction like that of the earth, which is very dense; but it possesses a compound and light colour, composed of the green and black colours, which is called by the special name of "the colour of the heaven." The black portion of it is due to the fact that its parts are more contracted than those of the other heavenly bodies. It does not, however, possess a true blackness, because some humidity is mixed with its composition, which imparts to it a kind of undefined greenness. We see that humidity is able to cause this colour, from what happens in rivers and pools, in which such a green colour occurs. Or we may state what is truer to the Book,[1] namely that the firmament takes this colour from the humidity that is above the heaven.

Some people have seen fit to affirm that this (colour) occurs because of the remoteness of the heaven. If this were so, however, why does not the air show the fixed stars which are in the heaven, and which are as remote as the heaven, with the same colour as the heaven? And why do not the sun itself and the moving stars change their colour slightly? If that is not the colour of the heaven, what colour then does it possess? Every body composed of the elements possesses a special colour. What is the colour of the heaven? Red? Lo, it is not seen to be red. White? Not at all. Yellow? No. Let us therefore believe what we see. Lo, it is impossible that a body should exist without colour. Indeed, we distinguish bodies by touch, by taste, by smell, by sound, or by colour.

[1] *Cf.* Gen. i. 7, etc.

A body emanating from the elements is by necessity distinguished by one of these. The elements cause the colours, as we showed above, by their antagonism and by the increase and decrease which come from them; and because the heavenly bodies possess colours identical with the colours of the bodies which we have with us, and the bodies which we have with us are composed of the four elements, the heavenly bodies are, therefore, composed also of the four elements. The heaven shows us the colour which it possesses, and not that which it does not possess.

Yet others have affirmed that the heaven is not a firm body, and that its colour is not its own, but that this same heaven is an expansion between the seven circuits of the seven moving stars and of the twelve Signs of the Zodiac; that this colour is an illusion of our vision, on account of the immense distance, and not a real colour, and that although it is always the same, it does not belong [1] to a body.

We answer that if this colour were the consequence of the weakness of our vision, it would have been necessary that when the sun shone in the daytime a change should have occurred in the colour, since it strengthens our vision by its light, and that in the night it should have had another colour, on account of the disappearance of the light and the consequent darkness of the air. Lo, we notice that people whose vision is weakened by (the darkness of) night, see a red thing black, but when the sun shines they see it red. If this takes place in real bodies, how much more would it not take place in things which are an illusion? Because, therefore, the colour of the heaven is the same by day and by night, without any change, it is, therefore, the colour of a thick body, like the bodies which we have with us; and as the colour of the bodies which we have with us suffers no change, so also that of the heavenly bodies suffers no change.

Further, if our vision imagined this colour on account of distance, it would have been necessary, when standing at the top of high mountains and looking over a distance of ten miles,
to imagine from this distance the colour of the heaven changed, 224

[1] Read *laiteh*.

however slightly; but since this does not occur, the colour (of the heaven) is not due to distance.

Further, it would have been necessary, when clouds collect in the heaven at a great distance from us, that on account of the weakness of our vision we should experience the illusion of this same colour of the heaven (in them). Lo, when clouds collect in the heaven, and one place is cleared of them, we see in that place in the heaven which is at a great distance from us, an emptiness between the clouds which has the colour of the heaven. This demonstrates that it is not distance that gives us the illusion of this colour, but that it is the very colour of the heaven.

Further, we see that some stars are fixed and some moving. In what would the fixed stars be fixed? In nothing? It is impossible. It is necessary that they should be fixed in something, and in this case, it is necessary that this thing should be a body like themselves. As they are not fixed in the moving stars—since the movement of the latter is opposite to them—and as there are no other than moving and fixed stars, they are therefore fixed in a third object, which is the firmament of the heaven, according to what the Prophet said: "He set them in the firmament of the heaven to give light."[1] Because the heaven is a fixed body,[2] and because there is no body without colour, and because colour is not itself a special composition, that colour belongs, therefore, to the heaven.

CHAPTER XXVI.

On the fact that the heavenly bodies have no reason, wisdom or soul.

Aristotle and many other ancient (philosophers) said that the stars have reason and wisdom, and that they are divine, and masters (of our destinies).[3] In this they strayed widely from the truth, as we will prove in the following manner. There are two movements, the voluntary and the compulsory. The voluntary movement is governed by *αὐτεξούσιον*,[4] that is to say, freedom

[1] Gen. i. 17. [2] Read *gushma.*
[3] See Aristotle, *De Caelo*, Book II, ch. 3, etc.
[4] Correct the text accordingly.

of choice. Freedom of choice gives birth to various movements in which are increases and decreases. Thus men are endowed with various movements which involve increases and decreases, but fire, earth, water and air are affected by compulsory movements.

It is known that men possess voluntary movements, from the fact that they are always modifying their movements. Indeed when they set out on a journey, with the intention of completing the whole distance, they may, when they reach half-way, (change their mind and) turn back, and this not forcibly. Further, they may move for one hour, and rest from movement for ten hours; or *vice versa*, they may move for ten hours, and rest for one hour, while their movements and their rest are not limited, nor do they emanate from the fight of the elements, but belong to the realm of (voluntary) cessation. This is due to the rational soul and the freedom of choice which men possess.

As to fire, earth, air and water, they possess compulsory movements which they are never able to modify. Indeed, fire has never gone down after it has risen up and completed half of its movement; nor has earth moved upwards. This is due to the innate compulsion of their nature.

Any being possessing voluntary movement is endowed with
free-will. As long as it is endowed with free-will, its ways and
movements are varied, it can check them of its own accord, and
it is rational; while any being that does not possess freedom of
choice and free-will is endowed with one movement which it
cannot change of its own accord. As such are fire, air, water, earth,
sun and moon, which possess one unchangeable and compulsory
movement which they cannot check; for instance, the sun and
the moon cannot, after having reached the middle of their course,
rest for one day or one hour, or go backwards. Their movement
does not emanate, therefore, from free-will or from freedom of
choice; and if it does not emanate from freedom of choice,
neither does it emanate from reason; and these beings are con-
sequently neither rational nor endowed with free-will, but are like
the elements, which possess their movement naturally and com- 225
pulsorily. That this is the case is known by the fact that some
of them are hot like fire, such as the sun and Mars, and some

of them are cold, such as Saturn and the moon. If they had a nature that was different from that of the elements, there might possibly have been a doubt about them. Though it may be said that a star goes backwards or burns, this happens from the antagonism of the elements and by compulsion, as it happens when water flees from fire or fire from water, and they turn backwards.

If, therefore, the stars are not rational or endowed with free-will and freedom of choice, how can they govern this world and grant life, death, wealth and poverty, like God, or guard from untoward accidents those who are on a journey, or cause one man to reign, and destroy another? How can a thing that has not brought to itself any change, increase or decrease, bring to others either benefit or injury? How can it give to those who are remote from it a thing which they do not possess, and either benefit or injure them?

They may be said to bring benefit and injury in the sense in which this is said of the elements. Indeed fire cooks food and gives heat, and in this way it brings benefit; but it brings injury when it scorches and burns. Water also brings benefit when it quenches thirst, but brings injury [1] when it causes drowning. Likewise earth brings benefit when we walk on it, dwell in it and derive [2] from it all the necessities of our daily life; but it brings injury when it falls upon us and kills us, causing our death. This applies to the stars which are above; so the sun causes benefit in heating and bringing fruits to maturity, but causes injury in scorching and burning. The same may be said of the other stars. But (the cherished conception) that they bring people to subjection at the time of their birth, or govern the world, or kill, or bring to life, is a vain hope. This type of action is a rational and wise one, which is not of the same nature as that of the elements; and because this action pertains to a rational and wise Being, and they are neither rational nor wise, it is not they that cause death, bring to life, give power, servitude, wealth and poverty; but these things pertain to another Being, who is God our Maker, who in His

[1] Read the verb with *Semkath* instead of *Ṣādhé*.

[2] Lit. "adorn."

incomprehensible wisdom gives kingdom to one, and wealth or poverty to another, as He sees fit.

Further, in what way could they bring power and wealth? Lo, we see that suddenly some kings reign, some rich men acquire wealth, and some men die; and to this may be added all the other changes that happen in this world. Since (the stars) are confined to the heaven and are not found in the earth, and since they are elemental bodies, how do they cause this man to reign and that man to be rich? We see that when God grants to somebody wealth of money or of other possessions, he does it through the movement of the transference of a body from one place to another; and in the same way, when a king wishes to render somebody wealthy he orders money or other possessions to be given to him, and this takes place through a servant carrying them from one place to another, until the receipt of them is acknowledged by the recipient. How, then, can stars, which are confined to the heaven, do this without coming down? If they were found both in the heaven and in the earth, this might possibly have happened, but since they are confined to the heaven this (power) does not belong to them, but belongs to the One who is both in the heaven and in the earth, and is unbounded. (A star) is from the nature of the elements, is not a creator, and is not found both in the heaven and in the 226
earth, while God is infinite, uncreated and one, and is truly capable of doing this, as the heaven is His throne and the earth His footstool.[1]

This compound syllogism may perhaps act as a natural rein to constrain you to leave your crooked path and walk in the straight path, and to affirm that the stars do not do these things of their own accord, but that it is God who does them; yet you may pretend that the prognostication of these events is found in the course of the stars, so that if a star reaches such a place or follows such a course, or is in conjunction with another star, it is a prognostication of death to a man who was born under it at such an hour or on such a day; or that another accident happening to (a star) is a prognostication of death to one man, of poverty to another, and of illness yet to another.

[1] *Cf.* Is. lxvi. 1.

Against this we will answer: while it is God who causes these things to happen, yet, for the purpose of enriching man with the knowledge of past, present and future events, He may have willed [1] to bring them about through the course of the stars, in the same way as through the prognostication of their course we know beforehand the night, the day and the seasons of the year. We may also say that we see a prognostication of death through their movement, in the same way as we know that a man will necessarily die when his head is cut off, that a living man will live if his head is not cut off, that another will be drowned when falling into water, and yet another will suffer injury from the bite of a reptile, *et cetera.* It is in this way that God may have placed in the course of the movements of the stars prognostications of death, life, wealth and poverty, while they themselves bring neither benefit nor injury. Lo, we know beforehand that when the sun follows in its course the southern path, it is winter, and when it is over us it is summer; that when in the winter no rain falls on the earth, the agriculturists and farmers will experience great want and poverty, but that when the earth is saturated with rain and no untoward accident happens, they will enjoy wealth; and that when a man never eats bread or drinks water or breathes, he will die. It is in a similar way that God may have placed a prognostication of events in the position of the movements of the stars. Such an opinion may carry some conviction, and everyone may choose what he wishes, as it is possible that through a foretoken which God placed in their movements we may obtain a foreknowledge of things, in accordance with what is written: God set them for signs and for seasons, and for days and months and years.[2]

Here ends the fifth discourse.

[1] Read the verb in masc.

[2] *Cf.* Gen. i. 14.

AGAIN BY THE ASSISTANCE OF GOD WE WILL WRITE THE SIXTH DISCOURSE.

CHAPTER I.

We will speak of the angels.

BECAUSE the angels do not fall under the senses like the heavenly and earthly species, we are not able to demonstrate their existence through the senses, but have to prove their existence by reasoning. We shall make use, as our guiding principles, of things which are perceived by the senses, and which lead us to the things which are beyond the senses.

We see that man has a soul which is rational and incorporeal, and which does not fall under the senses. This we have demonstrated above in short terms, and at length in a special treatise which we wrote *On the Soul*. Because we see that man is composed of two opposite principles, corporeal and incor- 227
poreal, we know that an incorporeal nature exists in the same way as a corporeal nature exists, of which two natures man is an intermediate composition. Indeed we see that there are bodies which have a separate existence, and which are dissimilar to the composition of man. As such are stones and soil. These are, however, similar to the body of man *quā* body, which may be cold, hot, *et cetera*. By analogy with this, it is necessary that an incorporeal being should be found with a separate existence, dissimilar to the soul of man, but similar to this soul in immortality and in incorporeality; and this is the nature of the angels, which is distinguished from the soul by the fact that the latter is imprisoned in a body and suffers with it through its union with it, and acts through it.

We will prove by the following reasoning that the angels do not suffer (from a body): if man is an intermediate being, he must be intermediate between two principles, the existence of which must be conceded. If this were not[1] so, there would have been neither an intermediate being nor any principle. It is imperative, therefore, that two principles should exist for this intermediate being, namely the corporeal and the incorporeal: the corporeal, such as fire, water and air, and the incorporeal, which is the nature of the angels.

(We will demonstrate our point) in another way: we see that some animate beings never know[2] God or praise[3] Him. As such are the irrational terrestrial animals, birds and fish. And there are some other animate beings who know God and praise Him, but not all of them or always. As such is man. Now it is necessary to suppose the existence of animate beings all of whom know God and always praise Him, as an opposite principle (διαμέτρον) to (the irrational animals). As such is the nature of the angels and the demons, who are part of them, and fell away only from His praise and not from the knowledge of His nature. As a counterpart of the necessary assumption of the existence of animate beings who are always irrational, we must assume the existence of animate beings who are always rational; and both should be known through intermediate beings who partake of some qualities of both, in the same way as there exist in all created things three kinds of occurrences:[4] the necessary, the impossible and the possible.

Because, then, the irrational animate beings, by the nature of their creation, are not able to know God or to praise Him, it is necessary to assume the existence of rational animate beings who know God and praise Him always, since we know the existence of intermediate animate beings, some of whom know Him and some of whom do not, and who do not always praise Him. Since we see and know these animate beings, some of whom know Him and praise Him, and some of whom do not, they point as with a finger to the existence of a nature of animate beings all of whom know God and praise Him always; and this is the nature of the angels. And because these always praise Him without becoming weary, they must be impassible; and because all of them—and not a part of them only—know Him, they are simple and all of them of one nature, and it would be a

[1] Insert the negative *lā*.
[2] Read the verb in *active* participle.
[3] Read *mshabbḥān*.
[4] Lit. "matter."

mistake to suppose that they emanate from antagonistic forces which cause suffering; and if they are simple, they are also imperishable and immortal. It follows from all this that the nature of angels does exist.

(We will demonstrate our point) in yet another way: if God is a king, and ruler of this world, He possesses servants between Him and the creation of the heaven and earth, which are composed of the elements; and these servants are the angels.
If they be His servants, they must be nearer to Him, and it 228
is necessary also that their nature should be higher[1] than that of the other beings who are known by name, in being impassible, immortal, simple, swift for service and spiritual. Lo, all the Books inspired by the Spirit proclaim this doctrine, and we must accept it by faith; but even if we were not men of faith, we should necessarily have to accept it from (arguments based on) the nature of things.

If you say that God can govern without servants by a sign only, we will answer: this is possible, as it was possible for Him to have composed the bodies without the four elements; but He willed it so, as not only all that He pleaseth He doeth in heaven and in earth,[2] but also in the way He pleaseth, and He was not pleased to create the bodies of animals without the elements, nor to create the world[3] by a sign only, nor to govern it without angels: "He made His angels spirits, and His ministers a flaming fire."[4]

CHAPTER II.

On the fact that the hierarchies of the hosts of angels are three, and these are subdivided into nine orders.[5]

We see that the beginning of everything that God created is ternary. As such are the heaven, the earth, and the intervening

[1] Read *rām*.
[2] Ps. cxxxv. 6.
[3] Text repeats "world."
[4] Ps. civ. 4.
[5] On the subject of this chapter, *cf.* the mediæval Latin adage: *omne trinum perfectum*. See also Aristotle, *De Caelo*, Book I, ch. 1: "For, as the Pythagoreans say, the world and all that is in it is determined by the number three, since beginning and middle and end give the number of an 'all,' and the number they give is the triad."

space, which are the three first compositions. The beginning of their composition is also ternary: essence, corporeality and incorporeality. Corporeality itself has three dimensions: length, depth and width. Even the names of bodies are ternary: themselves, their homonyms and their paronyms. Everything under heaven is also ternary: the object itself, what is outside the object, and what is inside the object. The composition of speech is also ternary: noun, verb, and that (part of speech) which refers to either of them, and which constitutes the third. Even a syllogism, constructed to show that something is true or not, is ternary: it is apodeictical, sophistical or dialectical,[1] and although some men saw fit to add two other (kinds of syllogism), these are really derived from the above three. Indeed a syllogism itself is composed of three propositions: the major, the minor and the middle.

The four mathematical sciences are also ternary: astronomy has three first movements: the circular movement of the sun and the rest of the stars, which takes place in the zenith [2] of the heaven; the lower movement which takes place in the southern region; and the middle movement which takes place between the two. The beginning of the geometrical movement also is directed towards three sides which complete the perfect composition of a triangle, and are not in the movement of a circle. Music also is composed of three movements of sounds: thick, thin and medium. Arithmetic also begins with a triad to form all numbers: an uneven number, an even number, and the sum of these two, which is the third. The composition of all animals is also ternary: two hands, and what comes between them; two eyes, and a nose. To these examples may be added others, on which there is no need to dilate. They illustrate a hidden mystery which testifies that there is one in three and three in one, and thus shows that the nature of the creation of the hosts of angels is one, and is known in three hierarchies, and that our God is one in three. If the Lord grants us life, we will write a particular and more specific discourse on this subject.

Although we said that the beginning of all division is ternary, yet innumerable other divisions sprang from it little by little.

[1] Correct the Greek word accordingly.

[2] Lit. "middle."

We have only wished to demonstrate that the first division of everything shows the number three, and because we see that this first division of everything which God created shows us 229
the number three, it is necessary that the beginning of the creation of the angels should also have been in three hierarchies, which He divided in a ternary way, and the sub-division of which He limited in a nonary way, like the rest of the created beings.

The cause of everything is one triune God who is one in three, and who, being without beginning and without end, cannot fall under any division. As to the created beings, it was at the beginning of the creation that they assumed the quality of being ternary-single; and because they began to be ternary-single, while the triune nature of God had no beginning, they acquired two other divisions which go beyond the ternary state: the angels a nonary state, and this visible world a denary state. It is clear that since the angels are the servants of God, some of them should be nearer to Him, some of them farther from Him, and some of them in an intermediate position between God whose dwelling-place is heaven, and His creation which is below.

Why did nine orders of angels come into being, and why did their division, proceeding from a ternary-single state, stop in a nonary state, and not reach a denary state like the division of our numbering, which proceeds from a single to a ternary state and from a ternary to a denary state?

We answer: The numbering found among created beings is derived from the elements through suffering, increase and decrease, while that of the angels takes place without suffering, increase or decrease. The angelic nature, beginning with a single-ternary state, and being without suffering, increase or decrease, reached a nonary state. It did not, however, go beyond the ternary-single state, for, although this ternary state, by being added three times to itself, reached a nonary state, its ternary quality was preserved in this addition of itself. It was distinguished from the ternary-single and single-ternary condition which had not been added to itself, only in the fact that it was added to itself in a ternary way.

As to the visible nature which we have with us, because

it possessed suffering, increase (and decrease), when it moved from a single-ternary state, not [1] only was it added three times to itself and thus reached a nonary state, but it went also beyond it towards a denary state, where it stopped. The angelic hosts, because they are created, were affected by an addition which does not affect the Creator, in the fact that their number was added three times to itself in a ternary way. The reason why the bodies (that we have with us) acquired a denary state,[2] in spite of their being added three times to themselves, is to be found in the fact that, at the beginning of all numbering, the number one moves first towards the number three, which is composed of uneven and even numbers.[3] Indeed numbering does not begin until the number two, which is even, follows the number one (making three); and it continues thus until the number three reaches the number nine, where the movement of mounting-up stops, according to the angelic division. It is after this that, in the composed beings, on account of suffering, an addition begins towards the number ten, as if the number nine wished to make a kind of link, in order to strengthen the numbering on account of its weakness, in accordance with the order required by the composed beings that these beings should be linked in all their compositions; and so that it might be linked, it reverted towards the number one in a circular way. The movement of numbering is thus completed in a kind of cycle. It is for this reason that the ancients invented, as a first sign for this number (ten) the (empty) space between the forefinger and the thumb, formed in a circular way.[4] Indeed when the numbers which we have with us reach a denary state they stop, and then turn back and mount up indefinitely.

This is the reason why the single-ternary state of the angelic hosts stops with a nonary state, while the order which we have with us stops with a denary state.

[1] Read *law* for *lewāth*. [2] Text is corrupt.

[3] This reasoning about the completeness of the number three, as embodying even and uneven numbers, is found also in the works of a contemporary of our author, the famous East Syrian or Nestorian Patriarch Timothy, whose residence was also in Baghdad. See my *Woodbrooke Studies*, Vol. II, pp. 63-65.

[4] This is an interesting explanation of the origin of the cipher.

On the function for which the angels were created.

The angels were created by God, as it pleased His will, in order to glorify and magnify His majesty. He was pleased that praise should be rendered to Him by the angels and the children of Adam, the former continually and without ceasing. 230
He made them a simple, swift and spiritual nature which does not experience weariness or fatigue, and He placed them in a higher sphere so that they might be near the glory of His majesty. Being Himself everywhere and infinite, He made them immaterial—not derived from the elements—and immortal, and divided them into three hierarchies, and caused these hierarchies to be enlightened one more than another, in proportion to its nearness to Himself, and one hierarchy to enlighten the hierarchy that came after it, and this the next, until the enlightenment reached the lowest order (of angels). This makes manifest to us His very high wisdom, and shows how in His bounty He granted them existence from non-existence, and the gift of abiding near the Godhead, the principle of life, and praising it continually. He made them also His servants and His messengers [1] to this world of ours, as [2] we have seen them sent to prophets and saints in order to fulfil His commandment. (He created them) also to receive the souls of men and take them where He should command them. They were made in three hierarchies in order to reveal in themselves the mystery of the Trinity in one God, creator of everything. Further, He created them so that, when He should come on the day of resurrection to raise the dead and judge the earth in righteousness, they might praise Him and glorify [3] Him while carrying the throne of His majesty and standing round Him in thousands and in tens of thousands. He will order them to take the wicked to the unquenchable fire and to the outer darkness of hell which has no end, and the just to the kingdom of heaven and to the everlasting life.[4]

[1] Read *meshtadrānê*.
[2] Text repeats.
[3] Read the verbs in the 3rd person masc. plural, *nezaiḥunai*(*hi*), etc.
[4] *Cf.* Matt. xxv. 46.

CHAPTER III.

On the fact that from the nature of things there will be an end to this world; and on the resurrection of bodies.

Dissolution takes place by necessity in all individual compositions of all species, at the time when the fight of the elements ceases in them, as happens, for instance, in the case of a certain man—Adam or Noah—or of a certain ox, horse, bird or pigeon. This individual dissolution points to the general dissolution. In the same way as the general composition of this world emanates from the elements, so also will be the case with its dissolution, when the fight of the elements from which this general composition emanates ceases. Since nothing new is found in the parts which is not found in the whole, except rationality; and since there is an individual composition and an individual dissolution; there must necessarily be a general composition and a general dissolution of this world. Indeed the cause which brings about the individual dissolution through the antagonistic powers is the same as that which will bring about a general dissolution through the antagonistic powers, and the same as that which made the first composition from the antagonistic powers. If the whole possessed over and above its part something more than rationality, there might have been room for question.

Why does individual dissolution precede general dissolution?

We answer that the dissolution of an individual composition takes place after the fight of the antagonistic forces with one another has ceased. When heat has little by little been defeated by cold, and humidity by dryness, the body which was composed of them dissolves. Lo, the dissolution of each body of animals and human beings takes place through death quickly—in the latter case at the age of about seventy years or eighty years,
231 or some other age [1]—because the antagonistic powers which are in them are found only in a small quantity. As to the general dissolution, it will take place in thousands of years, on account of the great quantity and energy of the antagonistic forces (in the world). The time (required for the dissolution) of a great quantity is long, and for that of a small quantity is short.

[1] Lit. "number."

Further, as the general composition took place[1] at one time and suddenly, while the individual composition takes place little by little, so also the general dissolution will take place at one time and suddenly, while the individual dissolution takes place little by little.

If the part should receive dissolution and the whole be imperishable, then the part could not be said to be composed of the same thing of which the whole is composed; for if the whole is imperishable, the part also should be imperishable. It follows that since the parts have an end, the whole will necessarily have an end. The only world which will not dissolve and have an end will be one possessing a simple nature in which there is no change brought about by the antagonistic forces; and because this world of ours is not simple, but composed of antagonistic forces, it will necessarily dissolve and end.

Further, if the beginning (of the created beings) as effected by the antagonistic forces is the beginning of their end, and their beginning and their end are thus the same thing, and if beginning and end are common to all of them, there will necessarily be a general end and dissolution, just as there is an individual end and dissolution.

Further, because the beginning of the general composition from the elements came about through their energy and vigour—since there was fight and antagonism between these elements in that very composition, with victory and defeat taking place in it—and because movement follows energy and vigour, while rest and cessation from movement follow defeat and weakness, rest and cessation from movement will, therefore, follow the weakness of the elements, and this means the dissolution and end of the world.

If you say: Because the general composition which emanates from the elements will suffer dissolution, the composition itself of the compound elements—fire, air, water and earth—will necessarily also suffer dissolution; and since in this case only the simple elements will be left, how will these simple elements be able to remain by themselves?—

[1] Ṛead *hewā*.

We answer: The bodies came into being from the compound elements after these compound elements became composed. The compound elements precede the bodies in priority of time,[1] while the simple elements do not precede the compound elements in this way, because they do not stand alone, but only in them. This is the reason why they have no priority or posteriority of time, but only of order. Because, therefore, the simple elements do not exist in themselves, but exist only in the compound elements—as we demonstrated in our discourse above [2]—the compound elements will not dissolve into the simple elements, as the latter are known only through the former as if they were their bodies. The compound and the simple elements,[3] together with the bodies, will, however, not perish and cease to exist, but will only undergo a change from their present existence to another kind of existence. In this way the rational bodies also, in dying,[4] will only undergo a change from their present existence to another kind of existence; and this is what we call "resurrection."

The elements will not cease from being hot, cold, humid and dry, because it is their nature to be so, as we have shown. Indeed if this nature ceases, what will remain of them? Nothing.
232 The heat, cold, humidity and dryness of the elements will, therefore, be preserved. They will only undergo a change. Their natural substance will not perish, but will be renewed, as the prophet said: "Thou renewest the face of the earth." [5]

What is this renewal and this change?

(The elements) will cease from fighting one another, and when they cease from fighting, they will have neither victory nor defeat; and because there will be neither victory nor defeat in them, there will be neither increase nor decrease; and consequently there will be, in the case of the bodies which emanate from them, neither generation nor corruption [6]; and wherever there is no generation and corruption, there is found no con-

[1] *Cf.* this statement with that of ch. XVII of the Second Discourse, where it is stated, perhaps with reference to the creative mind of God, that "there is no time intervening between the compound elements and the genera and species which emanated from them."

[2] See pp. 5 *sqq.*

[3] Delete the *Lamadh* at the beginning.

[4] Lit. "ceasing to exist."

[5] Ps. civ. 30.

[6] Read *ḥubbāla.*

ception, birth, dissolution or death. This is the change and renewal which will happen to the elements.

CHAPTER IV.

On the fact that renewal and change will affect[1] *the elements ; and on how it will happen that while heat, cold, dryness and humidity will be preserved in them, no fight will take place between them.*

The elements bring about two compositions, one of which is effected through their antagonistic powers, and the other through the balance and equilibrium established by their non-antagonistic powers. We have spoken of this above. The composition by means of the antagonistic powers takes place through the straight movement, which is the movement that governs our world ; that is to say, all the bodies which undergo dissolution, deterioration, increase and decrease from the fight of the antagonistic forces. The second movement, although emanating also from the antagonistic powers which fight and dissolve each other, is effected through the balance and equilibrium (between the elements) spoken of above, which the compositions of the bodies that we have with us contain in a small degree and quantity, and which constitute the life and health of the bodies. The heavenly bodies emanated from this second movement, through the predominance in them of a great measure of balance and equilibrium, on account of which they are devoid of increase and decrease, and are consequently imperishable, impassible and immortal.

How can the elements change their fight, struggle, suffering and antagonism for something opposite to these ?—

We answer : After the elements have fought in this world of ours a long period of time extending over myriads of years, now winning victory over one another, and now suffering defeat from one another, they will weaken and become weary and enfeebled. They will then remain in a quiescent state, and the balance and equilibrium which they possess to-day in a small measure, but which exists in the heavenly bodies in a great

[1] Delete the *Dalath* of *hāwé*.

measure, will predominate in them ; and after they have been defeated, have ceased from conflict, and have remained in a quiescent state, there will be in the bodies that emanate from them an end to increase, decrease, and other contingencies. After these bodies have, through the cessation of the struggle, rested and ceased to live, they will be renewed and will possess qualities contrary to their present nature, namely incorruptibility, immortality, impassibility, *et cetera*. As long as equilibrium is preserved in them, they will not be affected by increase or decrease, and thus the desire which wishes to fulfil a want will cease to exist in them ; and as long as there is no desire and no increase or decrease, there can be no struggle and no dissolution and death.

The heavenly bodies will also undergo a change when that small portion of passibility which they contain, and which emanates from the small measure of the straight movement which they possess, is conquered. They will then be wholly impassible. We possess an infallible testimony to the possibility of a composition from the elements being unchangeable—as will happen on the day of the resurrection—in the heavenly bodies. Indeed we have shown above how, in spite of the fact that they are from the elements, they are incorruptible, immutable and unageing.

It follows from all this that there will emanate from the elements another world which, after the antagonism (of the elements) has ceased to exist in it, will be incorruptible, everlasting and immutable.

CHAPTER V.

On the fact that there will be resurrection, and that this resurrection will affect men alone, and not other species of animals.

233 Since all men and all animals die and are dissolved, why will the resurrection affect men alone and not all other animals ?—

We answer : All irrational animals are composed of the single nature of the elements, while man is composed of two natures : the nature of the elements which is corporeal, and another nature which is incorporeal. As to the animals, because they emanate [1] from the elements only, they will dissolve into the elements. They will remain in these elements, will have the same renewal as the elements in general, and will not have a separate existence. Man, however, will have a special resurrection, because he is composed of two antagonistic natures, a corporeal and an incorporeal. In its composition with the body his incorporeal part is hidden in this world by his corporeal part, and is dominated by it, so much so that many believe that there is no incorporeal part in man, but that all of him is corporeal like the irrational animals. They think that all his actions in general are corporeal, and since the incorporeal action is noticed but little and seldom, they refer even [2] this action to the body.

When (these two natures) become separated from each other, and the body perishes and withers away, the soul, after having been loosed from its fetters and having gone out, waxes very strong and vigorous and desires to return to the body so that it may be avenged on it for what it had done to it in all the time when, before its separation from it, it was naturally united with it. Since the soul possesses towards the body a natural inclination like that of a magnetic stone towards iron, it will clothe itself with it, in accordance with the will of the Creator, will make manifest its action through it, and dominate it, as it was formerly dominated by it; and in the same way as it formerly suffered from it and was overcome by it, it will in its turn overcome it by the impassibility, immortality and incorruptibility which its nature possesses.

In the same way as when two gladiators have fought each other a long time and then have separated from each other a long time, (the defeated man) nurtures his strength and becomes stronger, more powerful and more vigorous after the combat (than he was before), resembling a man strengthening himself greatly against his adversary; and a natural thought having been borne in upon him inducing him to come again and fight against his adversary, he goes for him, and finding him weak and enfeebled, seizes him and binds him, in such a way that, on account

[1] Delete the negative *lau*. [2] Delete the negative.

of his weakness and feebleness, as contrasted with his own vigour, the adversary gives up all attempts at further struggle—in this same way will the soul, after it has been separated from the body which was fighting against it, wax strong and vigorous, return to it, join with it, defeat it, and bestow upon it benefits which its nature did not possess, and which are indissolubility, incorruptibility and immortality. The natural body dissolves, but will rise with the soul as a spiritual body.[1] It follows that the irrational animals, since they have no incorporeal nature to be united with their bodies on the day of resurrection, will not have resurrection. Only man has this.

Animals came into being to help man and serve him in this imperfect world, so that they might satisfy the necessities of his daily life, some of them acting as his mounture, and some others as his food, *et cetera.* When, however, man has risen, all passions, together with the desire for food, will cease to exist
234 in him, and so he will not be in need of the food and service which they afford him. If, therefore, they were created for him in this world in order to satisfy his necessities, and if he is not in need of anything in the next world, they will not rise on the day of resurrection, when the cause for which they had come into being has ceased to exist. God created nothing to no purpose or haphazardly.

(The point may be explained) in another way: God is a just Judge, who united the body to the incorporeal soul which is thus in this world joined with the body and suffers from it, while its own nature is impassible and immutable; and because God is just, and did not judge the body[2] and condemn it (in this world), He will judge it and condemn[3] it in the next world, in order to mete justice to both body and soul.

(The point may be explained) in yet another way: We see in this world fall and death, which, involving as they do the end of a state, imply the existence of a state which has no end. Death, which is an end, implies life which is unending; corruptibility, which is also an end, implies incorruptibility[4] which is unending; and fall implies a risen state which is

[1] *Cf.* I Cor. xv. 44.
[2] Read all the pronouns in masc.
[3] Lit. "give it death."
[4] Read *llā* instead of *lwāth.*

unending. In this world there exist death, corruption and fall, which are all of the nature of an end, while unending life, incorruptibility and resurrection are not found in it. There will, therefore, be another world in which there will be unending[1] life, resurrection, incorruptibility and immortality.

(The point may be explained) in yet another way: There are four contrasting propositions, the first of which is without beginning and without end, and the second also without beginning and without end; the third with beginning and with end, and the fourth with beginning and without end.[2] The first and second belong to God the Creator, and the third and fourth to His creation. We have already shown how the nature of God is without beginning and without end, and how the created beings found in this world in which we are had a beginning and will have an end; and since in this world in which we live only three of these propositions are known to exist, there must necessarily be (a world) in which the fourth, which is with beginning and without end will be found, and this is the next world, in which there are no antagonistic forces, no suffering, no fight, no dissolution and no death. It is said that it will be an angelic world, as the angels of God are in heaven. We will show this in the following diagram.

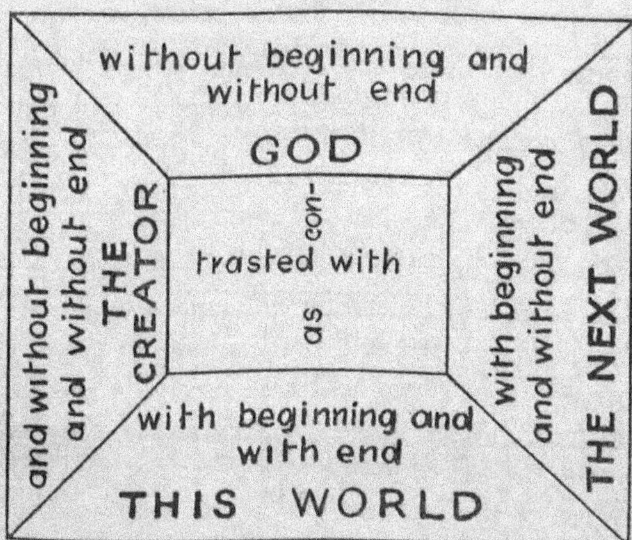

[1] Delete the *Wāu* before *dlā*.

[2] See diagram below.

(The point may be explained) in yet another way: We have shown that this world had a beginning and will have an end. Because something[1] will be left over from this beginning and this end, this will give rise to a different beginning. What is left over from this world will be the beginning of another. If this other beginning is different from the first one, it will necessarily take place through a change in the (elements) which brought it about. Since it will be different, its very difference will separate it from this one, to the extent of making it a new world.

(The point may be explained) in yet another way: The just have done good and the wicked evil, and both have died in this world without receiving the reward or punishment due to their deeds. There will, therefore, be another world in which there will be a rewarder of the just and a punisher of the wicked, so that justice may be done.

CHAPTER VI.

235 *On whether as people die in this world so they will rise on the day of resurrection: children as children, young men as young men, old men as old men, and tall men, short men, fat men, thin men, sick men, healthy men, et cetera, as such.*

We affirm that these are accidents which happen in this world on account of the increases, changes and deteriorations which take place in it. This is due to the fight and antagonism of the elements, and to the straight movements of which we spoke above. In the day of resurrection, however, because this antagonism will weaken, and the elements will desist from their fight, since in the place of the strong movement which they possess the medium balanced part will predominate in them, the bodies which emanate from them will also receive a medium balance and equilibrium. After they have received it, they will have, in the next world, the age of youth which has not the defect of the age of children or the increase which characterises the age of old men. If we believe in the Book, it will be the perfect age in which our Lord rose from the dead. Indeed till this age is reached, the

[1] *I.e.*, of the elements.

body receives increase in all its members, as by a natural reckoning we showed above. Because all the sufferings that arise out of antagonism, increase and decrease will cease in man, he will rise in one complete harmony in which there are no sufferings from increase and decrease, and he will consequently have no short stature, tall stature, fatness, thinness, or any incidental thickness. These occur from the deteriorations that affect him, and since these deteriorations which are the originating cause (of the above accidents) will disappear, the latter also will disappear, and in their place the nature of the body which had died [1] will rise in a perfect form.

If the body rises in a perfect form, it should be with all the members which make it perfect, such as hands, feet, head, *et cetera*, which do not emanate from suffering through increase and decrease, like tallness, shortness, fatness, thinness, *et cetera*, but from the natural substance of the elements found in all the species. The whole species possesses all the members, but no whole species possesses shortness, tallness, *et cetera*. These are, therefore, not essential, and will necessarily cease to exist, but the natural substance, together with all the members in which it is shown, will rise; for if the members which were formed by congealment from the natural substance of heat, cold, humidity and dryness do not rise, the natural substance itself will not rise; and if the natural substance does not rise, it will not be the man who died who will rise, but something else or nothing. The substance,[2] therefore, will rise with all the members which pertain to it.

If the man who does evil will receive punishment, and the man who does good will receive reward, it follows, since we do these things through our members, that we shall rise also with our members. With what shall we stand before the Judge? Will it not be with our feet? With what shall we present our apology? Will it not be with our mouth? With what shall we distinguish the sheep from the goats that are on the left hand? Will it not be with our eyes? The perfect body, therefore, consisting of its substance and all (the members) through which this substance is known, will rise.

[1] Lit. "fallen."

[2] Read οὐσία.

236 What about those who were righteous in their youth, and became unrighteous in their old age? Will the body of youth have its reward in their case, in the same way as in the case of those who committed unrighteousness in their youth and righteousness in their old age?

We answer: The body of young men and of old men is one and the same body, and there is no difference between them *quā* body. Socrates was one and the same man when he was young and when he became old, and his body did not change into that of Alcibiades [1] by the fact that it received a certain change from the quantitative increase which extends from childhood to old age. Since, therefore, it is the very same body *quā* body, whether it was still young when the man became unrighteous, and old when he became righteous, or whether it was young when he became righteous and old when he became unrighteous. A sin does not belong to youth or old age, to a small or to a great quantity, but to the body in general; and the body as such is always the same. Indeed a part of it does not become righteous while it is young, and another part become unrighteous in old age. If it were divided into parts there might have been room for doubt. In the same way as Socrates, when he was a young man, had an illness in his body, which was healed when he became old, and as this illness was in all his body in his youth, and this healing was in all his body in his old age, and he was not ill with some parts of his body in his youth, and (healed) with some others in his old age, but these parts were the very same parts when he was ill and when he was healed—in this same way the illness of sins is in all the body of a man in his youth, and not in one part of it only, and the healing of righteousness is in all of it in his old age, and not in one part of it only, as this state does not come upon the body from increases or decreases.

[1] Text erroneously "Helypiades," with a slight change of *Ḳōph* into *Wāu*.

CHAPTER VII.

On how the body will be dissolved into the elements after death; and on how on the day of resurrection the part which was dissolved will come and be composed into the body.

We affirm that [1] we have shown from the nature of things that the elements have in their nature the power to give birth to the bodies of animals, animal-plants and plants. We have shown also the reason why animals, to the exclusion of plants, have bones, flesh, veins, arteries, *et cetera.* We have also shown by a comparison between the movements of the bodies which we have with us and those of the heaven and of the stars, the reason why the latter came into being immutable and incorruptible. In the same way as in the first congealment the compositions of these bodies became incorruptible and immutable on account of their movement, so also when the antagonism (of the elements) is weakened [2] by the predominance of balance and equilibrium, they will bring about another composition emanating from the movement of heat, cold, humidity and dryness, this time acting in a balanced way, and not in such a way that one of them predominates over another. They will thus give birth to the bodies which had dissolved, and which will be exactly as they were, but without corruption. I believe, then, that a second composition will take place, and this is easily feasible, because the elements will preserve the power which they constantly possessed in this world, even if they have ceased from their antagonism.

The way in which that part of the bodies of men which had reverted to the elements and become one with them will be composed again, become a body and join with the soul, may be explained in the following way:

Each human body receives from the soul, by its union with it, a distinct individuality and particular characteristics. When the soul has moved away from the body, and the body has dissolved into the elements and become one with them—heat 237
with heat, cold with cold, *et cetera*—it will,[3] in becoming one

[1] Read the first letter as a *Dalath* instead of a *Wāu.*

[2] Change the verb accordingly.

[3] Remove the *Wāu.*

with them, preserve the same individuality that it possessed from the soul through its union with it. Indeed the body possesses in itself particular characteristics which, through a power unknown to the senses, will begin to act. This may be illustrated by the stone of Heracles [1] which, although possessing an identical nature with all other stones in coldness and dryness, possesses a little ferrous power, and is found in ferrous mountains. When iron is brought near to it, a power that is not perceived by the senses leaps out from the stone towards it, on account of the particular characteristics and individuality that it possesses from it, causing a mutual attraction. This characteristic is not found in the other black stones which (in appearance) resemble it in every way. The same thing will happen to the body of man, possessing as it does particular characteristics and powers which emanate from the soul on account of its union with it. When the soul, being immortal, acquires energy, and comes to the body to avenge itself on it and absorb it, the body will run [2] from the elements towards the soul, and the soul [3] towards the body, and the body will take from the elements the (individual) power which was kept in them, and which will match the soul and fit it, as an impression fits the stamp. The soul will not receive from the elements anything except that which belongs to it alone and not to another soul. In the same way as the impression of an image, when brought back to its stamp, fits it easily and immediately, on account of the likeness that it received from it, and in the same way as the impression does not fit another stamp which has no likeness to it, so also in the day of resurrection each soul, when coming to be reunited, will run towards the images and likenesses which it had impressed upon its body by its former union with it, and will impress itself again on it.

If you say: "How can you show me in the elements the individuality and the particular characteristics which each body took from its soul?"

We answer: I cannot show [4] you this through the senses in the same way as (I can show you that) in a triangle there are

[1] *I.e.*, the magnetic stone.
[2] Read the verb in masc.
[3] Read the pronoun in fem. *hī*.
[4] Lit. "This cannot be shown."

two equal sides, and in a circle a round circumference; nor can I show you through the senses the dissolution of the bodies into the four elements, nor how each of these elements receives what is of the same nature as itself, for this is known only mentally. It is in this same way that the individuality which the body possessed from the soul, and which was absorbed by the elements, is perceived only mentally. It is in a state of potentiality, and comes into action for the purpose of its union with its individual soul and not with another. As the (particular) power which is found in the earth, and which emanates from the elements, remains after the dissolution (of the particular body which it affects); and as when a grain of wheat or of barley or a bean, *et cetera*, falls into the earth, it draws to itself from the earth a power that is particular to itself; and as the particular power of each species of cereals is perceived only mentally, and not through the senses; and as one species does not receive the power that belongs to its neighbour—so also each soul coming to be reunited, will receive its individual body from the earth and from the other elements, and take out, for the act of reunion, the individuality possessed by them.

CHAPTER VIII.

In what does this world differ from the next world, and in what does it resemble it?

In this world the bodies (of men) are composed of the four elements, are joined with the soul, and experience generation,
corruption, increase, decrease, conception, birth, life, death, and 238
the fight of the elements with one another. The next world will resemble this world in that the bodies in it will be composed of the same elements, and will have life; but it will differ from it in that no dissolution will affect its composition, and no increase or decrease or conception or birth or death or fight of the elements will occur in it.

We prove this to be the case by two reasons, the first of which is connected with the Creator and the other with the movement of the elements. The reason connected with the Creator is His infinite goodness whereby He bestowed upon the Adamic race the gift of being in His image and likeness. Through this, man rose above all other creatures, which became his servants. This is the reason why He endowed him with something which does not perish and deteriorate, and in which that image and likeness is made manifest, namely the immortal and impassible soul. And He gave him also a passible and mortal body, and gave him in this world the things that pertain to this body, which are death and corruption. If there were always corruption, in what would the goodness of God have been known? And if there were no corruption, no death and no perishableness, we should not have known our salvation from our present life, corruption, death and perishableness. Indeed if we were immortal and imperishable, we might have believed that it was not the goodness of God that was the cause of this, and that it happened by a kind of natural necessity. When, however, there is a transformation from corruption to incorruption, from death to immortality, and from perishableness to imperishableness, we know that this transformation emanates from the providence of the goodness of God.

Further, if we had not been passible, we should not have enjoyed impassibility, as there would have been no contrast; and we find happiness in contrasts. We shall enjoy the next world as those enjoy good health who have experienced illness. It is with great wisdom, therefore, that the Creator made this world passible and mortal, and the next world impassible, immortal and imperishable. In proof of this we have the impeccable testimonies found in the Gospels and in all the other Books, about which we need not here weary our listeners, as they are clear and well-known to everybody, and are accepted by faith.

Our aim has generally been in all our works to demonstrate our statements from the nature of things, so that no one might be able to reject them. The testimonies taken from the Books are accepted without question by believers only, while non-believers do not accept them. For this reason we wrote a book containing ten syllogisms taken from the nature of things, which prove that Christ is both God and man, and we added to each of these natural syllogisms many testimonies from all the Books of the Prophets and from the Holy Gospel, which demonstrate also that Christ is both God and man. We based it on the method of the three laws: the natural, the written and the spiritual, a method that points to the Holy Trinity. We inserted in it many testimonies from the Books, because it was our aim to do so.

We further wrote a book *On Faith*, in which we demonstrated that there is a God, who is one in three and three in one; that He created everything from nothing; that we have by necessity to be baptised and to receive the body and the blood of Christ; that we have to worship towards the east; and that God has a Son and a Spirit, *et cetera*. We did not quote in this book any testimonies from Holy Writ, as the above points belong to those who believe, and believers accept them without difficulty.

We also wrote a book *On the Soul*, divided into twenty
chapters, which demonstrate its existence, its nature, how and 239
why it exists; that it is immortal, incorporeal and endowed with a movement that is perpetual and particular to itself; whether or not, and in what way, the different souls exist in a separate place before their bodies come into existence; whether the soul grows up with the body or grew up before it; the manner of its existence after the body has ceased to exist; and that it does not emanate from the elements, as some people have erroneously asserted; together with different other questions concerning the soul. In this book we did not refer to the demonstrations of the ancients concerning the soul, or to their opinions, nor did we deviate from the tenets of the Holy Church.

Our aim in all our books has been to demonstrate the truth by reasons taken from nature, and (to write) in a compendious way, so that the reader may obtain benefit without much trouble.[1] We have not quoted new theories set forth by others, thus resembling a man who takes pleasure in the crops of a field that does not belong to him, as you have already seen in the method adopted by us in this book. Indeed we do not believe that in it we have troubled anybody with the labour of others.

We also wrote before this book medical works *On the Causes*

[1] Lit. "through short terms."

of Fevers and *On Canine Hydrophobia.*[1] We wrote also *On Urine,*[2] and on other medical and non-medical subjects, according to the same method. These works, through the dissertations taken from the nature of things which they contain, through the compendious way in which the words are expressed in them, and through the moulds in which the opinions are cast in them, testify to themselves that they are the fruits of our own labour.

The second and natural reason mentioned above, which shows in what way this world differs from the next, and in what it is identical with it, is the following: God placed in the movement of the elements the power to become, through their mutual antagonism, the cause of a change which will be common to all of them, and to be a sign whereby rational beings may understand this change. This (future change) may be known from the fact that the elements are not one nature, but four natures different from one another. Indeed if they were one nature they would not have undergone any change; but because there are in them antagonistic changes common to all of them, these must necessarily give rise to a total reversal of [3] antagonism; and the natural and mutual antagonism and change (which they now possess) will bring also a change in this very antagonism and changeability.[4] The elements must, therefore, possess two different actions, having two categorically opposite principles, that is to say, the power to produce two worlds different from each other, each of them being contrary to the other in its general working. It follows that because there is here a general change and antagonism in the elements, that is to say, corruption, death,[5] suffering, fight and perishableness, there must necessarily be another kind of union of souls and bodies, which will be without corruption, death, suffering, fight or perishableness.

If you say that in this world heat possesses an antagonistic change in relation to cold, and humidity in relation to dryness, but that this change does not affect mutually all the four

[1] This work is found in the present MS. (Mingana, Syriac 559, ff. 36*b*-39*b*).

[2] All these books of the author are also mentioned in his work on *Canine Hydrophobia,* see *Catalogue of the Mingana Collection,* Vol. I, pp. 136-137.

[3] Read the word with a *Dalath* at the beginning instead of a *Wāu.*

[4] Lit. "will change that change."

[5] Read *Wāu* instead of *Dalath.*

elements, but is only partial and belongs to one or another of them:—

We answer: Corruption, fight and decomposition are due in this world to all the elements acting together, and if this is so, it is necessary, as we said above, that another change due to them all should exist, which will be without antagonism, and consequently without corruption and dissolution; and it is in this that the next world will be different from this one.

How can (the elements) possess such [1] an antagonistic change, affecting all of them together?

We answer: The elements, after having mutually waged a
great war lasting myriads of years among themselves, will 240
become weary; and after they have become weary, their movement will weaken, and after this weakening of their vigour, their antagonistic movement will quieten, and the balance and equilibrium of movements which is found in the elements, and of which we have spoken above, will hold sway over them.

As two fighters at the beginning of their fight are very strong and possess a powerful movement, but this movement weakens little by little in the course of the fight, in such a way that if they spend a long time in their fight this movement will weaken so much that they, having been deprived of their energy, will be completely unable to move—so also after the elements which we have with us have waged war for a long time, they will become weak and weary [2] and their antagonistic movement will quieten and cease to exist. Then general dissolution will set in, and the elements and everything that emanates from them will undergo a change and a renewal. When their antagonism has ceased, and the balance and equilibrium which are in them have been strengthened, a general renewal will [3] affect everything, and there will be a resurrection from death for all the Adamic race. After balance and equilibrium have reigned, there will of necessity be no increase, decrease or generation. Indeed desire takes place when there is something lacking which has to be filled, or when there is superfluity arising from addition which has to be cast out. And because (the next world) will have balance and equilibrium, without any increase or decrease, there

[1] Read *d-hāna*. [2] Read *methmaḥlīn*. [3] Delete the first *Wāu*.

will be neither dissolution [1] nor end to its composition. Indeed end occurs through dissolution, and dissolution through disintegration of parts, and (since this disintegration will not be found in the next world) it will have no dissolution. It follows that the next world will be devoid of dissolution, perishableness and suffering. This characteristic will differentiate this world from the next, but the next world will be identical with this in the fact that both are composed of the very same elements, so that the same elements which we have with us will also be found there.

The meaning of rising implies a state of having fallen ; the meaning of impassibility implies a state of having been passible ; the meaning of immortality implies a state of having been mortal ; and the meaning of imperishableness implies a state of having been perishable.

The body in union with the soul will, therefore, rise in the next world.[2]

CHAPTER IX.

On the fact that there will be heaven and hell, and on what heaven is, and what hell is, both from the Book and from the nature of things.

We find that in this world there are by necessity two kinds of happiness and two kinds of torment. What are they? One kind is the happiness and torment of the body, and the other kind is the happiness and torment of the soul. Food, drink, marriage, and the rest of such desires belong to the body, and its deprivation of them is its torment. On the other hand, meditation on God, and the knowledge of Him, together with the knowledge of new discoveries concerning the created beings and of the natural sciences, are the delight and happiness of the soul, and its deprivation of them constitutes its pain and torment. Because the happiness or the torment of the soul belongs to its own nature, and this nature will endure, and will not change,

[1] Read *sherāya*. [2] Lit. "there."

cease to exist, dissolve or end ; and because the nature of the body will dissolve and consequently its happiness will by necessity also dissolve, cease to exist and end, there will necessarily be a delight and a torment which will be unending and imperishable, corresponding with the nature of the soul ; and since this world, together with its desires, its happiness and its torment, will end, there will by necessity be another world which will not
end and perish. The happiness of that world will be the king- 241
dom of heaven, and its torment hell. And because meditation on God and the knowledge of Him, together with the knowledge of the discoveries of natural sciences concerning the created beings and the understanding of them are happiness, and ignorance of them is torment, these will endure in the next world. The kingdom of heaven will, therefore, be meditation on God and the knowledge of the nature of the created beings, while lack of them will be hell. It follows that there will necessarily be heaven and hell in the next world.

There will be a reward for the labours of the just, which is the kingdom of heaven, consisting in the delight of the Abrahamic table, a thing which eye hath not seen, nor ear heard, and which hath not entered into the heart of man.[1] There will also be hell, consisting in the torment with which the wicked will be requited for their evil deeds : a fire that is not quenched, a worm that dieth not,[2] and outer darkness.[3] The torment of the wicked will not be corporeal, because—since, as we have shown above, the happiness and delight of the body in this world come to it naturally through increase and decrease, by food and drink, and its punishment and torment by lack of food and clothing, and by pains and afflictions, and since we shall rise on the day of resurrection impassible, imperishable and immortal—we shall not be in need of the happiness and delight which emanate from food and drink, and which are caused by passibility. If we were in need of them we should still have been in this world and passible, and in this case in what would this world have been distinguished from the next? In nothing, if this were so. But because the things pertaining to the body will not rise with it, as the soul will

[1] *Cf.* I Cor. ii. 9. [2] *Cf.* Mark ix. 44 and 48.
[3] *Cf.* Matt. xxii. 13, etc.

overcome these things when it is reunited with the body and works in it the things pertaining to itself, there will be no happiness emanating from food and drink; and since this happiness emanating from food and drink will not be found in the next world, neither will the torment enamating from them be found in it. Indeed where [1] there is no food there is no deprivation of food, and where there is no antagonism to give rise to fight, there are no afflictions, suffering and corruption; and if these are not found there, what will be the happiness (of the just) and the torment of the wicked which will be found there, except meditation on God and on the created beings, and ignorance of them respectively, as we have learned?

Let us prove this point more fully by some short demonstrations from this world. Lo, we think of God as being the infinite, incomprehensible, unfathomable and bounteous Creator, who by His supreme wisdom created everything from nothing, and composed the various and innumerable genera and species, the heaven, the earth, the sun, the moon, the stars, and all creatures. When we do this we experience great happiness and delight. This delight is for us without satiety, and on many occasions it conquers the happiness of the body, which in this world often conquers the soul.

Further, when we enquire into the composition of the created beings, the differentiation of the species of animals, the cycle of the years of this world, the movement of the heaven, the course of the stars, the science of Arithmetic, Geometry and Music, and understand them to a small extent, we greatly rejoice and delight, and experience a happiness which outweighs the happiness of the things which relate to the body. Indeed we practise this kind of meditation naturally. Lo, we exercise our thoughts all day, from morning till evening, and find great happiness when our thought proceeds from one knowledge to another, and after understanding this one, it proceeds to yet another, and so on without ceasing; in such a way that if a man wished to put down
242 in writing every day the thought of his mind, he would be able to write a large book.

In the same way as a hungry man is urged [2] by his body

[1] Read *aiḳa*.

[2] Delete the first *Wāu*.

to eat, so also human nature is urged forcibly by the soul to enquire into things, and to know their causes. This is the reason why, when a man notices an assembly of men, he runs towards them to know the cause of their collecting together, even if his going there to enquire into their motives is to the detriment of the necessities of the body. This is the particular and natural food of the soul, and it will continue with it, and will not undergo any change. Its torment and pain, from which it greatly suffers, consists in its deprivation of knowledge; and when it wishes to understand, in accordance with its nature, the things pertaining to God and the things pertaining to the world, and cannot comprehend them, it becomes greatly distressed, even when the body is satisfied; and sometimes a man claps his hands together (in vexation), and often he refrains from eating, and his face assumes a sad appearance.

Very often when we wish to acquire the knowledge of an art or a science, such as building a house which we had intended to build, or fashioning a vessel, or making any other thing, and we fall short of the knowledge of our enterprise, we are stricken with much pain and grief; but when we perform it successfully we rejoice, jubilate and exult. It follows from this that the soul possesses a particular happiness belonging to its nature, and consisting in the comprehension of the knowledge of God and of the knowledge of the created beings, and (that it possesses) also a torment consisting in its deprivation of the knowledge of God and of the created beings. Its happiness in this world is small, because it is impeded by the body, as it is written: We know in part and we prophesy in part, and we see through a glass darkly, but then [1] perfectly and face to face.[2]

This may be proved as follows: When by our faith we comprehend a difficult point concerning God or concerning His providence towards the created beings, we rejoice and jubilate greatly, in such a way that we may die for it, in order not to give it the lie or deny it; and the more perfectly our point is established so that no one can deny it, the more we rejoice. Indeed if we rejoice when we acquire the knowledge of the building of houses, how much more shall we experience an

[1] Read *tamman*.

[2] *Cf.* I Cor. xiii. 9-13.

ineffable joy when we have acquired a perfect knowledge of the composition of this world, the greatness of the heaven and its nature, the right size and nature of the sun, moon and stars, the knowledge of the composition and differentiation of each species (of created things), the union of the soul with the body, and the structure of all the world? Will it not be a wonderful joy, of which we have no experience here, and which does not fall under any of the senses? Since, however, we have not experienced it we cannot possibly describe it.

When the resurrection takes place, the happiness of the body, which emanates from antagonism and which was an impediment to the soul, will cease to exist,[1] together with the torment with which it was affected, and which consisted in its deprivation of food. The body will rise incorruptible and devoid of any such passions, as it is written: We are sown in corruption and we shall rise in incorruption; it is sown a natural body and it is raised a spiritual body.[2]

When the bodies have risen in impassibility, then the soul
will conquer the body, as there is nothing to impede its light;
and it will absorb the body in the same way as it was itself
absorbed by the body in this world. When the desires and
passions of the body cease to exist, and nothing remains but
things pertaining to the soul, then we shall see clearly and face to
face. "To see clearly" means in this connection that we shall
243 see completely, that is to say, we shall know, as it is written:
I shall see the work of Thy hands, the heaven and the moon
which Thou hast ordained.[3] "To see" signifies here "to
know," for vision refers both to the soul and to the body.
Theory [4] is nothing else but the vision of the mind. In the next
world we shall know God and the created beings with the vision
of the soul; and it is this act that we shall perform in the future.
The sentence "I shall see" means that I have not yet seen, and
if I have seen I have seen only figuratively and not really, like a
shadow and not the true substance of face to face. This is
the happiness that is kept for the diligent.

As to the torment, it consists in a complete deprivation of

[1] Delete the *Wāu*.

[2] *Cf.* I Cor. xv. 42-43.

[3] *Cf.* Ps. viii. 3.

[4] In the original meaning of the word.

the knowledge of God and of the created beings, a thing contrary to the light of the kingdom (of heaven), and this is the outer darkness which stands before the knowledge of the soul and impedes it from rejoicing in the glorious and ineffable light of the Holy Trinity, one God. The sins and evil deeds which the wicked have committed will stand before the eye of the soul, and will blind it; and it will in consequence be affected with ἀμβλυωπία, that is to say, dimness of vision, which is an incurable and hopeless disease which acts like the thick matter of walls that impedes the light from penetrating and from illuminating those who are imprisoned in a house. This is the outer and the extreme [1] darkness.

If when we are deprived in this world of a certain knowledge, we experience great pain and anxiety, so that the pain conquers not only the soul but also the happiness of the body—how much more shall we not suffer in the next world an indescribable and insufferable pain when the nature of the soul shall stand in its naked reality without anything to impede or check it? The pain and torment that man will experience then will be more intense than the torment that he experiences here in a strong fire. He will burn there with an immaterial fire, which is not quenched, and which, because of the thought which will agitate it, kindle it and ignite it, will not cease from burning, in such a way that he will be constantly perturbed and never at rest; (and this fire will be like) a worm which will not die and to which there will be no end.[2] On account of its severity and insufferableness, when sinners compare that torment with the torment found in this world, they will long to be refreshed in the torment of this world. The torment and the happiness of the next world will be in proportion to its greatness, sublimity and power; and what is worse than anything else will be the lack of hope that that torment will ever come to an end. The fact that that torment will not come to an end will render it a double torment, than which there is none severer. As to the just, the holy and the righteous, when they rise [3] in the excellence of the angelic nature and in a glorious light—as it is written: All will rise up

[1] Change the *Nūn* into *Yodh*.
[2] *Cf.* Mark ix. 44.
[3] I read *d-ḳaimīn*.

like angels [1]—since there will be no opaque matter of sins which will dim (their vision) they will see [2] face to face,[3] through lamps the oil of which will not diminish or fail, but will shine, through the good works of these men, and enlighten the nature of their soul to contemplate and to receive the perfect knowledge of God, which is the true life, according to what is written: This is life eternal, that they might know Thee the only true God.[4]

The kingdom of heaven is truly life, and hell death,[5] and it is through life that the eye of the soul is enlightened to see God and delight in Him, as it is written: Blessed are the pure in heart, for they shall see God.[6] The knowledge of God is life, and life is the kingdom of heaven. Listen for this to the words of our Lord, who said: "These shall go into eternal life, and the others into everlasting torment."[7] The knowledge of God
244 is, therefore, life and the kingdom of heaven, of which it is written: Eye hath not seen nor ear heard, neither have entered into the heart of man, the things which God hath prepared for them that love Him.[8] This consists in a new life, a supreme delight and happiness, the greatness and the sublimity of which cannot be described with the tongue of the children of Adam. Indeed a thing that does not exist in this world cannot possibly have been heard by the ear, or seen by the eye, or have entered into the heart of man. Knowledge does not comprehend it, nor does man understand it, but God alone. If an eye had seen it or an ear heard it, it would have been in this world; and how could God have prepared for those who fear Him, that which might be possessed also by those who do not fear Him?

If somebody says here that God will grant the righteous [9] in the next world a variety of food and drink emanating from the elements, but not resembling that found in this world, because they will eat this food without suffering, and will drink without pain; and that in the same way the wicked will experience a torment emanating from the elements, but not resembling that found in this world:—

[1] *Cf.* Matt. xxii. 30, etc.
[2] Read *ḥāzain.*
[3] *Cf.* I Cor. xiii. 12.
[4] John xvii. 3.
[5] Delete the first *Wāu.*
[6] Matt. v. 8.
[7] *Cf.* Matt. xxv. 46.
[8] I Cor. ii. 9.
[9] Lit. "them." Read *lhōn.*

We will answer: Whatsoever God wills, He does; and if He wills to grant them something dissimilar from that found in this world, and emanating from the elements so that they may delight in it, He can do so easily. We cannot, however, prove whether this will be the case or not, as this surmise will be similar to another whereby one might say that God can easily create a third world from the elements; but we do not know whether God wills to do this or not. Indeed if He so wills, He can create not one world only, but ten worlds. We can only think of the thing which He has willed and created, and we see that He has created two natures in this world: a corporeal and an incorporeal. The corporeal nature will complete in this world all things pertaining to itself, and cease from action, and there will remain only the incorporeal nature, which will complete the things that pertain to itself in the next world, as we demonstrated above by a reasoning taken from the nature of things. Since there is no other nature outside these two natures, we can only prove the existence of two kinds of food, the first belonging to the body in this world, and consisting of a material food and drink, accompanied by suffering, corruption and death, and the second belonging to the next world, and consisting of the immaterial and subtile food of the soul which is the delight in (divine) theory, as we showed above. There is no other variety of food apart from these two. If the existence of a third world were proved to us, we might perhaps have been convinced of the existence of a third variety of food.

If the material and corruptible [1] food of this body of ours exists in the next world, there will also be natural desire; and if there is natural desire, there will also be passions, as the desire for food is nothing else but passions which wish to satisfy a want; and if there is [2] passion, there will also be corruption, increase to satisfy a need, decrease, digestion of food, and ejection of superfluities; and consequently there will also be fight between antagonistic powers, victory, defeat, dissolution and death. If this were the case, there would be no next world.

If you say that there will be reception of food, but without the above increases:—

[1] Delete the negative *lā*. [2] Read *hāwé*.

(We will answer) that this will not be reception (of food). Indeed what is reception, except an addition to something? Remove the reception of food, and you will remove also the increase that emanates from it on account of a decrease that had taken place, and there will remain no passion, corruption, dissolution or death. Believe in an increase, and you will necessarily believe also in decrease, dissolution and death. It follows then that there will be only one world and not two. The fact of our eating in a corporeal way involves the thick matter of food, the quantity of which diminishes from outside through transference, and is added to the body inside. If a body fed without an increase to itself, or if the food added to it were not digested, or if the digested food were not ejected by the excretory power, or if a change did not take place in the bowels, the odour would become malignant, and there would be suffer-
245 ing emanating from its great quantity, and sometimes also illness. The general consensus of opinion among rational men will not accept the assertion that the body could receive food without an increase to itself, together with the other consequences which we have just enumerated. If it does accept this assertion, it will be only by faith. Such an assertion would indeed resemble that of a man telling us: "This year or to-day, two and two make four; but next year or to-morrow, these two and two will make ten. You should accept this from us by faith." Or: "This one is a man to-day, but he will be a horse to-morrow. Do not doubt this, but accept it also by faith." A man can say anything he wishes in this way!

CHAPTER X.

On the fact that the next world, the heavenly kingdom and hell will have no end

Everything done in this world has a beginning and an end. This happens, as we showed above, because the antagonistic powers found in it fight with one another, as a consequence of which defeat occurs, parts suffer disintegration from their mutual composition, and an end takes place. As to the next world, because there will be no antagonism in it which would give rise to fight, there will be no defeat or victory, and consequently no disintegration of parts from their composition, but there will be perfect balance and equilibrium in which there will be no fight, victory or defeat, and consequently the parts will not suffer separation from one another. Because there will be no separation of parts, there will be no dissolution, and consequently no end and no death; and if there will be no end and dissolution of the composition, neither will there be an end and dissolution of the state to which this will give rise, that is to say, heaven and hell. The latter will have their course in the nature of that composition, and since that composition will have no end, neither will they.

Further, if the happiness of the next world consists in divine Theory and in the understanding of the knowledge of the created beings, since this Theory and understanding belong to the very nature of the soul, and the nature of the soul is not composed but simple, the next world, therefore, (will also be simple, and) will not suffer dissolution. Indeed how can a thing that is not composed suffer dissolution? The kingdom of heaven consists in the knowledge of God and in the understanding of the nature of the created beings; and if this kingdom of heaven will have no end, neither will its counterpart, which is hell, have an end, as it is written: Their fire shall not be quenched, neither shall their worm die.[1]

Further, thought and the knowledge of things belong to the nature of the soul, and we see in this world that the soul is never, even for a twinkling of an eye, without the things that pertain to the food of its nature, while the body sometimes feeds and sometimes not. If the soul, in this world in which it is bound with the fetters of the body and is kept captive as in a house, does not cease, in accordance with its nature, to feed on the divine and human knowledge, and has thus no end in the things pertaining to its nature, while the body has an end in the things pertaining to its nature, how much less in the next world will the soul have by its very nature rest or end in the

[1] Mark ix. 44.

things pertaining to its nature, namely the delight in the knowledge of the Holy Trinity and in that of the created beings, which constitute the kingdom of heaven!

Further, if there is an end to hell because it had a beginning, even if both heaven and hell began and will, therefore, end, how can the rationality of the soul end? Is it that its knowledge of God or its ignorance of Him will cease? If the natural characteristics which constitute its essence cease to exist, its very nature will cease to exist, and the words of our Lord, "These shall go into eternal life, and the others into everlasting torment"[1] will be belied. He did not say "to a temporary life or torment." The word "everlasting" means something that has no end, as the prophet said: For ever art Thou, O Lord.[2]

What shall we think about hell? Is it a kind of material fire and material worm? If this were so, and if we were to be
246 burnt with fire, we should have a dissolution and an end in it; this end would resolve itself into ashes, and these ashes would dissolve into the elements; and then one of two things would happen: either there would be another resurrection and another world, and yet another and another, and so on till the end; or we should remain in the elements without rationality, and the providence of God would be at fault.

If you say that we shall fall into the fire, but that we shall not be burnt, and that because we shall become incorruptible we shall not suffer injury from the fire:—

We answer: To burn (materially) without end is not found in the nature of things, but only in faith. The wicked, therefore, will burn in a fire which is not material, and which will have no end.

Some heedless people entertain a vain hope through which they wish to satisfy their passions, and by idle words corrupt sound opinions. They pretend that God, who is a just Judge, will requite us in the measure of time which we lived in this world in which we sinned. The time of the torment will, according to them, be measured by the period of life. They

[1] *Cf.* Matt. xxv. 46. [2] Ps. cxix. 89 (Syriac version).

pretend that as God is not unjust, there will be an end to His [1] judgment.

O wise men! You are comparing the method of that punishment with the one found in this world. The judges of this world strike anyone who strikes, and kill anyone who kills, *et cetera*. They inflict this punishment on the body, because it is in the body that people suffer and feel the pain which comes to them from the antagonistic (powers) that are found in the fight (of the elements). Since, however, in the next world we shall rise without suffering, corruption or death, what comparison would there be between this world and the next in the case of the killer who had to be killed? And would the one who had committed a hundred murders be killed a hundred times? Such a one would have to die and live again a hundred times! The same argument would apply to the case of a man who had struck another man, and would have to be struck in retaliation, according to the law: An eye for an eye. . . .[2]

He who does good is good, and he who does not do good is not good. On account of the goodness of God, it is necessary, if the welldoer is to endure, that the next world should be without end. Since He will not cease to be what He is, nor will He come to nothing, neither will the next world come to nothing. What despair could be greater than that which would arise if, after the created beings had delighted in their existence in the goodness of God, this delight were to be removed from them, and they were to come to the despair of non-existence?

Further, if the angels, souls and first elements are simple, how can a thing that is not composed suffer dissolution? How can it come to nothing, when it has no forms which can be separated? No (simple) created being will, therefore, come to nothing.

If you say that as that which is simple and devoid of composition came from nothing, it will also dissolve into nothing:—

[1] Text repeats "God."

[2] A lacuna. The copyist of our MS. informs us in a marginal note that the copyist of the MS. from which he was transcribing stated that a whole leaf was missing here. The lacuna, therefore, goes back to the archetype of the MS. from which ours was transcribed.

We will answer: That which is simple is wholly one. God also is wholly one and simple, but He is in addition infinite. He granted to the above created beings, by grace, the favour of being simple and wholly one, but finite. If His simpleness does not dissolve into nothing (neither will that of the above created beings, which are) one in their nature. It is He who granted to the simple creatures to be in their very nature one and not [1] to dissolve into nothing, and He having made them so, they will not dissolve into nothing.

Further, He granted to the simple and composed beings to exist naturally, and this by grace, since true existence belongs to Him, but He implanted in the composed beings dissolution, a thing which He does not possess, and which will endure in them as long as they exist in their state formed of antagonistic
247 principles. The simple beings are one naturally, and will not dissolve into nothing, while the composed beings will dissolve into the simple ones.

Nobody should, therefore, adhere to such a flimsy and baseless reasoning (as that which we have refuted).[2]

CHAPTER XI.

On the reason why God created this world terminable, and the next world interminable.

God—may His majesty be worshipped—is good, merciful and compassionate. He created this world by His goodness, and knew beforehand that this Adamic race which He made in His image and likeness would, as a consequence of its transgression, suffer labours, pains and afflictions; and because of His loving kindness and mercy He had pity on it, and wishing to release it from these labours, pains and afflictions, He made this world perishable and terminable, since interminability would have been detrimental to the Adamic race. Because man

[1] Add negative *lā*.

[2] The author seems to refer here to arguments brought forward by his adversaries, some of which were evidently contained in the missing leaf.

was from Him and for Him, He made this world terminable, so that after man—who is also from the world through his elemental parts which emanate from it—had escaped from afflictions and pains through the dissolution of these parts, he might, by His infinite and unlimited goodness, receive the benefit of being without end.

In the same way as by His mercy and grace He made this world terminable, as a deliverance from afflictions and passions, so also by His goodness He made another world in which there are no afflictions, labours or pains; and He made it to be without end so that man might delight in it in incorruption and impassibility, and without labour. He granted him in it ineffable blessings, and a happiness that is higher than all other happiness.

It has been shown that it is not to no purpose that God created the next world, and made it interminable, while He made this world terminable. In both of them [1] He is seen as the All-Wise, who for the benefit of man made the two worlds: the present terminable and the next interminable and without dissolution into nothingness, so that man might delight always in divine knowledge, in the ineffable Theory of the Holy Trinity one true God, and in the accurate knowledge of the created beings, a knowledge which is seen here only in the dark glass of a shadow, but there face to face,[2] and as accurately as created beings are able to bear.

Let us here put an end to this work.[3]

O Trinity, who art higher than all minds and intelligences, who art supreme above all and in all, who art infinite and uncircumscribed, one essence, one will, one lordship, one sovereignty and one provident and ruling Lord, I miserable one, beseech Thee, with the lifting up of the hands of my mind, with the abundant tears of my wretched soul, and with the sacrifice of my humble and faithful thanksgiving to Thee—and I would also implore here all those who derive benefit from the labour of my work to beseech with me—that as Thou hast rendered me worthy in this world of partial knowledge, and enabled me to contemplate Thee and the knowledge of the

[1] Read the word in masc. or *b-thartaihain*.

[2] *Cf.* I Cor. xiii. 12.

[3] Lit. "merchandise," as before.

created beings with the eye of my mind as through a glass darkly, and to explain their causes in the measure of human capacity, Thou wouldst in the next world render me worthy of perfect knowledge, in order to contemplate Thee and the created beings face to face, and so delight unerringly and unmistakeably in their true nature and causes; to be renewed in Thy knowledge with all the saints in the heavenly kingdom and in the unend-
248 ing and imperishable life; and to be worthy of the rest given to those who have endured labour and tribulation, while Thou deliverest me from the outer and extreme darkness of ignorance concerning Thee and the created beings, and from the complete perdition which will have no end. To Thee be glory and on us Thy grace in both worlds. Amen.

[The Colophon of the Old[1] Manuscript.]

Here the book ends with the assistance of God. It was written in the year 1532 (*of the Greeks :* A.D. 1221), *in the month of April, in the town of Caesarea, by the weak Basil the deacon, son of John, the head of the scribes of Melitene. All of it was accurately collated (with the original) by our pride Rabban Abu l'Ḥasan the physician—may God preserve him—in the month of April of the year* 1533 (*of the Greeks :* A.D. 1222), *also in Caesarea.*

[1] *I.e.*, the original from which the present MS. is derived.

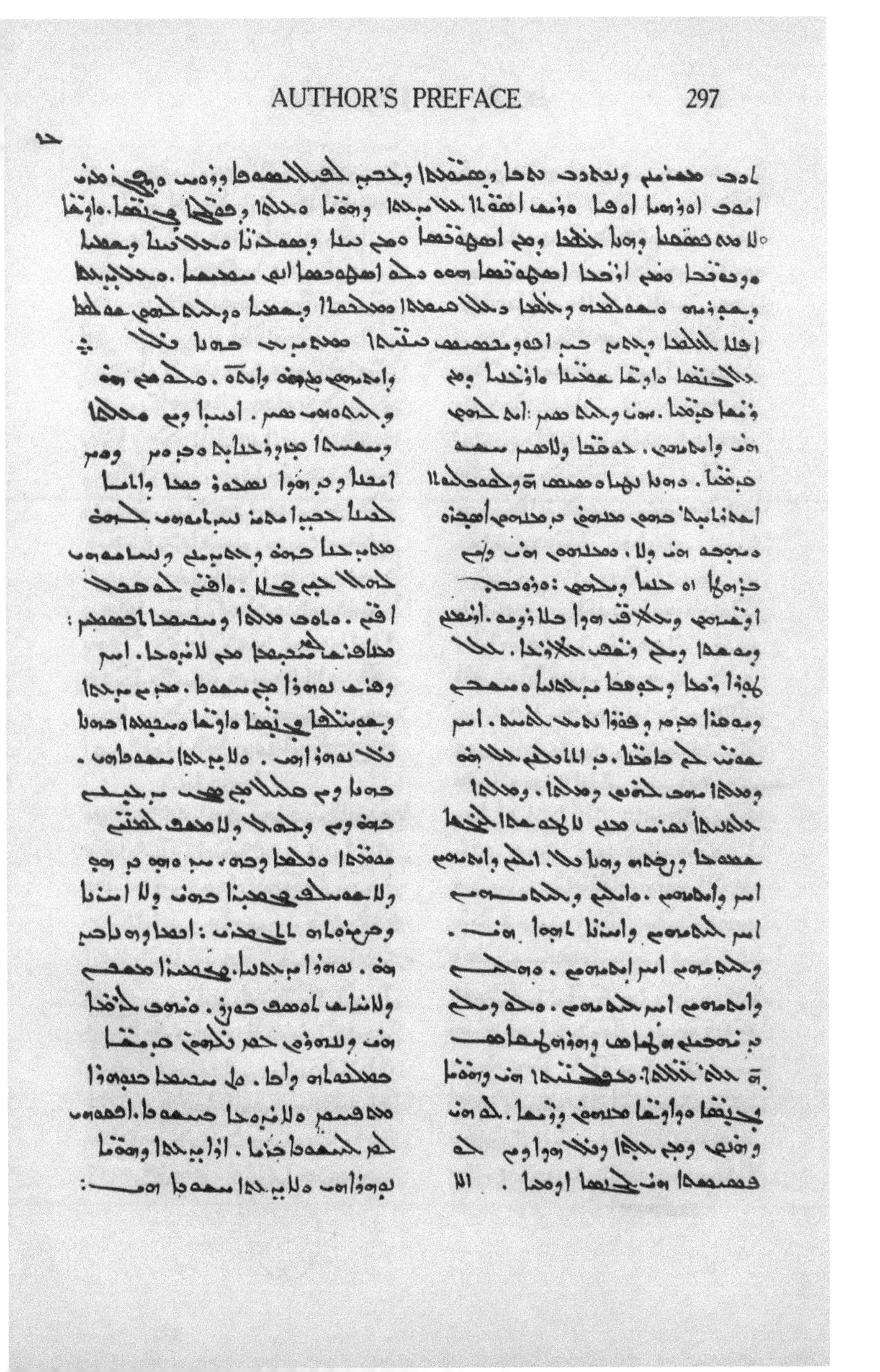

[illegible]

[illegible]

ܦܪܝܫ ܘܡܫܚܠܦܘܬܐ ܕܫܟܝܢܐ ܐܝܟ ܕܒܡܫܡܫܢܘܬܐ
ܐܣܝܐ ܕܝܢ ܐܦ ܡܝ̈ܐ ܗܠܝܢ ܠܗܘܢ ܐܝܟ ܕܗܘܝܢ
ܡܝ̈ܐ ܚܝ̈ܐ ܘܢܒܥܝܢ ܘܐܝܟ ܐܝܟ ܐܠܗܐ܇
ܒܫܘܝܢ ܘܡܫܬܚܠܦܝܢ ܘܡܢ ܚܕ̈ܕܐ ܗܘܝܢ
ܡܬܝܕܥ ܐܠܗܐ ܐܝܟܢܐ ܗܘܐ. ܘܡܝ̈ܬܐ ܘܚܝ̈ܐ
ܐܝܟ. ܡܕܝܢ ܡܢ ܟܠܗ ܘܟܝ̈ܢܐ ܘܡܝ̈ܐ ܡܢ ܟܠܗܘܢ
ܘܟܝܢܐ. ܘܠܐ ܗܘ ܕܢܫܬܚܠܦܘܢ ܠܐܠܗܐ ܡܝ̈ܐ
ܐܝܟ. ܘܡܝ̈ܐ ܡܫܚܠܦܘܬܐ ܘܫܟܝܢܐ. ܗܠܝܢ ܕܐܟܠܝܢ
ܚܝ̈ܐ ܘܟܠܗܝܢ. ܐܦ ܕܟܝܢܬܐ ܗܝ ܕܟܝܢܐ ܡܢ ܟܠܗ
ܘܟܠܗ ܘܡܝ̈ܐ. ܘܙܗܐ ܗܘ ܡܢ ܚܝܝܐ
ܘܢܦܫܗ. ܘܟܠܗ ܚܝܐ ܘܡܬܚܫܒܝܢ ܒܗ
ܚܙ̈ܝܢܐ ܘܐܠܗܐ ܗܘܝܢ. ܡܕܝܢ ܢܦܫܗ
ܒܐܪܝ ܗܘܐ ܚܐܘܪܐ ܗܘܐ ܙܡܢܐ ܘܡܢ ܚܝܠܐ
ܘܟܠܗ ܘܡܝ̈ܐ. ܐܝܟܢܐ ܘܡܫܬܟܚܐ
ܢܦܫܗ ܒܗܘ ܙܟܝܐ ܘܡܫܟܚܐ ܐܘܪܚܐ
ܡܦܝ̈ܣܐ ܠܐܢܫܐ ܡܢ ܕܟܝܢܐ. ܐܠܐ ܡܐ ܠܟܠܗܘܢ.
ܕܟܝܢܐ ܘܗܠܝܢ ܘܐܝܟܢܐ ܒܗܘ ܘܐܝܟܢܐ
ܐܣܝܐ ܘܟܠܗ ܐܠܗܐ ܘܒܝܬܐ

ܙܘܥܐ ܚܕܝܐ

ܡܛܠ ܟܠܗ ܘܐܣܛܘܟܣܐ ܦܫܝܛܐ ܘܡܪܟܒܐ
ܢܦܫܐ ܘܐܝܟܢܐ. ܘܐܢܫܐ ܐܝܬ ܦܪ̈ܨܘܦܐ ܐܦ
ܡܬܝܕܥܐ ܒܗܘ ܫܡܐ: ܐܣܛܘܟܣܐ ܒܪܢܫܐ
ܦܫܝܛܐ ܗܢܘܢ ܕܟܠܗܘܢ ܗܘ ܒܢܝ ܠܗܘܢ:
ܐܝܟܢܐ ܕܢܦܫܐ ܐܝܟ ܕܡܬܐ ܒܗܘ ܫܡܐ ܡܢ ܕܝܢ
ܗܢܘܢ ܕܝܢ ܡܪܟܒܐ ܢܦܩܝܢ ܐܝܟ ܦܪ̈ܨܘܦܐ.
ܦܫܝܛܐ ܕܝܢ ܣܡܝܡܘܬܐ ܚܡܝܡܘܬܐ ܐܝܟ ܕܐܝܬܘܗܝ
ܢܗܝܪܘܬܐ: ܡܪܟܒܐ ܕܝܢ ܢܘܪܐ: ܡܝ̈ܐ: ܐܪܥܐ: ܐܐܪ:
ܗܢܘܢ ܕܝܢ ܡܪܟܒܐ: ܠܐ ܡܫܬܒܩܝܢ ܠܟܠܗܘܢ ܫܡܐ.
ܘܡܬܚܙܝܐ ܐܝܟ ܠܟܠܗܘܢ ܒܗܘ ܐܠܐ. ܐܝܠܐ
ܒܗܘ ܦܫܝܛܐ. ܘܐܝܟ ܦܫܝܛܘܬܗܘܢ

ܦܫܘܛ ܠܟܠܗܘܢ ܐܝܟ ܕܐܠܗܐ ܦܫܝܛܐ
ܐܝܬܝܗ. ܦܫܝܛܐ ܠܟܠܗܘܢ ܘܐܪܥܐ ܘܐܐܪ ܠܘܬ
ܡܢ ܡܝ̈ܐ. ܘܗܘܝܐ ܠܟܠܗܘܢ ܐܐܪ. ܘܐܦ
ܗܘ ܡܫܟܚܝܢ ܦܫܘܛ ܠܟܠܗܘܢ ܘܐܝܟܢܐ
ܘܐܪܥܐ ܐܝܟ ܘܠܐ ܫܠܡܝܢ ܘܠܐ ܡܬܚܙܝܢ. ܡܕܝܢ
ܕܢܝܚܐ ܐܝܬ ܘܠܐ ܐܝܬܝܗ ܦܪ̈ܨܘܦܐ. ܐܝܟ ܟܠܗܝܢ ܐܦ
ܐܝܟܢܐ ܐܦ ܠܐ. ܘܐܬܚܙܝܘ ܘܐܝܟܢܐ
ܡܫܟܚܝܢ ܠܗܘܢ ܢܗܪܐ ܗܘܝܐ. ܕܝܢ ܐܝܟ
ܘܫܟܝܢܬܐ ܐܦܛܝܢ ܡܢ ܗܢܘܢ ܡܪܟܒܐ
ܠܗܢܘܢ ܦܫܝܛܐ. ܗܘܝܐ ܗܢܘܢ ܒܪܢܫܐ
ܡܪܟܒܐ ܡܢ ܐܪܒܥ ܡܟܝ̈ܢܬܐ ܐܝܬܝܗܝܢ
ܘܐܝܟܢܐ. ܢܗܪܐ ܡܢ ܣܡܝܡܘܬܐ ܘܡܟܝ̈ܢܬܐ
ܡܟܝ̈ܢܐ ܡܢ ܚܡܝܡܘܬܐ ܘܢܗܝܪܘܬܐ. ܘܗܢܘܢ
ܐܝܟ ܐܝܬܝܐ ܡܟܝܢܝܢ ܒ ܡܟܝܢܝܢ. ܟܠܗ
ܒܪ ܢܗܪܐ ܣܡܝܡܘܬܐ ܗܘܬ ܠܟܠܗܘܢ ܠܐ ܐܟܪܝܢ
ܗܘܝܢ ܘܡܪܟܒܐ ܗܘܬ. ܘܟܠܗܘܢ ܚܝܝܢ ܐܝܬܐ
ܠܒܗܘܢ ܗܘܐ ܠܐ ܐܦܪܝܢ ܗܘܝܢ ܘܡܪܟܒܝܢ ܗܘܘ
ܗܘܝܐ ܐܦ ܐܪܥܐ ܘܐܐܪ. ܐܠܐ ܡܬܚܙܝܐ ܘܡܫܝܚܐ
ܫܪܝܢ ܒܗܘܢ. ܦܪܝܢ ܠܗܘܢ ܡܪܟܒܐ
ܘܡܟܝܢ ܘܟܠܐ ܡܟܝ̈ܢܬܐ ܘܡܟܝܢ ܐܐܪ ܘܒܗ
ܦܠܓܝܢ ܘܫܪܝܢ ܘܗܘܝܢ ܡܟܝܢܐ ܘܟܠܗܘܢ
ܠܐܦܝ̈ܢ ܦܫܝܛܐ. ܡܢ ܟܠܗ ܘܦܫܝܛܬܐ ܐܝܟ.
ܦܫܝܛܐ ܒܪ ܐܝܟܢܐ ܗܘ ܡܐ ܘܟܠܗ ܒܟܠܗ.
ܗܘ ܕܝܢ ܐܝܟ ܠܗ ܘܠܐ ܦܫܝܛܐ: ܡܪܟܒܐ
ܕܝܢ ܗܘ ܘܗܘ ܘܐܝܟܢܐ ܡܢ ܐܪܒܥܬܐ ܗܘܬ ܘܐܪܥܬܐ
ܡܬܚܙܝܐ ܘܡܫܝܚܘܬܐ ܐܝܟ ܡܪܟܒܘܬܐ ܗܘܬ.
ܘܚܡܝܡܘܬܐ ܒܗܘ ܘܣܡܝܡܘܬܐ. ܘܡܫܝܚܘܬܐ
ܘܢܗܝܪܘܬܐ ܗܘܬ ܠܟܠܗܘܢ ܦܫܝܛܐ ܐܪܥܐ
ܐܣܛܘܟܣܐ

ܐܣܛܘܟܣܐ ܦܫܝܛܐ ܐܝܟ ܕܠܗ ܩܕܡܝܢ ܗܘܘ
ܐܬܪܐ. ܐܦܠܐ ܡܫܬܚܠܦܝܢ ܗܘܘ ܒܟܝܢܗܘܢ.
ܘܐܦܣܘܩ ܡܢ ܡܫܬܚܠܦܘܬܐ ܐܝܟ ܕܣܝܡܝܢ ܗܘܘ.
ܘܡܫܬܚܠܦܝܢ ܐܝܟ ܡܐ ܕܡܬܚܙܐ. ܐܠܐ ܡܛܠ ܕܡܪܟܒܐ
ܘܡܬܚܙܝܢ. ܗܘܐ ܗܟܝܠ ܕܡܢ ܡܪܟܒܐ. ܐܬܐ ܠܗ
ܡܕܡ ܐܚܪܝܢ ܐܝܟ ܗܘ. ܡܫܬܚܠܦܐ ܘܚܕܐ
ܗܘܐ ܕܝܘܠܕܐ ܘܡܬܢܝܢܘܬܐ ܘܡܪܟܒܐ ܠܗ ܡܢ ܙܘܓܐ -
ܣܛܘܟܣܐ ܐܦܪܝܐ ܘܡܬܚܙܝܐ ܠܗ ܗܘ
ܘܝܘܠܕܐ. ܘܫܘܚܠܦܐ ܘܛܒܥܐ ܠܗ ܗܘ ܘܡܬܒܥܐ.
ܡܢ ܐܝܟ ܐܝܬ ܦܫܝܛܐ. ܡܕܡ ܐܝܟ ܐܣܛܘܟܣܐ
ܦܫܝܛܐ. ܡܫܬܚܠܦ ܘܐܝܟ ܐܘܚܕܐ ܘܠܐ ܡܫܬܚܠܦܝܢ
ܒܟܝܢܗܘܢ. ܡܠ ܐܝܟ ܕܐܡܪ ܘܡܫܬܚܠܦܝܢ ܒܡܬܚܙܝܢ
ܠܡܫܬܚܠܦܘܬܐ. ܐܟܙܐ ܘܡܫܬܚܠܦ ܡܢ
ܠܡܫܬܚܠܦܘܬܐ ܘܡܪܟܒܐ. ܡܫܬܚܠܦܝܢ ܠܐ ܡܕܡ ܐܠܐ ܠܡܫܬܚܠܦܘ
ܡܕܡ ܕܠܗ ܐܬܐ ܐܝܟ ܐܣܛܘܟܣܐ ܦܫܝܛܐ
ܐܘ ܡܪܟܒܐ ܐܝܟ ܗܘ ܕܡܢ ܐܣܛܘܟܣܐ ܘܐܣܛܘܟܣܐ
ܦܫܝܛܐ ܗܘܝܐ: ܡܫܬܚܠܦܘܬܐ ܐܝܟ ܕܡܢ ܐܣܛܘܟܣܐ
ܦܫܝܛܐ ܘܡܬܪܟܒܐ ܐܘ ܡܫܬܚܠܦܐ ܘܡܬܦܠܓ
ܡܢ ܦܫܝܛܘܬܐ. ܦܫܝܛܘܬܐ ܕܝܢ ܐܝܟ ܗܘ ܕܐܣܛܘܟܣܐ
ܦܫܝܛܐ ܘܡܪܟܒܐ ܡܪܟܒܝܢ ܗܘܘ ܕܠܗܠ
ܠܡܫܬܚܠܦܘܬܐ. ܡܪܟܒܘܬܐ ܕܝܢ ܐܝܟ ܐܣܛܘܟܣܐ
ܦܫܝܛܐ ܘܡܬܚܙܝܐ ܘܡܬܦܠܓ ܡܢ ܙܘܓܐ.
ܙܘܓܐ ܐܝܟ ܐܣܛܘܟܣܐ ܦܫܝܛܐ ܘܐܫܬܝ
ܘܡܬܦܠܓ ܡܢ ܡܫܬܚܠܦܘܬܐ. ܗܘܝܐ ܕܝܢ ܐܦ
ܡܪܟܒܐ. ܝܘܪܐ ܕܝܢ ܐܣܛܘܟܣܐ ܫܚܠܦܐ ܘܡܫܬܚܠܦܐ
ܘܡܬܚܙܝܐ ܘܙܘܓܐ. ܘܐܚܪܢܐ ܡܫܬܚܠܦ ܐܘ ܦܫܝܛܐ
ܐܝܟ ܕܝܢ ܫܚܠܦܐ ܐܘ ܙܘܓܐ. ܗܢܘܢ ܕܝܢ
ܐܣܛܘܟܣܐ. ܗܢܘܢ ܕܠܡܬܚܙܝܢ ܗܘܝܐ. ܗܘ ܕܝܢ
ܘܡܬܦܠܓ ܡܢ ܦܫܝܛܘܬܐ ܡܫܬܚܠܦܘܬ ܡܫܬܚܠܦ
ܘܐܣܛܘܟܣܐ ܗܘ ܦܫܝܛܐ ܫܚܠܦܘܬܐ ܗܘ. ܐܬܐ

ܫܚܠܦܘܬܐ ܐܝܟ ܐܣܛܘܟܣܐ ܦܫܝܛܐ
ܘܡܫܬܚܠܦܘܬܐ ܘܡܬܦܠܓ ܡܢ ܦܫܝܛܘܬܐ :
ܗܢܘܢ ܕܝܢ ܘܐܦ ܐܣܛܘܟܣܐ ܕܗܠܝܢ ܐܬܪܐ
ܟܠܗ ܕܡܬܚܙܐ. ܡܕܡ ܡܫܬܚܠܦ. ܡܛܠ ܐܝܟ
ܕܐܡܪ ܕܦܫܝܛܐ ܡܫܬܚܠܦܝܢ ܠܡܪܟܒܐ. ܐܬܐ
ܐܝܟ ܗܘܘ ܡܕܡ ܕܡܬܚܙܝܢ. ܡܛܠ ܙܘܓܐ
ܠܙܘܓܐ ܕܡܬܚܙܝܢ ܗܘܘ. ܡܛܠ ܠܐ ܠܙܘܓܐ ܙܘܓܐ.
ܐܝܟܢܐ ܗܟܢܐ ܙܘܓܐ. ܐܘ ܡܬܚܙܝܢ ܘܦܫܝܛܐ
ܘܡܪܟܒܐ: ܘܐܡܪܝܢ ܕܙܘܓܐ ܠܗܘܢ ܐܝܟܢܐ
ܡܢ ܠܡܬܚܙܝܢܘܬܐ ܟܝܢܐ. ܘܐܬܪܐ ܘܠܐ ܡܬܚܙܝܢܘܬܐ
ܟܝܢܐ ܐܚܪܢܐ. ܐܠܐ ܡܬܚܙܝܢܘܬܐ ܕܟܝܢܐ ܠܡܬܚܙܝܢ.
ܡܬܚܙܝܢܘܬܐ ܟܝܢܢܝܬܐ ܗܘܝܐ. ܡܛܠ ܗܠܝܢ
ܙܘܓܐ ܕܡܢ ܡܪܟܒܬܐ ܕܡܬܚܙܝܢܘܬܐ ܟܝܢܐ
ܗܘܘ. ܘܫܘܝܢ ܡܪܟܒܬܐ ܕܡܢ ܙܘܓܐ ܐܬܪܐ
ܐܠܐ ܗܘܐ. ܡܬܚܙܝܢܘܬܐ ܟܝܢܐ ܐܚܕ ܘܡܬܚܙܝܢܘܬܐ
ܘܐܚܕ ܘܐܡܪ ܟܝܢܐ ܘܡܬܚܙܝܢܘܬܐ. ܡܪܟܒܐ ܗܘ
ܡܢ ܡܪܟܒܬܐ ܡܫܬܚܠܦܘܬܐ ܠܟܠܢ ܐܦܪܝܐ ܘܡܬܚܙܝܢ.
ܘܫܘܝܢ ܡܪܟܒܬܐ ܘܙܘܓܗ. ܡܬܚܙܝܢ ܠܙܘܓܗ
ܟܝܢܐ. ܡܫܬܚܠܦ ܘܡܢ ܐܬܪܐ ܐܠܙܘܓܗ ܐܘ
ܫܘܝܢ ܠܟܠܢ ܐܦܪܝܐ ܘܐܚܕܐ ܘܐܦ ܫܘܝܢ ܡܢ
ܐܬܪܐ ܠܟܠ ܡܬܚܙܝܢ ܘܠܘܬܗ ܙܘܓܐ. ܐܬܪܐ
ܘܡܢ ܟܠܗܘܢ. ܘܐܝܟ ܡܬܚܙܝܢܘܬܐ ܟܝܢܐ.
ܗܘ ܕܝܢ ܐܢܐ ܐܬܪܐ ܕܙܘܓܐ ܘܡܫܬܚܠܦܐ ܗܘ ܠܟܠܗ
ܘܠܟܠܗ ܠܗܘܢ ܡܪܟܒܬܐ ܦܫܝܛܬܐ
ܡܢ ܡܪܟܒܬܐ ܐܚܪ ܘܐܣܛܘܟܣܐ ܡܪܟܒܐ
ܡܢ ܦܫܝܛܐ. ܡܫܬܚܠܦ ܘܟܠܗ ܡܕܡ ܐܬܪܐ
ܘܡܢܗ ܐܠܙܘܓܗ ܦܫܝܛܐ ܘܐܚܕ ܟܝܢܐ
ܐܚܕ ܕܝܢ ܡܛܠ ܐܝܟ ܗܢܘܢ ܗܢܘܢ ܐܡܝܢܐ
ܐܝܟ ܗܢܘܢ ܗܢܘܢ ܒ ܠܟܠܗܘܢ ܐܡܝܢܐ.
ܡܛܠ ܕܐܡܪ ܐܡܪܐ: ܡܛܠ ܠܟܠ ܡܢ ܙܘܓܐ

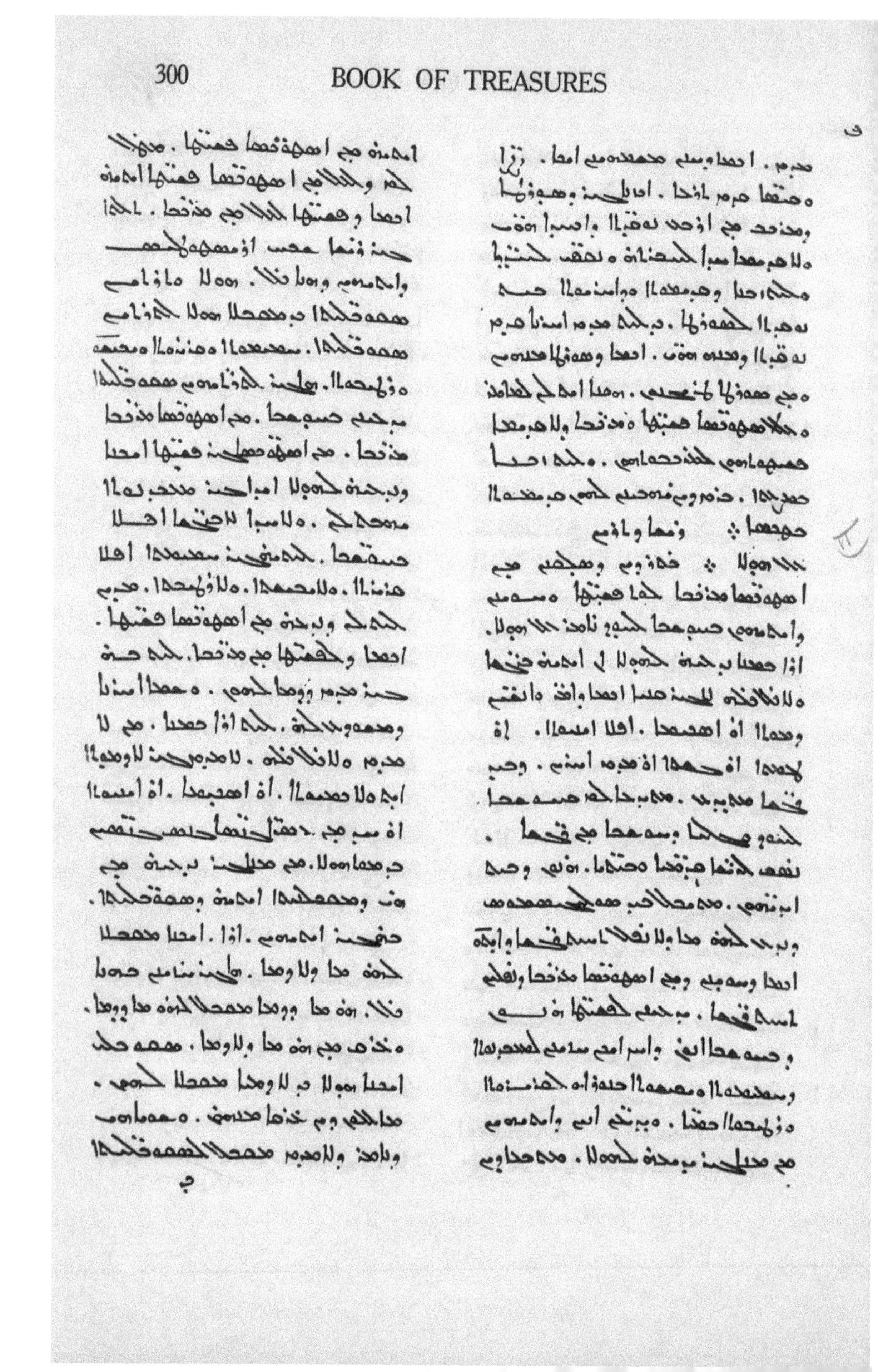

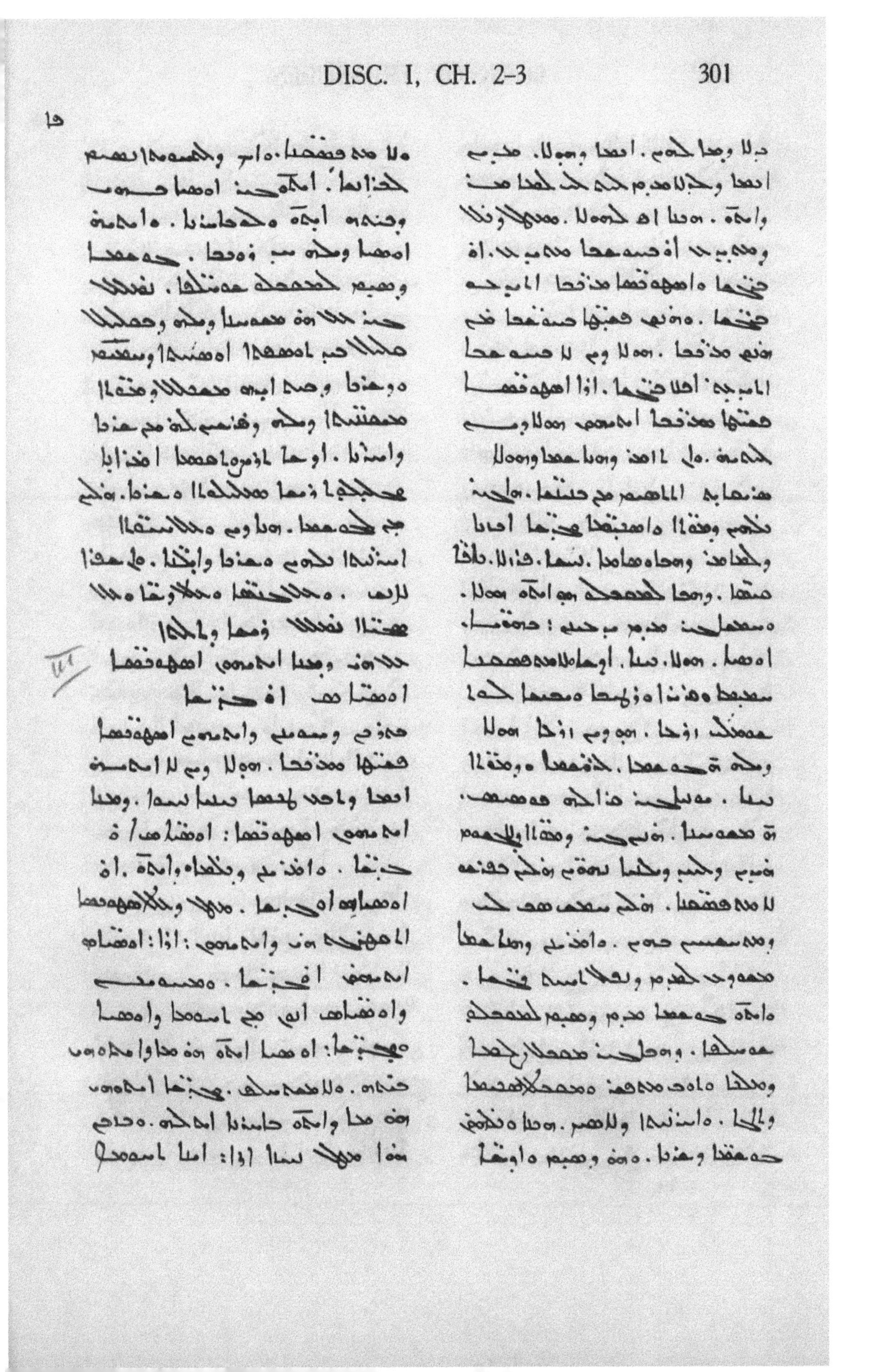

[illegible]

[illegible]

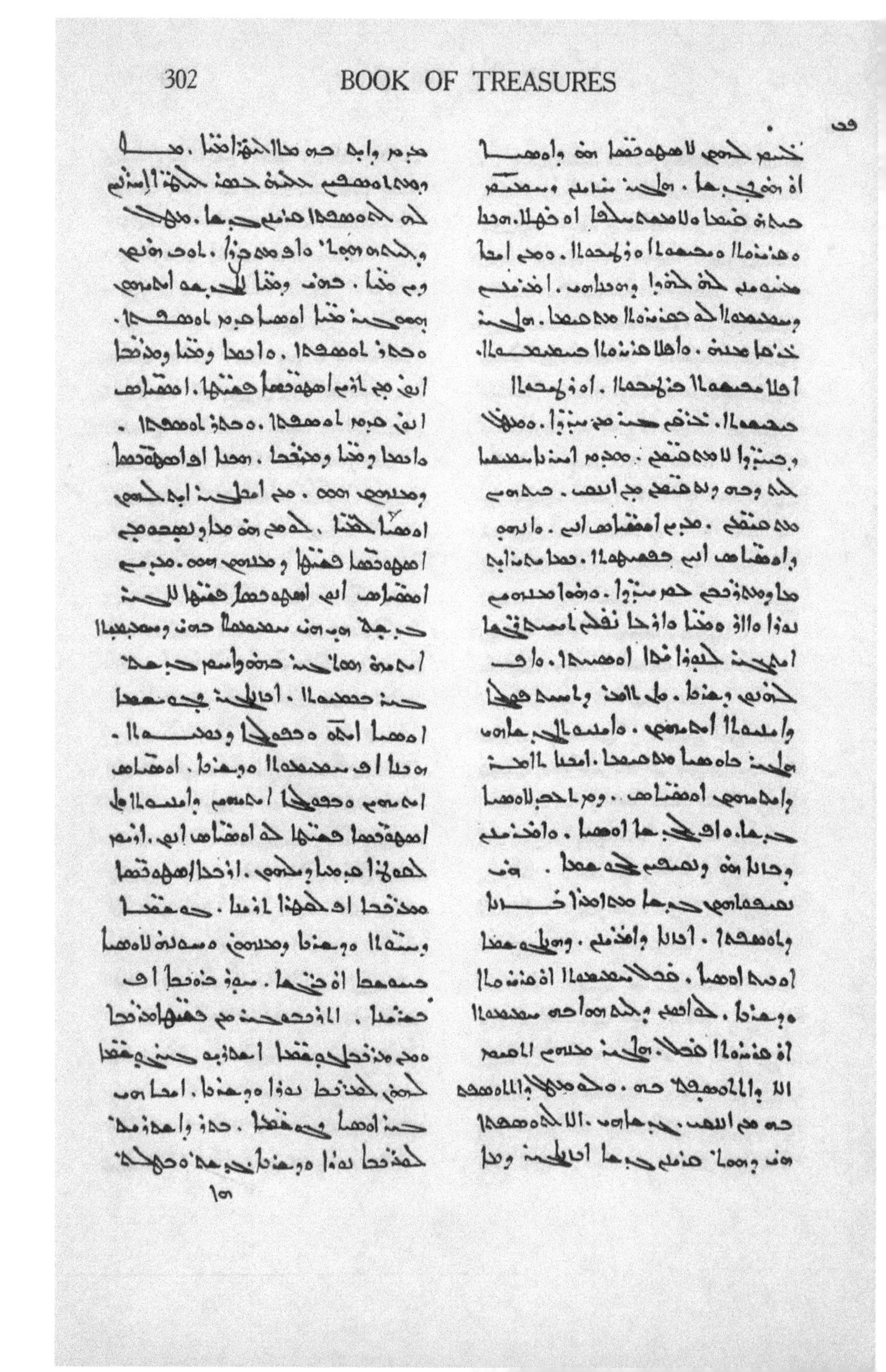

[illegible]

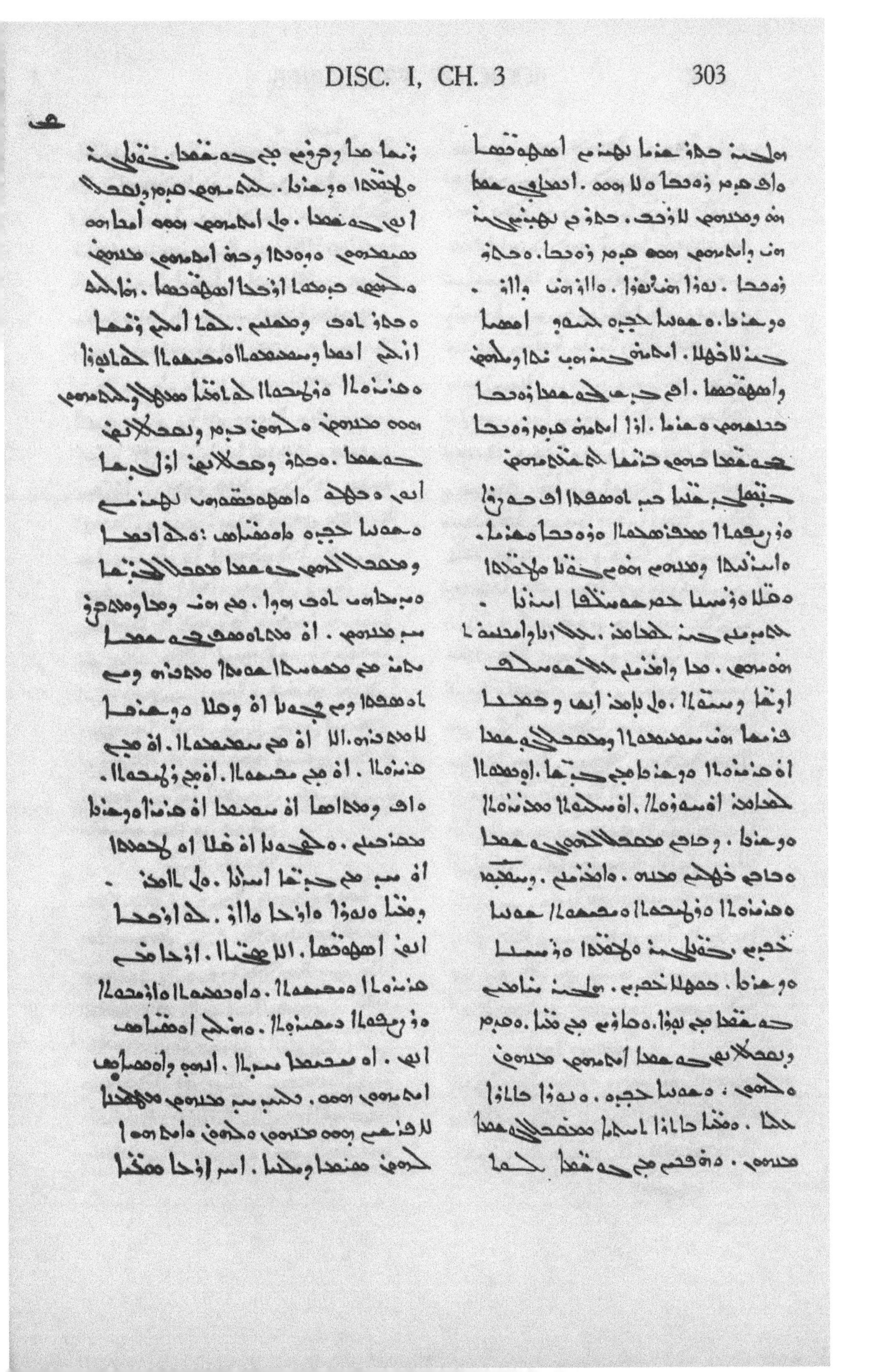

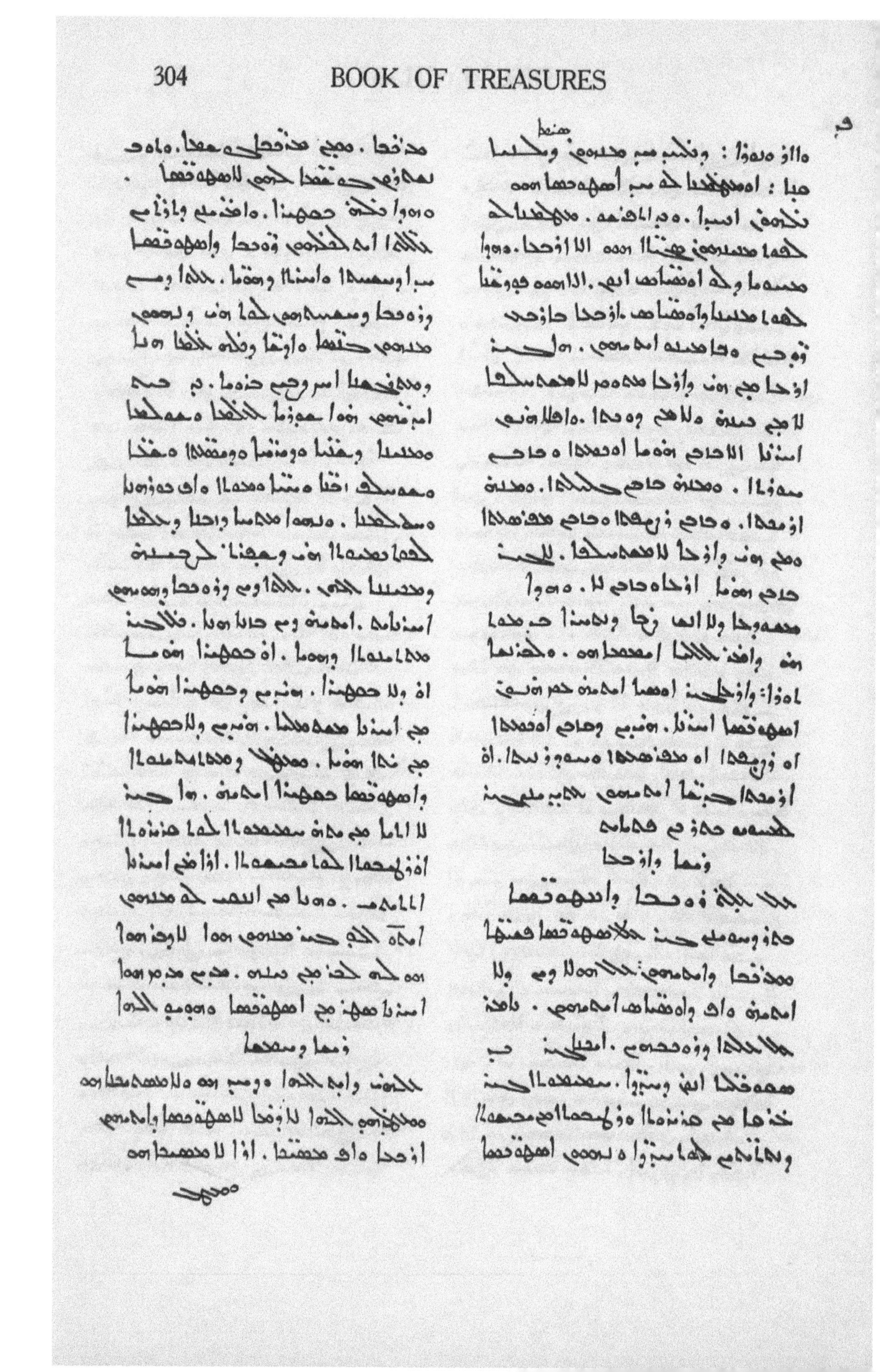

[illegible]

[illegible]

ܘܡܛܠ ܕܠܐ ܡܫܬܡܠܝܢ ܗܘܘ ܡܢܗ. ܟܠܠ ܡܫܬܡܠܝܢ: ܒܗ ܥܡܐ. ܐܠܐ ܫܘܡܠܝܐ ܕܠ ܒܗܠܝܢ ܡܘܠܟܢܐ ܐܠܗܝܐ. ܗܘ ܕ ܗܘ ܐܝܟ ܒܒܢܘܗܝ. ܘܗܘ ܗܕܐ ܕܗܘ ܕ ܐܝܟ ܒܒܢܘܗܝ ܗܘܐ. ܡܢ ܗܘ ܘܐܦ ܡܢ ܗܘ. ܗܘ ܡܫܡܠܝܢ ܕܡܢ ܗܢܐ ܐܘܪܚܐ ܐܦ ܒܗܝ ܕܝܠܗ. ܒܗ ܡܢ ܗܘ ܡܬܡܠܝܢ ܐܠܐ ܫܠܝܡܝܢ ܒܗ ܕܝܠܗ ܕܡܢ ܒܗ ܐܝܟ ܐܡܪܢܐ ܡܫܡܠܝܐ ܒܗ ܡܢ ܗܘ. ܡܫܡܠܝܢ ܕܗܘ ܡܢ ܗܘ ܒܐܝܕܗ. ܗܘܐ ܕܠܐ ܗܘܐ. ܒܗ ܘܡܢ ܟܠ ܒܬܘܪܬܐ ܕܟܠ ܕܢ ܦܘܩܕܢܐ ܗܘܐ ܟܠܐ ܕܩܕܡܐ ܕܙܘܥܝܐ. ܐܠܐ ܐܡܪܢܐ ܘܐܦ ܟܠܐ ܗܘ ܡܢ ܗܘ. ܠܝܬ ܕܟܠ ܠܐ ܦܓܪܐ ܐܝܟ. ܐܠܐ ܡܫܟܚܝܢ ܒܗ ܕܢܦܫܐ ܘܕܡܢ ܕܡܢ ܕܦܓܪܐ ܐܝܟ ܐܢ

ܪܫܐ ܕܫܬܐ

ܟܠܗܘܢ ܕܝܢ ܕܐܣܟܝܡܐ ܕܐܝܬܘܗܝ ܐܣܟܝܡܐ ܕܦܓܪܐ ܘܕܢܦܫܐ ܘܕܪܘܚܐ ܐܝܟ ܟܠ ܘܚܕܐ ܚܕܐ ܐܦ ܫܘܡܠܝܐ. ܘܟܠ ܘܚܕܐ ܫܘܡܠܝܐ. ܐܦ ܫܘܪܝܐ ܚܕܐ. ܘܟܠ ܘܠܐ ܚܕܐ ܚܕܐ ܠܐ ܚܕܐ ܫܘܡܠܝܐ. ܘܟܠ ܚܕܐ ܫܘܡܠܝܐ ܠܐ ܚܕܐ ܫܘܪܝܐ. ܘܡܛܠ ܕܐܣܟܝܡܐ ܐܝܟ ܕܟܠܗܘܢ ܚܕܐ ܚܕܐ ܡܫܬܟܚܐ ܐܦ ܒܬܪܬܝܢ. ܐܦ ܐܝܟ ܕܟܠܗܘܢ ܫܘܡܠܝܐ. ܠܐ ܐܝܟ ܕܟܠܗܘܢ ܫܘܡܠܝܐ ܐܘ ܫܘܪܝܐ. ܟܕ ܐܦ ܐܝܟ ܐܝܟ ܕܐܝܬ ܐܝܟ ܘܠܐ ܫܘܪܝܐ. ܐܠܐ ܕܟܠܗ ܦܓܪܐ ܥܠ ܦܓܪܐ ܐܝܟ ܕܢܦܫܐ ܐܝܟ ܘܢܦܫܐ ܘܐܦ ܗܘ ܕܐܝܬܘܗܝ ܫܘܡܠܝܐ ܗܘ ܢܦܫܐ ܘܟܠ ܚܕܐ ܚܕܐ. ܡܫܡܠܝܢܐ ܘܦܓܪܢܝܬܐ ܘܢܦܫܢܝܬܐ ܘܪܘܚܢܝܬܐ. ܥܠ ܗܢ ܕܝܢ ܘܟܠܗܘܢ

ܡܫܬܡܠܝܢ ܗܘܘ ܚܝܐ. ܘܫܘܡܠܝܐ ܟܠܗ ܘܐܝܬܘܗܝ ܕܡܢ ܗܘ ܕܦܓܪܐ ܠܐ ܗܘܐ. ܗܘ ܗܘܐ ܘܐܝܬܘܗܝ ܗܘ ܡܫܡܠܝܐ ܗܘ. ܘܗܘܐ ܐܝܟ ܐܝܬܘܗܝ. ܥܠ ܟܠ ܘܢܦܫܐ ܘܦܓܪܐ ܟܠ ܗܘ ܡܫܡܠܝܢ ܡܢ ܗܘ ܫܘܡܠܝܐ ܘܐܦ ܗܘܐ ܒܗ ܕܠܐ ܗܘܐ ܐܦ ܗܘ ܐܝܟ ܚܝܐ ܘܡܫܡܠܝܐ ܐܝܟ ܚܝܐ ܕܐܝܟ ܐܝܬܘܗܝ ܢܦܫܐ. ܐܠܐ ܡܟܝܠ ܗܘ ܕܝܢ ܐܝܟ ܘܡܢ ܗܘ ܐܦ ܗܘ ܐܝܟ ܗܘܐ ܕܠܐ ܚܝܐ ܘܗܘ ܥܠ ܐܝܟ ܗܢܐ ܐܝܟ ܐܦ ܒܡܫܡܠܝܢ ܘܐܝܬܘܗܝ ܫܘܡܠܝܐ. ܐܦ ܟܠ ܚܕ ܚܕܐ ܕܡܢ ܫܘܡܠܝܐ ܚܕܐ. ܐܦ ܡܢ ܗܘ ܒܝܬ ܫܘܡܠܝܐ ܐܦ ܚܕ. ܥܠ ܡܢ ܗܘ ܦܓܪܐ. ܐܠܐ ܐܝܟ ܟܠ ܡܫܡܠܝܢ ܐܝܟ ܐܝܬܘܗܝ ܚܝܐ. ܐܝܟ ܚܝܐ ܕܡܢ ܗܘ ܘܐܦ ܗܘܐ ܘܐܦ ܫܘܡܠܝܐ ܘܡܫܡܠܝܐ. ܘܐܦ ܗܘ ܕܝܢ ܟܠ ܕܝܢ ܡܫܡܠܝܢ ܘܫܘܡܠܝܐ. ܘܐܝܟ ܐܝܬܘܗܝ ܡܫܡܠܝܐ ܘܟܠܗ ܡܢ ܗܘ. ܘܐܝܟ ܐܝܬܘܗܝ ܟܠ ܘܐܦ ܫܘܡܠܝܐ. ܘܐܝܟ ܗܢܐ ܟܠܗ ܘܢܦܫܐ ܘܦܓܪܐ. ܐܦ ܐܝܟ ܐܝܬܘܗܝ ܢܦܫܐ ܘܦܓܪܐ ܘܐܝܟ ܢܦܫܐ ܕܦܓܪܐ. ܘܐܦ ܗܘ ܕܝܢ ܟܠ ܢܦܫܐ ܘܦܓܪܐ ܘܪܘܚܐ. ܐܝܟ ܐܝܬܘܗܝ ܟܠ ܢܦܫܐ ܐܝܟ ܦܓܪܐ

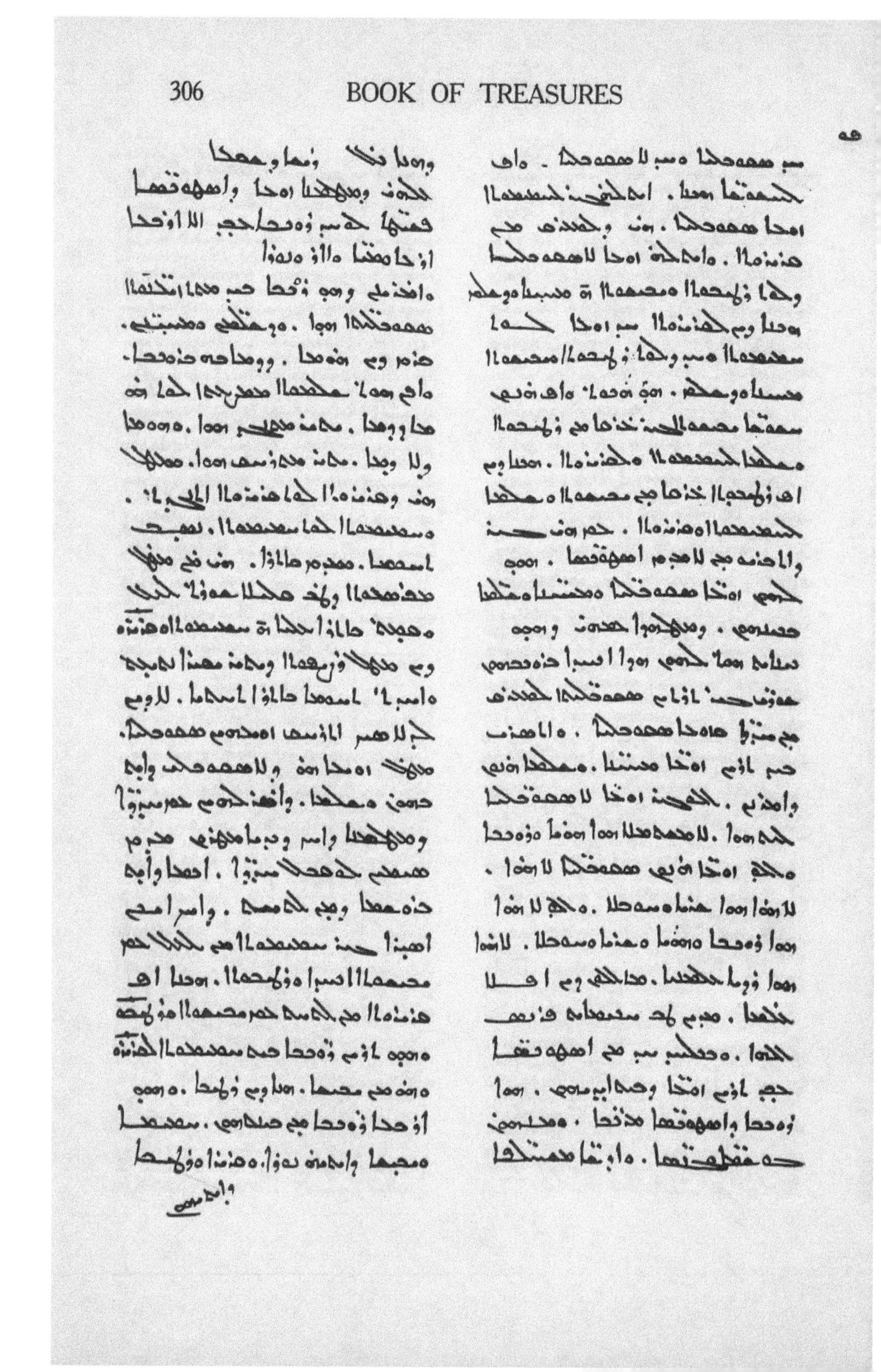

[illegible]

[illegible]

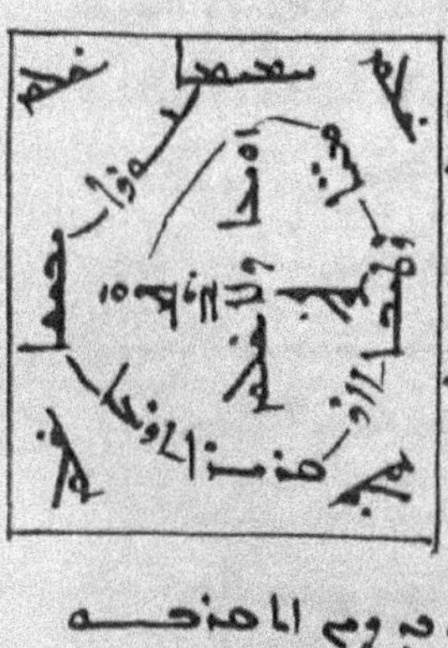

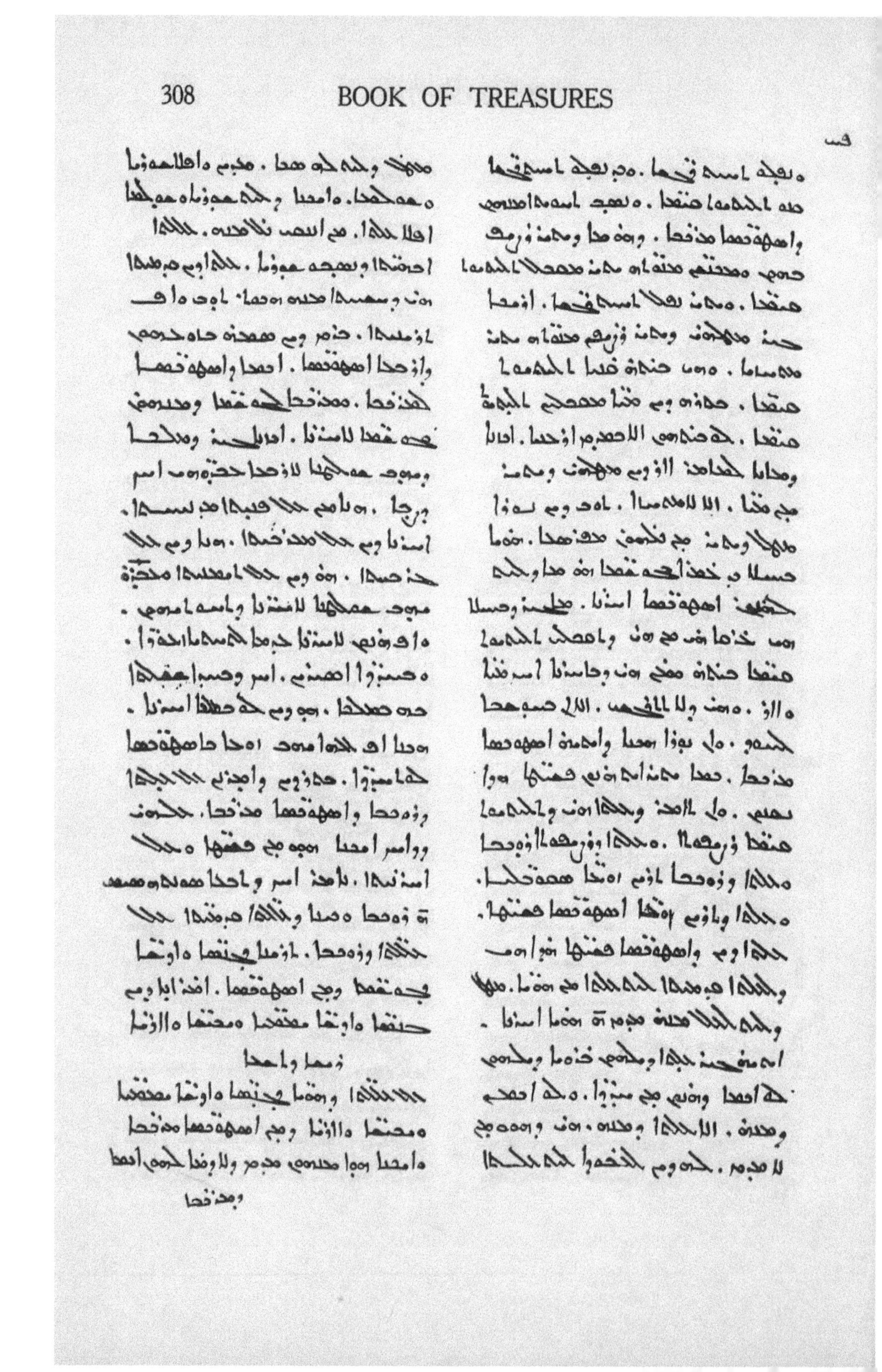

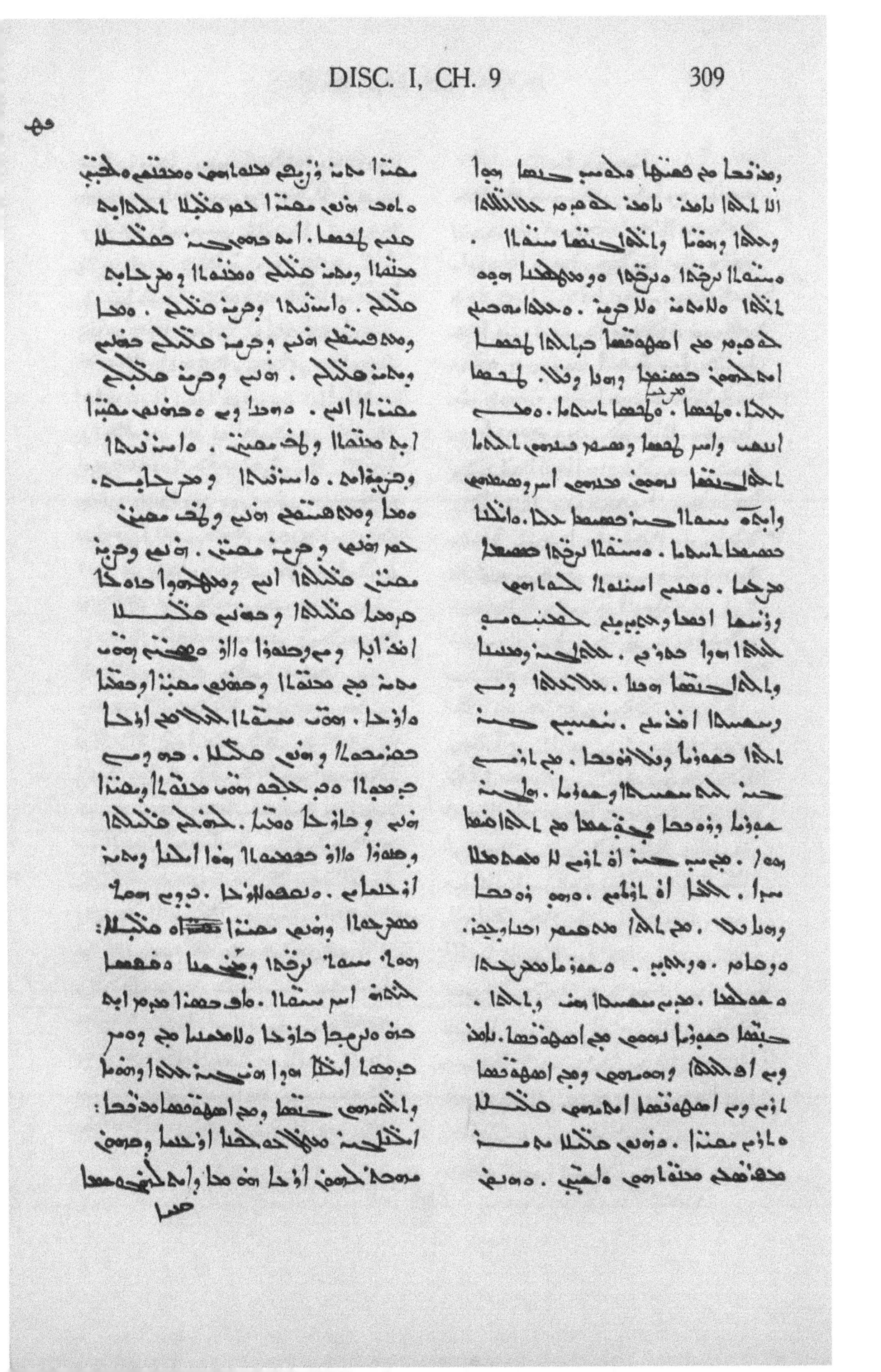

[illegible]

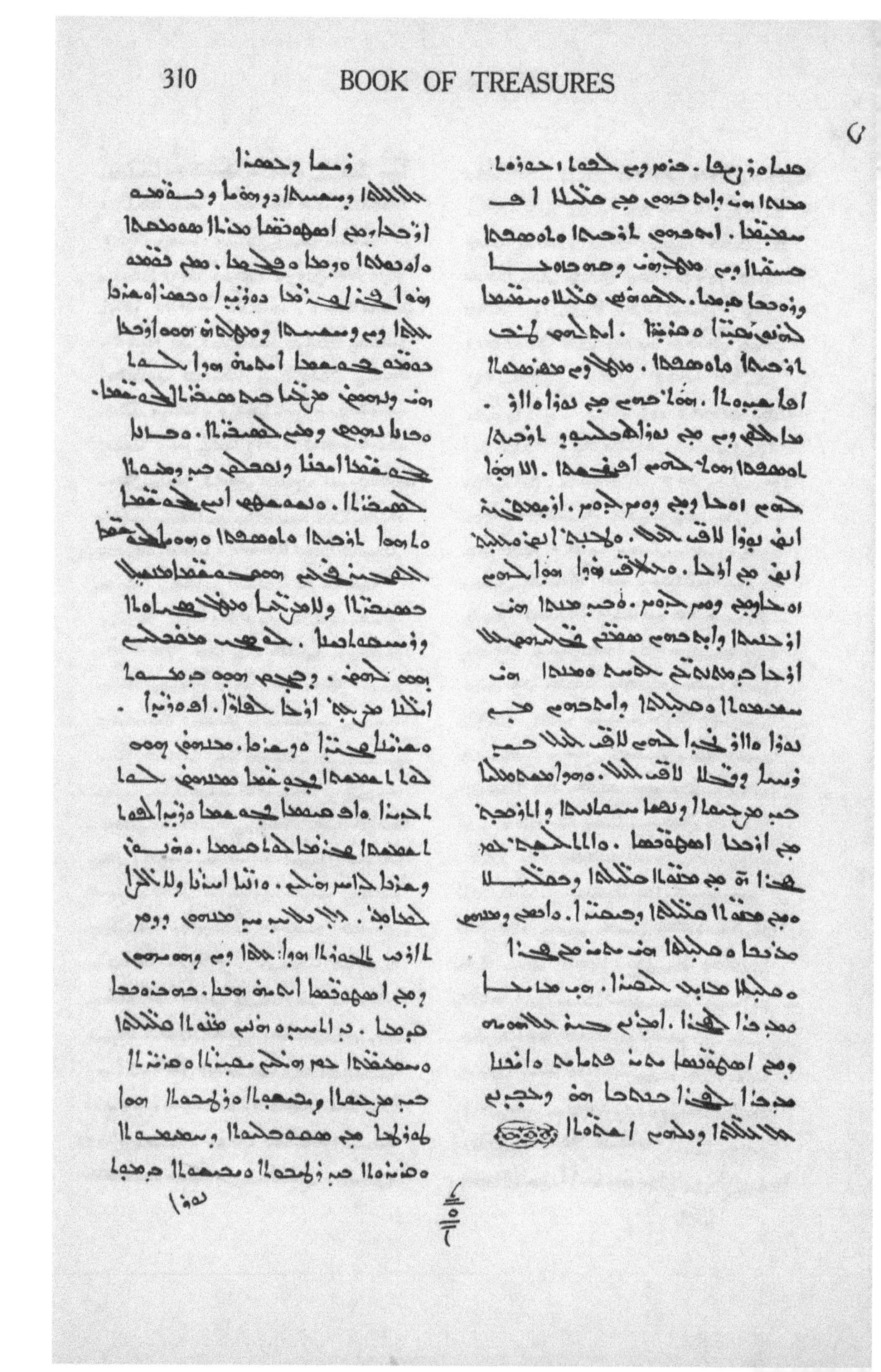

[illegible]

ܢܓ

ܘܡܛܠ ܡܢܐ ܡܬܡܛܝܢ ܐܠܝܢ ܟܠܗ ܡܢܗ
ܒܫܘܝܘܬܗ . ܘܐܬܝܢ ܘܡܬܡܠܟܢ ܒܗ
ܘܗܘ ܡܢܐ ܘܐܝܟܢܐ ܐܝܬܝܗܘܢ ܕܡܬܦܪܫܢ
ܡܢܗ ܐܬܝܢ ܘܗܘܝܢ ܒܥܢܢ̈ܐ ܘܪܘܚ̈ܐ ܘܡܛܪ̈ܐ
ܘܡܫܡܫܢܘܬܐ . ܘܗܘ ܡܢܐ ܘܐܝܟܢܐ ܗܘܐ ܒܗܘܢ
ܘܡܫܬܡܫ ܘܦܠܚ ܗܘܐ ܡܛܠܬܗ ܕܟܝܢܐ . ܘܡܢ
ܗܘ ܡܢܐ ܘܐܝܟ ܗܘܐ ܒܗܘܢ ܘܡܛܠ ܕܗܘܢ ܗܘܐ
ܠܐܪܥܐ ܘܫܡܫܐ . ܘܐܬܪܚܩ ܡܫܡܫܢܐ ܕܗܘܝܐ
ܒܫܡܫܐ . ܡܢ ܗܕܐ ܠܚܘܝܐ ܘܐܝܟ ܕܬܩܢܝܢ
ܠܐܝܟ ܕܡܬܩܢܝܢ ÷ ÷ ÷ ÷ ÷ ÷ ÷ ÷ ÷ ÷ ÷ ÷

ܪ̈ܫܐ ܕܫܒ ܒܫܡܫ

ܘܡܛܠ ܕܠܝܬ ܐܠܐ ܕܝܢ ܪܘܚܐ ܒܥܘ̈ܡܩܐ
ܝܕܝܥ ܗܘ ܠܡܐ ܕܗܘܐ ܠܘ ܐܦ
ܡܫܡܫܢܐ ܟܠ . ܐܝܟܢܐ ܐܦ ܐܪܥܐ . ܐܠܐ
ܗܘ ܐܝܟܢܐ ܡܫܡܫܢܐ ܘܐܦ ܡܫܡܫܢܐ ܕܚܝܐ
ܡܛܠܠܐ ܗܕܐ ܐܝܟ ܡܫܡܫܢܘܬܐ ܕܡܫܡܫܢܐ
ܗܕܐ ܕܠܐ ܘܠܗܘܢ ܘܠܚܝܘܬܐ ܘܠܡܫܡܫܢܘܬܐ
ܘܠܚܝܐ ܘܠܡܫܡܫܢܐ . ܘܡܫܡܫܢܐ ܗܘ ܕܡܢ
ܠܟܠ ܗܘܐ ܠܟܠ ܚܝܐ ܘܡܫܡܫܢܘܬܐ . ܗܘ ܕܝܢ
ܕܠܗܘܢ ܠܟܠ ܚܝܐ ܘܡܫܡܫܢܘܬܐ . ܘܗܘ ܕܠܚܝܐ
ܘܠܡܫܡܫܢܘܬܐ ܘܠܚܝܐ ܘܠܡܫܡܫܢܐ ܠܟܠ
ܚܝܐ ܕܦܠܚܘܬܐ ܘܡܫܡܫܢܘܬܐ . ܗܘ ܕܝܢ ܕܡܫܬܪܪ
ܡܢ ܗܘܐ ܕܠܗܘܢ ܕܟܠ ܡܫܬܪܪܐ ܘܐܝܟ ܐܡܪ
ܝܕܥܝܢ ܗܘ ܡܫܡܫܢܘܬܐ ܡܫܡܫܢܘܬܗ
ܘܡܠܟܘܬܗ ܕܟܠ . ܘܡܫܬܒܚ ܗܘܐ
ܗܘܐ ܕܐܦ ܠܟܠ . ܘܐܦ ܡܢܐ ܕܡܫܡܫܢܘܬܐ
ܡܫܬܪܪܝܢ ܘܡܫܡܫܢܝܢ ܘܡܬܒܝܢ ܗܘܐ ܕܐܦ
ܠܗܘܢ . ܐܦ ܕܝܢ ܕܒܢܝܢ ܦܠܚܘܬܐ ܘܡܫܡܫܢܐ
ܘܐܝܬܝܗ ܟܠܗ ܕܫܡܫܐ ܘܠܟܠ ܡܫܡܫܢܐ
ܘܐܪܥܐ ܕܝܢ ܡܫܡܫܢܘܬܐ ܟܠܗ ܫܡܫܐ ܠܗ

ܟܣܝܐ . ܘܡܢ ܡܬܚܙܝܢ ܒܚܙܘܐ . ܘܡܫܬܡܠܝܐ
ܘܡܫܬܠܡܢܐ ܒܗ ܡܠܐ ܘܡܚܝܐ ܠܚܘܫܒܐ ܐܪܥܝܐ
ܦܠܚܝܢ ܗܘܐ ܫܘܪܝܐ . ܡܛܠ ܕܐܝܬܝܗܘܢ ܟܠܗ
ܗܘܝܐ . ܐܪܥܝܢ ܦܪܨܘܦܐ ܘܡܫܡܫܢܐ
ܡܢ ܕܡܬܚܙܝܢ ܡܫܡܫܢܘܬܐ ܘܢܘܪܐ ܠܦܪܨܘܦܬܐ
ܘܐܪܥܐ ܕܝܢ ܡܢ ܕܚܙܝܐ ܘܡܫܡܫܢܘܬܐ ܟܪܝܐ ܦܪܨܘܦܬܐ
ܘܐܪܥܐ ܡܫܡܫܢܐ ܟܠܗ ܡܢܐ . ܟܠ ܦܪܨܘܦܬܐ
ܕܟܣܝܢ ܟܪܐ ܠܚܫܟܢܘܬܐ . ܗܟܢܐ ܕܝܢ ܡܫܡܫܢܐ
ܘܡܫܬܪܪ ܦܪܨܘܦܬܐ ܘܡܢܐ ܡܢ ܡܫܡܫܢܘܬܐ
ܘܢܘܪܐ ܟܪܝܐ . ܦܠܚܘܬܐ ܘܡܢܐ ܟܠܗ ܐܦ .
ܡܢܐ ܐܝܟ ܘܐܝܬܘܗܝ . ܡܫܡܫܢܘܬܐ ܘܐܦ ܕܝܢ
ܦܪܨܘܦܬܐ ܘܡܢܐ ܟܪܝܐ ܟܠܗ ܢܘܪܐ . ܢܘܪܐ ܕܝܢ
ܡܢ ܘܐܝܬܘܗܝ ܡܢ ܦܪܨܘܦܬܐ . ܘܠܗܠܝܢ ܟܠܗܘܢ
ܡܛܠ ܡܢ ܡܪ ܐܡܪܝܢ ܘܡܫܡܫ ܠܟܠ ܕܐܝܬܘܗܝ
ܟܠܗ . ܡܢ ܟܪܝܐ ܠܟܠܗܘܢ ܘܦܠܚܐ
ܟܠ ܡܫܡܫܢܘܬܐ ܘܐܪܥܐ . ܘܚܫܟܐ ܐܝܬܘܗܝ
ܟܠ ܦܪܨܘܦܬܐ ܘܟܠ . ܘܗܘܐ ܗܘܐ ܫܘܪܝܐ
ܟܠ ܐܝܟ ܟܐܢܘܬܐ ܗܘܐ ܟܢ ܘܟܠ ܦܠܚܐ
ܡܫܬܪܪܝܢ . ܗܘ ܕܝܢ ܗܘܐ ܡܢ ܐܪܥܐ
ܟܠ ܢܘܪܐ ܡܢ ܢܘܪܐ ܟܠܗ ܐܦ ܘܡܢ ܐܦ
ܟܠ ܡܢܐ . ܡܢ ܡܢܐ ܟܠ ܐܪܥܐ . ܟܪܝܘܬܐ
ܦܪܨܘܦܐ ܕܡܢ ܠܟܠܗ ܘܡܬܚܙܝܐ ܒܥܘ̈ܡܩܐ
ܗܘܐ ܡܥܒܕ ܡܢ ܐܣܛܘܟܣܐ ܘܐܝܟ ܟܠܗܘܢ
ܡܫܡܫܢܝܢ ܐܬܪ̈ܘܬܐ ܡܢ ܐܣܛܘܟܣܐ
ܠܗ ܡܫܡܫܢܝܢ . ܘܗܘܐ ܗܘܝܐ ܡܫܡܫܢܐ
ܟܪܝܐ ܦܪܨܘܦܐ . ܘܡܫܬܪܪ ܡܫܡܫ ܕܦܪܨܘܦܐ
ܗܘܐ ܐܪܥܐ ܘܡܫܡܫ ܫܘܪܝܐ ܟܠܗ ܚܝܠܐ
ܘܡܢ ܐܣܛܘܟܣܐ . ܘܟܠܗ ܐܦ ܡܫܡܫܢܝܢ
ܐܝܟܢܐ ܗܘܐ ܟܠܗ ÷

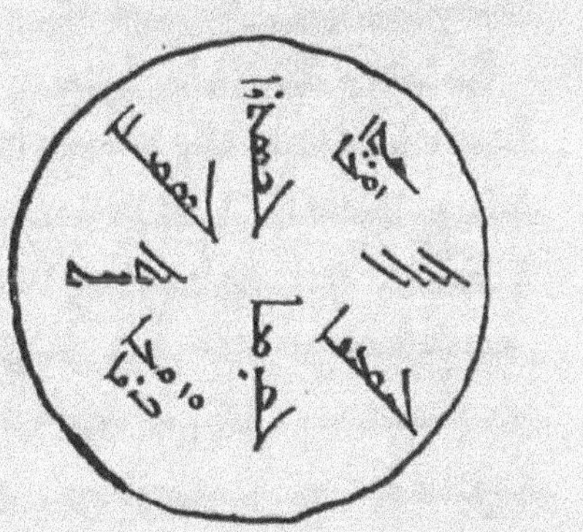

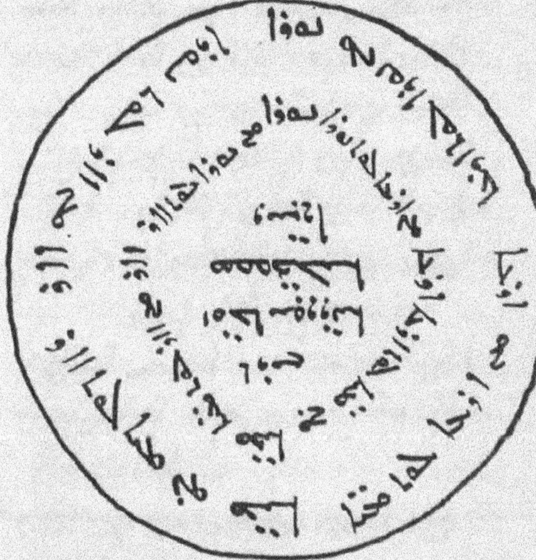

[illegible]

ܢ

ܒܢܝܢܐ ܕܓܒܠܐ ܘܡܬܒܪܢܐ ܠܓܫ̈ܡܐ ܠܗܿܝ ܐܘܣܝܐ
ܕܗܘ̈ܝܬܐ. ܘܡܬܢܛܪܐ ܒܗ ܕܚܝܐ ܘܡܬܡܪܝܬܐ.
ܡܬܢܫܡܐ ܡܬܡܠܝܐ ܕܐܐܪ ܡܥܠܐ ܠܗ ܐܘܪܐ
ܘܚܝܐ. ܘܢܘܗܪܐ ܕܪܘܚܐ ܚܝܘܬܢܝܬܐ. ܫܪܝܪܐ
ܒܗ ܐܘܪܐ ܕܓܘܫܡܐ ܦܓܪܢܝܐ. ܘܡܫܡ̈ܫܢܐ
ܒܗ ܡܫܡܫܢܬܐ ܕܡܬܢܛܪܬܐ ܘܡܬܚܠܡܐ ܠܗܘܢ
ܩܢܝܐ ܠܐܪ̈ܓܫܐ ܘܠܡܥܒܕܐ. ܘܚܝܐ ܡܬܚܙܝܐ
ܗܘ ܡܢ ܒܪܢܫܐ ܕܚܝܐ ܒܓܘܫܡܐ.
ܠܒܬܪ ܕܚܠܦܝ̈ܢ ܗܠܝܢ ܡܬܚܙܝܢ ܡܢ
ܚܝܐ. ܡܢ ܒܬܪ ܕܡܬܠܒܠܒܐ ܠܐ
ܡܬܡܠܟܢ ܗܘܘ ܠܡܣܬܒܪܢܘܬܗܘܢ ܒܗ
ܢܦܫܬܐ ܐܢܫܝܬܐ ܘܨܒܝܢ. ܚܝ̈ܠܐ ܕܝܢ
ܘܗܘܝܢ ܕܡܢ ܐܣܛܘܟܣ̈ܐ ܐܪܒܥܐ ܗܢܘܢ ܒܢܝܐ
ܒܪܢܫܐ ܡܢ ܕܡܫܟܚܝܢ ܒܥܘ̈ܡܩܐ ܐܣܪܢܐ
ܓܘܫܡܐ ܡܢ ܗܢܘܢ ܐܝܟ ܘܡܬܒܢܝܐ. ܒܓܘܗ
ܐܘܪܚܐ ܘܡܬܦܫܛܐ ܒܟܘܠܗܝܢ ܐܣܛܘܟܣ̈ܐ
ܐܪ̈ܒܥܐ. ܘܐܝܬܝܐ ܕܡܬܩܪܝܐ ܕܒܪܐ ܘܐܘܪܚܐ
ܦܪܝܫ ܡܢ ܚܝ̈ܐ. ܡܛܠ ܡܫܡܫܢܘܬܐ ܕܢܦܫܐ
ܒܡܠܟ. ܢܦܫܐ ܡܢܐ ܘܡܬܦܪܫܢܐ
ܒܥܒܕ̈ܐ ܕܐܪܚܐ ܡܢ ܚܝܐ ܡܥܠܝܢ ܒܝ
ܦܪ̈ܝܫܝܢ ܠܒܢܝܐ ܚܝܐ ܡܬܩܪܝܐ ܡܟܝܠ ܘܡܟܝܠ.
ܡܢܐ ܕܝܢ ܡܛܠ ܐܘܪ̈ܚܬܐ ܕܚܠܝܢ ܒܗܘܢ
ܒܗ ܕܒܗܘܢܐ ܗܘ ܕܐܝܬܝܗܿ: ܒܗ ܚܝܐ
ܡܬܚܙܝܐ ܐܝܟ. ܘܡܬܩܪܝܐ ܡܠܐ ܘܦܪܫܢܝ
ܠܡܬܚܙܝܢܘܬܗ ܕܒܗܘܢܐ ܚܘܒܐ. ܘܡܬܕܝܢܐ
ܒܗ ܚܝܘܬܐ ܢܦܫܝܐ ܘܚܝܐ ܗܘ̈ܝܢ. ܡܛܠ
ܡܬܩܪܝܐ ܕܗܘܢ. ܘܡܢ ܦܓܪܢܝܬܐ. ܗܘ ܕܝܢ
ܦܪܨܘܦܐ ܘܕܡܘܬܐ ܕܦܓܪܐ ܕܒܪܢܫܐ. ܡܢ ܕܡܬܬܩܢ
ܡܛܠ ܕܚܝ̈ܐ ܒܗܘܢ ܡܫܡܫܢܘܬܐ. ܠܟ ܠܐܦ̈ܝܗܘܢ
ܡܬܩܪܝܢ ܡܫܡܫܢܐ. ܘܐܪ̈ܓܫܐ ܒܟܠ

ܚܝ̈ܘܬܐ. ܘܐܦ ܗܘܐ ܐܣܛܘܟܣܐ ܒܟܠܗܿ.
ܡܛܠ ܗܘܐ ܚܝܘܬܢܝܐ. ܒܗܘܢܐ ܡܫܡܫܢ̈ܐ
ܡܢܐ. ܒܗ ܕܝܢ ܗܘ ܡܫܡܫܢܝܢ ܕܟܠ
ܚܝܘܬܐ ܐܝܟ ܕܐܝܬܝܗܘܢ. ܘܢܦܫܐ ܕܝܢ ܘܡܬܦܫܛܐ
ܘܡܬܒܪܐ. ܡܛܠ ܕܒܡܠܬܐ ܡܫܡܫܢܝܬܐ ܘܡܠܟܐ
ܒܗܘܢ ܚܝܐ ܡܫܡܫܢܐ ܗܘ ܒܗܘܢܐ ܡܟܝܠ

ܕܢܦܫܐ ܕܐܪ̈ܓܫܐ ܒܓܘܫܡܐ

ܡܬܚܫܚܢܐ ܒܝܢܬ ܚܕܘ̈ܡܐ ܗܘ ܘܗܘܦܟܐ
ܕܝܢ ܘܡܬܦܫܛܐ ܒܟܠܗ ܕܗܘܐ ܒܝܢܬ ܚܝܘܬܐ
ܘܗܘܦܟܐ ܘܡܬܦܫܛܐ ÷

ܒܟܠܗ ܕܝܢ ܕܡܫܡܫܢܘܬܗܘܢ. ܗܢܐ ܗܘ
ܒܝܢܬ ܚܝܘܬܐ. ܒܗ ܗܘ ܕܚܝܐ
ܒܡܠܟ ܡܬܚܙܝܐ ܕܗܘܝܢ ܐܝܟܢܐ ܕܡܬܦܪܫ
ܒܟܠܗ ܘܢܦܫ ܒܓܘܫܡܐ ܘܠܐ ܚܝܐ
ܢܦܫܐ ܡܛܠ ܕܡܫܡܫܢܘܬܐ ܡܫܡܫܢܘܬܐ. ܘܢܦܫܐ
ܕܝܢ ܒܗ ܗܘ ܕܢܦܫܐ ܡܫܡܫܢܘܬܐ ܠܟܠܗ
ܒܓܘܫܡܐ ܕܒܗܘܢܐ ܠܒܣܪܐ. ܘܡܬܢܛܪܐ ܒܗ
ܗܘ ܕܐܝܬܝܗܘܢ ܒܡܫܡܫܢܘܬܗ ܕܝܢ ܪܘܚܐ
ܚܝܘܬܐ ܕܟܠܗ ܐܦ ܐܣܛܘܟܣܐ ܒܗ
ܕܒܗܘܢܐ ܡܬܒܪܝܐ ܗܘ ܕܝܢ ܒܓܘܫܡܐ. ܒܡܫܡܫܢܐ
ܡܛܠ ܕܒܗ ܒܓܘܫܡܐ ܠܓܘܫܡܐ ܐܦܣܪ ܗܘܝܢ
ܗܘܐ ܒܐܪܐ ܒܪܢܫܐ. ܒܕ ܐܟܠܝܢ ܗܘܘ ܦܪܘܫܐ
ܠܐܪ̈ܒܥ ܗܘܘ ܡܢ ܢܦܫܐ. ܘܡܬܢܛܪܐ
ܒܓܘܫܡܐ. ܒܐܝܠܝܢ ܕܚܝܐ ܗܘ ܒܟܠܗ.
ܘܠܐ ܐܘܕ ܡܬܩܪܝܐ ܗܘܐ ܒܐܝܠܝܢ ܕܡܬܚܙܝܢ ܗܘ
ܘܠܐܘܕ. ܒܟܠ ܒܐܪܐ ܒܪܢܫܐ ܐܝܬܝܗܘܢ ܗܘܘ.
ܡܬܩܪܝܢ ܗܘܘ ܡܫܡܫܢܐ. ܡܛܠ ܕܪܘܚܐ ܗܘܐ
ܠܐ ܡܬܚܙܝܢ ܗܘܘ ܐܦ ܒܗ ܦܓܪܐ ܓܘܫܡܐ
ܕܒܗ ܕܝܢ ܕܡܫܡܫܢ ܘܡܢ ܐܣܛܘܟܣ̈ܐ ܐܪܒܥܐ
ܗܢܘܢ ܒܟܠܗ ܒܗ ܗܘ ܒܐܪܐ ܠܓܘܫܡܐ.

ܡܛܠ

ܘܢ

[illegible]

ܪܫܐ ܕܚܡܫܬܥܣܪ

[illegible]

ܪܫܐ ܕܫܬܬܥܣܪ

[illegible]

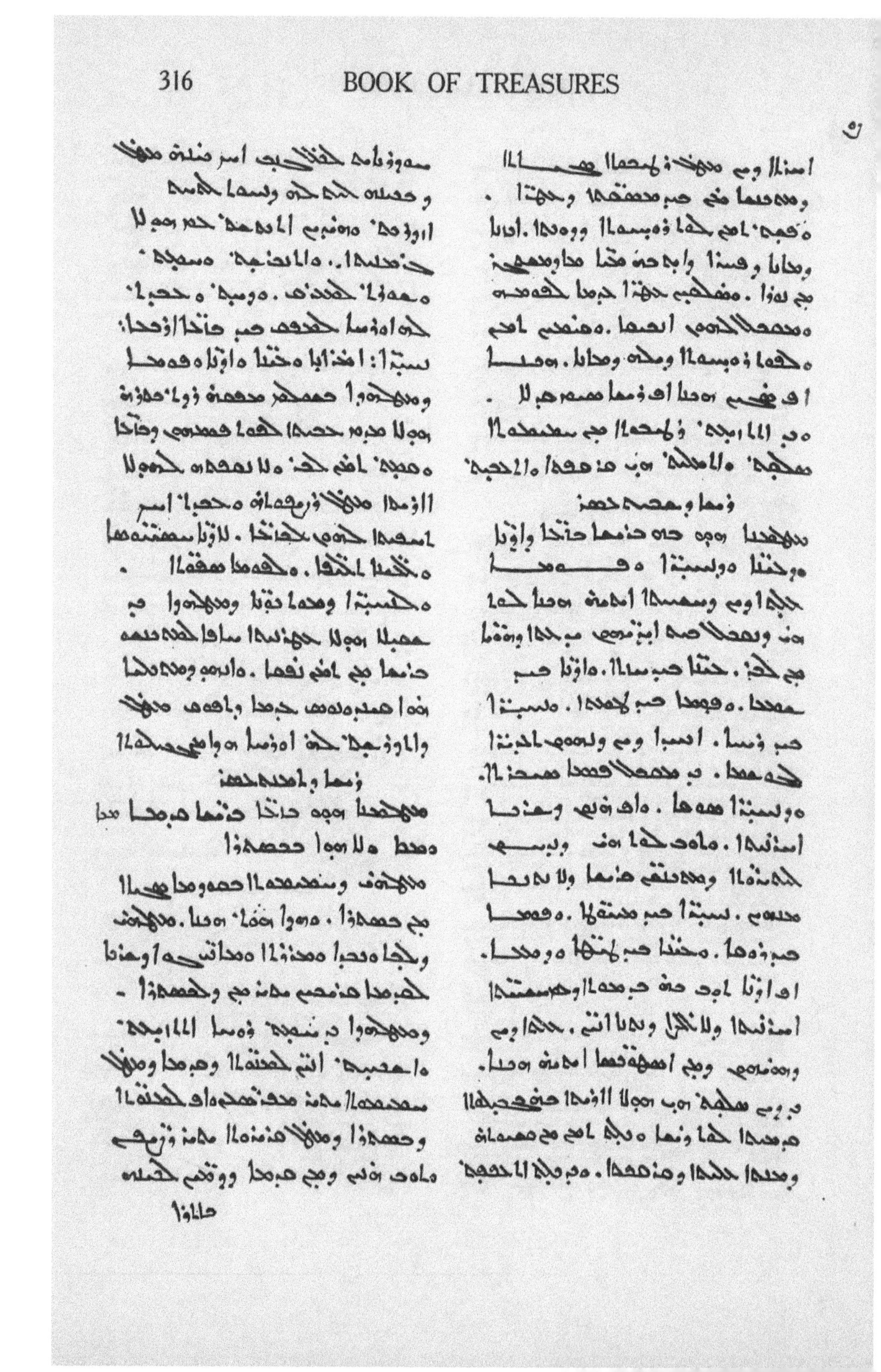

[illegible]

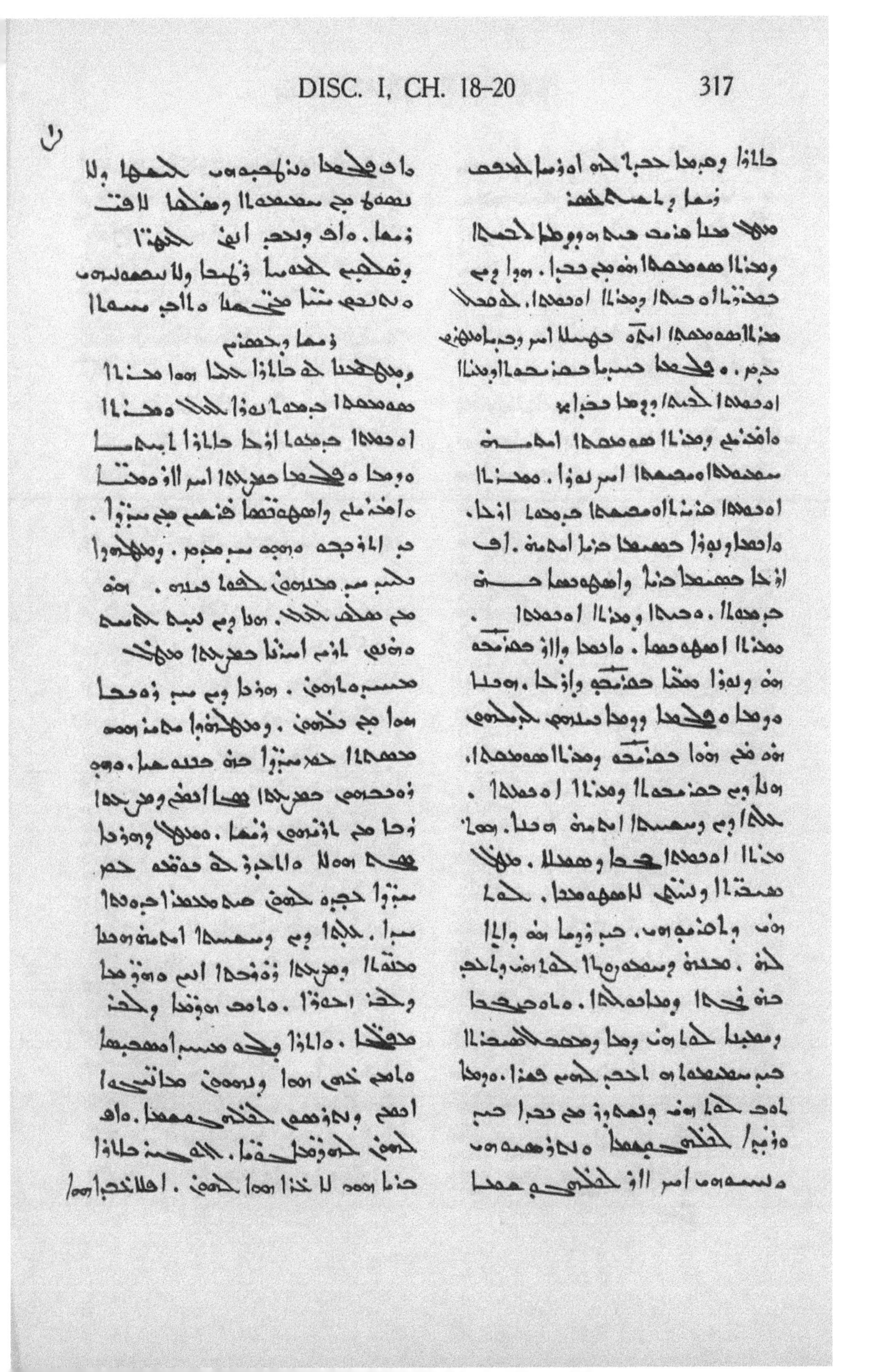

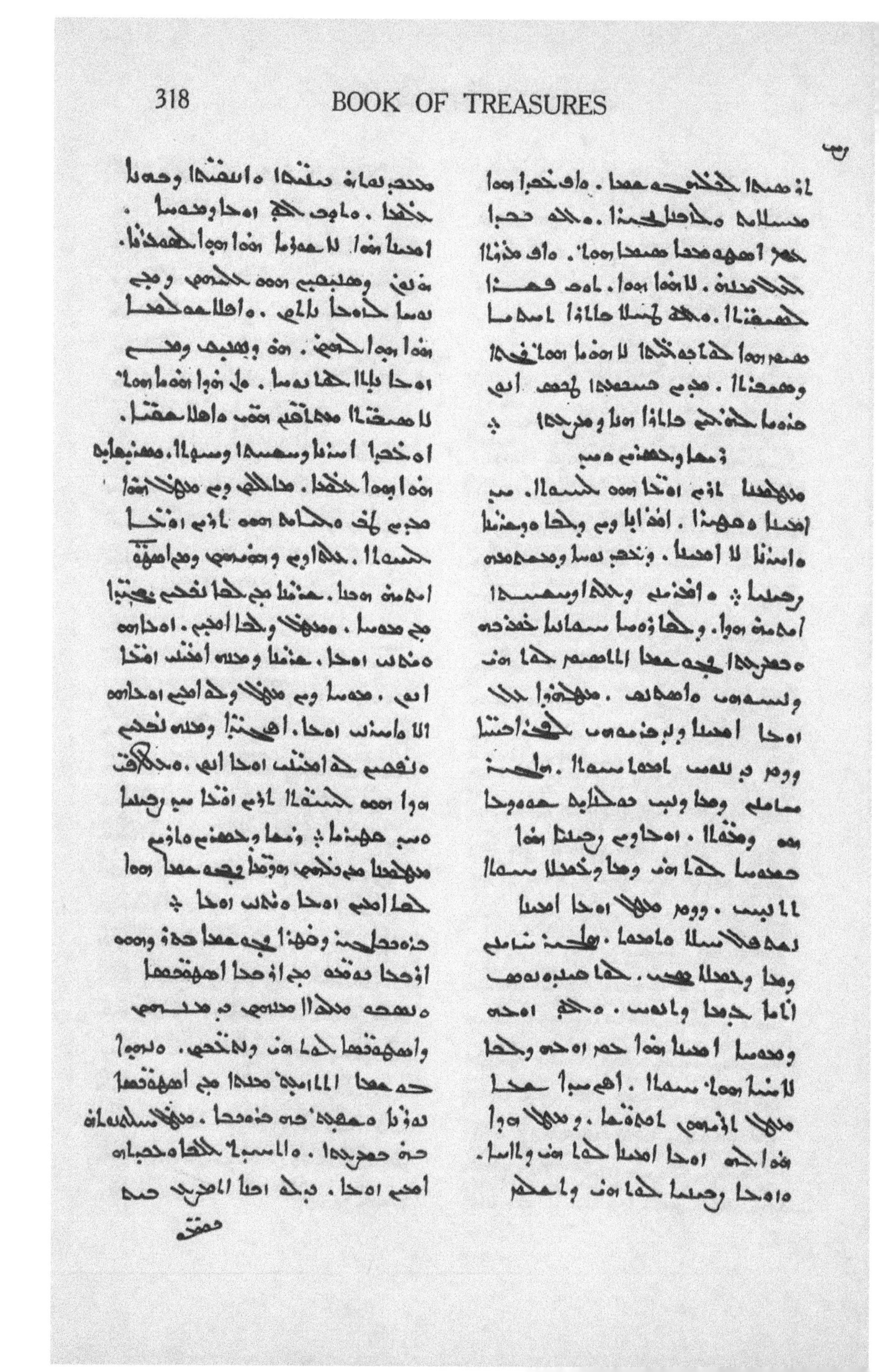

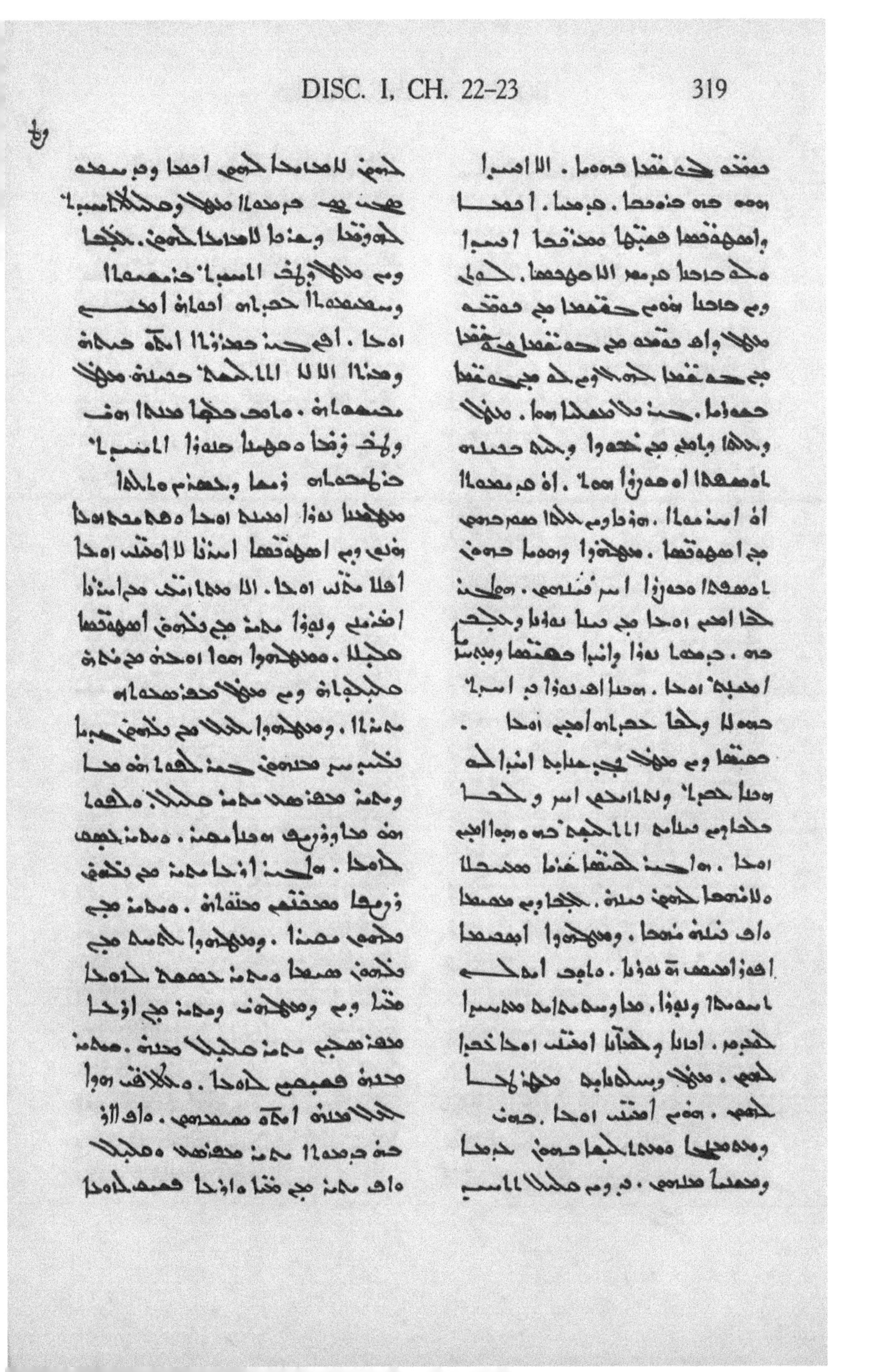

ܘܐܦ ܥܠ ܡܠܘܗܝ ܐܡܬ̈ܐ ܗܢܐ ܕܡܢ ܐܦ
ܗܘܐ ܕܝܢ ܡܢ ܢܟܝܦܘ ܐܣܛܘܟ̈ܣܐ ܡܟܝܠ
ܘܡܬܗܘܐ ܕܝܢ ܡܢ ܢܟܝܦܘ ܡܬܦܪܫܢܐ
ܡܬܪܫܡܢܘܬܐ ܘܡܬܚܙܝܢܐ ܘܗܢܐ ܕܝܢ
ܡܬܝܕܥ. ܗܘ ܐܝܟ ܗܘܐ ܕܡܢ ܢܦܫܗ.
ܗܘܝܘ ܐܡܝܢܐ ܗܘܐ. ܒܝܕ ܟܠܗ ܣܓܝ̈ܐܬܐ
ܡܛܠ ܐܣܛܘܟ̈ܣܐ ܐܚܪ̈ܢܐ ܕܟܠ ܐܝܟ.
ܡܬܬܝܗܒܐ ܡܛܠ. ܘܒܗܝܢ ܡܥܒܪܢܐ ܟܠܗ
ܐܬܬܠܝ ܡܢ ܗܘܢ. ܘܡܢܐ ܕܐܝܬ ܠܗ ܐܢܫ
ܗܘ ܡܛܠ ܕܒܗܘܢ ܡܬܦܪܫܢܘܬܐ ܐܝܬܝܗ ܟܠܗܝܢ
ܘܡܬܚܙܝܢܐ ܗܘܐ ܠܟܠܢܫ ܡܢ ܡܬܚܙܝܢܘܬ ܕܝܢ
ܡܢ ܡܬܦܪܫ. ܡܛܠ ܕܒܗܘܢ ܐܝܟ ܕܡܬܚܙܝܢ
ܠܗܕܐ. ܗܘ ܕܡܬܚܘܐ ܟܠܗܘܢ ܗܘܝܘ ܕܡܢ
ܡܢܗ. ܗܘ ܕܝܢ ܕܐܣܛܘܟ̈ܣܐ ܐܚܪ̈ܢܐ ܡܬܚܙܝܢ
ܡܬܚܙܝܢܐ. ܗܘ ܕܝܢ ܠܐ ܡܬܚܙܝܢܐ ܒܗܘܢ
ܟܠܗ ܕܗܘܐ ܐܘܪܐ ܡܫܬܡܫܢܐ. ܘܒܗܘܢ
ܗܪܡܢܐ. ܘܒܗܘܢ ܡܬܚܙܝܢܐ. ܘܒܗܘܢ ܕܐܝܬܝܗ
ܪܘܚܐ ܕܡܢ ܐܝܟ ܗܘܐ ܘܐܪ. ܗܘܐ ܕܡܢ ܠܐܬܪܐ
ܒܗܘܢ ܗܪܡܢܐ ܘܒܗܘܢ ܕܐܝܬܝܗ. ܡܛܠ ܕܗܘܐ
ܒܗܘܢ ܡܫܬܡܫܢܐ ܡܢ ܗܘ ܕܐܝܟ ܗܘܐ
ܡܢ ܕܢܗܘܢ ܡܬܚܙܝܢ ܗܘܐ. ܗܘܐ ܕܝܢ ܡܛܠ
ܗܘ ܕܠܐ ܡܬܚܙܝܢܐ ܡܫܬܡܫܐ. ܠܐ ܗܘܐ ܡܫܬܡܫܢܐ
ܡܬܚܙܝܢ ܕܟܠܐ ܠܗܘܢ. ܘܡܛܠ ܗܢܐ
ܢܦܫܗ ܗܘܐ. ܐܝܟ ܕܐܝܬ ܠܟܠܗܘܢ ܐܣܛܘܟ̈ܣܐ
ܐܚܪ̈ܢܐ ܗܘܐ. ܗܘ ܕܝܢ ܗܘܐ ܢܦܫܗ ܗܘܐ
ܡܛܠ ܕܟܠܗܘܢ ܐܝܬ ܘܐܝܟܢܐ. ܘܗܘܝܘ ܐܝܟ ܗܘܐ
ܕܗܘܐ. ܗܘ ܕܡܢ ܐܚܪ̈ܢܐ ܠܗܘܢ ܐܝܬ ܐܢܫ
ܕܡܢ ܐܝܟ ܗܘܐ ܐܡܝܢܐ ܗܘܐ ܡܛܠ ܕܟܠܗ
ܟܠ ܡܫܬܡܫܢܘܬܐ ܘܟܠܗ ܐܝܬ ܕܒܗ ܗܘܐ
ܕܟܠ ܗܘܐ ܐܝܟ ܡܛܠ. ܘܗܘܐ ܠܐ

ܡܬܚܙܝܢ ܕܒܗ ܟܠܗ ܐܣܛܘܟ̈ܣܐ
ܐܚܪ̈ܢܐ ܕܦܠܓܝܢ ܐܝܟܢܐ ܘܠܐܣܛܘܟ̈ܣܐ ܗܢܘܢ
ܐܚܪ̈ܢܐ ܗܘܐ ܕܒܡܬܚܙܝܢ ܘܐܪܟܐ
ܡܬܚܙܝܢܐ ܡܬܚܙܝܢ ܐܝܟ ܗܘܐ ܗܘ. ܠܐ ܡܬܚܙܝܢ
ܟܠܗ ܘܢܘܪܐ
ܘܐܦ ܕܡܢ ܘܡܝܐ ܘܡܢ ܡܬܚܙܝܢܘܬܐ ܕܗܘܐ
ܐܚܪ̈ܢܐ ܕܝܢ ܪܘܚܐ ܘܪܘܚܐ. ܐܠܐ ܕܡܢ
ܡܬܚܙܝܢܐ ܘܡܬܚܙܝܢ ܠܗܘܢ ܟܠܐ ܡܫܬܡܫܝܢ
ܠܬܚܘܡܗ ܗܘܐ ܡܫܬܡܫ ܠܗ ܡܬܚܙܝܢ ܡܛܠܗ
ܐܡܝܢܐ ܕܗܘܐ ܐܪܥܐ ܘܡܫܒܚ ܠܗ ܟܠ ܐܝܟ
ܐܘܦܝ ܐܝܟ ܡܕܡ. ܘܐܦ ܡܬܚܙܝܢ ܕܗܘܝܐ
ܢܦܫܐ ܒܗ ܟܠܗ. ܡܛܠ ܕܟܠܗ ܗܘܐ
ܘܢܦܠܐ ܗܘܐ ܘܠܗ ܘܡܫܬܡܫ ܟܠ ܗܘܐ
ܕܟܠܐ ܘܡܫܬܡܫ ܐܦܠܐ ܡܢ ܟܠܗܝܢ
ܡܢ ܕܡܢܐ ܟܠܗ ܢܦܫܐ. ܡܬܚܙܝܢ ܗܘ ܐܝܟ
ܕܟܠܗ. ܡܫܬܡܫܢܐ ܟܠܗ ܐܝܟ ܕܗܘܐ ܐܦܠܐ ܐܝܬ
ܡܕܡ ܡܫܬܪܪ ܗܘܐ ܪܘܚܐ ܒܝܕ ܡܫܬܡܫܢܐ
ܘܡܬܚܙܝܢ ܡܢܗ ܠܟܠܗ ܗܘܐ. ܡܬܚܙܝܢܐ
ܠܐ ܗܘܐ ܠܗ ܗܘܐ ܐܡܝܢܐ. ܒܗܘܢ ܘܡܫܬܡܫܢ
ܘܐܝܟܢܐ ܡܢ ܐܣܛܘܟ̈ܣܐ ܡܫܬܡܫܢ ܠܗ ܘܠܐ ܗܘܐ.
ܐܡܝܢܐ ܒܟܠ ܟܠܗܘܢ ܐܪ̈ܥܐ
ܗܘܐ. ܕܝܢ ܗܘܐ ܗܘ ܡܬܚܙܝܢ ܒܗ ܡܢ
ܕܡܢ ܗܘܐ ܡܫܬܡܫܢ ܕܢܗܘܐ ܟܠ ܡܫܬܡܫܢܐ
ܘܡܫܬܡܫܢܘܬܐ ܗܘ. ܐܝܟܢܐ ܕܡܢ ܐܪܥܐ
ܢܘܪܐ ܘܡܫܬܡܫܢ ܘܡܫܬܡܫܢ
ܡܬܚܙܝܢܐ ܕܟܠܗܘܢ ܡܫܬܡܫܢܐ ܐܝܟ ܘܡܫܬܡܫܢܐ
ܒܟܠܐ. ܐܦܠܐ ܘܐܘܪܐ ܡܫܬܡܫܢܐ ܡܛܠ
ܟܠܐ ܘܐܝܬܝܗܝܢ ܗܘܐ ܗܘܐ ܒܢܝܢܐ. ܘܡܛܠ
ܡܫܬܡܫܢ ܡܢ ܡܬܚܙܝܢܘܬܐ ܘܒܗܘܢ ✣ ✣ ✣ ✣
ܘܐܦܪܝܫ ܕܟܠܗ ܘܡܫܬܡܫܢܐ ܐܡܝܢܐ ܢܘܪܐ
ܡܫܬܡܫ

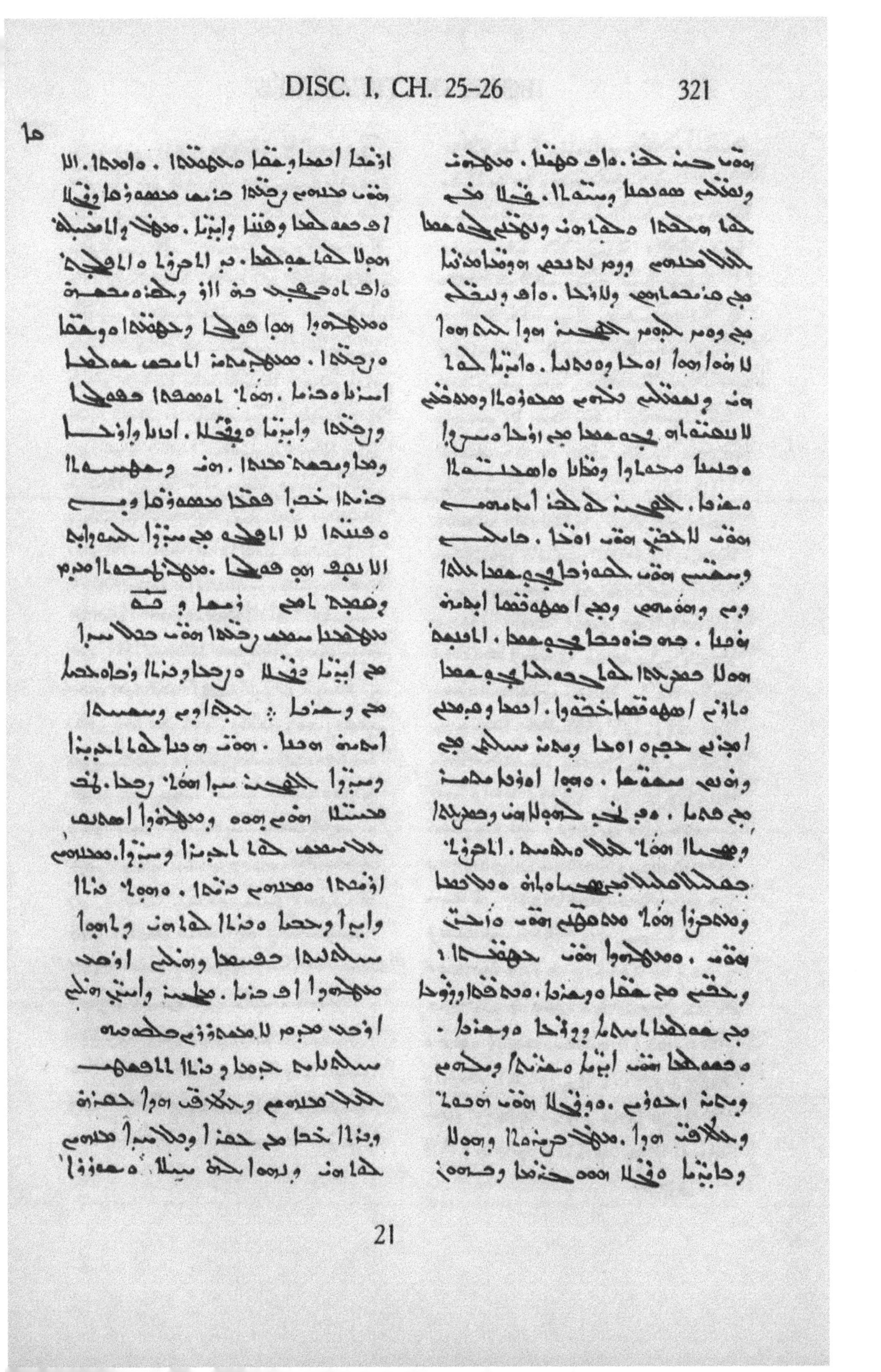

ܡܐ

[illegible]

[illegible]

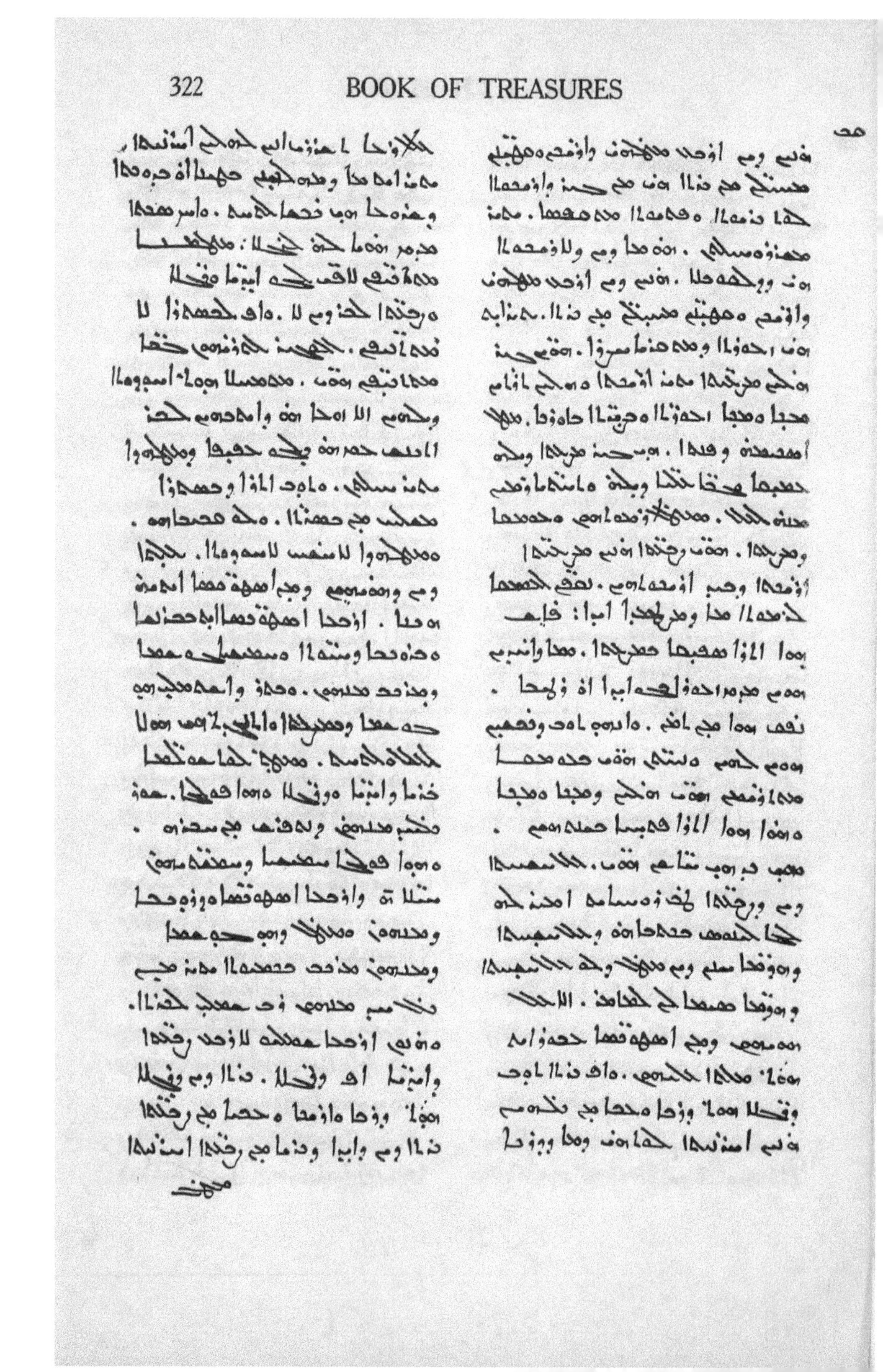

[illegible]

ܡܛܠ ܗܘܐ ܕܐܝܟܐ ܗܘ ܕܢܦܫܗ ܢܣܒ
ܒܓܘܫܡܐ ܡܢ ܗܝܘܠܐ ܡܬܪܐܡ. ܡܛܠ
ܡܬܪܐܡܘܬܗ ܕܗܝܘܠܐ ܘܫܘܫܠܐ ܘܐܒܢܘ
ܕܢܦܫܗ ܘܚܒܝܫܘܬܗ ܠܓܘܫܡܐ ܕܐܪܥܢܐ
ܘܚܝܐ ܒܕܘܟܬܐ ܗܝ ܕܩܕܡܐ ܕܦܓܪܐ
ܘܠܐܦ ܠܗ. ܡܫܡܫ ܠܗ ܘܡܬܪܐܡ. ܐܝܟ
ܕܒܗ ܗܘܐ ܘܗܘܠܐ ܐܦܝ. ܗܢܘ ܕܝܢ ܕܓܢܒܪܐ
ܘܠܐܦ ܠܗ. ܡܛܠ ܕܡܬܪܐܡܘܬܗ ܘܗܘܠܐ ܐܦܢ
ܠܐ ܢܦܩ ܘܡܬܦܪܫ ܡܢ ܦܓܪܐ ܡܬܪܐܡܘܬܗ
ܘܡܬܪ ܪܘܚܐ. ܡܛܠ ܪܘܚܐ ܕܗܘܝܐ.
ܦܓܪܐ ܕܝܢ ܘܐܝܟܢܐ. ܡܛܠ ܕܡܬܪܐܡ ܗܘܐ ܘܐܠܐ ܐܝܟܢܐ
ܒܐܝܪܐ ܡܢ ܦܓܪܐ ܡܛܠ ܕܫܡܫܐ ܐܝܟܐ
ܘܠܐ ܗܘܐ ܠܒܫܬܐ ܘܫܡܫܬܐ ܗܘ
ܗܟܢܐ ܠܐ ܢܫܒܘܩ ܡܢܗ ܐܠܗܐ ܐܝܟ
ܕܠܐ ܕܦܓܪܐ ܘܒܠܐ ܗܘܐ ܗܘܐ ܘܪܘܚܐ.
ܡܢ ܕܓܢܒܪܐ ܗܢܘ ܐܝܟܢܐ. ܗܢܘ ܕܓܢܒܪܐ
ܘܐܝܟܢܐ ܗܘܐ ܘܐܦܢ ܡܬܪ ܪܘܚܐ ܘܡܬܪ
ܡܢ ܓܢܒܪܐ. ܐܬܪܐ ܕܦܓܪܐ ܡܛܠ ܕܫܡܫܬܗ
ܐܝܟ ܐܡܪ ܡܬܪ ܡܢ ܕܦܓܪܐ ܕܝܢ ܐܝܟ
ܠܓܢܒܪܐ ܡܬܪ ܡܢ ܕܦܓܪܐ. ܠܐ ܗܘܐ ܐܦ
ܢܦܫܐ ܘܪܘܚܐ ܗܘܐ ܘܒܪܘܚ ܡܛܠ ܕܫܡܫܬܗ
ܘܐܝܟ ܠܓܪܐ ܡܢ ܘܒܓܢܒܪܐ ܘܢܦܫܐ ܘܗܘܝܐ
ܡܛܠ ܕܗܘܠܐ ܘܒܓܪܗ ܗܟܢܐ ܡܬܪܐܡ

ܪܫܐ ܕܟܚ

ܡܛܠ ܡܢܐ ܢܦܩ ܒܓܘܫܡܐ ܠܓܢܒܪܐ
ܡܬܚܙܐ ܒܫܡܫܬܐ ܕܝܢ ܠܐ ܀
ܐܠܐ ܐܠܗܐ ܒܓܘܫܡܐ ܗܢܘ ܕܟܝܢܐ
ܡܬܪ ܡܢ ܕܝܢܐ ܪܘܚܢܝ. ܘܗܘܐ ܗܢܘ.
ܗܟܢܐ ܡܛܠ ܘܒܪܘܚ ܡܬܪܐ ܕܓܘܫܡܐ
ܐܝܟ. ܡܛܠ ܪܘܚܢܝܐ ܘܡܬܪܐ. ܘܪܘܚܗ

ܘܒܬܪ ܕܝܢ ܡܫܡܫ ܒܪܘܚܐ ܘܦܓܪܐ
ܪܘܚܢܝܐ ܐܝܬ. ܡܛܠ ܘܪܘܚܐ ܡܬܚܙܐ ܠܗ
ܡܬܪܐܡܘܬܐ ܠܓܘܫܡܐ ܒܪܘܚܐ ܐܘ ܠܓܘܫܡܐ
ܘܒܪܘܚܐ ܐܝܟ ܐܠܗܐ ܘܗܘ ܘܠܐ ܕܦܓܪܐ
ܘܐܦ ܡܢ ܪܘܚܢܝܐ ܘܡܬܪܐܡܘܬܐ ܘܐܦ
ܓܢܒܪܐ ܡܢ ܓܘܫܡܐ ܘܡܢ ܕܝܢܐ ܡܬܪܐ
ܡܬܚܙܝܢ. ܗܢܐ ܡܫܡܫܐ. ܡܬܪ ܕܝܢ ܡܛܠ
ܡܫܡܫ ܘܪܘܚܐ. ܠܐ ܓܒܪܐ ܡܫܡܫܐ ܠܓܘܫܡܐ
ܐܘܫܡܫܬܐ. ܒܪܘܚܐ ܕܝܢ ܒܪܘܚ ܒܗ ܠܓܘܫܡܐ
ܡܛܠ ܘܪܘܚܢܝܐ ܘܫܡܫܬܐ ܕܓܢܒܪܐ ܒܗ ܦܓܪܐ
ܡܢ ܕܓܢܒܪܐ. ܡܬܚܙܝܢ ܪܘܚܢܝܐ ܠܫܡܫܬܐ
ܓܘܫܡܐ ܘܫܡܫܬܐ ܡܬܪܐ. ܡܛܠ ܗܘܐ
ܠܓܘܫܡܐ ܡܫܡܫ ܐܝܬ. ܠܓܢܒܪܐ ܕܝܢ ܠܐ
ܡܫܡܫܐ. ܡܛܠ ܕܒܪܘܚܐ ܡܢ ܪܘܚܢܝܐ
ܗܢܘܢ ܗܘܐ. ܘܐܦ ܟܠܐ ܠܓܢܒܪ. ܘܐܦ ܒ
ܓܢܒܪܐ ܗܘ ܕܒܟܠ ܓܘܫܡܐ ܘܫܡܫܐ ܀

ܪܫܐ ܕܟܛ

ܡܬܚܙܝܢܐ ܒܗ ܒܓܘܫܡܐ ܕܪܘܚܐ ܘܐܘܪܚܐ ܠܟܠ
ܗܘܐ ܪܫܐ ܠܟܠܐ ܘܫܘܡܠܝܐ ܘܪܘܚܐ ܘܐܘܪܚܐ
ܕܢܦܫܗ ܒܓܘ ܠܐ ܠܓܢܒܪܐ ܘܫܡܫ ܦܓܪܐ
ܒܡܬܚܙܐ ܠܡܬܚܙܝܢ ܡܬܪܡ ܀
ܓܢܒܪܐ ܕܝܢ ܘܡܫܡܫܢܐ ܐܡܬܪܐ ܗܘܐ. ܗܘܐ
ܪܫܐ ܠܟܠܐ ܡܛܠ ܕܗܘ ܐܡܬܪܐ ܡܫܡܫܐ
ܦܓܪܐ. ܘܡܫܡܫ ܗܢܝܢ ܗܘܝܐ. ܘܡܛܠ ܗܘܐ
ܡܫܡܫܐ ܕܗܘܐ ܗܘܐ. ܟܠ ܡܫܡܫ ܘܡܬܪܐܡ
ܡܬܪ ܡܢ ܢܦܫܗ ܗܘܝܐ ܕܒܓܘܫܡܐ.

ܡܒ

ܘܐܣܛܘܟܣܐ ܕܟܠ ܘܐܣܛܪܢܐ ܗܘ ܣܟܠܐ
ܡܢ ܕܐܡܪ ܐܣܛܘܟܣܐ ܘܣܕܪܐ ܡܬܬܠܝܢ
ܘܢܦܠܢ ܘܒܚܠܦ ܚܕܝܗܘܢ. ܐܠܐ ܡܬܕܒܪܝܢ
ܡܢ ܫܠܝܛܐ ܘܡܫܒܚܐ ܒܟܠ ܡܬܚܙܝܢܐ. ܘܡܫܡܫܢܐ
ܫܠܝܚܐ ܡܢܗܘܢ ܒܟܠ ܐܬܪܘܬܐ ܘܫܥܕܘܗܝ
ܘܡܬܚܙܝܢܐ ܡܫܡܠܝܢ ܐܣܛܘܟܣܐ ܒܡܠܬܗ. ܘܐܠܘ
ܒܪܐ ܕܡܐ ܡܢܗܘܢ ܡܬܕܒܪ ܗܘܐ. ܥܕܡܐ ܠܡܣ
ܡܫܬܟܚ ܗܘܐ ܡܢ ܐܘܟܠܘܣ ܕܫܡܝܐ ܘܒܗ
ܘܗܘܝܢ ܘܒܗ ܘܗܘܝܢ ܠܢܗܪܐ ܘܐܘܟܠܘܣ ܡܬܚܡܝܢ
ܗܘܐ ܕܡܝܐ ܘܡܛܠ ܐܣܛܘܟܣܐ ܒܐܬܪܗ ܘܠܢܗܝܪܘܬܗ
ܡܬܡܛܐ ܒܗܘܝܐ ܠܥܠܡܐ. ܐܘܦ ܗܘ ܐܣܛܘܟܣܐ
ܘܢܘܪܐ ܐܬܪܗ ܚܕ ܚܕ ܢܘܗܪܗ ܡܢ ܕܐܡܪܝܢ
ܢܬܬܠܐ. ܕܠܐ ܢܚܒܠ ܒܗ ܡܕܡ. ܐܣܛܘܟܣܐ
ܕܝܢ ܘܐܣܛܪܐ ܡܬܘܣܟܝܢ ܒܐܬܪܘܬܐ.
ܠܐ ܦܣܩ ܒܫܡܝܐ ܗܘܐ. ܒܕ ܐܡܪ ܡܢ ܕܐܡܪ
ܢܘܪܐ ܕܒܘܫܡܝܐ. ܠܐܬܪܗ ܗܘ ܕܡܬܥܒܕܝܢ
ܡܐ ܘܡܬܬܒܪ ܒܐܬܪܘܬܐ ܒܟܠ ܐܢܫܐ ܚܢܢ
ܗܘܝܐ. ܘܦܣܩܢ ܗܢܘܢ ܘܗܘܝܢ ܡܢ
ܒܫܡܝܐ ܘܩܪܒܝܢ ܒܗܘܢ. ܐܣܛܪܐ ܕܝܢ
ܡܬܚܙܝܢ ܒܗܘܢ ܘܗܘ ܫܡܫܐ ܗܘ ܘܣܗܪܐ
ܪܒܘܬܐ ܠܐܢܫܐ. ܘܨܒܝܢ ܒܗ ܐܝܟ ܕܢܚܙܐ
ܣܕܪܐ ܡܢ ܐܣܛܘܟܣܐ ܐܚܪܢܐ ܦܬܝܪ ܡܢ ܚܒܘܪܝܐ
ܡܬܚܙܝܢ ܗܘܘ ܘܠܐ ܦܣܩ ܒܫܡܝܐ ܒܟܠܗ ܘܡܢ
ܘܗܘܝܢ ܘܡܢ ܐܣܛܘܟܣܐ ܐܚܪܢܐ ܗܘܐ.
ܡܢ ܕܐܡܪܝܢ ܕܒܬܪ ܫܡܝܐ ܗܘܐ ܐܬܝܕܥܘܬܐ
ܒܗܘܢ. ܘܒܟܠܗܘܢ ܗܘܐ. ܡܬܝܕܥ ܕܢܗܘܐ
ܒܗܘܢ. ܒܕ ܗܘ ܐܬܝܕܥ ܠܐܦ ܟܠ
ܥܡ ܐܬܪܘܬܐ. ܘܡܫܬܟܚ ܗܘܐ ܠܐ ܚܡܝܡܐ
ܒܡܬܡܠܝܐ ܟܠ. ܘܒܠܐ ܐܝܬܘܗܝ ܒܗܝܢ
ܐܘ ܒܡܬܡܪܐ. ܐܘ ܒܗܘ ܡܢ ܒܥܠܡܗ ܗܘܐ

ܘܡܐ ܡܬܝܕܥܢܐ ܘܠܐ ܐܬܝ ܒܣܓܝ ܡܢ ܒܒܬܝ
ܡܬܬܠܝܢ ܘܡܢ ܕܢܘܪܐ ܗܘܘܬܐ. ܐܬܪܐ ܕܝܢ
ܘܐܘܪܐ ܒܠܐ ܗܘܝܐ. ܗܘܡܢܐ ܗܘ ܐܬܪܐ
ܐܠܐ ܕܚܒܝܟ ܒܥܡܪܐ. ܒܥܡܪܐ ܕܠܐ ܡܫܚܠܦ
ܠܐܘܪܐ ܠܐܦ ܐܪܥܐ. ܠܐܝܟܐ ܕܚܒܝܟ.
ܗܘܠܐ. ܐܦ ܒܐܬܪܐ ܐܚܪܢܐ. ܒܕ ܐܬܥܒܕ
ܒܗ ܗܘܐ ܘܐܬܚܙܝ ܠܐܦ ܟܠ. ܐܬܪܐ
ܐܘ ܕܚܒܝܫ ܗܘܐ ܒܡܥܒܕܬܐ. ܘܡܬܚܫܒ
ܡܢ ܒܗ ܡܬܒܠܒ. ܐܝܟ ܕܗܘ ܒܪܝܬܐ
ܘܡܛܠ ܒܗܘܠܐ ܘܐܘܪܐ ܒܟܠ. ܘܒܐܬܪܐ
ܐܝܟ ܘܠܐ ܐܬܥܒܕ ܘܢܫܘܦ ܒܗܝܢ ܡܬܝܕܥ
ܡܢ ܕܠܐ ܒܝܬ ܡܬܚܙܝܢ ܘܗܘܝܐ
ܡܬܬܠܝܢ ܘܢܘܪܝܬܐ ܘܒܗ ܘܠܢܫ ܒܗܘܢ
ܗܘܐ. ܘܡܢ ܒܪܝܐ ܘܡܛܠ ܐܣܛܪܐ. ܐܬܪܐ
ܒܪ ܘܡܬܐ ܘܡܕܝܢܬܐ ܡܐ ܕܢܣܒ ܟܠ
ܘܗܘܝܐ ܘܐܝܟ ܒܗ ܡܬܐ. ܘܢܫܒܐ ܗܘ
ܢܘܪܐ ܚܝܐ ܒܗܘܢ ܘܡܬܐ. ܘܡܢܐ
ܕܥܡܪܐ ܠܗ ܡܢ ܕܘܟܬܐ ܐܘ ܡܬܚܠܦ ܒܗ ܐܬܪܐ
ܡܠܝܢ ܘܡܬܐܠܦ ܡܢ ܗܘ ܕܐܡܪܝܢ ܘܡܬܐ. ܘܐܡܪ
ܘܠܒܗ ܢܘܗܪ ܒܗܘܢ. ܘܐܝܟ ܘܡܢ ܐܡܪ
ܢܘܗܪ ܒܗܡܘܢ ܟܠ. ܡܬܝܕܥܐ ܡܢ ܡܬܐ
ܘܒܪ ܡܬܥܒܕܐ ܐܣܛܘܟܣܐ ܟܠ
ܡܢ ܡܬܐ. ܗܘܝܐ ܐܦ ܐܘ ܒܟܠ ܡܢ ܐܬܪܐ
ܐܝܟܢܐ ܒܗ ܡܬܐ ܒܗ ܐܬܪܐ ܒܟܠܐ. ܘܒܢ
ܕܝܢܐ ܘܡܛܠ ܐܣܛܪܐ. ܘܗܝܢܐ ܘܐܦ
ܒܡܬܐ ܡܐ ܕܢܦܩ ܠܡܬܐ ܘܡܢܗܐ
ܐܘ. ܒܡܬܐ ܕܝܢܐ ܡܬܬܠܐ ܢܚܒ ܐܣܛܪܐ
ܡܬܝܕܥܐ. ܐܝܟ ܒܐܬܪܐ ܐܚܪܢܐ. ܘܡܐ ܒܐܬܪܐ
ܒܟܠ ܐܝܟܢ ܘܐܘ ܡܬܩܢ ܒܗ ܡܢ ܡܢ
ܗܘܪܡܐ ܒܗ ܡܬܐ. ܒܕ ܡܢ ܒܗ ܡܢ ܟܠ

ܒܢ

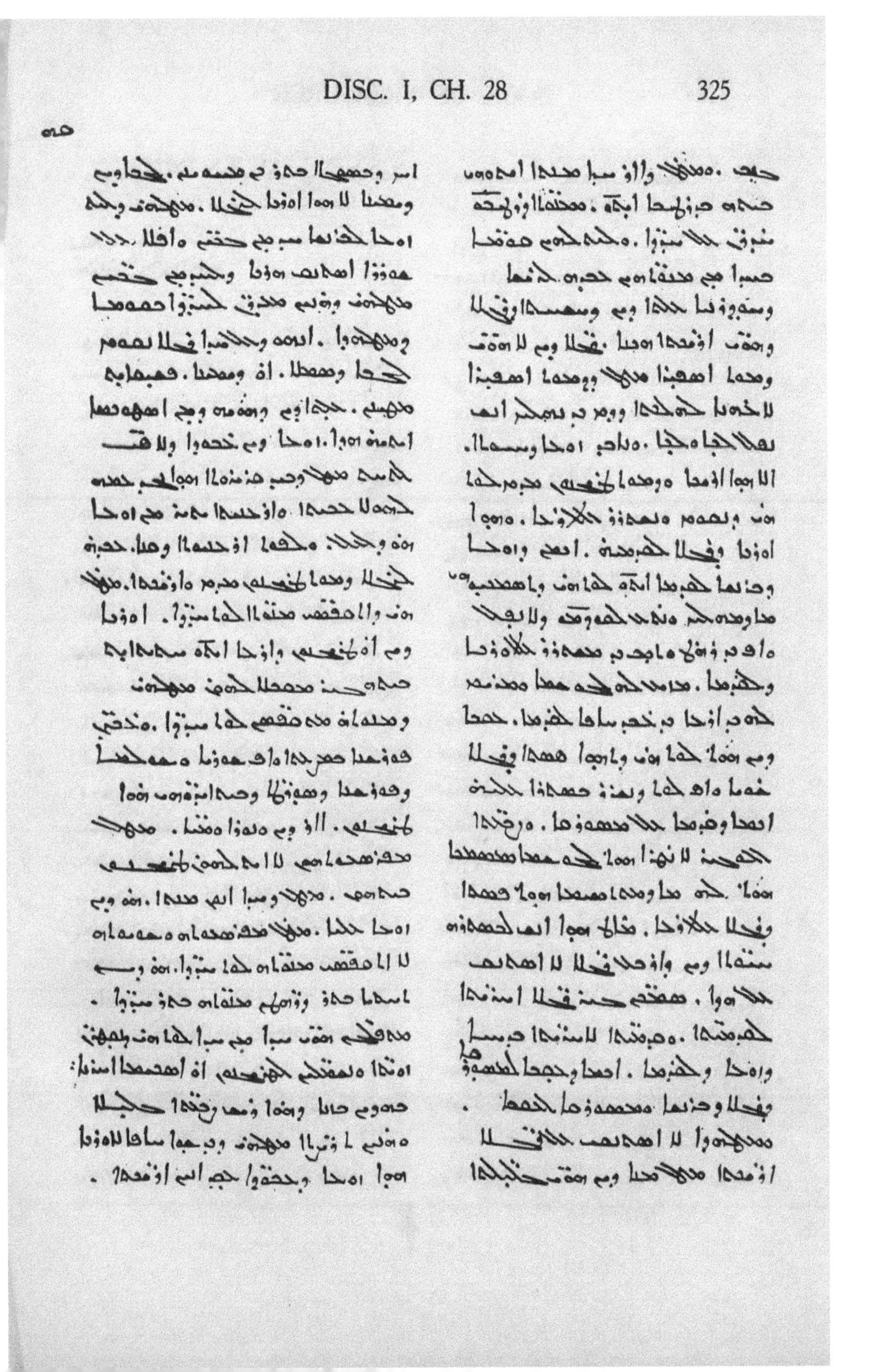

[illegible]

[illegible]

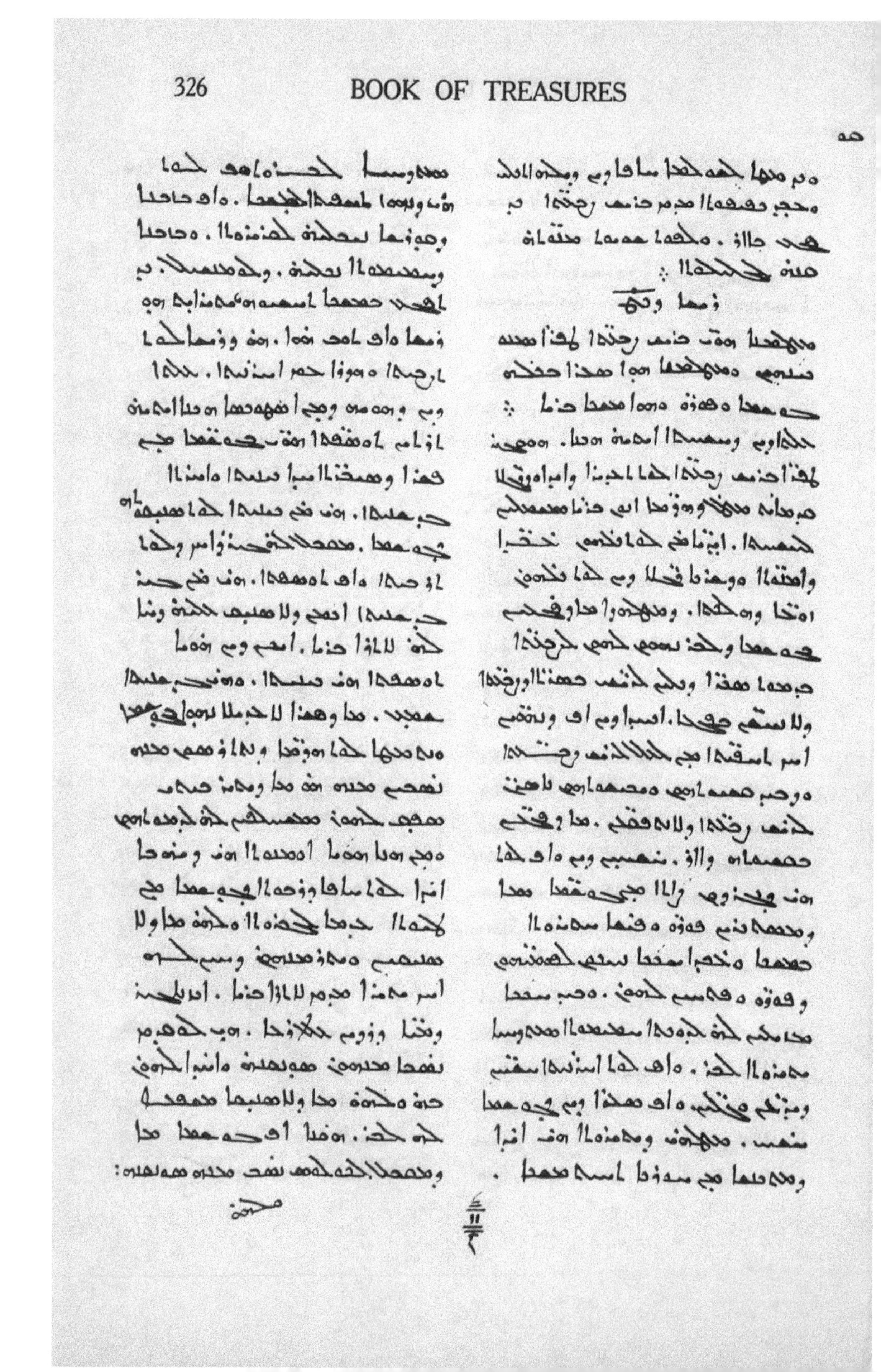

ܗ

ܡܛܠ ܗܢܐ ܠܐ ܡܫܒܩ ܕܢܐܬܝܗ ܠܐܬܪܐ
ܒܢܐ ܡܢ ܗܕܐ ܗܘܐ ܡܬܩܪܝܐ . ܘܠܥܡܪܢ
ܐܝܟ ܠܟܠܗܘܢ ܐܢܫܐ ܘܚܝܘܬܐ . ܘܐܝܟ
ܕܡܬܦܪܫܝܢ ܕܗܐ ܠܫܘܠܡܬܐ ܢܦܫܐ
ܠܐܘܪܚܐ . ܠܐܬܪܐ ܕܝܢ ܦܫܝܛ ܕܠܗ
ܠܟܠܗܘܢ ܡܫܠܡܗ . ܐܫܬܠܡܬ ܢܟܪܝܬ
ܘܡܠܟܐ ܐܦܝܢ . ܐܝܟܐ ܕܒܠܬܐ ܘܫܡܝܐ
ܐܘ ܢܟܝܪ ܡܢ ܗܘܢܐ ܕܒܢܝܢܫܐ ܕܗܝܐ
ܠܡܬܪܘܬܐ ܦܝܫ ܕܚܝܘܬܐ . ܐܫܬܘܕܥܬܐ
ܠܡܠܟܐ ܕܡܢ ܦܪܐ . ܫܡܝܐ ܕܟܠܐ
ܠܡܫܬܥܠܝܘ ܠܡܫܬܪܝܐ ܕܝܗܒ ܟܕ ܦܪܐ .
ܘܐܦܢ ܗܘܪܐ ܕܝܢ ܠܦܪܐ ܘܡܢ ܡܢ ܗܢܐ
ܟܠܐ . ܗܘ ܕܢܦܘܩ ܠܗ ܟܠܗ ܕܟܠܗ
ܠܦܘܪܬܐ . ܗܢܘ ܕܝܢ ܗܘ ܫܘܝܐ ܠܦܓܪܐ
ܒܘܪܟܬܐ . ܘܡܫܒܚܐ ܠܡܪܢ ܠܒܘܪܟܬܐ
ܗܘܬ ܠܗ ܡܬܪܘܬܐ ܕܡܫܘܕܥܐ ܡܢ
ܫܠܝܛܘܬܗܘܢ ܕܗܘܪܡܐ . ܡܛܠ ܕܡܫܠܡ
ܗܘ ܕܒܟܠܚܕܐ ܠܟܠܚܕ ܦܝܣܐ ܐܡܪܝܢ
ܡܬܪܘܬܐ ܘܡܢ ܕܗܘܝܢ ܒܗܘܪܢܐ ܕܐܢܫ
ܐܢܫ ܡܬܪܘܬܐ ܕܝܢ ܗܘܝܐ ܡܢ ܗܘ ܦܓܪܐ
ܪܘܚܢܐ ܡܬܪܝܢܐ ܡܫܬܪܪܐ . ܡܛܠ
ܕܢܦܫܐ ܠܫܠܝܛܘܬܗܘܢ ܕܦܓܪܐ ܐܬܬܐ
ܟܕ ܗܘܐ ܪܒܝܐ . ܐܦ ܡܛܠ ܕܐܬܐܪ
ܕܢܐ ܗܘܐ ܡܬܪ ܐܬܦܫܛ ܘܗܘܐ ܠܗ
ܦܓܪܐ ܡܫܬܪܪܐ ܕܢܦܫܗ ܕܗܘܐ ܦܓܪܐ
ܕܝܢ ܗܘܐ ܡܬܚܙܐ ܕܪܘܚܢܬܐ ܗܘܬ ܡܬܕܒܪܐ
ܠܗ ܡܫܬܐ ܘܡܫܬܪܪܐ ܡܢ ܫܡܫܘܬܐ ܗܘ
ܘܡܬܦܫܛܐ ܡܢ ܐܬܪܐ ܠܟܠܗ ܐܬܪ
ܕܢܐ . ܘܟܕ ܦܓܪܐ ܡܬܚܙܐ ܡܠܟܗ
ܠܪܘܚܢܬܐ ܡܫܬܪܪܐ ܘܡܢܗܪܐ . ܘܡܫܬܠܛ

ܡܢ ܟܠܝ ܡܬܪܘܬܐ ܠܗܝ ܡܢܗܪܐ ܘܟܠܝ ܒܘܪܪܐ
ܒܫܡܝܐ ܡܫܬܡܠܝܢ ܘܫܦܝܪܐ . ܗܢܘ
ܡܬܪܘܬܐ ܕܫܠܝܛܘܬܐ . ܡܛܠ ܗܪܟܐ ܘܡܢܝܢ
ܕܢܦܫ ܒܦܓܪ . ܘܠܗܘܢ ܒܫܡܝܐ ܘܠܟܠܗܘܢ
ܗܢܘܢ ܡܫܬܢܝܐ . ܦܓܪܗ ܕܝܢ ܗܘܐ ܟܠܗ
ܒܟܠܗܐ ܘܡܫܬܪܪ ܒܫܘܠܡܐ ܗܘ ܟܠܐ
ܕ ܐܬܚܕܝ ܒܡܬܪܘܬܐ ܘܫܡܝܐ ܠܗܐ ܡܢܗܪܐ
ܘܐܬܪܘܡ ܐܦ ܡܢ ܟܠܗ ܫܘܪܒܗܘܢ ܕܠܗܐ
ܡܢܗܪܐ ܠܦܓܪܗ ܡܢ ܟܠܗܘܢ ܠܐܦܝ ܠܟܠ
ܘܟܕ ܚܙܘܪܐ ܟܠܗ ܡܫܬܥܒܕܢܐ ܗܘ ܕܚܫܚܬܐ
ܘܡܚܝܠܐ . ܐܦ ܦܓܪܗ ܕܐܬܦܪܣ ܟܠܗ
ܒܡܬܪܘܬܐ . ܠܦܪܐ ܕܝܢ ܗܘܐ ܡܢ ܡܬܪܘܬܐ
ܕܓܪܡܐ ܘܫܡܫܢܘܬܐ ܡܫܬܪܪܐ . ܕ
ܐܬܘܚܕ ܠܟܢܫ ܪܘܚܢܬܐ ܡܬܠܒܫܢܘܬܐ
ܘܪܡܐ ܠܦܓܪܐ . ܕܠܗܘܢ ܐܝܟ ܒܗܘܢ ܦܓܪ
ܫܡܝܬܐ ܕܒܫܘܠܡ ܦܓܪܐ ܘܪܘܚܢܬܐ ܠܐ
ܦܓܪܐ ܕܒܫܘܠܡܐ ܡܫܬܡܠܝܐ ܘܪܦܝܢ
ܠܓܒܪܐ ܡܚܝܠܐ ܐܦ ܠܚܫܝܫܘܬܐ
ܟܠܗ ܫܘܝܐ ܘܡܫܬܘܐ ܠܐ ܡܢܗ ܐܬܪ
ܐܝܟ ܦܓܪܐ . ܐܠܐ ܡܬܠ ܐܢܫ ܡܫܬܠܛܬܐ
ܓܒܪܐ ܘܫܦܝܪܘܬܐ . ܠܐܘ ܕܝܢ ܦܓܪ
ܦܓܪܐ ܡܫܬܠܛ ܫܠܝܛܐ ܘܟܠܗ
ܘܐܘܦ ܡܫܬܠܛ ܕܡܬܪܘܬܐ ܐܝܟ ܕܠܐ ܐܬܪ
ܒܫܡܝܐ ܘܦܓܪܐ ܐܪܥܐ . ܗܘ ܕܝܢ ܒܡܬܪܘܬܐ
ܐܦܠܐ ܫܡܝܐ ܗܘܐ ܦܓܪܐ ܡܫܬܠܛ ܕܪܘܚܢܬܐ
ܗܘ ܡܬܪܬܐ ܕܐܬܘܚܕ . ܗܘܐ ܕܝܢ ܫܡܝܐ
ܟܕ ܗܘ ܕܪܘܚܢܬܐ ܗܘ ܕܐܬܐܪ ܟܕ ܗܘܐ
ܕܫܡܝܐ ܐܬܐܪ ܠܗ ܡܬܪ ܠܫܡܫܬܐ ܘܟܠܗ
ܘܐܬܚܫܒ ܘܐܬܚܕܝ ܡܬܪܘܬܗ ܟܠܗ
ܡܢܗܪܐ . ܘܐܬܚܫܒ ܗܘܐ ܫܡܝܐ . ܐܝܟܢܐ

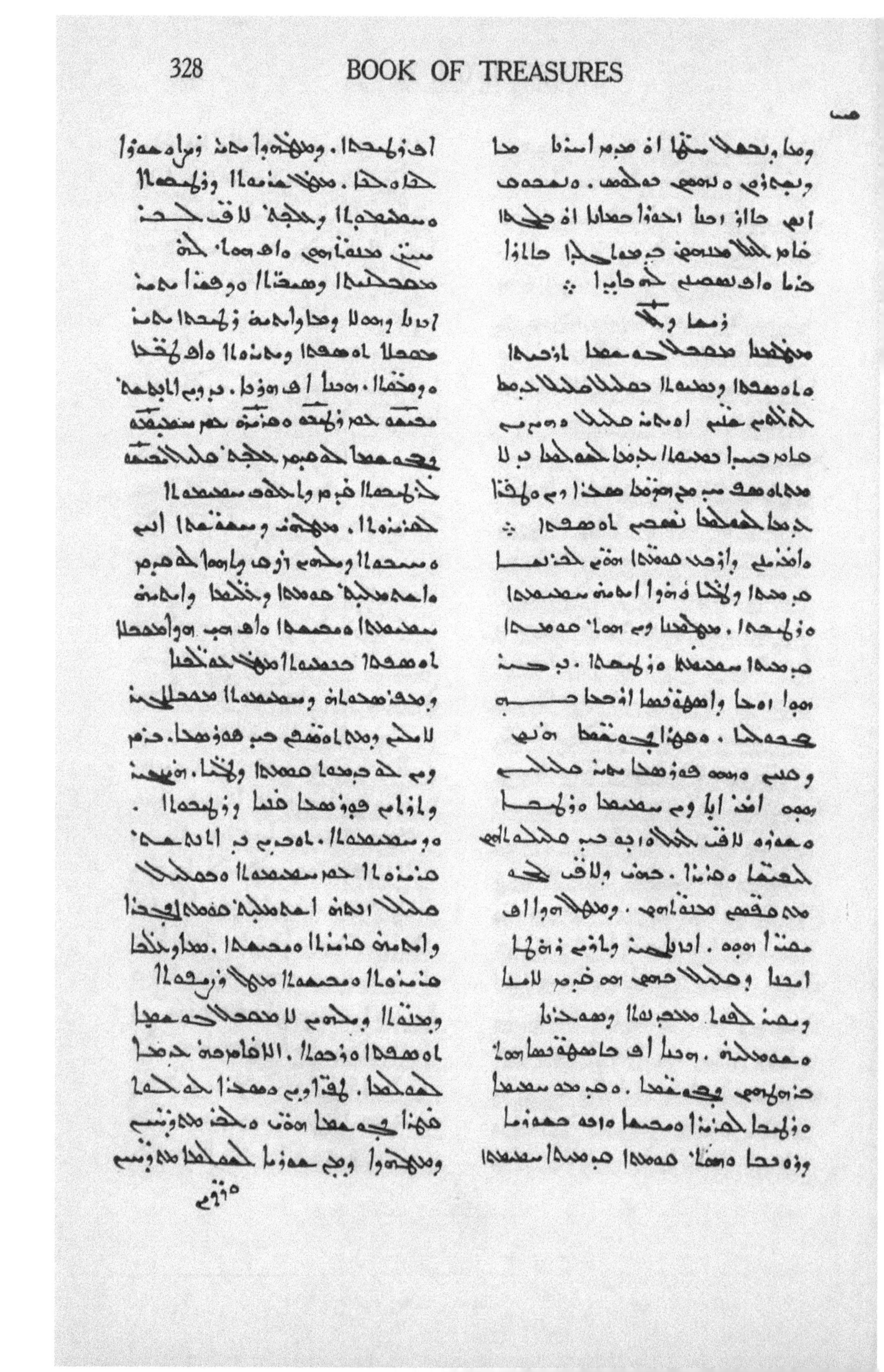

[illegible]

ܘܪܘܡܐ ܘܠܦܩܕ ܗܟܢ ܕܝܢ ܡܢ ܡܬܡܪܢܘܬܐ
ܗܢܐ ܚܕ ܥܡܠܐ ܕܠܐܢܫܐ ܗܘ ܡܬܐ ܚܣܝܒ
ܡܢ ܐܒܗܬܐ. ܐܝܟܐ ܕܡܬܡܪܢܘܬܐ ܗܢܐ ܢܨܚܐ
ܗܢܐ ܕܝܢ ܥܡܠܐ ܚܡܝܐ ܘܚܡܝܐ ܚܡܝܪܘܬܐ
ܘܐܦ ܠܐܢܫܐ ܠܐܢܫܘܬܐ. ܗܢܐ ܕܝܢ
ܚܝܠܐ ܚܡܝܐ ܐܘ ܚܕ ܚܝܠܐ ܡܬܚܫܚ ܗܘܐ ܠܐ
ܡܬܚܫܚܢܐ ܗܘ ܚܝܠܐ ܕܝܢ ܕܚܫܚܐ
ܐܡܬܝ ܗܘܝܐ. ܗܘܢܐ ܐܡܬܝ ܟܠܗ ܚܝܠܐ
ܐܘ ܚܫܚܐ ܘܐܢܫܐ ܘܚܘܪܢܐ. ܘܡܬܚܫܚ
ܗܘܐ. ܗܢܐ ܗܢܐ ܟܠܗ ܚܫܚܐ ܕܒܝܫܐ
ܟܕ ܚܝܐ ܒܟܢܘܬܐ ܡܚܫܚܬܐ ܕܒܝܫܐ
ܠܚܫܚܘܬܐ. ܐܠܐ ܟܠ ܘܐܚܝܕ ܡܬܚܫܚܢܘܬܐ
ܘܐܢܫܐ. ܟܕ ܠܐ ܚܫܝܚ ܡܬܚܫܚ ܗܘܐ ܡܢ
ܐܦ ܡܢ ܗܢܐ ܗܘܐ ܟܠ ܚܫܚܐ ܕܚܝܐ ܘܢܘܬܐ
ܘܠܐ ܚܝܘܬܐ ܗܢܐ ܗܘܐ

ܪܫܐ ܕܠܐ

ܡܬܚܫܚܢܐ ܚܫܚ ܐܝܟܢܐ ܕܚܫܚܬܐ ܠܐ ܢܫܒܘܩ
ܘܡܫܚܢܘܬܐ ܐܚܝܕ ܗܘܐ ܡܬܚܫܚܢܐ ܡܚܫܚܢܐ
ܚܘܫܚܐ ܘܗܘܪܡܐ ܥܢܝܐ ܐܚܝܕ ܟܠܗ ܚܫܚܢܐ.
ܗܟܢܐ ܕܝܢ ܠܐ ܗܘܐ ܗܢܐ ܐܠܐ ܚܫܚܬܐ ܡܫܚܢܐ
ܡܚܫܚ ܡܬܚܫܚ ܗܘ ܚܫܚܘܬܐ ܠܚܫܚܬܐ ܘܐܦ
ܟܠܗ ܡܚܫܐ ܘܟܠܗ ܚܫܚܬܐ ܚܫܚܬܐ.
ܘܐܫܬܡܫ ܕܐܬܚܫܚ ܚܫܚܬܐ ܐܝܟ ܟܠܗ ܚܫܚܐ
ܘܚܫܚܘܬܐ ܚܕܐ ܡܢ ܟܠܗܝܢ. ܘܐܚܝܕܐ ܡܢ
ܐܚܫܚܢܐ ܘܚܫܚܬܐ. ܗܢܐ ܡܢ ܕܡܢ ܟܠܗܝܢ
ܠܐܢܫܐ ܘܐܚܫܚܬܐ ܐܘ ܚܘܪܢܐ. ܡܬܚܫܚܢܘܬܐ
ܠܚܫܚܢ ܡܢ ܚܫܚܘܬܐ ܚܫܚܬܐ
ܘܠܐ ܘܡܚܫܚ. ܡܟܝܠ ܕܚܫܚ ܡܢ
ܘܠܗ ܚܫܚܬܐ ܟܠܗ. ܐܝܟ ܕܝܢ ܘܐܦ ܗܘ
ܡܠܐ ܕܡܫܚܢ ܡܫܚܠܐ ܚܫܚ ܟܠܗܘܢ

ܘܚܘܒ ܕܢܦܫܐ. ܚܡܝܠܐ ܡܫܠܡܢܘܬܐ ܕܡܫܝܚܐ
ܠܟܠܗܘܢ. ܗܢܐ ܘܡܫܚܢ ܗܘܝܐ ܘܚܢܢ
ܗܟܢܐ ܚܫܚܬܐ ܘܐܢܫܐ ܠܟܠܗ ܕܝܢ ܡܫܚܢܐ
ܚܝܠ ܗܘ ܕܝܢ ܗܘܘ. ܘܟܠܗ ܟܠܗ ܗܘ ܐܚܝܐ
ܚܫܚܐ ܘܐܢܫܐ. ܚܫܚܢ ܟܠܗܘܢ ܠܟܠܗܘܢ
ܚܫܚܐ ܕܠܐ ܚܘܪܢ ܚܝܠ ܗܘ ܘܟܕ ܗܘ.
ܗܢܐ ܐܦ ܠܟܠܗܘܢ ܕܟܠ ܡܚܫܚ ܚܫܚܬܐ
ܘܠܐ ܦܠܚܝܢ ܕܟܠܗܘܢ ܡܬܚܫܚܝܢ ܚܫܚܢ
ܕܟܠܗ ܚܫܚܬܗܘܢ. ܘܗܘܐ ܕܟܠ ܡܚܫܚܐ
ܡܫܚܢܐ ܚܫܚ ܘܟܕ ܐܚܝܕ ܘܚܫܚ ܠܚܫܚܢܐ
ܚܫܚܢܐ ܠܟܠܗܘܢ. ܡܫܚܢܘܬܐ ܡܚܫܚܢ ܚܫܚܐ
ܘܟܕ ܚܫ ܠܟܠܗܘܢ ܘܡܫܚ. ܐܡܬܝ
ܐܦ ܐܚܝܕ ܠܟܠ ܚܫܚܐ ܘܚܫܚܢ ܐܚܝܕ
ܘܠܐ ܐܬܚܫܚ ܡܫܚܢ. ܟܕ ܗܘ ܕܝܢ ܡܫܚܢ ܚܫܚܐ
ܡܫܚܠܐ ܡܫܚ ܢܫܚܢ ܚܫܚܐ ܘܚܫܚܢܢ
ܕܡܢ ܐܘܟܠܐ ܐܫܚܬܢܐ ܚܫܚܐ ܚܫܚܢܐ
ܡܫܚܢ ܢܫܚܢܝܢ ܚܫܚܬܐ ܘܚܫܚܢܘܢ ܐܘܚܢܐ
ܘܡܫܚܢܘܬܐ. ܟܠܗ ܗܢܐ ܘܡܢ ܕܐܚܝܕ ܟܠܗ
ܪܘܚܐ ܘܡܫܚܢܘܬܐ ܡܫܚܠܦܬܐ ܘܚܫܚܝܢ
ܡܢ ܚܫܚܢܐ ܘܚܫܚܬܐ ܚܫ ܐܘܚܫܚܢ
ܘܐܫܚܢܬܐ ܚܫܚܬܐ ܐܝܬܝܗ ܘܚܫܚܬܐ
ܘܚܫܚܬܐ ܘܚܫܚܐ. ܟܕ ܡܫܚܢܘܢ ܘܐܬܚܫܚ
ܡܢ ܐܚܫܚܢܐ ܚܫܚܬܐ ܐܡܬܝ ܐܚܫܐ
ܒܟܠܗ ܚܫܚܐ ܘܠܐ ܚܫܚ. ܡܫܚܢ ܘܚܫܚܢ
ܡܢ ܡܫܚܢܐ ܚܫ ܐܘܚܐ ܘܐܫܚܬܐ ܚܫܚܬܐ
ܘܡܫܚܢܐ ܒܟܠܗ ܚܫܚܐ ܘܚܫܚܬܐ ܐܡܬܝ
ܐܘܚܐ ܒܟܠܗ ܚܫܚܐ ܘܠܐ ܚܫܚ. ܡܫܚ ܘܚܫܚ
ܡܢ ܡܫܚܢܐ ܘܚܫܚܢ ܡܫܚܢ ܘܢܫܚܢܘܢ
ܡܢ ܚܫܚܢܐ ܚܫ ܐܫܚܬܐ. ܗܢܐ ܐܘ
ܡܫܚܢܐ. ܚܫܚܬܐ ܐܡܬܝ ܡܫܚܢܐ ܒܟܠܗ ܚܫܚܐ

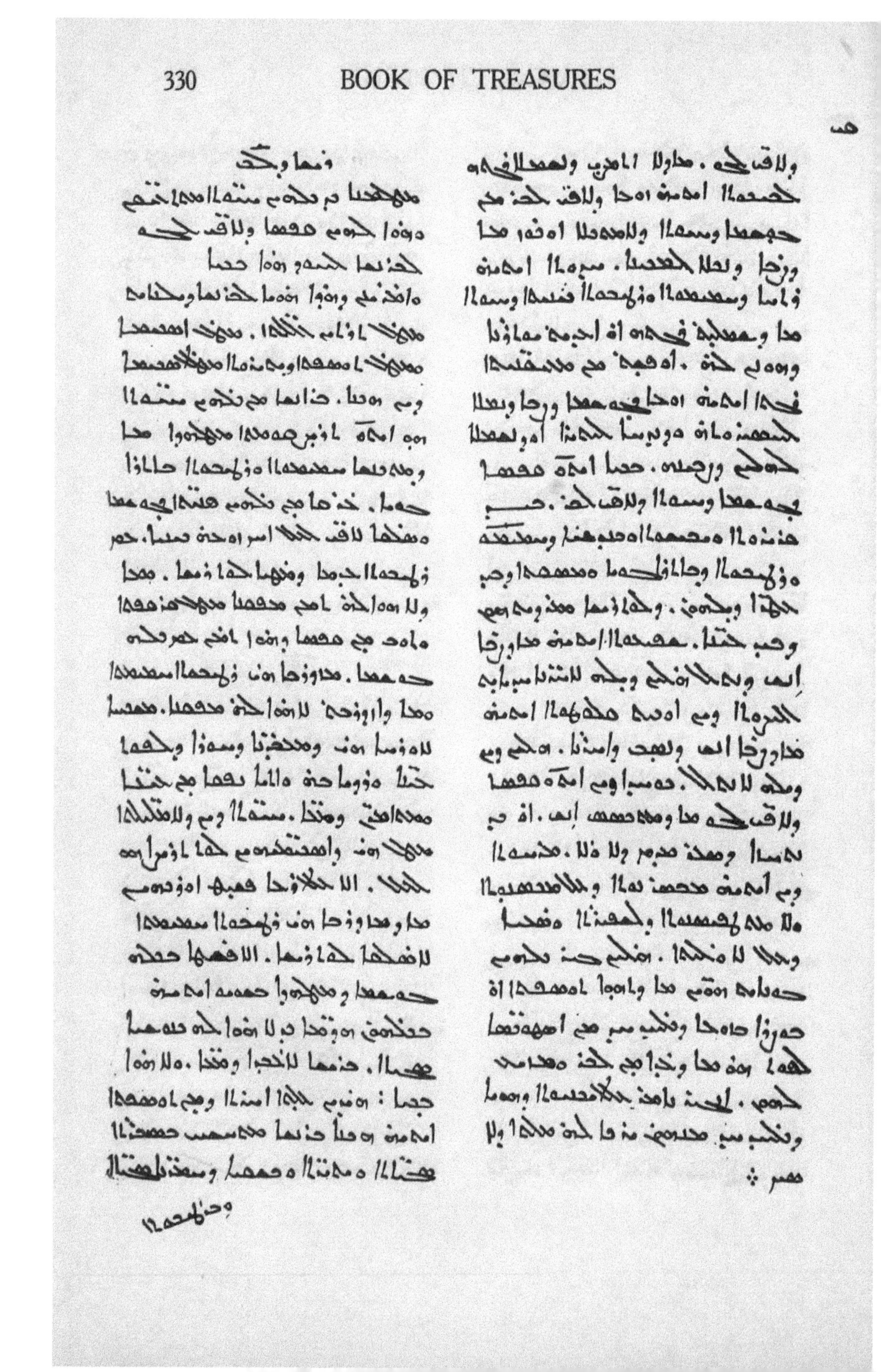

[illegible]

[illegible]

ܗܐ

ܘܒܡܫܡܫܢܘܬܐ ܕܩܘܕܫܐ ܘܐܟܬܒܐ ܗܘ ܗܕܐ
ܘܠܐ ܡܬܡܫܟܢ ܒܗ ܡܫܝܚܘܬܐ ܐ ܡܢ ܠܒܫܐ
ܘܡܫܬܠܡ ܗܘܐ ܠܗ ܡܫܡܫܢܐ ܡܬܡܛܐ ܒܗ ܡܫܡܫܢܘ
ܘܕܡܫܡܫܢܘܬܐ ܐܚܪܬܐ ܕܡܢ ܘܥܠܡܐ ܘܕܡܫܡܫܢܐ
ܘܦܠܚܐ ܠܐܠܗܐ. ܡܛܠ ܕܡܬܕܪܟܐ
ܡܢ ܡܫܡܫܢܐ ܘܐܘܟܠ ܘܠܐ ܦܓܪܢܝܐ ܡܫܡܫܢܐ
ܘܟܕ ܡܢ ܗܪܟܐ ܡܫܡܫܢܘܬܐ. ܦܠܚܝܢ ܡܫܡܫܢܐ
ܠܐܠܗܐ. ܡܫܬܠܡ ܗܘܐ ܟܕ ܡܢܗܪܐ ܘܟܝܢܐ
ܪܘܚܐ ܗܝ ܡܫܡܫܢܘܬܐ ܒܗ ܟܕ ܡܢܗܪܐ ܘܐܪܙܐ
ܐܘ ܦܘܡܐ ܐܘ ܘܠܫܢܐ. ܐܡܪܝܢ ܕܡܢ
ܟܠܗܝܢ ܪܘܚܐ ܡܫܡܫܢܘܬܐ ܡܢ ܪܘܚܐ ܕܡܢ
ܟܢܫܐ ܕܟܢܫܐ. ܘܡܫܬܠܡ. ܡܫܬܡܫܢܘܬܐ
ܘܥܠܝܐ ܡܫܡܫܢܘܬܐ ܗܝ ܕܡܫܬܠܡ ܒܗܘܢ
ܘܡܢ ܚܝܐ ܕܡܫܡܫܢܘܬܐ ܠܦܠܚܐ ܚܝܠܐ
ܘܐܪܙܐ ܟܠܗ ܟܢܫܐ. ܘܐܝܟ ܟܢܫܐ ܡܢ
ܡܢ ܐܪܙܐ ܘܦܘܡܐ ܘܠܫܢܐ ܡܫܡܫܢܝܢ
ܡܫܬܡܫܢ ܗܘܐ ܠܐܦܘܬܗܘܢ ܡܫܡܫܢܐ ܘܟܪܝܐ
ܢܦܫܝܢ ܒܗ ܒܦܘܡܐ. ܘܐܡܪ ܐܦ
ܒܪܘܚܐ ܗܘ ܡܢܗܝ. ܟܕ ܕܝܢ ܕܟܠܐ ܟܢܫܐ
ܠܡܫܡܫ ܪܘܚܐ ܗܘ ܡܫܡܫܢܘܬܐ ܐܝܟ ܕܡܫܬܡܫ
ܟܢܫܐ ܡܫܡܫܢ ܟܠܗ ܗܘ ܡܫܡܫܢ ܕܟܠܐ
ܒܗ ܕܡܫܡܫܢܘܬܐ ܟܠ ܦܘܡܐ ܘܠܫܢܐ. ܘܠܦܘ
ܟܕ ܡܫܡܫܢܝܢ ܪܘܚܐ ܘܗܘܐ ܡܛܠ ܕܟܠܗ
ܐܝܬ ܠܗ ܟܢܫܐ. ܟܠܗ ܐܪܙܐ ܕܡܢ ܦܓܪ ܠܐ ܐܢܫ
ܡܫܬܡܫܢ ܗܘܐ ܕܝܢܐ ܘܡܫܡܫܢ ܕܡܫܬܠܡ
ܒܪܘܚܐ ܗܝ ܡܫܡܫܢܘܬܐ ܡܢ ܡܫܡܫܢܐ. ܗܟܢ
ܕܝܢ ܕܟܠܐ ܡܫܡܫܢܐ ܡܫܬܠܡ ܡܫܡܫܢܘܬܐ. ܗܘ
ܘܡܫܬܡܫ ܡܢ ܡܫܡܫܢܘܬܐ. ܡܛܠ ܕܐܝܬ ܒܗ
ܟܠܗ ܪܘܚܐ ܘܐܝܬ ܒܗ. ܡܫܬܡܫܢܐ ܕܡܢ
ܗܢܐ ܡܫܡܫܢܐ ܠܐ ܦܠܚܝܢ ܟܠܗ. ܡܫܬܡܫܢ

ܢܦܫܝܢ ܘܡܫܡܫܢ ܟܠܐ. ܘܐܡܪܝܢ ܕܡܢܐ
ܘܡܫܡܫܢܐ ܟܠܗ ܟܢܫܐ ܡܢ ܪܘܚܐ ܘܡܫܡܫܢܘܬܐ .
ܘܡܛܠ ܡܫܡܫܢܐ ܠܡܫܡܫܢܘܬܐ ܘܡܫܡܫܢܘܬܐ ܠܐܦܝ
ܟܠܗ. ܡܫܡܫܢ ܡܫܡܫܢܐ ܟܠܗ ܕܝܢ ܟܢܫܐ
ܟܠܐ ܪܘܚܐ. ܡܛܠ ܡܫܡܫܢܘܬܗ ܠܐ ܡܫܡܫܢܐ
ܠܗ ܕܐܝܬܘܗܝ ܟܕ ܡܢܗ. ܘܐܦܠܐ ܐܝܬ ܟܕ
ܢܦܫܬܐ ܡܫܡܫܢܐ ܠܗ ܕܟܕ ܪܘܚܐ ܡܫܡܫܢܐ
ܢܦܫܘܢ ܘܡܫܡܫܢ ܗܘܐ ܟܠܗ ܘܡܫܡܫܢ
ܟܠܐ. ܡܛܠ ܕܡܢ ܪܘܚܐ ܡܫܡܫܢܘܬܐ ܬܫܡܫ
ܡܫܡܫܢܘܬܐ ܘܡܫܡܫܢܐ ܘܡܫܡܫܢ ܟܠܗ ܗܘ
ܘܠܐ ܦܘܡܐ ܠܦܘܡܐ. ܗܢܝܢ ܟܢܫܝܢ
ܡܫܡܫܢ ܘܚܝܐ ܘܪܘܚܐ ܡܫܡܫܢܘܬܐ. ܐܡܪܝܢ
ܘܟܕ ܦܘܡܐ ܘܠܫܢܐ ܡܫܡܫܢܐ ܘܡܫܡܫܢܘܢ
ܟܠ ܡܫܡܫܢܐ. ܐܦܘܢ ܡܫܡܫܢ ܡܫܡܫܢܘܬܐ
ܡܫܡܫܢܐ ܘܐܦ ܡܫܡܫܢ ܡܫܡܫܢܐ ܠܡܫܡܫܢܘܬܐ
ܡܫܡܫܢܐ. ܡܫܬܡܫ ܘܡܫܡܫܢܐ ܡܫܡܫܢܘܬܐ
ܘܪܘܚܢܘܬܐ ܟܠܗ ܠܗܘܢ ܟܕ ܡܫܡܫܢܐ ܡܫܡܫܢܘܬܗ
ܒܪܘܚܐ ܟܕ ܡܫܡܫܢܘܬܐ ܘܡܫܡܫ ܡܫܡܫܢܐ
ܘܡܫܡܫܢ ܠܐ ܬܪܬܝܢ ܟܠܗ

ܪܝܫܐ ܕ ܠܓ

ܥܠ ܡܫܡܫܢܐ ܕܪܘܚܐ ܘܟܠܗ ܘܡܫܡܫܢܝܢ
ܡܢ ܐܘܪܚܐ ܕܡܫܡܫܢܘܬܐ ܘܐܡܪܝܢ ܗܟܢ ܡܢ
ܐܡܫܡܫܢܐ ❖ ܥܠ ܡܫܡܫܢܘܬܐ ܕܡܢ
ܕܟܠܗ ܘܡܫܡܫܢܐ ܘܡܫܡܫܢܘܬܐ ܘܡܫܡܫܢܐ ܡܫܡܫܢܘܬܐ
ܕܪܘܚܐ ܕܡܫܡܫܢܐ ܘܡܫܡܫܢܘܬܐ ܘܡܫܡܫܢ
ܡܫܡܫܢܘܬܐ ܡܢ ܕܡܫܡܫܢܐ ܘܡܫܡܫܢ ܠܟܠܗ
ܚܕ ܡܢ ܐܘܪܚܐ ܕܡܫܡܫܢܘܬܐ ܐܦ ܒܗ ܘܡܫܡܫ
ܠܟܠܗܘܢ ܡܫܡܫܢܐ ܦܓܪܗ. ܐܠܐ ܡܫܡܫ
ܡܢ ܡܫܡܫܢܐ ܡܫܡܫܢܘܬܐ ܠܟܠܗܘܢ ܘܡܫܡܫܢܘܬܗܘܢ
ܐܘ ܕܡܫܡܫܢܘܬܐ ܠܡܫܡܫܢܐ: ܘܡܫܡܫ ܚܕ ܡܢ ܐܘܪܚܐ

ܫܠ

ܡܬܬܪܝܢܘܬܐ ܕܚܝܘܬܐ ܗܢܐ. ܘܐܦ ܥܒܕܝܢ
ܘܡܬܚܙܝܐ ܐܝܟ ܦܪܨܘܦܐ. ܘܒܗ ܡܬܬܪܝܢܘܬܐ
ܒܟܠܚܕ ܡܢ ܐܪ̈ܒܥܐ ܡܢ ܐܝܬܝܐ ܘܡܢܗܘܢ
ܘܐܣܛܘܟܣ̈ܐ ܐܝܬܝܗܘܢ. ܘܡܫܘܕܥ ܠܢ ܥܠܗ
ܗܘܐ ܥܒܘܕܘܬܐ ܕܒܪܝܬܐ ܡܢ ܡܬܠ ܕܟܝܢ
ܡܫܘܕܥܐ ܥܠ ܦܫܝܛܘܬܐ. ܘܡܢ ܪܘܚܢܝܘܬܐ
ܡܫܡܫܢ ܐܦܐ ܘܐܦ ܪܘܚܢܝܘܬܐ ܕܫܡܫܐ
ܒܗ ܒܪܝܘܬܐ ܥܡ ܡܫܡܫܢܘܬܐ. ܘܪܕܝܐ ܐܣܝܐ
ܡܫܘܕܥܝܢ ܘܡܢ ܐܣܛܘܟܣ̈ܐ ܐܝܟ. ܐܡܟܝܠ
ܥܕܢܐ ܗܘܐ̈ ܣܓܝܐܐ ܘܦܫܝܛܐ ܘܕܘܟܝܬܐ ܘܟܝܢܐ
ܒܒܪܝܬܐ. ܐܦ ܒܥܘܗܕܢܐ ܘܒܟܠܟܠ. ܘܠܐ ܫܪܝܐ
ܒܬܪܝܢ ܡܫܘܕܥ ܐܣܛܘܟܣ̈ܐ ܘܥܠܬܐ ܕܡܢ
ܒܗ ܠܗܘܢ ܒܬܪܝܢ ܚܕܐ ܗܘܘ ܕܝ ܗܘܐ.
ܗܘܝܢ ܡܢ ܠܥܠ ܘܥܒܕܐ ܗܘܐ ܐܝܟ ܠܡܦ
ܠܐ ܡܫܘܕܥܢܐ ܕܚܕܐ ܥܠ ܘܠܚܝܠ ܠܥܠܡܠܡ ܣܓܝ.
ܠܡܥܝ ܠܡܫܡܫ. ܠܡܥܒܕ ܠܡܫܘܕܥܘ ܐܣܘܪܝܢܗ
ܚܕܐ ܘܡܢ ܡܫܒܚ ܕܐܝܬܝܗ ܗܘ ܠܐܘܪܐ ܡܫܒܚܝܢ
ܢܚܬ ܫܒܝ ܣܥܫܥܘܬܐ. ܡܫܒܚܝܢ ܫܒܝ
ܦܪܝܫܬܐ ܘܬܪ̈ܬܝ. ܘܬܒܥ ܠܫܡܫܐ ܘܠܝܠܝܐ
ܘܫܡܫܐ ܘܒܣܬܐ. ܥܡ ܗܢܝ ܐܣܛܘܟܣ̈ܐ
ܐܬܕܝܢ ܒܗ ܗܘܕܐ ܠܥܠܡ ܠܥܠܡܝܢ ܦܪܘܫܐ
ܘܡܫܬܟܚܐ. ܘܡܢ ܡܫܘܕܥܢܐ ܩܢܝ ܗܘ ܗܠܝܢ
ܟܬܒ ܐܘܠܐ ܘܣܒ ܗܘ ܐܠܗܐ ܕܠܐܘܪ ܟܠ
ܡܫܚ ܡܫܒܚܐ. ܐܡܪܐ ܘܡܠܠ ܡܫܘܕܥܐ
ܠܐܘܪܐ ܡܫܒܚܐ. ܡܫܒܚ ܐܦ ܡܬܐܡܪ
ܘܐܦ ܕܡܬܐܡܪ ܐܦ ܡܫܒܚ ܡܫܘܕܥܢܐ
ܘܡܢ ܕܝ ܡܬܐܡܪ ܐܦ ܡܫܒܚ ܡܫܘܕܥܢܘܬܐ.
ܠܥܠܡ ܦܬܝ ܡܢܘܬܗ ܥܘܠܡܐ. ܐܦ ܡܫܝܢܐ
ܐܦ ܡܫܘܕܥܐ ܠܐܘܪܐ. ܒܥܠܡܝܢ ܡܢ ܡܬܐܡܪܝܢ
ܒܪܝܬܐ ܘܗܦܟܝܢ ܒܟܠܗ ܘܡܫܝܢܐ

ܗܦܟܐ ܫܡܝܐ. ܘܢܗܪܐ ܕܡܫܒܚ ܠܡܫܡܫܢܘܬܐ
ܟܫܦܝܪܐ ܒܪܝܐ. ܒܪܝܗ ܫܡܫܐ ܢܗܝܪܐ ܘ
ܚܕ ܡܢ ܬܪܝܢ ܡܢ ܚܕ ܕܐܠܗܐ ܡܢ ܣܘܪܝܐ ܕܝܢ
ܐܣܛܘܟܣ̈ܐ ܡܫܘܕܥܝܢ ܕܝܢ ܚܕ ܕܗܘܐ
ܐܝܬܘܗܝ ܒܗ ܗܘܐ ܗܘ. ܒܗ ܗܘܐ ܕܟܠ
ܘܗܠܝܢ ܡܫܡܫܝܢ ܠܟܠ ܘܚܝܘܬܐ ܐܦܪܝܢ.
ܡܠܐܟܐ ܐܦܝܢ ܕܐܠܗܐ ܗܘ ܕܒܪܐ ܒܚܟܡܬܐ
ܘܐܦ ܕܢܗܘܘܢ ܡܫܘܕܥܝܢ ܘܡܬܬܪܝܢܘܬܐ ܚܕ
ܐܢܘܢ ܕܝܢ ܡܫܡܫܝܢ ܐܣܛܘܟܣ̈ܐ ܡܢ ܚܕ. ܘܗܘܐ
ܡܬܬܪܝܢܘܬܐ ܘܡܫܟܚܐ ܡܫܒܚܐ ܘܠܗ ܚܕ
ܡܫܡܫܢܘܬܐ ܘܡܫܟܚܐ. ܐܦ ܐܬܪܘܬܐ ܐܦ
ܘܦܪܝܫܬܐ ܐܦ ܡܫܘܡܠܝܐ ܡܬܬܪܝܢ. ܐܠܐ
ܥܠܡܐ ܘܡܫܡܫܢܐ ܘܡܫܟܚܐ. ܡܢ ܗܘܐ
ܥܠ ܥܠܡܝܢ ܡܫܡܫܢܘܬܐ ܘܡܫܟܚܐ. ܐܦܝܢ
ܥܠ ܘܡܫܘܕܥ ܘܡܫܒܚܐ ܗܘ ܐܝܟ ܡܢ
ܗܦܟ ܠܐܘܪܐ. ܗܘܐ ܐܦ ܡܫܬܪܝܢܘܬܐ ܘܡܫܘܕܥ
ܒܗ ܡܫܘܕܥ ܕܐܝܬܘܗܝ ܘܠܐܘܪ ܗܘ ܒܪܝܬܐ
ܡܫܬܪܝܢܘܬܐ ܐܝܬܝܗ. ܘܐܦ ܐܬܝܐ ܒܟܠ
ܘܕܝܢ ܗܘ ܐܡܪܐ ܫܡܥܗ ܗܘ ܒܫܘܪܝܐ
ܒܪܝܬܐ. ܘܫܡܫܐ ܘܒܣܬܐ. ܒܗ ܚܕܐ
ܘܡܫܟܚܐ ܠܐܘܪܐ ܠܥܠܡ. ܘܐܦܪܝܢ ܘܠܫܡܝܐ
ܚܕ ܡܢ ܐܪ̈ܒܥܐ ܠܐܣܛܘܟܣ̈ܐ ܐܝܟ ܠܗ ܡܫܡܫ
ܘܒܛܝܪܐ ܡܢ ܐܣܛܘܟܣ̈ܐ ܘܠܗܠܝܢ ܗܘܢܐ
ܗܘܐ ܠܟܠܗܘܢ ܚܕܐ ܐܣܛܘܟܣܐ ܐܦ ܗܘܐ
ܘܚܕܐ ܘܡܫܟܚܐ ܘܚܕ ܡܫܬܟܚܐ. ܐܠܐ ܡܫܒܚ
ܘܡܫܟܚܝܢ ܡܫܬܪܝܢܘܬܐ ܡܢܗ ܠܐܘܫܒܚܐ
ܡܫܡܫ ܘܒܪܝܬܐ ܡܫܬܪܝܢ ܒܗܘܐ. ܐܡܟܝܠ
ܘܠܐ ܗܘܐ ܦܫܝܛܐ ܡܢ ܚܕܪܐ. ܘܡܫܟܚܐ ܟܠ
ܚܕ ܚܕ ܡܢ ܠܐܘܫܒܚܐ ܗܘܘ ܐܦܪܝܢ ܡܫܡܫܢܘܬܐ
ܗܘܝܐ ܐܦ ܐܬܪܘܬܐ ܫܡܘܕܥܐ ܗܘܘ ܠܟܠ

ܚܝ

ܚܕ ܚܕ ܡܢ ܐܘܪ̈ܚܬܐ. ܐܝܟ ܕܠܒܪ ܕܢܐܬܘܢ
ܠܟܠܚܕ ܚܕ ܡܢ ܐܘܪ̈ܚܬܐ ܒܝܘܩܢܐ ܕܡܬܪ
ܥܩܒܝ ܡܫܝܚܘܬܗ ܕܡܢ ܚܘܒܗ ܡܫܘܕܪܐ
ܘܡܐ ܕܒܪܢܫ ܒܪܘܚ ܗܘܐ ܚܙܐ. ܘܡܐ ܕܡܬܝܩܪ
ܡܫܪܝܐ ܗܘ. ܘܟܕ ܒܪܘܚܐ ܐܬܓܠܝ ܘܐܬܚܙܝ
ܡܫܠܡܐ ܗܘܐ ܠܐܘܪ̈ܚܬܐ ܡܢ ܚܕܪ̈ܐ. ܘܠܗܘܢ
ܡܫܡܠܝܐ ܕܟܠܗܘܢ ܗܢܘܢ ܐܦ ܡܫܡܠܝܐ
ܕܡܫܘܕܪܐ. ܕܒܫܘܪܝܐ ܡܫܬܘܕܥ ܐܝܟ ܗܘ
ܗܢܘ. ܐܝܟܢܐ ܐܘܪ̈ܚܬܐ ܕܒ̈ܢܝܐ ܐܝܬ ܠܒ
ܘܡܬܚܙܝܐ ܗܟܢ ܡܫܬܡܠܝܢ ܡܫܡܠܝܢܘܬܐ
ܡܢܗ ܕܡܫܝܚܘܬܐ ܡܫܘܕܪܢܘܬܐ. ܘܗܟܘ ܠܟܠ
ܚܕ ܚܕ ܡܢܗܘܢ ܕܡܬܚܙܐ ܕܠܐ ܕܢܥܒܕ ܒܪܘܚܗ
ܗܢܘܢ ܕܒܟܝܢ ܗܢܘܢ ܐܦ ܚܕܬܐ. ܐܦ
ܕܐܦ ܡܢܗܘܢ ܕܡܢܗܘܢ ܘܐܝܟܢܐ ܗܢܘܢ
ܡܢܗܘܢ ܗܢܘܢ ܘܡܫܡܠܝܢܐ ܡܬܬܩܢܝܢ.
ܘܡܢܗܘܢ ܡܬܩܢܝܢ ܡܢܗܘܢ ܡܬܬܟܝܢ
ܡܢܗܘܢ ܡܢܗܘܢ. ܘܗܘܐ ܡܫܡܠܝܢܘܬܐ ܐܝܟ
ܠܟܠܚܕ ܡܢܗܘܢ ܕܒܟܝܢܗ ܡܬܚܙܐ
ܘܚܝܠܐ ܕܒܗ. ܗܢܘܢ ܕܝܢ ܗܢܘܢ ܘܡܫܬܡܠܝܢ
ܕܒܟܝܢܗܘܢ ܗܘ ܕܒܟܝܢܐ ܘܟܠܗܘܢ ܠܟܝܢܗܘܢ
ܘܗܢܘܢ ܡܫܡܠܝܢܐ ܘܦܫܝܛܐ ܥܒܕܝܢ ܠܟܝܢܗܘܢ
ܠܟܝܢܐ ܘܡܢܗܘܢ ܡܢܗܘܢ ܠܟܝܢܗܘܢ ܘܡܫܬܡܠܝܢܐ
ܡܢܗܘܢ ܡܫܡܠܝܢܐ. ܘܡܫܡܠܝܢܐ ܗܢܘܢ
ܘܡܫܡܠܝܐ ܘܦܫܝܛܐ. ܢܥܒܕ ܠܟܠ ܚܕ ܚܕ ܒܟܝܢܐ
ܘܡܫܡܠܝܢܐ ܦܫܝܛܐ ܥܒܕܝܢ ܠܟܠ ܒܟܝܢܐ. ܐܝܟ
ܗܘܝܐ ܗܘܐ ܘܥܒܕܝܢ ܐܝܟ ܕܥܒܕܝܢ ܒܟܝܢܐ
ܘܗܢܘܢ ܡܫܡܠܝܢ ܐܝܟܢܐ ܕܗܘ ܟܠܗܘܢ.
ܘܐܝܟ ܕܥܒܕܝܢ ܒܟܝܢܐ ܘܡܫܡܠܝܢ ܘܡܫܡܠܝܢ
ܐܝܟܢܐ ܕܡܫܡܠܝܢܐ ܘܐܦܠܐ. ܘܐܝܟ
ܘܗܢܘܢ ܘܦܫܝܛܐ ܘܡܫܡܠܝܢܐ ܡܢܗܘܢ

ܐܝܟܢܐ ܕܐܝܟ ܡܢܗܘܢ. ܘܐܝܟܢܐ ܘܡܫܡܠܝܢ
ܘܡܫܡܠܝܢ ܘܡܫܡܠܝܢ ܡܢܗܘܢ. ܘܟܠ ܒܟܝܢܐ
ܐܝܟܢܐ ܕܡܫܡܠܝܢܘܬܐ ܘܐܝܟܢܐ ܘܡܫܡܠܝܢ ܘܡܫܡܠܝܢ
ܘܠܐ ܢܥܒܕ ܒܟܝܢܐ. ܐܦܠܐ ܐܝܟܢܐ. ܘܐܝܟ ܐܢܫ
ܟܠܐ ܘܡܫܡܠܝܢܘܬܐ ܡܫܡܠܝܢܐ. ܟܕ ܡܫܡܠܐ
ܘܒܟܝܢܗ ܚܕ ܗܘ. ܘܐܝܟܢܐ ܘܡܫܡܠܝܢ ܘܡܫܡܠܝܢ
ܡܢ ܗܢܘܢ ܘܡܫܡܠܝܢ ܡܫܡܠܝܢ ܡܫܡܠܝܢ ܘܡܫܡܠܝܢ
ܡܫܡܠܝܢ ܠܟܝܢܐ. ܘܡܫܡܠܝܢ ܒܟܝܢܐ ܗܘܐ .
ܗܘܐ ܘܡܫܡܠܝܢ ܠܟܝܢܐ ܗܢܘܢ ܡܫܡܠܝܢܐ.
ܘܐܝܟ ܕܐܡܫܡܠܝܢܐ ܐܝܟ ܠܟܠ ܡܫܡܠܝܢܐ
ܠܟܝܢܐ. ܡܢܗܘܢ ܘܡܫܡܠܝܢ ܘܐܦ ܠܗܘܢ ܘܗܢܘܢ
ܠܐ ܡܫܡܠܝܢ ܠܗ ܠܗܘܢ. ܕܒܟܝܢܗ ܡܫܡܠܝܢ
ܟܗܢܘܬܐ ܘܡܫܡܠܝܢܘܬܐ ܐܦ ܡܫܡܠܝܢ ܐܝܟܢܐ
ܘܡܢ ܟܠܗ ܗܘܐ ܐܝܟ ܕܠܗܘܢ ܠܗܘܢ ܟܠܗܘܢ
ܡܫܡܠܐ ܥܒܕܘܗ ܘܗܢܘܢ ܘܡܫܡܠܝܢ ܘܟܕ ܡܫܡܠܝܢ
ܡܫܡܠܝܢܐ ܕܒܗ ܕܡܫܡܠ ܠܐ ܐܦ ܠܟܝܢܗ ܐܝܟ
ܢܦܫܗ . ܗܘ ܗܘ ܕܒܗ ܒܐܦܝ ܡܫܡܠܝܢ ܘܡܫܡܠܝܢܐ
ܘܡܫܡܠܝܢ. ܐܝܟ ܟܠܗ ܡܫܡܠܝܢ ܘܡܫܡܠܝܢܐ
ܡܫܡܠܝܢ ܟܝܢܐ ܡܢ ܡܫܡܠܝܢ ܡܫܡܠܝܢ ܡܫܡܠܝܢ
ܡܢ ܕܟܠܗܘܢ. ܘܡܫܡܠܝܢ ܕܥܒܕܘ ܡܫܡܠܝܢ
ܠܟ . ܘܡܫܡܠܝܢ ܗܢܘܢ ܡܫܡܠܝܢ
ܟܝܢܐ. ܗܢܘܢ ܕܝܢ ܘܡܫܡܠܝܢ ܠܐ ܡܫܡܠܝܢ
ܗܘܠܐ ܘܡܫܡܠܝܢ ܘܐܝܟܢ ܘܡܫܡܠܝܢ ܠܟܝܢܐ
ܡܫܡܠܝܢܐ ܐܝܟ ܘܐܦ ܡܫܡܠܝܢ. ܘܡܫܡܠܝܢ
ܡܫܡܠܝܢ ܐܝܟ. ܡܫܡܠ ܘܐܦ ܠܗ ܡܫܡܠܝܢ
ܠܡܫܡܠܝܢ ܡܫܡܠܝܢ. ܘܦܫܝܛܐ. ܗܢܘ ܕܝܢ
ܐܝܟܢܐ ܡܫܡܠ ܡܫܡܠܝܢ ܒܟܝܢܗ. ܘܐܦ
ܡܢ ܡܫܡܠܝܢ ܡܫܡܠܝܢ ܘܥܒܕܘ ܡܫܡܠܝܢܐ
ܘܡܫܡܠܝܢ ܡܐ ܘܡܫܡܠܝܢ ܠܟܠ ܟܝܢܐ
ܟܝܢܐ ܟܕ ܡܫܡܠܝܢ ܠܗ ܕܢܫܡܠܝܢ. ܐܠܐ

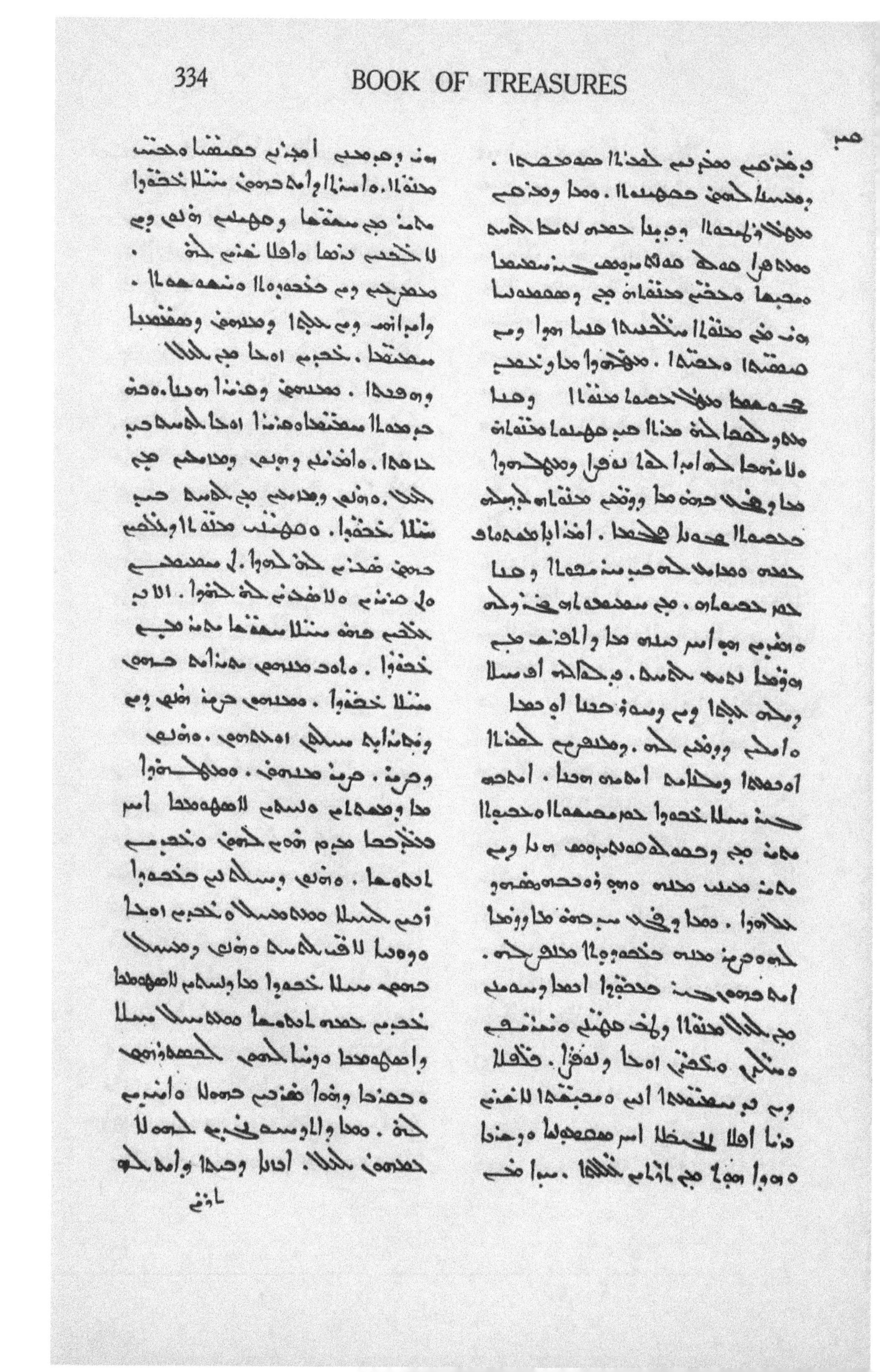

[illegible]

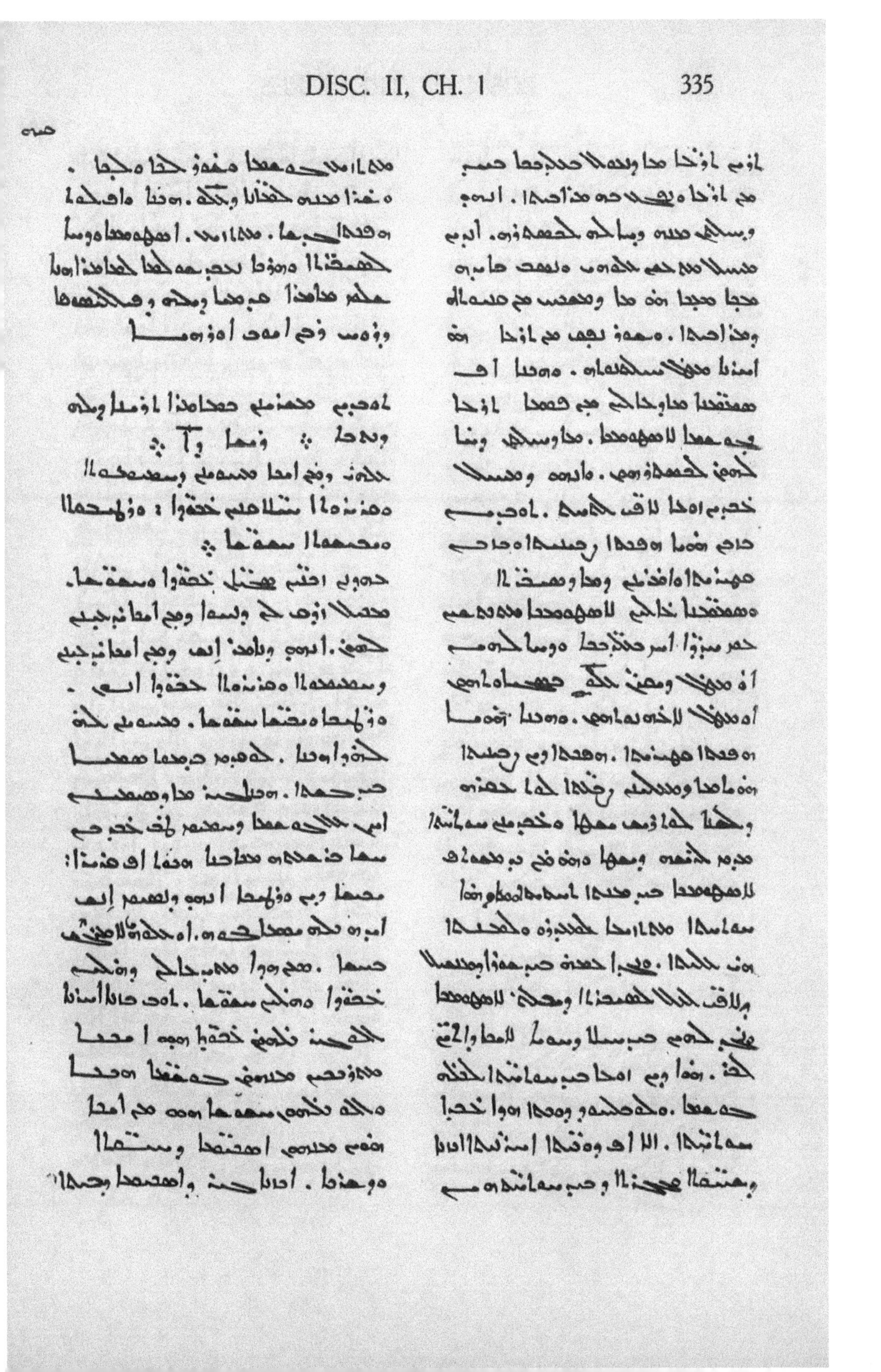

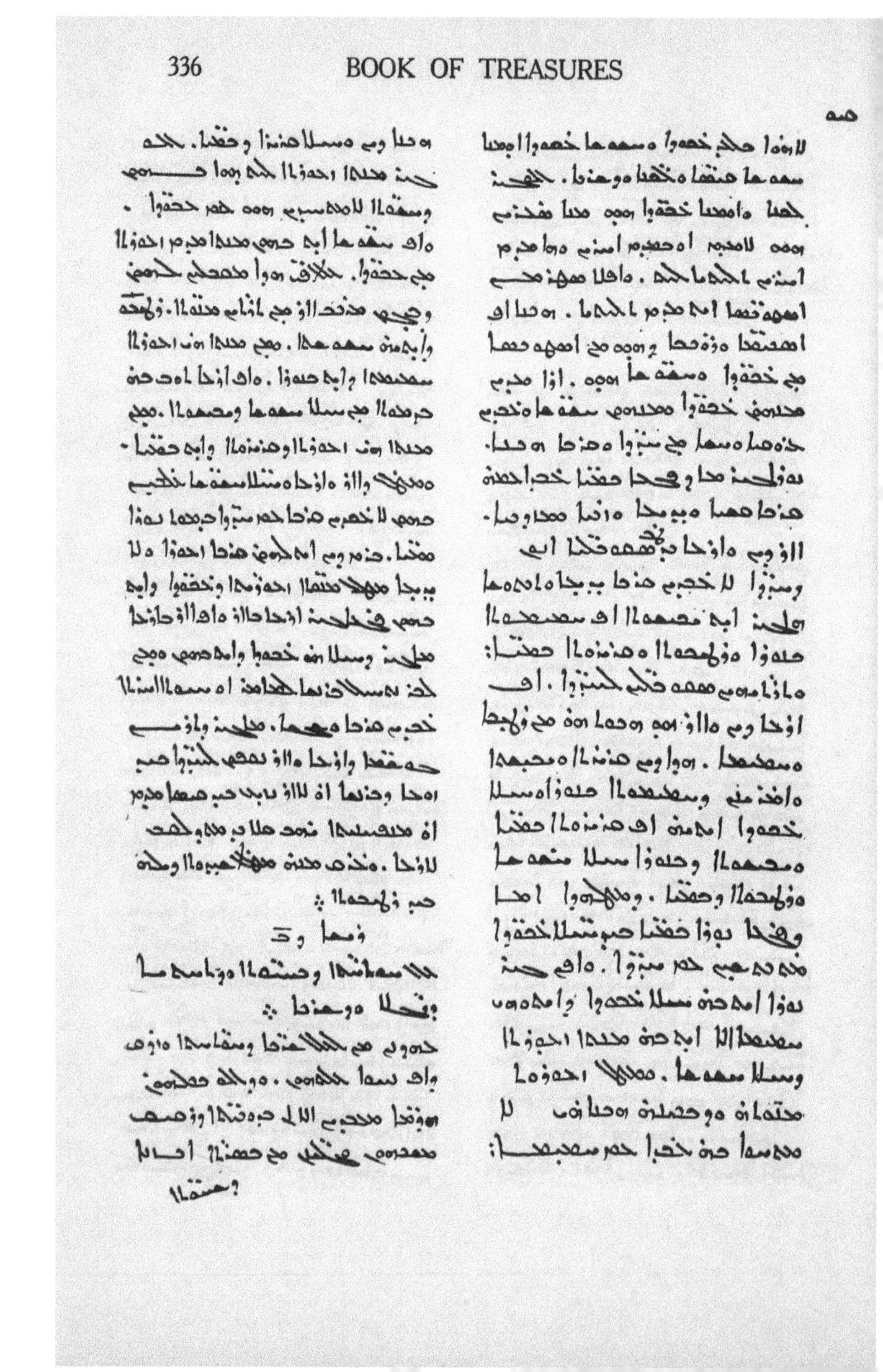

ܩܗ

[illegible]

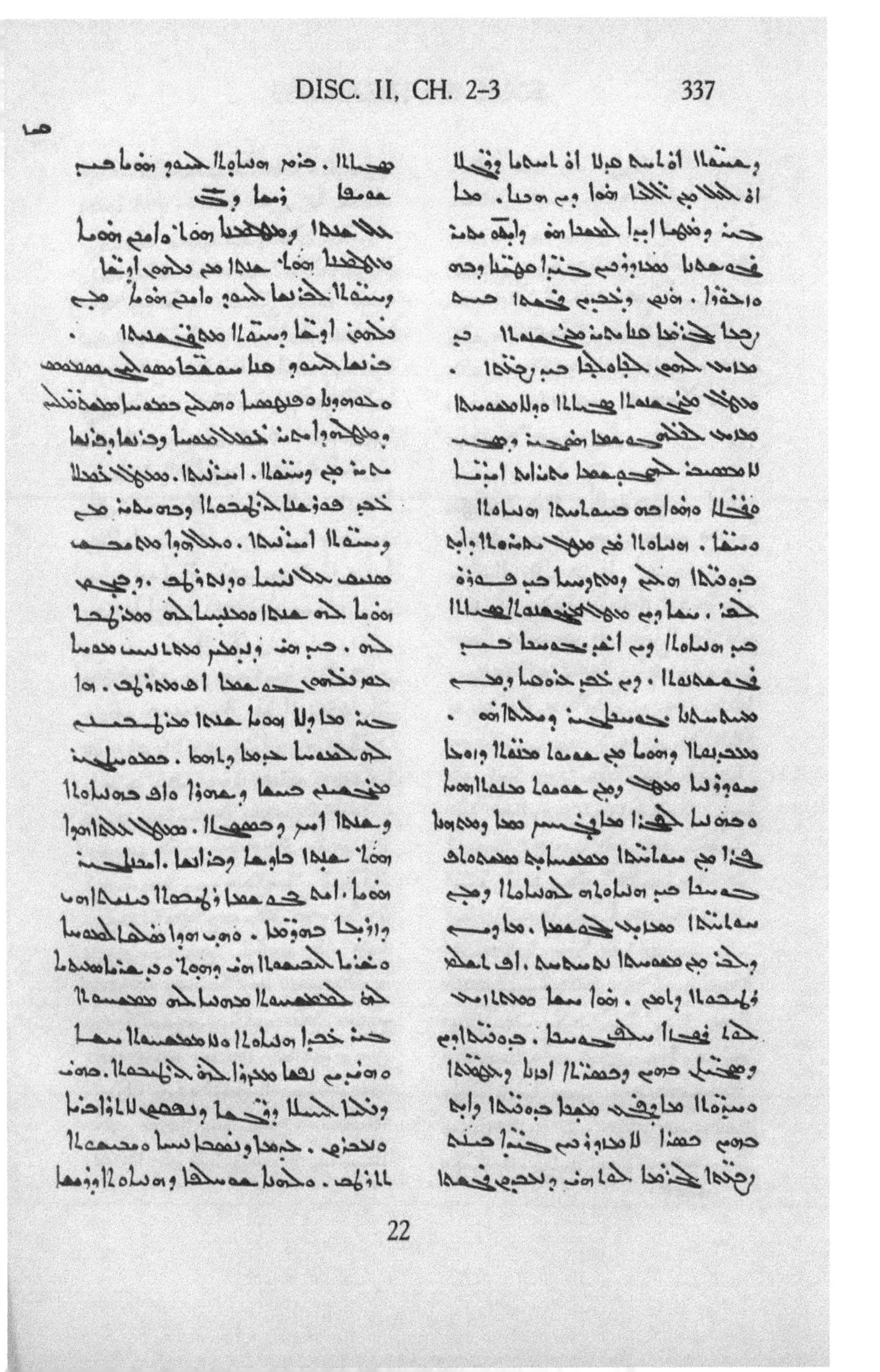

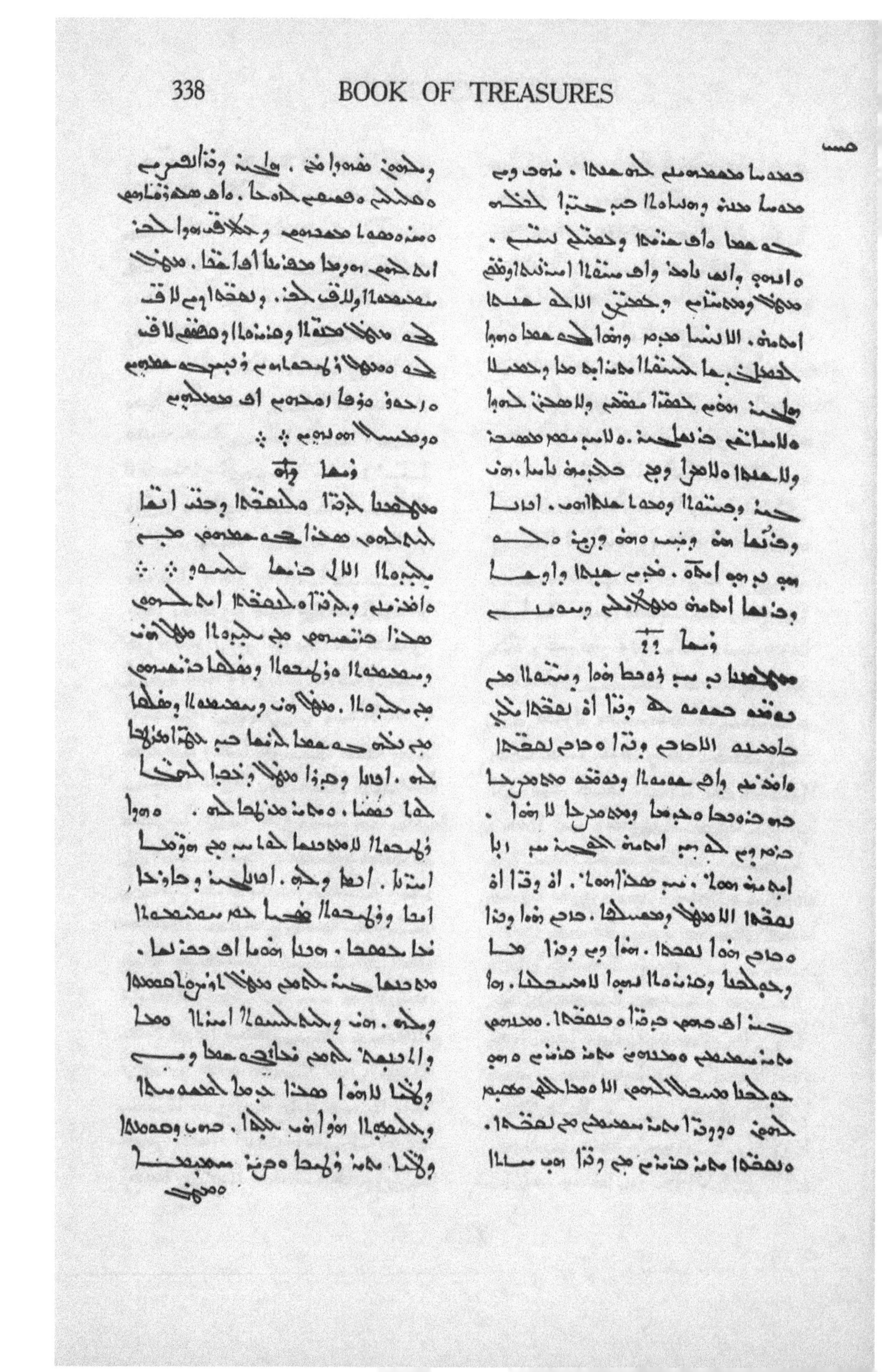

[illegible]

ܣܛ

ܘܡܛܠ ܕܒܠܐ ܦܫܝܛܘܬܐ ܠܗܝܡܢܘܬܐ ܠܐ
ܗܘܐ ܒܗܘܢ ܡܫܪܐ. ܡܛܠܗܢܐ ܕܒܟܠܗܠܝܢ
ܡܠܠܐ ܟܪܙܐ ܦܫܝܛܘܬܐ ܘܡܢܚܐ ܠܗܝܡܢܘܬܐ
ܐܡܝܢܐ ܕܠܐܢܫ ܒܗܘ ܡܘ. ܢܚܐ ܗܟܢܐ ܐܠܗܘ ܡܗܝܡܢ
ܐܝܠܝܢ ܕܐܪܥܐ ܡܢܐ ܕܒܝܬܐ ܠܡܠܬܗ ܡܗܝܡܢܐ
ܘܐܬܪܗܛ ܥܘܦ ܠܐܦܝܐ ܕܪܗ ܒܚܘܒܐ ܒܪܝܐ
ܘܒܠܚܘܕ ܒܠܒܗ ܥܡܠܐ ܘܣܓܝ ܠܗ ܠܗܝܡܢܘܬܐ
ܕܒܗ ܘܐܬܡܛܝܐ ܒܗܘܢ ܡܫܡܫܢܘܬܐ. ܘܐܡܪ
ܠܗ ܐܪܡܝܐ ܠܟܠܐ. ܗܟܢܐ ܐܦ ܒܫܘܡܠܝܐ
ܘܢܦܫܐ ܠܐܦܝܐ ܡܫܪܐ. ܒܪ ܡܢ ܒܫܘܡܠܝܐ ܘܢܒܠܥܐ
ܡܐܢ ܕܐܬܐܬܩܢ ܡܫܡܫܢܘܬܐ ܒܟܢ ܐܝܢ
ܢܦܫ ܡܫܪܐ. ܘܡܬܦܫܩ ܢܦܫ ܒܪܘܚܐ
ܕܪܘܚܐ ܘܡܬܦܫܩܢ ܘܡܢ ܐܡܬܝܬ
ܡܫܡܫܢܘܬܐ ܘܗܝܡܢܘܬܐ. ܐܝܟܢܐ ܕܥܡܗܘܢ ܐܬܐ ܐܦ
ܘܡܢܗ ܒܟܠ ܗܢܘ ܡܢ ܕܡܬܐܡܪܐ ܡܫܡܫܢܘܬܐ
ܒܟܠ ܡܠܠܐ ܒܦܢܝܬܗ ܡܫܪܐ ܢܚܐ ܡܫܪܐ ܀

ܪܝܫܐ ܘ̄

ܥܠ ܕܡܢܐ ܕܗܘܐ ܒܟܪ̈ܝܐ ܀

ܗܘܐ ܕܝܢ ܕܡܢܐ ܠܟܪ̈ܝܐ ܠܒܢܝ̈ܐ ܕܝܢ ܘܠܢܫܐ
ܡܬܩܪܒܝܢ ܠܐ ܗܘܐ ܡܫܬܟܚܝܢ ܗܘܘ .
ܒܡܘܬܐ ܕܢܫܐ ܡܢ ܕܡܛ ܦܫܝܛܐ ܒܗܘܢ
ܕܟܢ ܒܪܐ ܐܪܝܐ ܠܐ ܗܘܝܢ ܘܗܝܡܢܘܬܐ. ܘܐܦ
ܡܫܝܚܐ ܕܢܦܫܬܐ ܡܢ ܕܦܫܝܛ ܒܪ̈ܘܚܐ ܐܢܫܐ
ܗܘܐ ܘܡܬܦܫܩܢܐ ܗܘܐ ܠܗ. ܘܡܢ ܡܫܝܚܐ ܕܒܗܘܢ
ܥܡ ܠܗܘܢ. ܘܐܦ ܕܒܪܐ ܒܡܢ ܡܫܬܡܠܝܢ ܠܒܢܝ̈ܐ
ܠܗ ܒܗܘܢ ܡܫܡܫܢܘܬܐ ܗܘ ܕܝܢ ܕܡܬܡܣܟܢܐ
ܒܗ ܦܫܝܛܘܬܐ ܡܬܚܙܝܐ ܡܫܪܐ. ܠܐ ܕܝܢ
ܗܘܐ ܠܗܘܢ. ܐܠܐ ܡܫܪܐ ܠܗ ܗܝܡܢܘܬܐ.
ܘܒܝܕ ܡܐ ܗܘܐ ܕܗܘ ܡܠܐܟ ܡܢ ܗܢܘ ܕܠܝܬ ܘܡܠܐ
ܘܐܬܐܡܪ ܡܫܡܫܢܘܬܐ ܒܟܠܢ ܕܟܘܢ ܗܘ

ܕܒܪܝܐ ܗܘܐ ܘܐܬܚܙܝ ܢܚܐ ܗܟܢܐ ܠܟܘ ܡܗܝܡܢ
ܘܐܦ ܗܘܐ ܒܗܘܢ ܕܡܢܐ. ܘܡܢܗ ܕܡܫܝܚܐ
ܕܢܦܫܐ ܕܦܫܐ ܒܠܚܘܕ ܠܐ ܗܘܐ ܠܗܘܢ ܕܡܢܐ
ܘܐܦ ܓܝܪ ܗܘܐ ܒܗܘܢ ܡܫܪܐ ܐܝܟܢܐ ܕܟܘܠ
ܐܠܐ ܒܪܝܐ ܗܘ ܡܢܐ ܕܡܬܦܫܩ ܪܘܚܗ ܕܟܪ̈ܝܐ
ܡܬܐܡܪܐ ܒܡܫܝܚܐ ܡܫܡܫܢܘܬܐ ܐܝܟܢܐ
ܕܟܘܠ ܐܡܪ ܡܠܠܐ ܡܢ ܕܠܒܢܝܐ. ܘܡܫܝܚܐ
ܘܡܠܠܐ ܐܡܪܗ ܠܐܦܝܐ ܠܐܬܪܐ ܐܝܟܢܐ ܐܡܪܐ
ܐܢܫܐ ܘܡܠܠܐ. ܘܐܡܪ ܡܢ ܒܪܝܐ ܡܫܬܡܠܝܢ ܥܡܠܐ
ܘܗܘܐ ܒܗܘܢ ܕܟܘܠ ܗܘܐ. ܒܟܪ̈ܝܐ ܕܝܢ ܠܐ ܗܘܐ
ܡܬܩܪܒ ܘܡܐܟܠ ܒܗܝܡܢܘܬܐ ܘܡܫܡܫܢܘܬܐ
ܘܐܬܬܦܫܩ ܒܡܫܡܫܢܘܬܐ ܗܘܐ. ܗܘܐ ܠܟܪ ܘܒܝܐ
ܠܡܬܡ ܒܟܪ̈ܝܐ ܐܝܟܢ ܦܫܝܛܐ. ܡܢܐ ܕܐܬܚܙܝ
ܡܫܡܫܢܘܬܐ ܣܠܩܐ ܘܐܬܬܦܫܩ ܒܦܬܓܡܐ
ܟܐܢܐ ܘܒܪܐ ܘܐܬܝܠܕ ܒܗܝܡܢܘܬܐ ܘܒܦܘܡܐ
ܘܡܫܝܚܐ ܒܡܬܢܝܬܐ ܘܒܦܘܡܐ ܢܚܐ ܡܫܪܐ
ܠܡܬܒܗ ܐܝܟܢ ܘܢܦܫ ܡܢ ܒܗܝܡܢܘܬܐ
ܘܒܦܘܡܐ. ܘܗܘ ܕܝܢ ܒܟܠܠܐ ܡܠܠܐ ܟܪ̈ܝܐ
ܡܫܡܫܢܘܬܐ ܘܢܦܫܐ ܠܗ ܠܒܝܐ ܐܝܟܢ
ܦܫܝܛܐ. ܘܐܬܐ ܗܘ ܕܡܬܚܙܐ ܠܗܝܡܢܘܬܐ ܘܐܦ
ܡܬܬܪ ܡܫܬܪܝܐ ܠܐܦ ܠܗܘܢ. ܡܢܐ ܕܡܬܚܙܐ
ܠܒܗ ܐܝܟܢܐ ܦܢܝܐ ܐܡܪ ܒܠܒܗ ܠܗ
ܠܐܪܐ ܕܐܝܟܢ ܠܗܘܢ. ܒܟܐܢܐ ܕܣܓܝܐܐ ܘܡܢܐ
ܢܗܘܐ ܠܟܪ̈ܝܐ. ܠܒܢܝܐ ܕܝܢ ܠܐ ܠܗ ܗܘ ܕܢܗܘܘܢ
ܟܪ̈ܝܐ ܒܗ ܡܫܝܚܐ ܘܣܝܡܐ ܦܝܫܝܢ ܡܢ
ܢܦܫܐ. ܘܐܦ ܡܢ ܒܢܝܐ. ܘܐܦ ܕܟ̈ܪܝܐ ܗܘܘ
ܘܢܦܫܐ ܒܠܚܘܕ ܘܢܗܘܐ ܣܓܝ ܠܒܢܝܐ
ܐܝܟ ܕܝܠܗ ܕܟܪܝܐ. ܡܫܝܚܐ ܕܝܢ ܘܟܪ̈ܝܐ
ܢܦܫܐ ܗܘܘ ܘܡܬܐܡܪܝܢ ܢܦܫܐ ܕܟܪ̈ܝܐ ܗܘܘ
ܦܫܝܛܐ ܠܒܢܝܐ. ܡܢ ܕܝܢ ܡܫܝܚܐ ܒܗܘܢ

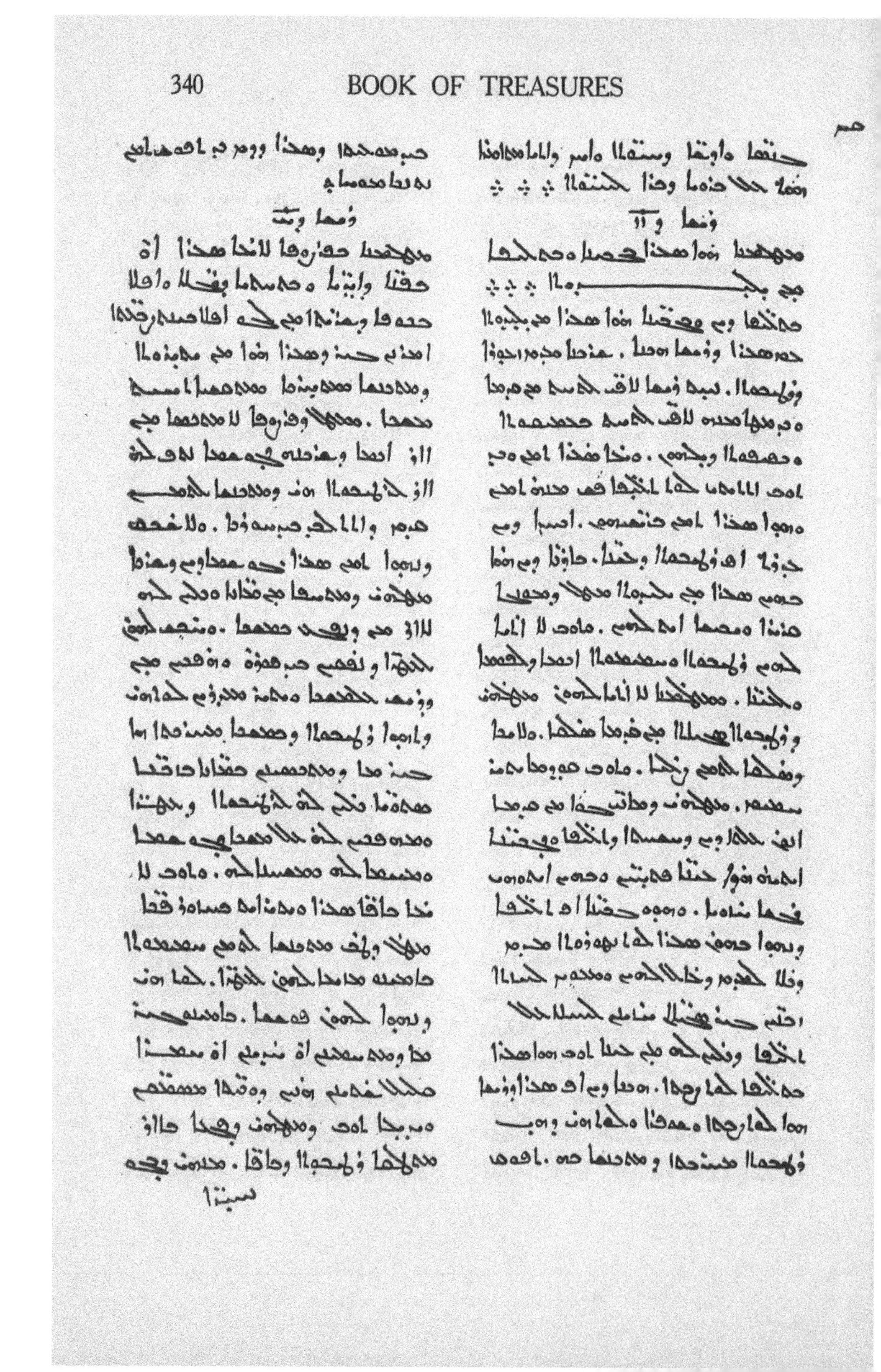

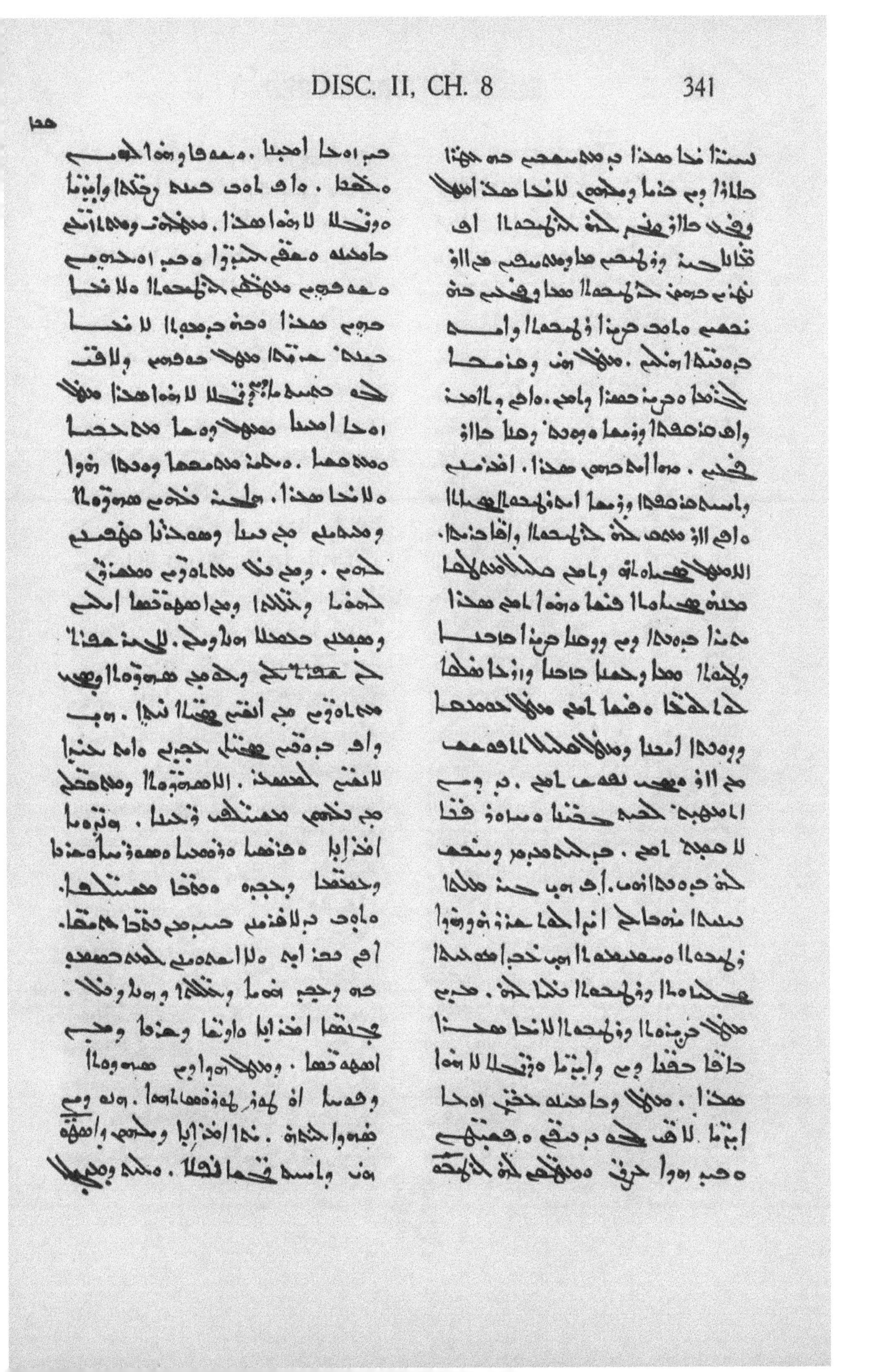

ܫܡ

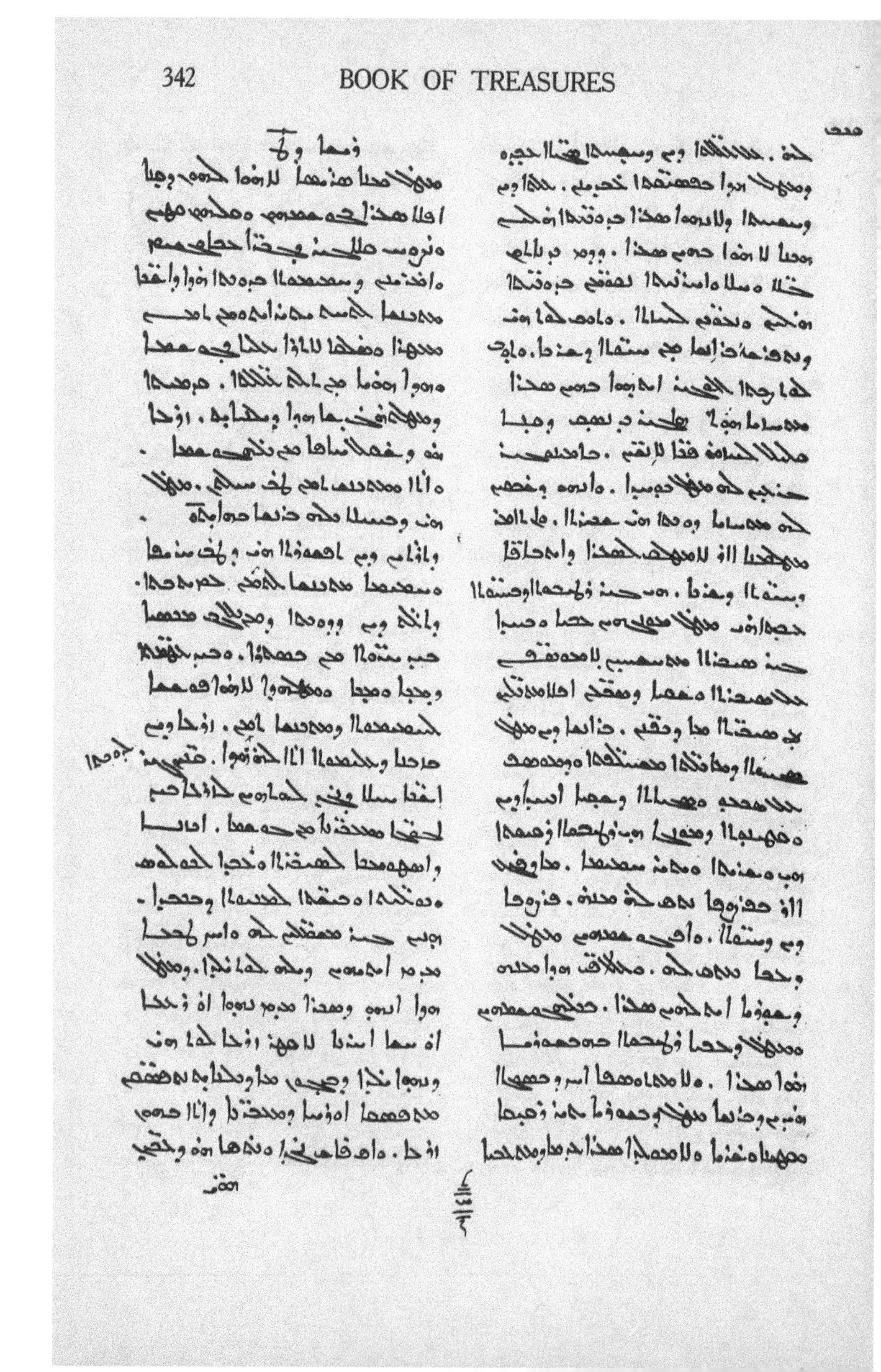

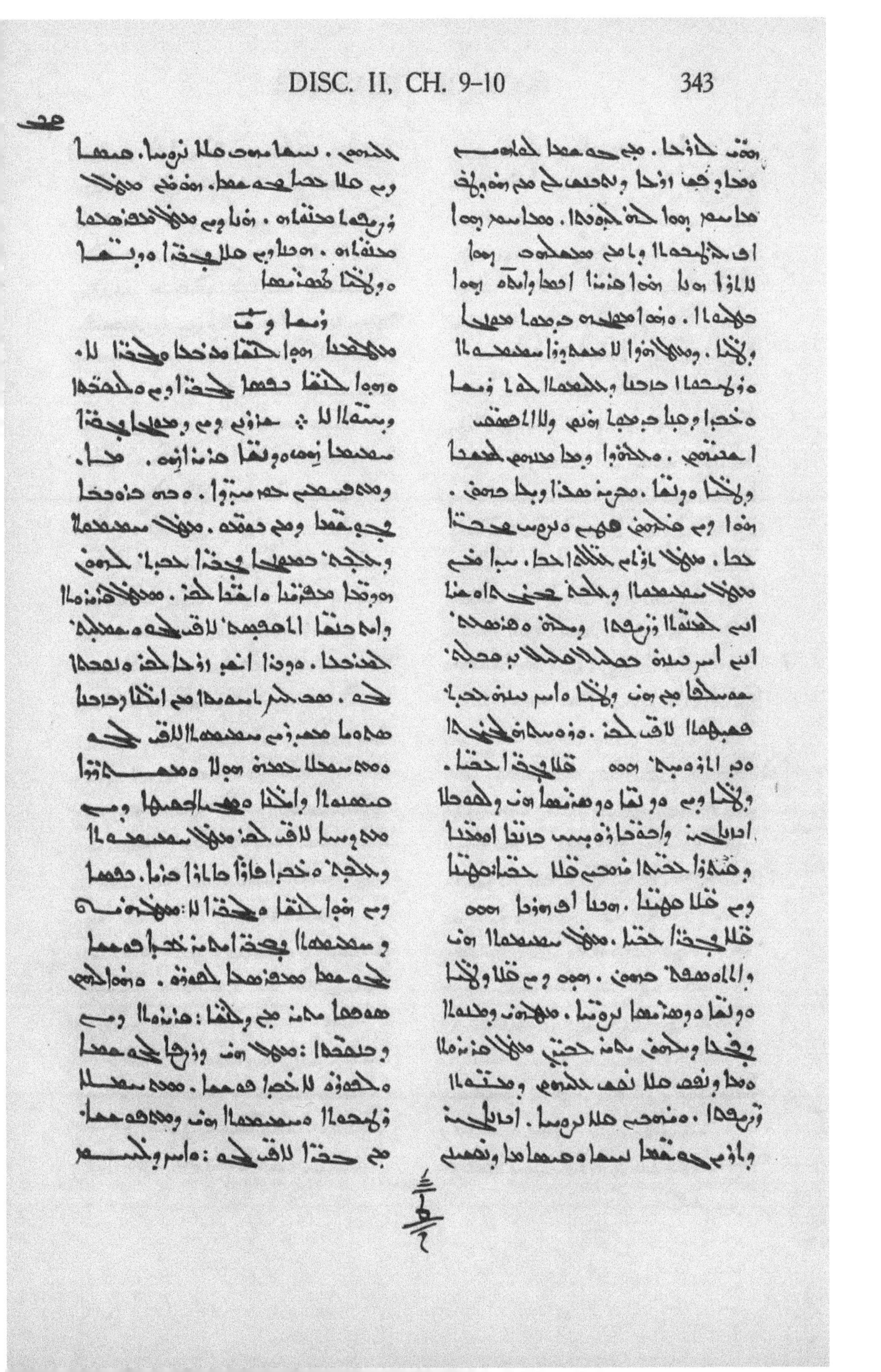

[illegible]

ܠܒܢ̈ܝ ܐܢܫܐ ܠܒܪܗ ܘܡܬܚܙܐ ܘܐܬܬܣܝܡ
ܒܢ̈ܝܐ. ܐܝܟ ܕܐܬܦܫܛܬ ܗܢܐ ܐܟܘܬܗ.
ܡܢܗ ܒܢ̈ܝܐ ܐܝܟ ܕܢܦܩܝܢ ܡܢ ܕܡܣܝܐ ܠܐܬܪ
ܗܝܡܢ ܒܣܝܐ ܐܝܟܢܐ ܡܢܗܘܢ ܘܠܐ ܐܬܟܪܝܗ
ܠܟܝܢ ܡܢ ܦܘܪܫ ܠܬܘܪܝܐ ܡܬܝܚܕܝܬܐ
ܠܢܦܫܬܐ ܕܝܢ ܕܚܝܠܬܐ ܠܐ ܗܘܬ ܒܟܝܢܐ. ܡܛܠ
ܕܡܬܢܛܪܝܢ ܒܟܝܢܐ ܒܟܝܢܐ. ܘܠܐ ܐܦ
ܡܬܦܢ̈ܝܢ ܠܛܠܝܘܬܐ. ܘܐܝܟ ܕܡܬܪ ܒܢ̈ܝܢ
ܘܐܦ ܡܬܚܠܦܝܢ. ܡܛܪ ܡܢܗ ܘܫܡܐ ܡܢ
ܕܢ̈ܐ. ܘܡܠܐܟ̈ܐ ܗܘܐ. ܠܐܝܟܐ ܕܒܝܢ ܡܛܪܬܐ
ܕܐܝܟ ܠܗ. ܐܠܐ ܗܢܐ ܕܦܩܕ ܠܒܢܝܢ
ܠܟܝܢ. ܢܦܫܐ ܕܝܢ ܡܬܚܙܝܢ ܕܒܪܢܫܐ
ܒܟܝܢ ܡܢܗܪܬܐ ܡܬܝܚܕܝܬܐ ܡܢ ܚܝ̈ܐ
ܐܦ ܒܥܩܒܬܐ ܗܠܝܢ ܐܦ ܢܝܚܝܢ ܙܒܢܐ
ܗܘ. ܡܢܗ ܒܢ̈ܝܐ ܒܟܝܢ ܡܬܪܘܬܐ ܡܢ ܦܝܢ
ܢܦܫܬܐ ܡܢ ܕܚܝܠ ܠܒܪ. ܘܐܦ ܐܬܦܪܫܬ
ܡܢܐ ܡܢ ܕܟܝܢ ܗܢܐ ܠܒܢܝ ܕܡܢ ܠܒܪ
ܐܬܚܙܝܬ ܕܗܘܠ ܕܡܟܢܐ ܢܦܫܐ. ܘܐܝܟ ܡܢ
ܕܟܝܢ ܗܘܐ ܡܬܚܙܐ ܒܟܝܢܐ. ܐܦܢܝܢ
ܕܗܘܐ ܠܗ ܡܬܪܘܬܐ ܗܢܐ ܘܠܐܬܪܐ ܠܟܢܗ
ܗܝ ܫܪܝܬܐ ܘܠܟ ܡܢܗ ܕܡܣܝܐ ܐܟܚܕܐ ܒܢ̈ܝܐ.
ܐܠܐ ܗܢܐ ܗܘܐ ܠܟܐ ܡܛܠܬܐ ܠܢܦܫ ܡܬܚܠܦܢ
ܘܗܢܐ ܒܟܠ ܗܢܐ ܙܘܥܐ ܠܟܐ ܐܬܪܥܝܬ. ܕܝ
ܡܢ ܐܬܦܫܛܬ ܐܝܟ ܕܒܬܪܗ ܘܒܝܢܗ
ܘܫܘܬܐ ܡܢ ܢܦܫܗ ܗܘ ܐܟܘܬ ܒܢܝܢ
ܘܠܟܠܒܝܢ ܡܢ ܗܘܢ̈ܟܐ ܡܢܐ ܘܡܠܬ
ܐܬܦܫܛܬ. ܐܬܦܫܛܬ ܕܗܘ ܠܦܫܛܐ ܠܟܢ
ܒܦܫܛܗ. ܘܐܝܟ ܕܡܢ ܦܫܛܐ ܡܬܚܙܐ
ܡܬܒܥܐ ܚܝܐ ܦܫܛܐ ܗܘܐ. ܗܘ ܚܝܐ ܡܢܐ
ܘܗܢܐ ܒܟܠ ܡܢ ܢܦܫܗ ܗܘ ܕܚܝ̈ܐ

ܘܒܝܬ ܡܬܪܘܬܐ ܕܚܝܐ ܘܡܢ ܐܬܦܫܛܬ ܠܗܘܢ
ܗܘܠܐ ܠܟܐ ܐܬܦܫܛ. ܡܛܠ ܕܐܝܬ ܒܐܬܪ
ܒܗܠܝܢ ܕܒܗ ܙܘܥܐ ܐܬܐ ܐܝܟ ܕܡܠܟܐ
ܠܟܐ ܐܬܚܙܝܬ. ܡܛܠ ܕܐܬܦܫܛ ܡܢ
ܫܠܝܐ ܠܐ ܐܝܟ ܐܬܐ ܗܝ ܗܘܠܐ ܠܟܐ
ܡܬܪܘܬܐ ܕܐܬܚܙܝ ܠܟܢ. ܫܠܡܐ ܕܝܢ
ܕܫܘܠܛܢܐ ܐܬܦܫܛܬ ܗܘܝܐ. ܗܘ ܒܐܬܪܐ
ܕܡܠ ܫܘܠܛܢܐ ܐܬܦܫܛܬ ܒܟܠ ܗܘܢ̈ܝܢ ܠܒܢ
ܕܢܦ̈ܫܐ ܢܦܫܬܐ ܠܟܐ ܦܫܛܐ ܕܒܢܝܢܐ
ܕܢܦܫܐ ܙܘܥܐ ܡܬܚܙܐ ܐܝܟ ܕܦܫܛ ܠܒܢܝܢܗ
ܘܠܒܢ ܒܢܝܢ ܒܙܘܥܐ ܗܘ ܕܒܢ̈ܝܐ ܘܐܝܟ ܦܫܛܬܐ
ܘܐܬܦܫܛܬ ܢܦܫܬܐ ܘܗܝܢܐ ܐܦ ܠܢܦܫܬܐ ܘܒܢ̈ܝܐ
ܘܫܘܬܐ ܘܫܪܝܐ. ܢܦܫܗ ܒܢܝܢ ܐܬܦܫܛܬܐ
ܘܡܠܝܢ ܠܟܠܗܘܢ ܐܝܟ ܡܬܝܚܕܝܢ ܠܟܠ
ܒܙܘܥܐ. ܘܡܛܠ ܦܫܛܐ ܘܒܝܬܗ ܠܐ ܐܬܬ
ܕܡܬܚܙܝܢܐ ܘܫܡܐ. ܦܫܛܐ ܒܙܘܥܐ ܘܒܢܗ
ܐܝܟܢܐ ܘܡܢܗ ܢܦܫܐ ܡܬܚܙܐ ܠܐ ܐܬܬ ܐܬܪܐ
ܐܝܟ ܕܠܒܝܢ ܦܫܛܐ ܘܒܝܢ ܡܬܡܠܝܐ ܘܗܢܐ
ܡܢܐ ܘܡܬܚܙܝ ܒܗ ܕܢܐ ܐܝܟܢܐ ܕܐܬܦܫܛܗ
ܘܐܬܚܙܝܬ ܒܟܐ ܐܬܪܥܝܬ ܘܒܡܫܠܡܐ ܕܒܪ
ܠܐܬܪܥܝܬ ܡܢ ܐܬܦܫܛܬܐ ܘܡܬܪܘܬܐ ܗܢܐ
ܘܦܫܛܐ ܦܢ ܕܡܢ ܘܐܝܟܗ ܒܟܝܢܐ. ܘܡܛܠ ܕܒܙܘܥܐ
ܗܘ ܕܢܐ ܘܐܬܚܙܝ ܒܗ ܠܟܐ ܗܘܝܐ
ܡܬܦܢܝܐ ܠܟܢ ܘܐܦܢ ܒܟܦ ܦܫܛܬܐ ܠܟܐ
ܡܢܙܘܐ. ܠܟܐ ܗܢ ܕܐܬܚܙܝ ܦܫܛܢܐܝܬ
ܘܐܬܦܫܛ ܗܘܐ ܘܦܠܢ ܒܢܝܢܗܘܢ ܠܟܠܢܫܐ
ܘܠܟܝܢܐ. ܘܗܘܐ ܦܢ ܒܙܘܥܐ ܘܡܠܝܢ
ܘܡܬܪܘܬܐ ܠܐܬܪܐ. ܗܘܐ ܙܒܢ ܕܐܬܦܫܛ ܘܐܬܚܙܝ
ܘܐܬܦܫܛ ܘܐܬܪܐ ܀ ܙܘܥܐ ܕܬܪܝܢ
ܡܬܚܙܝܢ ܡܢܦܫܬܐ ܘܡܬܪܘܬܐ ܒܦܫܛܗ ܘܡܠܟܗ

ܦܘܡ

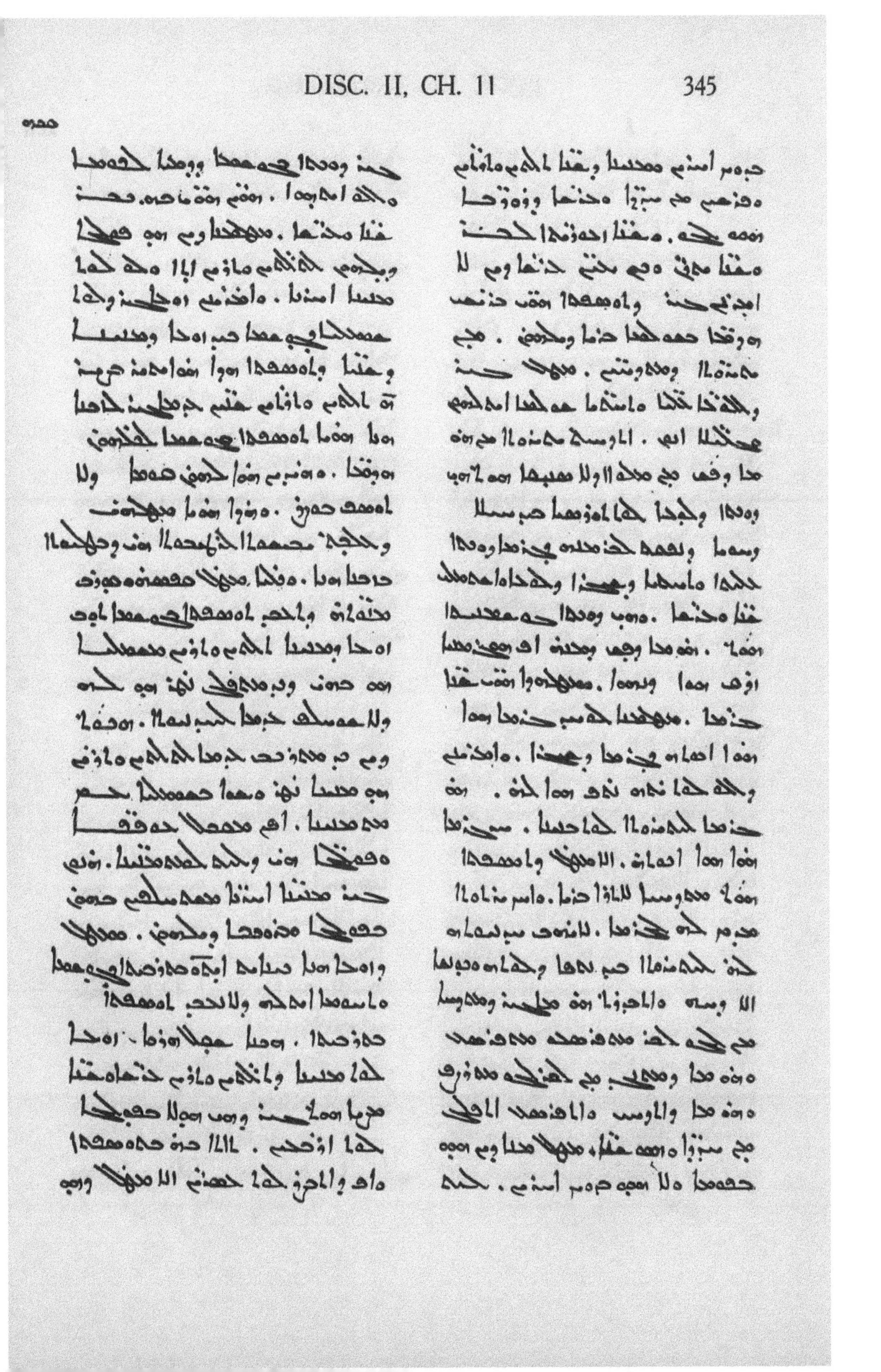

[illegible]

[illegible]

ܩܡܗ

ܚܝܐ ܕܪܘܚܐ ܕܐܬܦܫܛܬ ܒܥܘܡܩܐ ܬܘܒ
ܐܫܬܟܚܘ ܒܒܪ̈ܢܫܐ. ܗܟܢ ܠܚܝܘܬܐ ܕܒܪ̈ܢܫܐ
ܘܢ̈ܦܫܬܐ ܗܝ ܕܗܝ. ܗܟܝܠ ܕܒܪ̈ܢܫܐ
ܡܬܚܠܠ ܗܘܘ ܠܐܦ̈ܝ ܟܠܗ. ܘܒܪ̈ܘܚܐ ܐܬܬܣܝܡ
ܘܢ̈ܦܫܬܐ ܕܒܗܘܢ ܗܘܬ ܘܠܐܦ̈ܝ ܟܠ ܐܢܫ ܗܘܐ
ܗܘܬ ܕܢܦ̈ܫܬܐ ܕܒ̈ܢܝܐ ܗܢܝܐ ܘܟܠܗܘܢ. ܡܢ
ܗܝ ܕܗܘܬ ܘܐܝܬܝܗ ܡܢ ܫܡܫܐ ܘܡܬܚܙܝܢ
ܫܡܫܐ ܐܠܐ ܙܘܥܐ: ܗܘܐ ܕܒܠܚܘܕ ܡܢܗ
ܘܡܢ ܦܫܛܐ ܗܘܬ ܕܒ̈ܢܝܐ ܒܓܘܗ ܘܟܠܗ
ܗܢܐ ܐܝܟ ܡܬܚܙܐ ܗܢܘܢ ܐܬܬܚܙܝ ܗܘܐ
ܘܣܒܝܪ ܚܢܢ ܒܓܘܗ. ܡܬܪܐ ܬܘܒ ܕܡܢ
ܗܘܐ ܘܟܠܐ. ܡܬܚܙܝܐ ܗܘܘ ܒܪ̈ܢܫܐ
ܙܪܥ ܡܢ ܟܠܐ. ܐܝܟ ܕܝܢ ܦܪ̈ܨܘܦܐ ܡܢܝܢܝܢ
ܘܙܝܘܐ ܘܢܗܝܪܐ. ܒܪ̈ܢܫܐ ܕܝܢ ܠܐ ܐܫܬܪܝ ܒܓܘܗ
ܘܐܬܪܡܝ ܕܒܪܝܐ ܘܐܬܚܙܝ ܠܡܫܬܡܫܘ ܐܝܟܢܐ
ܗܘ ܘܙܗܝܪܐ. ܟܕ ܕܝܢ ܒܪܝܬܐ ܠܗܘܢ ܒܪܢܫܐ
ܕܝܠܗ ܙܩܝܦܐ. ܘܙܩܦܐ ܡܬܐܡܪ ܕܐܬܦܪܫ
ܡܬܚܫܒܐ ܕܢܦܫܬܐ ܟܝܢܐ ܡܬܚܙܝܐ ܐܬܐ
ܘܡܬܚܙܝܐ ܡܢ ܗܝ ܡܠܟܘܬܐ ܟܝܢܐ ܡܫܬܡܫܝܢ
ܐܦ ܟܝܢܐ ܐܚܪܝܐ. ܡܬܚܙܐ ܕܟܝܢܐ ܘܕܝܠܗܐ
ܫܡܫܐ ܐܝܟ ܕܒܢܘܗܝ ܕܗܘܠܐ ܪ̈ܘܚܢܝܐ ܘܕܝܠܗ
ܡܢܐ ܘܐܬܦܫܛ ܗܘܠܐ ܕܒ̈ܢܝܐ ܡܬܚܫܒܐ
ܐܦܢܝܬܐ ܕܒ̈ܢܝܐ ܕܟܠܐ. ܗܘ ܕܝܢ ܕܝܢ
ܐܦ ܡܬܦܪܫܬܐ ܗܝ ܕܒܗ ܡܬܚܙܝܐ ܘܡܬܚܫܒܐ
ܟܠܗ ܡܢ ܚܕ ܐܝܬܐ ܟܠ ܟܝܢܐ ܘܩܢܘܡܐ
ܘܗܘܝܐ ܕܐܝܬܝܗ ܗܘ ܗܝܘܠܐ ܕܡܬܚܫܒܐ ܗܝ
ܘܡܬܚܫܒܐ ܗܝ ܒܟܠܗ ܒܟܝܢܐ ܠܐܦܝ ܟܠ
ܒܠܐ ܡܬܚܫܒܐ ܟܢܝܫ ܐܝܟ ܡܬܐܡܪ ܩܢܘܡܐ ܕܟܠܗܘܢ
ܡܬܚܙܝܐ ܘܠܐ ܗܘܐ ܐܝܟ ܚܕ. ܘܗܟܢܐ
ܡܢ ܕܒ̈ܢܝܐ ܘܢܦܫܐ. ܘܡܬܒܪܝܐ ܗܘܬ ܕܗܘܐ ܕܐܬܚܙܝ

ܡܢ ܚܝܐ ܟܠܐ ܘܟܠܗܝܢ ܒܟܠܗܘܢ ܘܠܐ
ܡܬܚܙܝܐ ܟܕ ܗܘܠܐ ܐܝܬܝܗ. ܠܚܝܢܐ
ܢܦ̈ܫܢ ܕܒܢܝܐ ܘܢܦܫ ܙܪܥܐ ܕܒܗܝ
ܐܬܚܙܝܬ ܘܗܘܝܐ ܒܗܘܢ ܡܬܦܪܫܢܐ ܠܟܠ
ܗܘ ܐܝܟ ܡܬܚܙܐ ܗܘ ܘܢܦܫ ܒܗ ܡܢܗ
ܘܟܠܐ. ܡܢܗ ܡܢ ܡܬܚܫܠܐ ܘܠܐ ܐܬܒܪܝ
ܡܢܗ ܕܒ̈ܢܝܐ ܡܬܚܫܒܬܐ ܟܝܢܐ ܘܟܠܗܝܢ
ܢܦܫܐ ܘܢܦܫܐ ܕܗܝܐ ܕܟܠܗ ܘܟܠܗܝܢ ܘܐܦ
ܚܝܐ ܗܢܘܢ ܘܟ̈ܝܢܐ ܡܬܚܙܝܢ ܡܬܚܫܒܬܐ ܘܡܬܚܫܒܬܐ
ܘܟܠܗܘܢ ܗܘܘ ܙܡܢܝܢ ܣܓܝܐܐ ܘܠܐ ܡܬܦܪ̈ܫܢ
ܐܝܟ ܡܕܝܢܬܐ. ܘܡܢܗ ܡܢ̈ܝܢ ܢܦ̈ܫܢ ܟܠܐ
ܒܟܝܢܐ ܘܒܪ̈ܢܫܐ ܐܢܫܐ ܡܬܚܫܒܬܐ
ܘܕܝܠܗ ܟܠܐ ܘܡܬܚܫܒܬܐ ܘܕܝܠܗ ܒܪ̈ܢܫܐ
ܕܝܢ ܠܐ ܡܬܚܙܝܢ ܒܓܘܗ ܟܠܐ ܡܬܚܙܝܐ ܘܟܝܢܐ
ܟܝܢܐ ܕܒܗ ܒܚ̈ܝܐ ܐܦ ܡܢ ܡܬܚܙܝܐ
ܡܬܚܙܝܐ ܘܟܠ ܗܘܠܐ ܦܫܝܛ: ܠܗܘܢܐ
ܘܗܘܢܐ ܕܙܪܥܐ ܗܘ ܘܟܠ ܐܝܟ ܡܢܗ
ܘܡܬܚܙܝܐ ܘܡܬܚܫܒܐ ܒܟܝܢܐ ܘܟܠܗܝܢ ܐܢ̈ܫܐ
ܒܪ̈ܢܫܐ ܕܝܢ ܟܠܗ ܘܟܠܗ ܘܟܠܗܝܢ ܟܝ̈ܢܐ
ܗܝܐ ܗܘܬܐ ܠܟܠܗܘܢ ܘܗܘܬܐ ܗܝ
ܘܡܢܐ ܡܬܚܙܝܐ ܘܗܘܝܐ ܡܬܚܫܒܐ
ܒܗ ܙܡܢܐ ܗܘ ܘܟܠܗܘܢ ܘܟܠܗ
ܟܝܢܐ ܚܝܠܐ ܙܪܥܐ ܗܘ ܘܟܝܢܐ. ܘܐܦ
ܡܬܚܙܝܐ ܕܗܝ ܗܘܬܐ ܗܢܐ. ܘܐܬܐ
ܘܐܝܟ ܟܕ ܠܐ ܡܬܚܫܒܐ ܡܬܚܫܒܬܐ ܒܓܘܗ
ܕܗܘܬܐ ܗܘ ܕܟܠܐ ܠܐ ܡܬܚܙܝܐ ܡܬܚܫܒܐ
ܠܐ ܡܬܚܙܝܐ ܠܐܦܝ. ܐܝܟ ܕܝܢ ܕܡܬܚܙܝܐ
ܐܡܪܗ ܗܘܐ. ܒܗܕܐ ܕܝܢ ܟܝܢܐ ܠܟܠܗܘܢ
ܟܠܗ ܗܘ ܘܒܟܠ ܡܬܚܫܒܬܐ ܘܡܬܚܫܒ
ܐܝܟ ܘܟܠܗ ܐܝܟ ܟܠܗܘܢ ܘܟܠܗܘܢ
ܡܬܚܫܒ

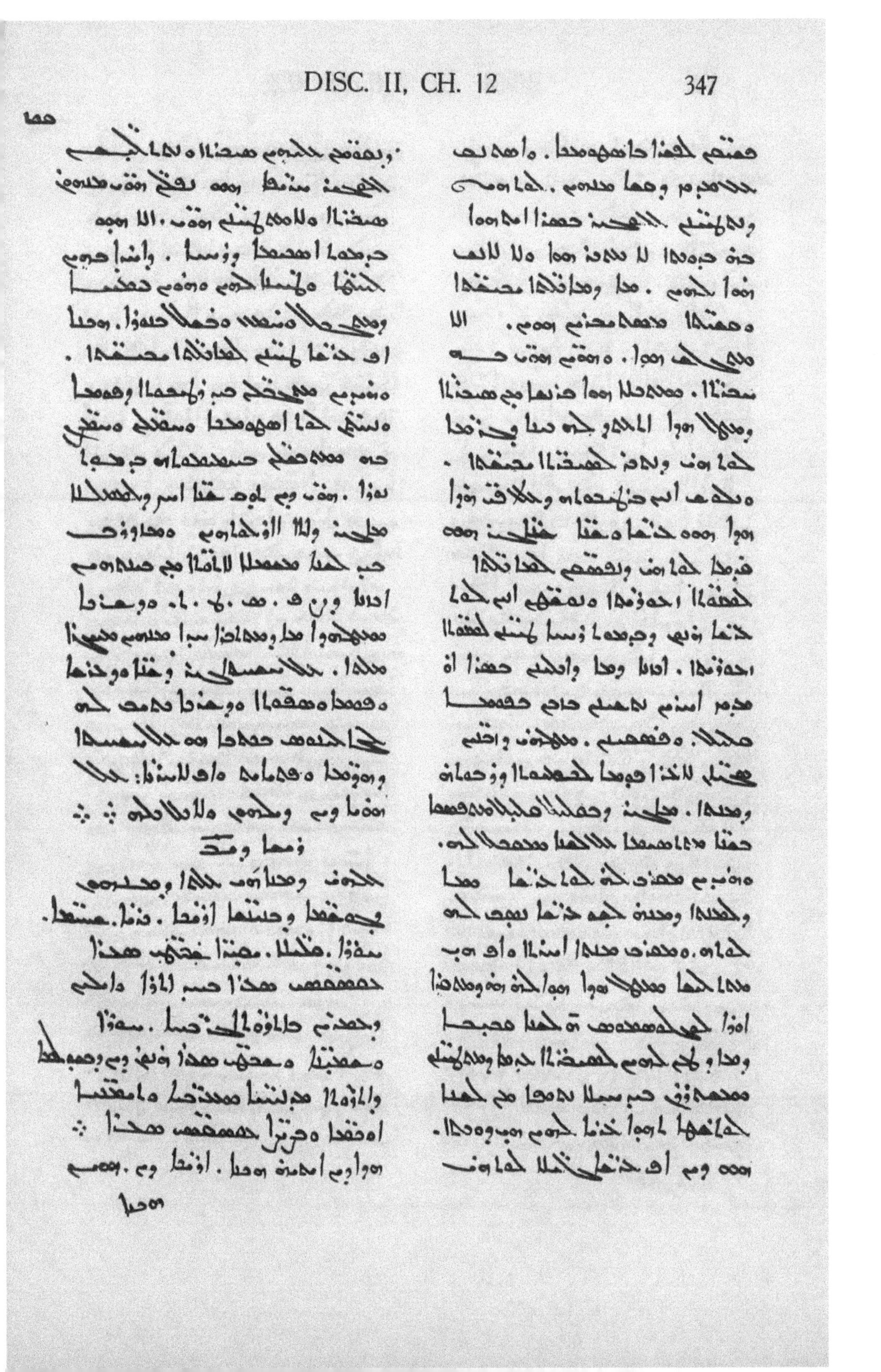

[illegible]

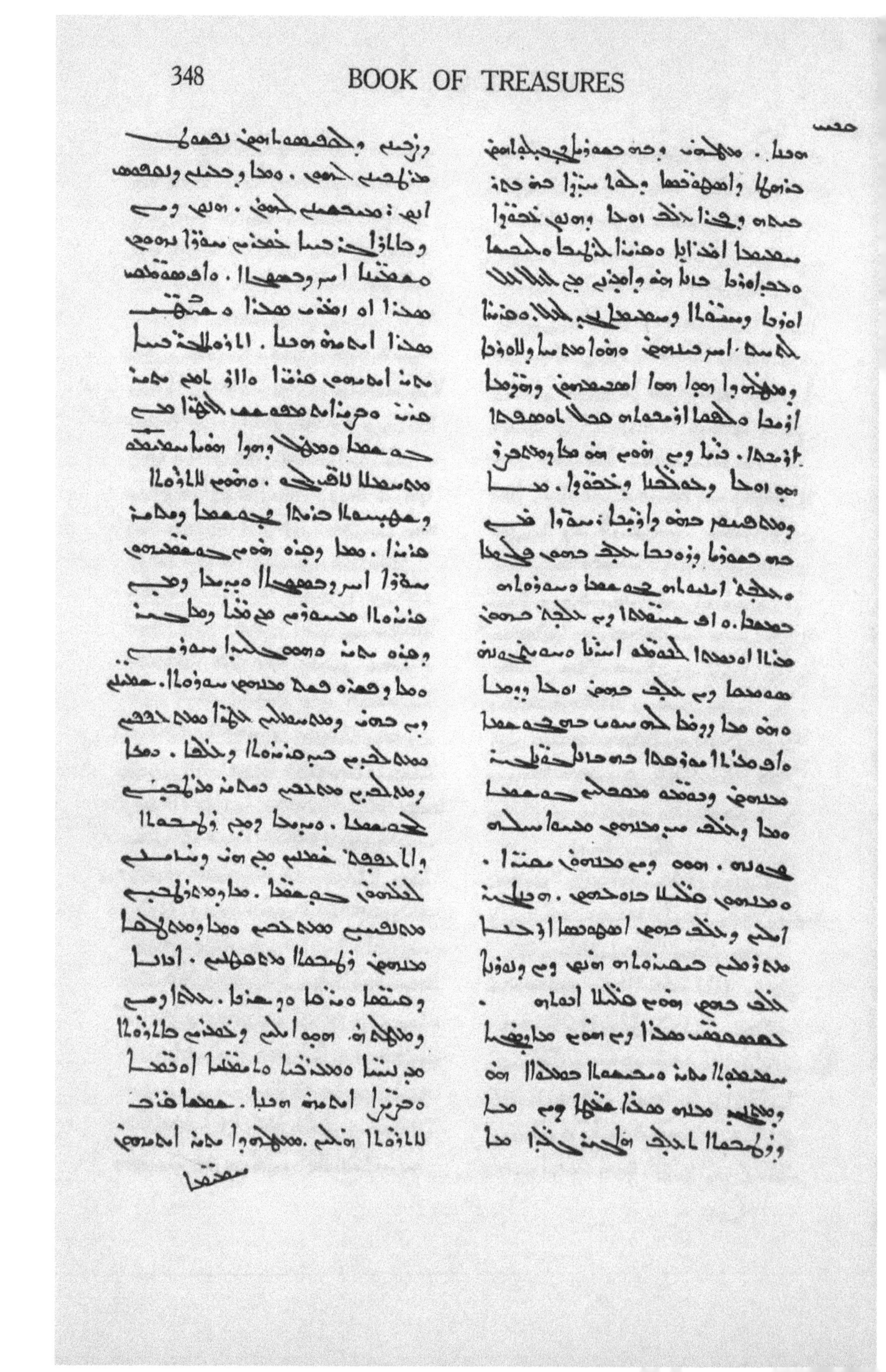

[illegible]

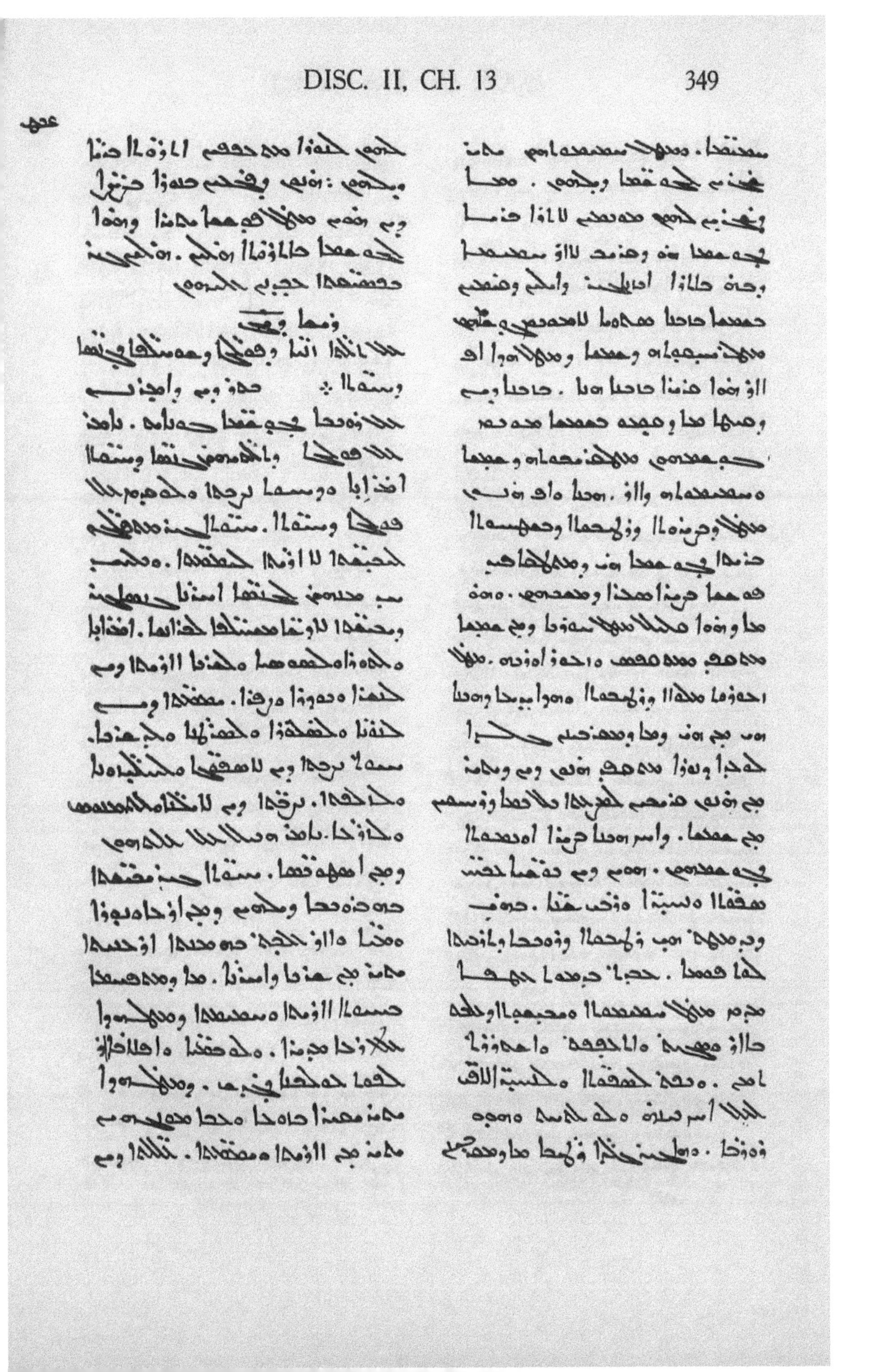

ܥܠܐ

ܕܒܗ ܟܬܒܐ ܣܗܕܝܢ ܒܟܪܘܙܘܬܐ ܐܘܟܝܬ ܡܢ
ܒܗ ܒܪܘܚܐ ܩܕܝܫܐ܆ ܒܗܘ ܡܫܡܫܢܘܬܐ
ܐܡܝܢܬܐ܆ ܕܡܢ ܟܕ ܗܘܬ ܐܘܪܫܠܡ ܘܡܩܕܫܐ
ܘܟܗܢܐ ܕܝܗܘܕܝܐ ܘܩܘܪܒܢܘܗܝ ܕܒܢܝ ܒܩܪܐ.
ܘܡܬܒܨܪܐ ܗܘܬ ܒܫܘܠܛܢܗ ܕܐܢܫܐ ܗܘܬ. ܘܪܘܚܐ
ܩܕܝܫܐ. ܘܒܗܘ ܙܒܢܐ ܐܬܐ ܠܟܠ ܒܗ
ܥܠܡܐ ܕܐܬܐ ܟܕ ܡܢ ܟܕ ܗܘ ܡܫܝܚܐ ܘܐܘܪܚܐ
ܕܠܟܠܢ ܗܘܐ ܗܘܐ ܕܡܢܗ ܠܐܬܪ ܘܒܗܘ
ܠܒܒ ܘܦܘܡ. ܕܡܫܬܠܡܐ ܕܡܢ ܟܠܗ ܗܘܬ ܗܘܐ
ܐܡܪܝܢ. ܗܘܐ ܠܗܘܢ ܟܢ ܠܐܝܩܘܢܬܐ ܕܡܫܬܐ
ܒܗ ܒܣܐܪܐ ܦܪܝܫܐ ܕܟܝܢܐ ܗܘܬ ܠܟܠܐ
ܚܕ ܚܕ ܡܢܗܘܢ ܠܟܠܗ ܒܝܘܡܬܐ ܗܘܬ.
ܒܪܘܟܐ ܕܫܢܬܐ. ܘܟܠܚܕ ܚܕ ܡܢܗܘܢ
ܠܟܠܗ ܕܟܝܢܐ ܫܠܡ ܘܐܘܟܝܬ ܒܗܘ
ܟܠܡܟܘܢ ܗܘ ܕܘܝܐ ܘܫܒܩ ܡܫܟܚܐ
ܚܕ. ܐܘܪܚܐ ܩܕܝܫܐ܆ ܚܕ ܕܟܝܢܐ
ܐܝܟܐ. ܟܕ ܐܝܟܐ ܡܬܒܨܪܐ. ܘܕܘܟܝܐ
ܢܦܫܐ ܘܡܫܝܚܐ܆ ܠܟܪܣܛܝܢܐ. ܘܡܫܬܡܠܝܐ
ܟܡܐ ܐܘܪܚܬܐ. ܐܘܪܚܬܐ ܒܟܪܣܛܝܢܘܬܐ
ܐܡܪܬ ܗܘ ܕܟܢܫܬܗ ܐܝܟܢܐ ܐܡܪܗ
ܐܘܣܛܘܟܣܐ ܠܟܘܡܫܚܬܐ ܕܟܝܢܐ ܘܐܘܪܚܐ
ܠܗ ܒܟܢ ܚܕ ܡܢܗܘܢ ܦܚܡܐ ܡܢ
ܐܘܪܟܝܗܘܢ ܕܟܝܢܐ ܕܡܢ ܒܗ ܠܟܠܐ
ܕܚܕ ܡܢܗܘܢ܆ ܒܗܘ ܦܚܡܐ ܘܐܘܪܚܐ
ܘܟܠܗܝܢ ܗܘܐ ܕܟܝܢܐ ܒܐܘܪܚܐ ܡܬܒܨܪܐ
ܐܝܟܐ ܕܦܪܝܫ ܐܡܪܝܢ ܗܘܢܐ ܐܦ
ܠܦܘܪܫܢܐ ܫܡܗ ܐܬ ܟܕ ܡܢ ܐܬܘܗܝ. ܐܦ
ܦܢܐ ܠܦܘܪܫܢܐ ܟܕ ܡܢ ܡܫܠܘܬܐ ܘܠܫܘܚܠܦܐ
ܡܫܬܒܚܐ ܗܘܐ ܕܡܢ ܦܪܫܝܢ ܒܗܘܢ ܒܐܝܠܝܢ
ܒܗܠܝܢ ܕܐܬܘܗܝ ܫܡܗܐ. ܘܐܚܪܢ

ܒܗ ܒܐܘܪܚܐ ܗܘ ܕܙܘܕܩܐ ܕܟܝܢܐ ܘܐܘܪܚܐ
ܡܫܝܚܝܐ ܗܘ ܕܡܢ ܡܫܝܚܐ ܗܘܬ ܡܫܬܠܡܐ܆
ܠܐܢܫܐ. ܘܗܘ ܕܟܪܘܙ ܫܠܝܚܐ܆ ܠܟܪܣܛܝܢܐ
ܘܠܐܘܪܚܬܐ. ܘܗܘ ܕܡܫܝܚܐ ܗܘܬ ܗܘܬ
ܡܟܝܠ ܠܐܘܣܛܘܟܣܐ ܕ ܗܘ ܟܕ ܕܐܬܐܗ
ܡܕܒܪ܆ ܘܐܘܪܚܐ ܠܐܢܫܐ. ܐܝܟܢܐ ܕܡܫܬܠܠܐ
ܘܒܫܪܪܐ ܕܡܢ ܐܘܪܚܬܐ ܐܡܝܢܝܢ. ܘܐܦ
ܒܗܘܢ ܡܫܬܠܡܢܘܬܐ ܒܗ ܕܡܢ ܒܐܢܐ ܘܐܦ
ܒܣܘܪܐ ܕܟܝܢܐ ܗܘܘ ܠܐܢܫܐ ܦܪܘܩܐ. ܒܗܕ
ܕܡܢ ܗܘ ܠܐܡܪ ܟܠ ܗܘܝܐ ܘܡܫܝܚܐ
ܕܐܘܪܚܐ ܘܡܠܘܗܝ ܒܫܪܪܐ. ܘܠܗ ܡܫܝܚ ܟܠ
ܡܫܝܚܐ ܐܘܪܚܐ ܘܗܘ ܡܫܝܚܘܬܐ. ܒܐܝܠܝܢ ܐܡܪܢܐ
ܘܫܘܫܠܐ ܘܐܘܪܚܐ ܘܒܫܪܪܐ. ܒܗ ܦܪܘܩܐ
ܐܡܪܝܢ ܠܟܠܗ ܕܡܫܝܚܐ ܟܠܗܐ ܘܟܠ
ܢܟܣܝܢ ܫܡܗܬܐ ܡܫܝܚܐ ܗܘܐ ܟܪܝܟܐ ܟܕ ܡܢ
ܡܫܝܚܐ. ܘܟܠܐܘܪܚܐ ܡܢܗܘܢ. ܘܟܠ ܟܠܐܝܬܐ
ܘܦܘܠܓܐ ܐܝܟ ܫܢܝܬܐ. ܘܢܟܣܝܢ ܦܪܫܝܢ ܕܒܟܠ
ܟܕ ܒܝܕ ܒܝܕܘܬܐ ܡܢܗܘܢ. ܠܐܝܩܘܢܬܐ
ܡܟܠܐ ܘܒܝܢ ܡܫܝܚܐ. ܦܪܝܫ ܕܡܢ ܐܘܪܚܐ
ܦܘܠܓܐ ܘܒܫܪܪܐ ܕܡܢ ܡܫܝܚܘܬܐ ܐܝܟ ܟܠܗܝܢ ܢܟܣܝܢ
ܡܫܬܒܚܬܐ ܘܟܠ ܕܡܢ ܡܫܝܚܐ. ܐܝܟܢܐ ܕܫܡܝܫܬܐ
ܡܢ ܐܝܟܘܪܘܬܐ ܘܡܦܪܫܐ ܠܐܡܪܬܐ ܗܘ ܡܫܬܪܪܐ
ܡܬܦܠܓܐ ܘܐܘܪܚܬܐ. ܘܒܪܝܟ ܡܢ ܢܦܫܬܐ
ܘܡܢܝ ܕܒܫܪܐ ܡܫܝܚ ܕܗܘܪܐ ܒܐܝܟܢܐ

ܢܦܫܐ ܕܡܢ

ܟܠܗܘܢ ܕܡܫܬܠܡܢܐ ܡܢ ܢܦܫܬܐ ܒܪܝܬܐ
ܠܢܦܫܗ ܕܐܝܬ ܡܫܬܡܠܝܢ. ܘܢܦܫܬܐ ܕܟܠܗ
ܒܗ ܡܫܬܟܚܐ ܐܝܟ ܠܟܠܗܝܢ ܢܦܫܬܐ ܕܡܢ ܘܐܦ
ܡܢ ܡܫܝܚܝܢ ܦܪܫܝܢ ܘܟܠܗ ܘܕܟܠܗܝܢ ÷
ܘܐܡܪܝܢ ܕܦܪܫܝܢ ܡܢ ܐܘܪܚܐ ܢܦܫܐ ܡܢ ܒܪܝܐ

ܡܚ

ܗܠ

ܡܢ ܓܘܫܡܐ ܕܡܢ ܐܣܛܘܟܣܐ. ܘܡܢ
ܢܦܫܐ ܡܠܝܠܬܐ ܕܠܐ ܡܝܘܬ. ܘܠܐ ܐܣܛܘܟܣܝܐ
ܗܘܝܢ: ܚܕܐ ܕܡܢ ܐܣܛܘܟܣ̈ܐ ܡܪܟܒ .
ܓܘܫܡܐ ܗܘ ܕܐܣܒ ܦܓܪܐ ܢܦܠ ܐܠܗܐ
ܟܠܗ ܓܘܫܡܐ ܗܘ ܕܝܢ ܕܠܐ ܡܝܘܬ. ܡܬܚܙܐ
ܘܠܐ ܡܝܘܬ ܗܘ. ܘܠܐ ܢܦܠ ܠܐܣܝܪ ܦܓܪܐ
ܐܝܟ ܟܠܗ ܓܘܫܡܐ. ܘܐܦ ܡܕܝܢ ܘܚܙܘܐ
ܘܐܒܗܬ. ܘܗܫܝܢ ܢܦܩ ܠܗ ܘܫܟܢܗ .
ܡܣܬܟܠ ܕܦܓܪ̈ܐ ܠܐ ܡܬܚܙܝܢ. ܐܦ ܕܚܙܐ
ܣܘܓܐ ܚܝܐ ܘܐܒܗܬ. ܘܟܠܗܝܢ ܡܣܘܡܢ
ܡܢ ܡܬܚܠܦܢܘܬܐ ܗܕܐ ܘܠܐ ܡܬܚܙܝܢ ܚܫܚܘܬ
ܦܓܪܐ ܘܡܢ ܐܝܟ ܗܢܘܢ ܟܕ ܒܪܝܢ ܡܫܘܬܦܘܬܐ
ܠܟܠܗܘܢ ܕܐܝܬ ܢܦܫܐ ܘܠܓܘܫܡܐ ܚܝܐ ܒܪ̈ܢ
ܚܝ̈ܐ ܒܓܘܫܡܐ ܠܡܫܘܕܥܘ ܘܠܟܠ ܡܐ
ܘܡܬܦܢܝܢ. ܡܚܐ ܘܡܬܚܠܦܢ ܘܠܐ ܡܬܦܢܝܢ
ܐܝܟ ܚܝ̈ܐ ܗܢܘܢ ܐܝܬܝܗܘܢ ܒܗ ܡܢ ܟܪ̈ܝܬܐ
ܘܚܝܝܢ ܒܟܠܗ ܗܘܝܐ ܘܚܝ̈ܐ ܠܟܠܗ
ܟܪ̈ܝܬܐ ܘܫܪܝܢ ܒܟܠܗ ܗܘܝܐ ܘܚܝ̈ܐ
ܠܗ ܟܪ̈ܝܬܐ. ܐܦ ܟܕ ܡܢ ܗܘܝܐ ܘܟܪ̈ܝܬܐ
ܘܚܝܝܢ ܐܣܛܘܟܣܐ ܘܡܠܝܐ ܘܢܦܫܐ ܒܓܘܗ
ܗܟܢ ܕܝܢ. ܡܬܚܙܐ ܕܐܣܛܘܟܣܐ ܐܝܟ ܠܗܘܢ
ܘܠܗ ܗܘܝܐ ܐܝܟ ܕܒܓܘܫܡܐ ܐܣܝܪܐ
ܘܚܝܝܢ ܕܝܢ. ܐܦܠܐ ܗܘܝܐ ܐܝܟ ܠܗܘܢ
ܐܝܬܝܗܘܢ ܚܝܐ. ܡܕܝܢ ܐܣܝܪܐ ܗܘ: ܡܢ
ܗܝܐ ܒܓܘܫܡܐ ܐܝܬܝܗܘܢ. ܘܠܒܪ ܚܝ̈ܐ
ܕܓܘܫܡܐ ܡܬܚܙܝܢ ܘܢܦܫܐ ܚܝܐ ܗܘܝܢ .
ܗܟܢ ܕܝܢ ܠܐ ܡܬܚܙܝܢ. ܡܬܚܙܐ ܗܘ ܕܠܐ
ܡܬܪܓܫ ܡܢ ܡܣܬܟܠܢܘܬܐ ܘܡܬܪ̈ܓܫ ܡܣܬܘܢ
ܗܢܘܢ ܕܓܘܫܡܐ: ܡܕܝܢ ܚܝ̈ܐ ܘܢܦܫܐ
ܘܠܐ ܒܓܘܫܡܐ ܐܝܬܝܗܘܢ ܐܝܟܢܐ ܕܚܝ̈ܐ ܕܝܢ

ܡܣܬܟܠ ܕܒܪܢܫܐ ܐܝܬܝܗܘܢ ܚܝ̈ܐ ܐܝܟ ܠܗ. ܡܕܝܢ
ܗܘ ܡܢ ܠܒܪ ܕܐܣܝܪܐ ܘܠܒܪ ܕܒܓܘܫܡܐ. ܘܗܘܝܐ
ܕܠܒܪ ܡܬܪܟܒܐ ܡܢ ܐܣܛܘܟܣܐ ܡܬܚܙܝܢܐ
ܘܢܦܠܐ ܠܗ ܐܝܟ ܦܓܪܐ ܡܢ ܐܝܟ ܟܠܗܘܢ ܦܓܪ̈ܐ
ܗܘ. ܘܗܫܝܢ ܒܓܘܫܡܐ ܠܐ ܡܬܪܟܒܐ ܗܘ ܕܡܢ
ܐܣܛܘܟܣܐ ܘܠܐ ܡܬܚܙܝܢܐ ܘܠܐ ܢܦܠܐ ܠܐܣܝܪ
ܦܓܪܐ. ܡܕܝܢ ܘܟܠܗ ܠܓܘܫܡܐ ܡܬܚܙܝܢܐ
ܐܦܐ ܡܢ ܠܐܢܫ ܚܝ̈ܐ ܡܪܟܒ ܒܪܢܫܐ ܡܢ ܢܦܫܐ
ܘܠܓܘܫܡܐ ܘܡܢ ܦܓܪܐ ܘܐܝܟܢܐ ܒܓܘܫܡܐ. ܘܗܟܢ
ܠܐ ܗܫܝܢ ܡܬܚܙܝܢܐ. ܠܒܪ ܕܡܢ ܟܠ ܕܡܬܚܙܝܐ
ܡܢܐ ܘܡܢ ܚܝ̈ܐ ܚܝܝܢܘܬܗܘܢ ܡܬܚܙܝܢ ܘܚܝܝܢ
ܒܓܘܫܡܐ ܗܫܝܢ ܠܗܘܢ. ܟܪ̈ܝܬܐ ܘܚܝܝܢ
ܘܗܘܐ ܟܠ ܕܡܢ ܡܬܚܙܝܢ ܒܟܠܗܘܢ. ܗܘ
ܟܕ ܡܬܚܙܝܐ ܘܡܢ ܟܪ̈ܝܬܐ ܐܝܬܝܗܘܢ. ܘܗܘܐ
ܠܡܬܚܙܝܢܘ ܡܕܝܢ ܚܝ̈ܐ ܘܢܦܫܐ ܗܘܝܢ:
ܗܘ ܕܠܐ ܗܘ ܕܐܝܬܝܗܘܢ. ܘܗܘ ܡܐ ܕܡܬܚܙܝܢ
ܕܚܝܬܐ ܡܕܝܢ ܒܟܠܗ ܠܟܠܗܘܢ ܐܦ ܗܟܢܐ ܠܓܘܫܡܐ
ܘܚܝ̈ܐ ܒܓܘܫܡܐ ܡܬܚܙܝܢ ܠܗ ܕܒܪܢܫܐ ܗܘ ܒܪ ܗܟܢ
ܘܢܦܫܐ ܗܘܝܢ. ܡܕܝܢ ܘܡܬܚܙܝܢ ܚܝ̈ܐ
ܒܓܘܫܡܐ ܗܘܐ ܐܝܟ ܗܘ ܘܚܝܝܢ ܒܟܠܗ
ܘܒܓܘܫܡܐ. ܡܬܚܙܝܢ ܘܚܝܝܢܘܬܐ ܠܟܠܗܘܢ
ܚܝ̈ܐ ܕܒܓܘܫܡܐ ܠܐܝܟ ܒܢܦܫܘܬܐ ܗܘ ܘܐܝܬܝܗ
ܡܬܚܙܝܐ ܠܐܣܛܘܟܣܐ ܐܘ ܡܬܚܙܝܢ ܠܐ ܡܬܚܙܝܢ
ܒܪܝܬܐ ܘܡܬܚܙܝܢܘܬܐ ܚܝ̈ܐ ܘܠܒܪ. ܠܟܠܗ
ܠܗܘܢ ܒܪܢܫܐ ܡܕܝܢ ܘܟܕ ܒܢܦܫܐ ܒܓܘܫܡܐ
ܡܬܚܙܝܢ ܠܗܘܢ ܢܦܫܐ ܘܠܓܘܫܡܐ .
ܐܝܟ ܕܐܝܬ ܚܝܝܢ ܠܒܪܢܫܐ ܘܐܝܬܝܗ ܡܬܚܙܝܢܘܬܐ
ܡܣܬܟܠܢܘܬܐ ܡܬܚܙܝܢܐ. ܡܪܝܐ ܗܘ ܘܡܬܚܙܝܢܐ
ܠܓܘܫܡܗ ܐܝܬܝܗ ܘܚܝܝܢ ܠܗ: ܘܡܬܚܙܝܢܐ
ܗܘ ܠܓܘܫܡܗ ܚܝܐ ܕܢܦܫܐ ܗܘ ܘܡܢ

ܡܟ

ܐܣܛܘܟܣܐ ܡܫܡܫܢܝܐ. ܘܡܛܠ ܗܢܐ ܐܦ ܐܐܪ
ܘܡܝܐ ܘܢܘܪܐ ܘܐܪܥܐ. ܘܐܬܡܢܥܘ
ܡܫܡܫܢܘܬܐ ܕܟܠܗ ܥܠܡܐ ܡܢ ܗܪܟܐ. ܘܕܠܐ
ܠܐ ܡܫܟܚܐ. ܘܗܕܐ ܦܫܝܛܐ ܗܘܬ ܡܢ ܥܘܡܩܐ
ܘܐܣܛܘܟܣܐ. ܗܕܐ ܕܠܐ ܐܢܫ ܚܙܐ ܚܙܝܐ ܗܘܬ
ܚܙܝܬܐ ܕܝܠܗ ܕܗܘܝܐ ܗܘ: ܘܡܢ
ܗܘ ܕܗܘܐ ܕܢܦܩ ܠܚܙܝܬܐ ܚܙܝܐ ܗܘܬ
ܡܢܟ. ܠܐ ܗܘܬ ܦܫܝܛܐ. ܗܘܐ ܕܢܦܩ ܚܙܝ
ܚܙܝܬܐ ܗܘܬ ܦܫܝܛܐ. ܘܗܝ ܕܡܬܚܙܝܐ ܕܝܢ
ܢܦܩ ܚܙܝܬܐ ܘܡܬܚܙܝܐ ܦܫܝܛܐ ܡܟܝܠ
ܡܫܡܫܢܝܢ ܠܚܙܝܬܐ. ܘܡܢ ܕܡܢ ܦܫܝܛܝܢ
ܡܫܡܫܢܐ ܐܝܬܝܗ ܕܦܫܝܛܐ ܘܡܫܟܚܐ ܕܐܪܥܐ
ܘܐܪܥܐ ܡܫܡܫܢܐ ܡܫܡܫܢܐ ܘܠܐ ܡܫܟܚܐ
ܕܝܢ ܢܘܪܐ ܦܫܝܛܐ ܡܫܡܫܐ ܐܣܛܘܟܣܐ
ܠܐ ܗܘܬ ܗܘܬ ܡܫܟܚܐ ܕܡܫܡܫܢܘܬܐ. ܠܐ
ܡܫܡܫܢܐ ܡܬܚܙܝܐ ܠܚܙܝܬܐ. ܐܦ ܡܢ
ܕܝܢ ܐܢܫܐ ܡܫܡܫܢܐ ܘܐܝܬ ܢܘܪܐ ܘܐܐܪ
ܘܗܘܐ ܠܐ ܡܫܡܫܝܢ ܐܢܫܐ ܐܪܥܐ ܡܬܚܙܝܢ ܕܡܢܗܘܢ
ܡܢ ܢܦܫܗܘܢ ܕܝܢ ܡܫܡܫܢܝܢ ܐܝܟ ܡܫܡܫܢܐ
ܘܡܝܐ ܘܐܪܥܐ. ܕܢܘܪܐ ܕܝܢ ܗܘܐ ܘܗܘܐ
ܗܘܬ ܚܙܝܬܐ. ܕܝܠܗ ܕܗܘܐ ܚܝܐ ܐܢܫܐ
ܘܡܫܡܫܢ ܠܐܣܛܘܟܣܐ. ܕܢܘܪܐ ܕܝܢ ܐܝܬ
ܠܗ. ܘܐܦ ܗܘ ܕܐܪܥܐ ܐܬܪܐ ܡܫܡܫܢܝܬܐ
ܗܘܐ ܡܫܡܫܢܘܬܐ ܗܘ ܡܢ ܕܡܫܡܫܢܐ
ܠܡܫܡܫܢܘ. ܐܢܫܐ ܕܡܫܡܫܢܘ. ܐܦ
ܡܢ ܕܝܢ ܐܢܫܐ ܐܝܬ ܠܐܣܛܘܟܣܐ ܕܟܠܗܝܢ
ܠܗܘܢ ܘܐܝܬܘ ܢܘܪܐ ܡܫܡܫܢܐ ܘܐܐܪ ܕܝܢ.
ܐܡܪܝܢ ܗܢܘܢ ܕܟܠܗܝܢ ܚܙܝܬܐ ܕܝܠܗ
ܐܢܫ ܡܢ ܢܘܪܐ. ܘܦܫܝܛ ܠܗ ܡܫܡܫܢܐ
ܠܗ ܕܝܢ ܡܫܡܫܢ ܠܗ ܢܘܪܐ ܘܡܢ

ܐܣܛܘܟܣܐ ܗܘ. ܠܚܙܝܬܐ ܡܢ ܐܘܪܚܐ
ܕܚܙܝܬܐ ܕܝܢ ܐܣܛܘܟܣܐ ܠܚܙܝܬܗ
ܘܡܫܡܫ ܡܢܗ ܐܦ ܡܫܡܫܢܐ ܡܫܡܫܢܘܬܗ
ܐܠܐ ܒܢܘܪܐ ܚܙܝܐ ܕܠܐ ܢܦܫ ܐܘ
ܡܫܡܫ ܚܙܝ ܗܘܐ ܗܘܬ ܗܘܝܐ. ܡܫܡܫ
ܕܚܝܐ ܐܢܫܐ ܐܝܟ ܠܗ. ܡܫܡܫ ܡܢ ܗܘ
ܕܐܣܛܘܟܣܐ ܕܚܝܐ ܡܫܡܫܢ ܘܗܘ
ܡܫܡܫ. ܘܐܦ ܕܐܣܛܘܟܣܐ. ܘܐܢܫ
ܕܚܙܝܬܐ ܐܢܫ ܡܢ ܢܘܪܐ ܠܗ ܕܐܝܬ
ܘܢܦܫܗ ܠܗ ܡܫܡܫܢܐ ܠܗ. ܘܕܝܢ
ܠܗ ܡܫܡܫܢ ܡܢ ܠܗ ܢܘܪܐ ܘܡܢ
ܐܣܛܘܟܣܐ. ܐܦ ܚܝܐ ܡܢ ܠܐܣܛܘܟܣܐ
ܘܐܣܛܘܟܣܐ. ܐܝܟ ܢܘܪܐ ܘܚܙܝܬܐ ܠܚܝܐ
ܐܣܛܘܟܣܐ ܘܡܫܡܫ ܘܡܫܡܫܢ ܠܗ.
ܘܐܝܬ ܐܢܫܐ ܘܡܢ ܡܫܡܫ ܘܐܝܟ ܡܬܚܙܝܐ
ܚܙܝ ܠܐܣܛܘܟܣܐ ܢܦܫ ܐܢܫܐ ܘܡܫܡܫܢܐ
ܐܡܪ ܗܘ ܠܗ. ܚܙܝܬܐ ܘܟܠܗܝܢ ܢܘܪܐ
ܘܡܢ ܚܙܝܬܐ ܢܘܪܐ ܡܫܡܫܢ ܘܟܠܗ
ܡܫܡܫ ܕܝܢ ܐܝܟ ܢܘܪܐ. ܘܡܫܡܫ ܢܦܫܗ
ܠܗ ܢܦܫܗ ܚܙܝܬܐ ܕܢܘܪܐ ܘܐܝܬ ܠܚܙܝܬܗ
ܡܫܡܫܢܐ ܕܝܢ ܡܫܡܫܢܘܬܐ ܚܝܐ
ܘܢܦܫܐ. ܚܙܝܬܐ ܕܝܢ ܗܘ ܕܐܝܟ ܡܫܡܫܢܐ
ܢܘܪܐ ܚܙܝܬܐ ܗܘ ܠܡܫܡܫܢܘܬܗ ܡܢ ܢܦܫܗ
ܚܝܐ ܘܐܪܥܐ ܦܠܓܐ. ܘܡܫܡܫܐ ܐܪܥܐ ܦܠܓܐ
ܡܫܡܫ ܐܝܟ ܗܘܐ ܘܢܘܪܐ ܡܫܡܫܢ
ܘܡܢ ܕܐܝܬ ܢܘܪܐ ܡܫܡܫܢ ܚܙܝܬܐ. ܡܢ
ܠܗ ܡܫܡܫܐ ܘܐܣܛܘܟܣܐ ܢܦܫܐ ܡܫܡܫܢܐ
ܡܫܡܫ ܚܝܐ ܘܢܦܫܐ ܡܫܡܫܢܐ ܗܘ. ܐܝܟ ܠܐ
ܚܙܝܬܐ. ܚܙܝܐ ܠܐܦ ܟܠܗܘܢ ܘܡܢ ܚܙܝܬܗ
ܘܚܙܝܬܐ. ܡܫܡܫ ܠܗ ܠܐܦ ܟܠ ܐܣܛܘܟܣܐ
ܐܣܛܘܟܣܐ

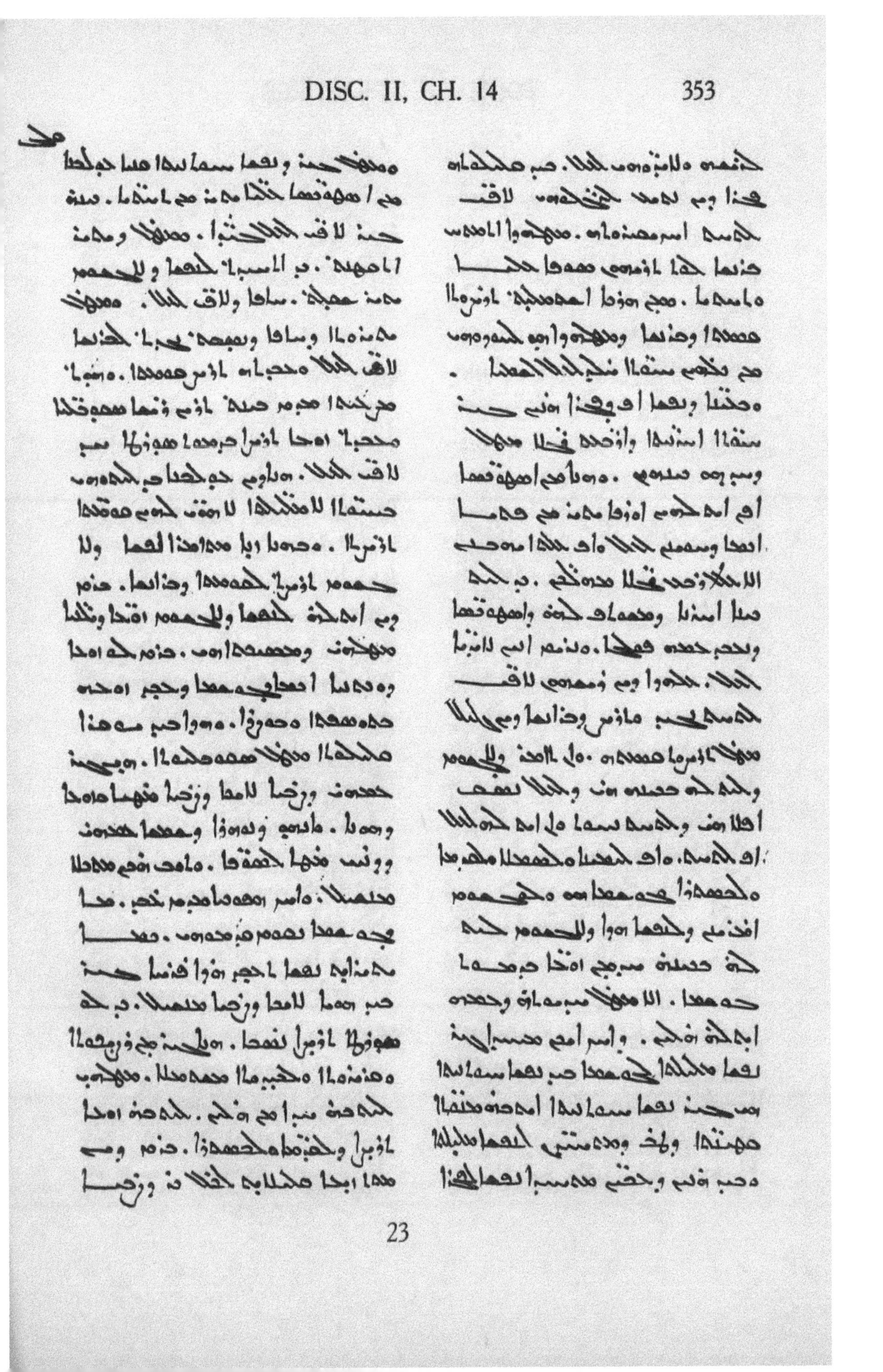

[illegible]

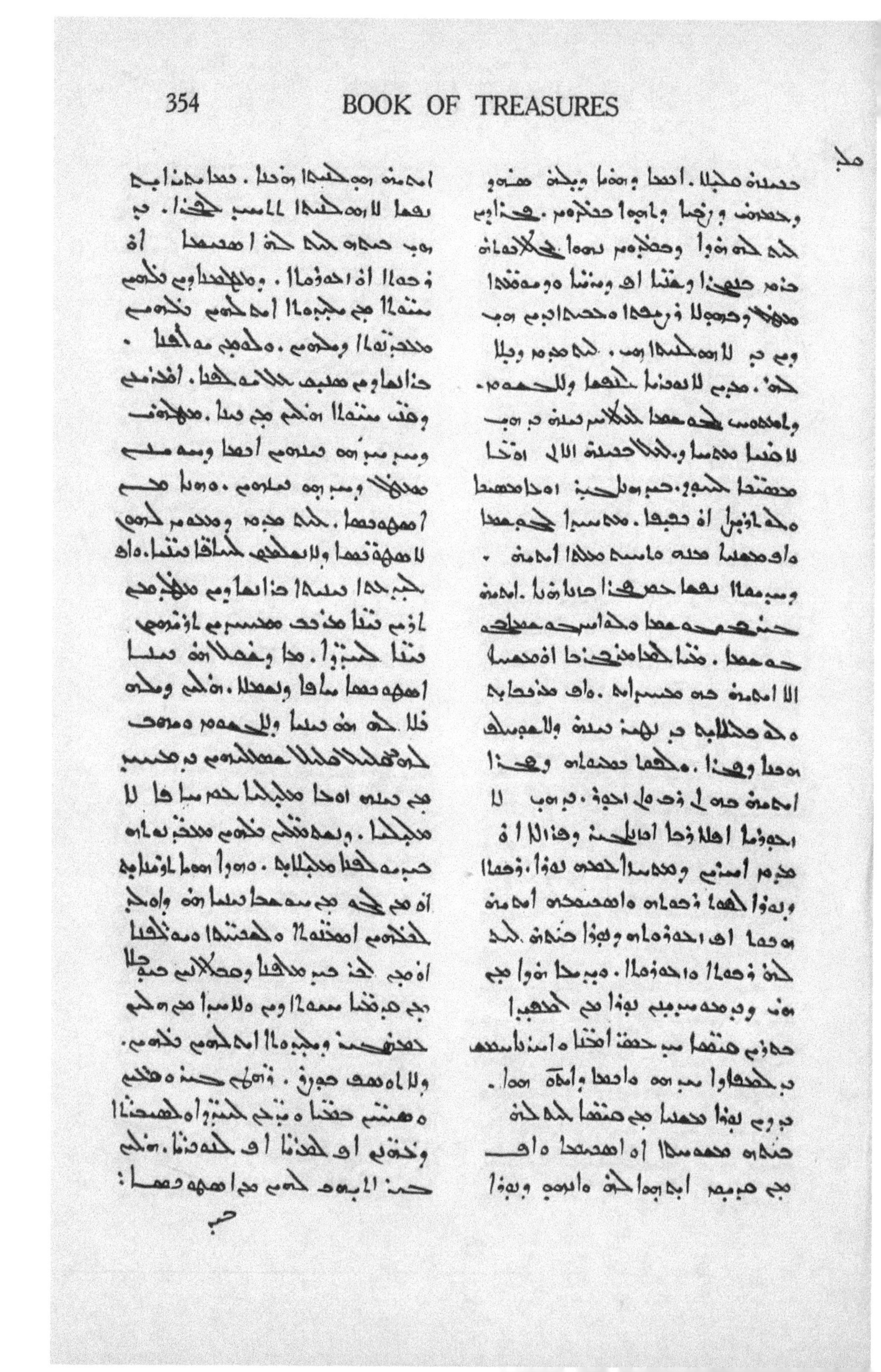

[illegible]

[illegible]

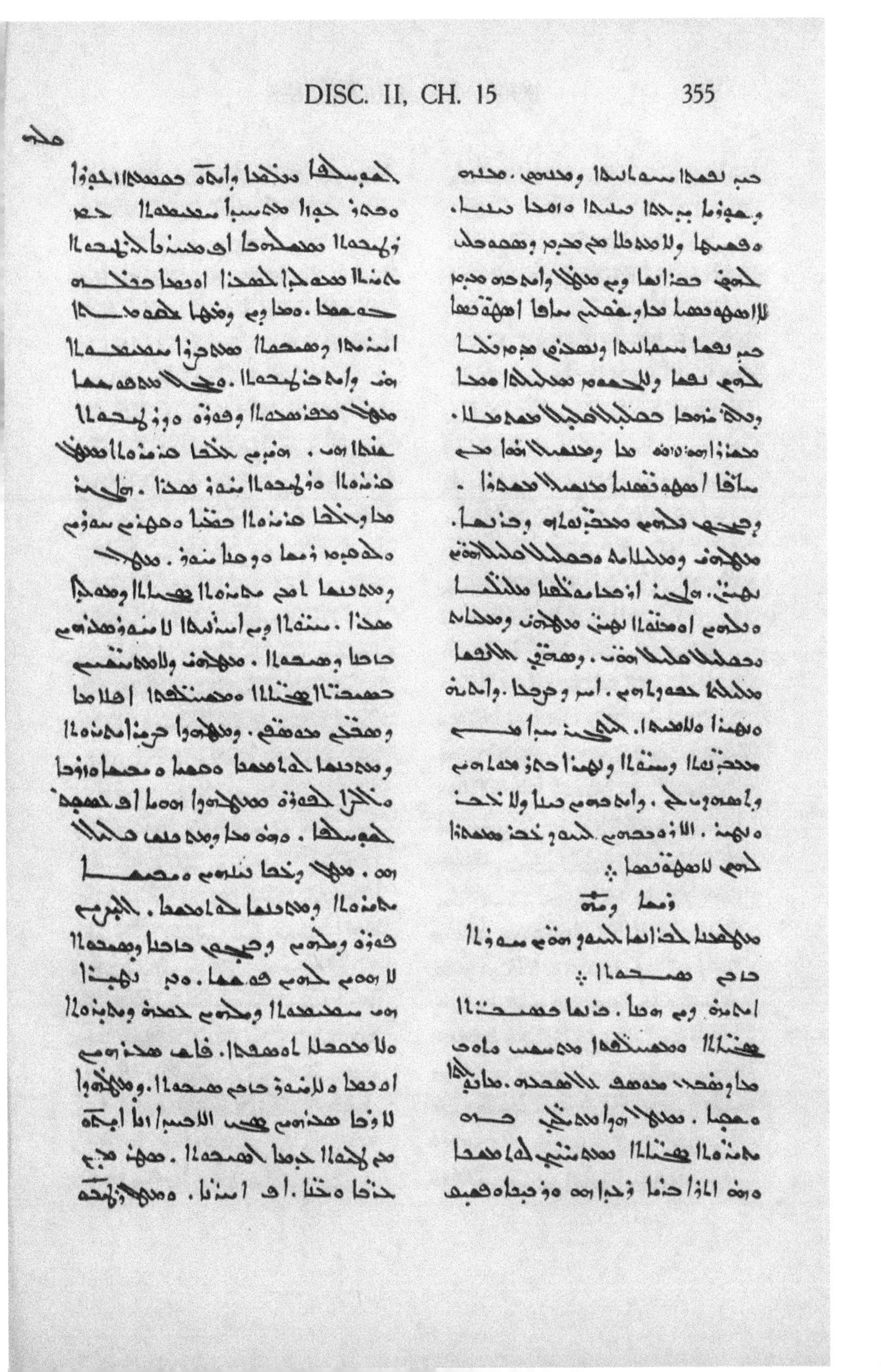

ܘܡܫܡܫܢܘܬܐ ܕܚܝܐ ܢܦܫܢܝܐ ܡܬܡܢܝܐ ܡܢ ܕܫܪܝܐ
ܗܘܐ ܠܗܘܢ ܫܘܪܝܐ ܒܡܫܩܝܬܐ ܕܠܒܗܘܢ .

ܪܝܫܐ ܕܡ̄ܗ

ܡܛܠ ܗܘ ܕܟܠܗ ܕܒܗ ܟܠܗܘܢ ܚܝ̈ܘܬܐ ܚܕ
ܐܝܬ ܠܗܘܢ: ܫܘܫܡܐ ܐܘ ܛܘܪܐ ܐܘ ܟܠܝܬܐ
ܐܘ ܠܒܐ ܐܘ ܟܒܕܐ ܘܫܪܝܐ ܕܒܗ ܟܠܗܘܢ
ܡܬܠܘܢ ܕܐܣܛܘܟܣܐ ܐܪ̈ܒܥ ⁘
ܒܗ ܒܪܘܟܐ ܚܕܡܐ ܕܗܘܐ ܡܢ ܐܣܛܘܟܣܐ
ܗܢܘܢ ܠܐܣܛܘܟܣܐ ܘܒܘܪܪܐ ܡܢ ܟܠܗ:
ܡܫܘܚܠܦܘܬܗܘܢ ܕܒܚܝ̈ܘܬܐ. ܠܠܒܪ ܡܢ ܗܕܐ
ܗܘܐ ܕܝܢ ܐܝܟ ܕܗܘܐ ܡܬܚܙܐ. ܡܛܠ
ܕܠܘ ܡܢ ܚܕ ܐܣܛܘܟܣܐ ܗܘܘ ܟܠܗܘܢ
ܚܕ ܗܘܐ ܐܣܛܘܟܣܐ ܘܟܠܗ ܗܘܐ ܒܗܘܢ
ܡܫܡܫܢܐ ܘܠܐ ܐܬܡܫܚܬܐ ܘܒܘܪܪܐ. ܟܕ
ܓܝܪ ܗܘܬ ܕܝܢ ܐܝܬܝܗ ܗܘܬ . ܘܗܘܝܐ
ܐܝܬܝܗ ܫܡܝܢܐ ܘܚܝ̈ܘܬܐ ܡܬܩܢܝܐ ܘܡܬܩܢܝܢܘܬܐ
ܘܐܪܥܐ ܘܚܝ̈ܘܬܐ ܢܦܫܢܝܬܐ ܘܐܝܟܢܐ. ܡܛܠܗܢ
ܘܗܢܘܢ ܡܬܢܨܝܐ ܘܠܐ ܦܫܛܐ ܡܬܚܫܚܢܘܬܐ ܡܢ
ܐܣܛܘܟܣܐ ܘܒܘܪܪܐ ܒܗ ܒܪܘܟܐ ܚܕܡܐ ܠ
ܡܢ ܐܣܛܘܟܣܐ ܠܗܘܢ ܡܬܓܒܪ ܗܘܐ ܗܘ
ܡܢܝܢܘܬܐ ܒܗܘܢ . ܘܒܬܪ ܕܠܗܐ ܠܒܪ̈ܝܬܐ
ܟܠܗ̈ܝܢ ܚܝ̈ܘܬܐ ܘܢܦܫܐ ܢܦܫܬܐ. ܘܐܢܫ ܡܢܗܘܢ
ܘܡܢ ܟܠ ܕܐܦ ܗܘܐ ܒܘܚܠܦܐ ܐܝܢܐ
ܘܡܛܠ ܕܡܠܐ ܡܢ ܗܘ ܚܕܡܐ ܠܟܠܗܝܢ ܡܢ
ܡܢ ܓܢܣܐ ܕܐܬܐ ܒܫܘܚܠܦ ܢܦܫܐ. ܗܘ
ܓܝܪ ܦܘܩܕܐ ܕܢܗܘܐ ܡܬܚܫܚܢܐ ܘܡܬܚܫܚܢܘܬܐ
ܘܐܪܥܐ ܘܒܪܝܬܐ ܐܝܟܢܐ ܠܐܘܡܢܘܬܐ
ܪܘܚܐ. ܘܫܡܝܐ ܢܦܫܐ ܕܟܠܗܝܢ ܘܐܣܛܘܟܣܐ
ܘܫܠܡܘܬܐ. ܘܗܘ ܡܬܚܫܚܢܐ ܠܐܘܪ̈ܚܐ ܘܠܐܦܝܣ
ܟܕܝܢܐ ܘܟܠܗܘܢ ܘܫܘܫܐ ܘܡܢ ܦܘܩܕܢܐ .

ܗܘ ܡܫܡܫܢܘܬܐ ܠܘܬ ܠܐܘܪ̈ܚܐ ܘܠܐܦܝܣܛܘ̈ܠܐ
ܠܡܦܩܕܐ ܠܡܦܪܢܣܘ ܘܒܫܪܝܐ. ܘܗܘ ܐܘܪܚܐ
ܠܒܪܐ ܡܘܬܐ ܕܒܘܪܪܐ ܘܒܫܪܝܐ. ܘܟܠܗܝܢ
ܡܢ ܐܘܪ̈ܚܐ ܟܠܗܘܢ ܦܘܩܕܢܐ ܕܢܝܐ ܠܡܫܪܝܘܗܝ
ܘܦܘܠܚܗ ܘܡܘܣܦ ܘܐܦ ܐܘܪܐ ܠܟܘܪܐ
ܠܐܘܒܕܢܐ. ܘܡܘܬܐ ܠܚܝܘܬܐ ܡܫܬܠܛܢܐ
ܡܫܪܐ. ܗܢܐ ܕܝܢ ܪܫܐ ܐܚܪܝܢ ܪܫܐ. ܗܟܢܐ.
ܒܢܪܐ ܠܡܦܩܕ ܕܟܠܦܐ ܘܦܘܢܝܐ ܕܫܪܝܐ ܒܗ
ܒܢܝܐ. ܐܝܟܢܐ ܕܡܫܝܚܐ ܒܪܘܚܐ ܕܡܢ ܠܗ ܡܢܗ
ܘܐܝܟܢܐ ܕܢܗܘܐ ܡܫܬܠܛܢܐ ܡܢ ܐܣܛܘܟܣܐ .
ܡܬܚܙܝܐ ܠܒܢ̈ܝܐ ܐܢ̈ܫܐ ܗܘܝܐ ܢܦܩܘ ܐܦ
ܠܒܢ̈ܝܐ ܕܡܬܪܘܢܝܢ ܘܐܦ ܠܟܘܪܐ ܠܐܘܪ̈ܚܐ
ܡܢ̈ܝܐ ܐܘ ܟܠܗܘܢ ܦܘܩܕܢܐ. ܘܠܟܠܗܘܢ
ܡܢ ܠܗܘܢ ܦܘܩܕܢܐ ܐܝܟ ܗܘܐ ܦܘܠܚܢܐ ܐܝܟܢܐ
ܡܬܚܫܚܢ ܗܘܐ ܗܘܐ. ܦܘܩܕܢ ܕܝܢ ܕܠܗܘܢ
ܚܝ̈ܘܬܐ ܡܫܡܫܢ ܒܪܢܫܐ ܘܒܫܪܐ ܘܗܘܝܐ
ܘܦܪܝܫܐ ܕܒܫܪܐ. ܘܒܠܚܘܕ ܐܝܟܢܐ
ܡܢ ܟܠܗܘܢ ܒܗ ܢܝܚܘ ܘܢܦܫܬܐ ܗܢܘܢ ܡܫܡܫܢܘܬܐ
ܘܦܠܓܐ ܘܪܘܚܐ ܐܝܟܢܘܬܐ ܘܦܪܘܫܘܬܐ ܘܡܫܡܫܢܘܬܐ
ܐܪܥܝܘܬܐ ܘܚܝ̈ܘܬܐ ܢܦܫܢܝܬܐ ܒܗ ܒܡܕܡ ܕܬܬ
ܡܬܚܙܝܐ ܘܡܘܬܐ ܐܣܛܘܟܣܐ ܒܗܘܢ ܘܐܝܬܝܗܘܢ
ܠܟܠܗܝܢ ܡܢ ܒܢ̈ܝܐ ܡܫܬܠܛܢܐ ܡܕܡ
ܒܗܢܐ ܘܡܬܚܫܚܢܐ . ܘܐܦ ܠܐܘܪ̈ܚܐ ܡܫܬܠܛܢܐ
ܡܕܡ ܡܫܬܠܛܢܐ ܕܒܗܘܢ ܕܟܠܗܝܢ ܚܝ
ܡܬܚܫܚܢ ܦܪܝܫܐ. ܡܢ ܐܝܟܢܐ ܘܐܝܟܢܐ
ܡܫܡܫܢܐ ܡܢ ܠܐܘܪܐ ܡܢ ܫܪܝܐ. ܘܗܘܝܐ ܗܘܬ
ܡܢ ܒܪܘܟܐ ܚܕܡܐ ܘܐܣܛܘܟܣܐ ܚܕ ܡܬܡܘܬܐ .
ܦܢ ܕܡܬܚܘܬܐ ܘܡܫܡܫܢܘܬܐ ܘܦܪܝܫܘܬܐ ܘܦܘܠܚܢܘܬܐ
ܘܡܬܚܘܬܐ ܕܒܗ ܡܫܡܫܢܘܬܗܘܢ ܗܘܬ
ܠܟܠܗܝܢ ܡܢ ܐܘܪ̈ܚܐ ܠܐܣܛܘܟܣܐ ܘܒܘܪܪܐ

ܡܢ

[illegible]

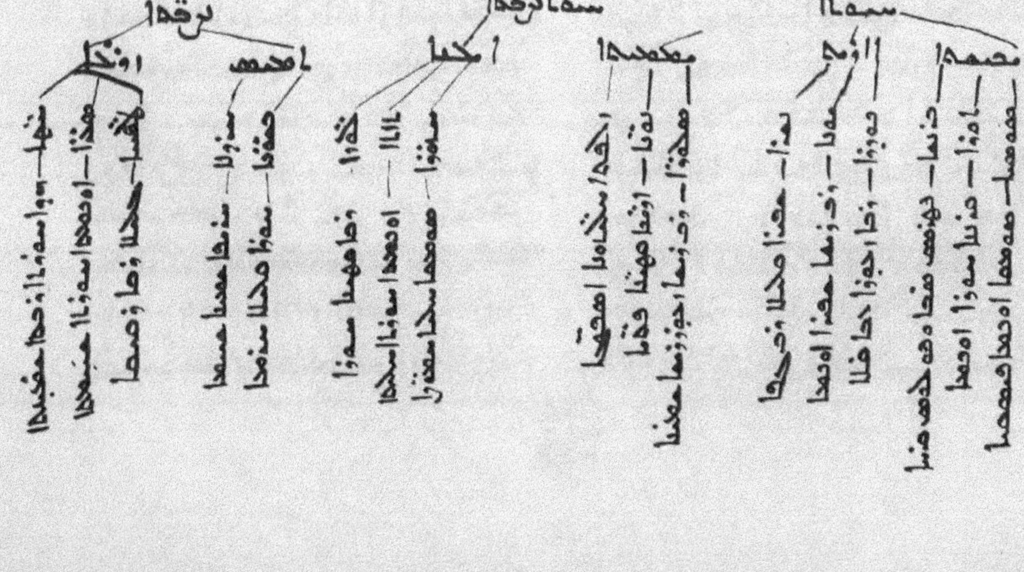

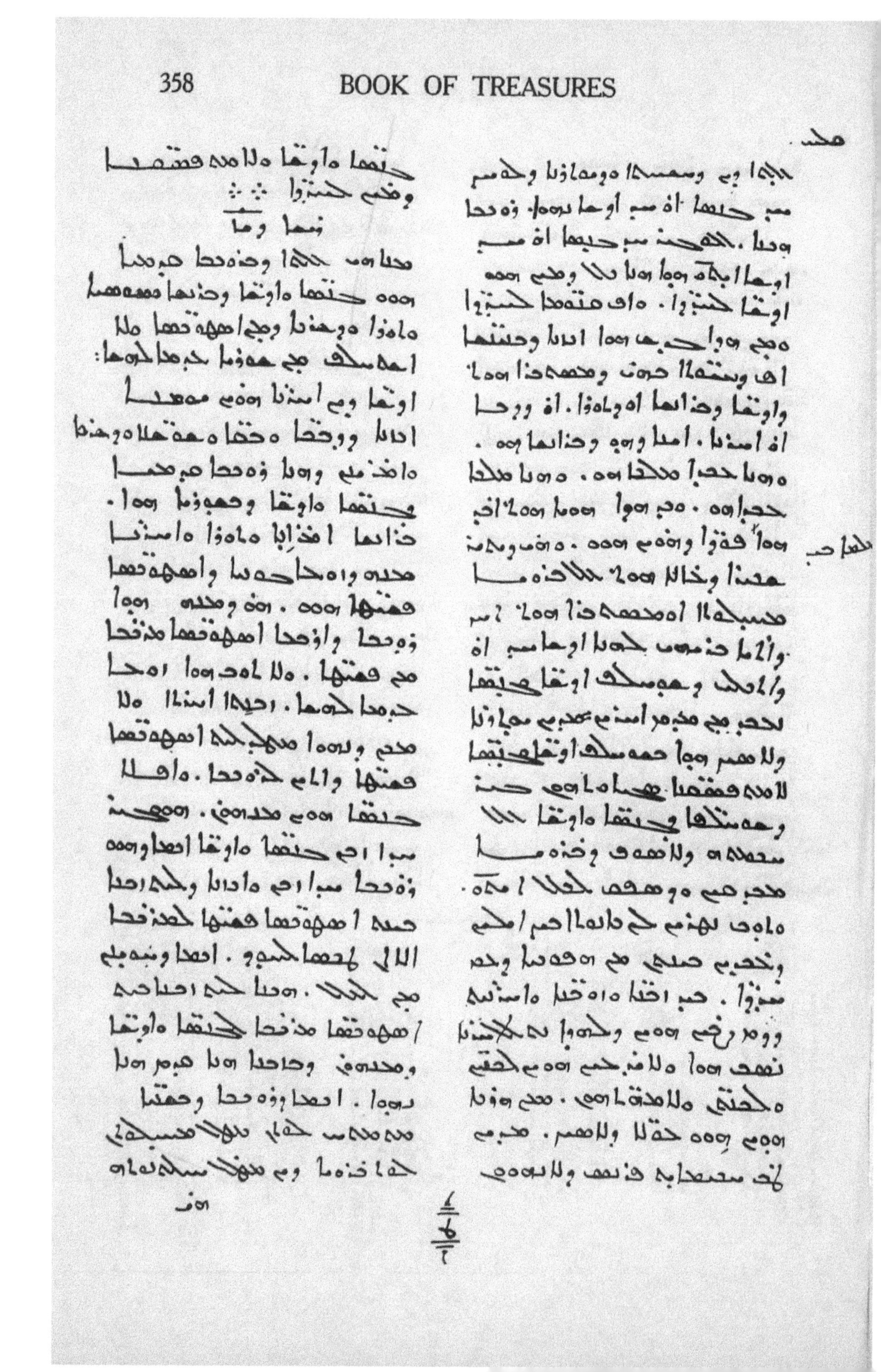

[illegible]

[illegible]

ܗܘ ܕܐܣܒܪܐ. ܘܗܘ ܡܫܘܕܥܢܐ ܠܐ ܡܬܚܙܝܐ
ܡܫܘܡܠܐ ܕܝܢܐ. ܒܣܘܪܝܐ ܠܡ
ܚܕ ܒܟܠܗ ܫܡܝܐ ܘܐܪܥܐ ܘܗܘܐ ܐܬܩܪܝܬ
ܒܣܘܪܝܐ ܘܠܐ ܒܣܘܪܝܐ ܫܡܝܐ. ܘܒܣܘܪ ܫܡܘܪܝܐ
ܐܪܥܐ. ܒܗܠܝܢ ܦܘܪܫܢܐ ܘܡܬܚܙܝܢܘܬܗ ܠܝ
ܡܘܡܐ ܦܫܝܛܐ ܘܐܪܡܝܐ ܘܥܒܪܝܐ ܗܘܐ ܡܚܕܐ
ܚܢܢ ܕܝܠܗ ܠܠܫܢܐ ܗܘܐ ܘܡܬܟܬܒ
ܠܗܘܢ ܟܠܗܝܢ ܕܝܠܗ ܠܗ ܗܘܬ. ܘܐܦ ܐܢܐ
ܘܡܘܕܥ ܠܐ ܗܘܝܢ ܐܣܛܘܟܣܐ ܡܪܟܒܐ
ܡܢ ܦܫܝܛܬܐ. ܒܪ ܡܢ ܐܡܬܝ ܗܘܘ
ܦܫܝܛܬܐ ܒܟܠܗܘܢ ܐܢܫܐ ܕܐܡܬܝ ܗܘܘ
ܗܟܢܐ ܐܦܠܐ ܐܪܒܥܐ ܗܢܘܢ ܡܪܟܒܐ ܡܢ ܐܣܛܘܟܣܐ
ܐܢܫܐ ܕܗܘܘ ܒܥܒܪܐ. ܐܡܬܝ ܕܡܬܚܙܝܢ ܒܗܘܢ
ܘܐܬܦܪܫܘ ܠܗ ܗܘ ܕܒܪܘܗܝ ܐܝܟ ܠܟܠܗܘܢ
ܚܫܘܒܐ ܐܢܫܐ ܕܗܘܐ ܠܟܠܗܘܢ ܒܡܘܪܐ
ܡܕܝܢ ܐܦܠܐ ܗܢܘܢ ܡܟܠܗܘܢ ܐܪܒܥܐ ܗܘ ܚܕ
ܗܢܘܢ ܕܡܬܚܙܝܢ ܡܟܠܗܘܢ ܘܐܪܒܥܐ ܡܢܗ
ܥܡܘܪܐ ܘܓܘܫܡܐ ܠܥܠܡܐ. ܗܢܘܢ ܡܟܠܗܘܢ
ܘܐܣܛܘܟܣܐ ܐܪܒܥܐ ܕܢܘܪܐ ܘܡܝܐ ܘܥܦܪܐ
ܘܐܐܪ ܕܢܦܩܝܢ ܠܟܘܢ. ܐܦ ܡܢ ܕܐܡܪܢ
ܐܣܛܘܟܣܐ ܕܟܠܗ ܡܕܡ ܘܐܪܒܥܬܗܘܢ ܐܝܟ
ܠܗܘܢ ܠܐܣܛܘܟܣܐ. ܡܛܠ ܕܐܦ ܒܐܝܠܝܢ ܡܢ
ܚܕ ܡܢܗܘܢ ܘܠܐ ܐܬܡܣܟ. ܘܐܢܫ ܠܐ ܗܘܐ ܡܢܗ
ܕܒܪ ܘܐܬܡܣܟܘ ܘܐܬܚܠܛܘ ܘܐܬܡܙܓܘ
ܐܪܒܥܬܗܘܢ ܘܙܘܓܐ ܦܓܪܐ ܘܗܘܘ ܡܢ ܐܣܛܘܟܣܐ
ܒ ܢܦܫܢ ܒܢܝܢܫܐ. ܡܫܟܚܝܢ ܕܐܝܬ ܒܟܠ
ܡܠܝܢ ܡܢ ܗܢܐ ܠܗܢܐ ܒܪܢܫܐ ܠܗܘܢ
ܒ ܢܦܫܢ ܗܘܘ ܦܓܪܐ ܘܢܦܫܐ ܫܢܝܐ ܘܒܢܝܗܘܢ
ܐܝܟ ܡܫܟܚܐ ܠܗ ܗܘܢ ܡܢ ܚܕ ܐܪܥܐ
ܠܗ ܐܢܫܐ ܕܡܢ ܚܕ ܟܠܗ ܗܘ ܗܘ ܓܢܣܐ

ܕܥܘܪܐ ܢܦܫܐ ܚܕܐ ܠܟܠܗܘܢ. ܘܠܐ ܐܣܛܘܟܣܐ
ܡܢ ܐܣܛܘܟܣܐ ܚܕ ܡܫܡܫܢܐ ܘܚܝܐ. ܘܗܘܐ
ܠܗ ܦܓܪܐ ܡܬܬܙܝܥܢܐ ܘܚܝܢܐ. ܐܘܟܝܬ ܠܓܘܫܡܢ
ܘܗܢܘܢ ܡܫܟܚܢܐ ܡܢ ܐܣܛܘܟܣܐ ܐܬܡܣܟܘ
ܡܢ ܐܣܛܘܟܣܐ ܗܘ ܕܟܠܗ ܘܐܬܡܣܟܘ
ܘܐܬܚܠܛܘ ܘܐܬܡܙܓܘ. ܐܟܙܢܐ ܕܗܘܝܢ
ܘܚܝܐ ܘܪܘܚܐ ܘܡܝܐ ܡܬܚܠܛܝܢ ܗܘܘ ܠܐ ܗܘܐ
ܠܗܘܢ ܡܬܥܠܝܢ ܟܕ ܚܢܢ ܡܢ ܗܘܘ ܡܢ ܡܫܡܫܢܐ
ܓܝܪ ܗܝ ܕܗܘܝܐ ܘܦܘܠܚܢܐ. ܘܡܢ ܕܡܐ
ܗܘܐ ܘܡܬܝܠܕܐ ܗܘ ܡܢ ܗܘ ܕܫܡܝܐ ܘܐܪܥܐ
ܘܡܢ ܡܪܟܒܐ ܡܫܡܫܢܐ ܘܐܬܚܠܛܘ ܗܢܐ. ܠܐ ܗܘܘ
ܡܢܗ ܠܡܬܚܙܝܐ ܘܐܦܠܐ ܐܟܠܐ ܟܕ ܡܕܐ ܘܡܫܡܫܢܐ
ܠܐ ܚܫܝܫܐ ܦܘܪܫܢܐ ܘܢܦܫܐ ܐܝܟ ܕܐܝܬ ܡܢ
ܘܚܫܝܫܐ ܢܦܫܐ ܪܘܚܐ ܡܢ ܚܝܢܐ ܘܐܬܦܠܓܘ ܚܫܝܫܐ
ܐܣܛܘܟܣܐ ܕܡܢ ܐܪܥܐ ܢܦܫܐ ܘܗܘ ܐܪܥܐ ܡܚܕܐ
ܗܘ ܐܣܛܘܟܣܐ ܘܗܢܐ ܚܝܠܐ ܪܘܚܐ ܕܐܬܐܡܪ ܠܗ
ܡܬܚܙܐ ܢܦܫܐ ܢܦܫܢܝܐ ܡܢ ܐܪܥܐ ܘܐܬܦܠܓܘ
ܗܢܝܢ ܕܚܫܝܫܝܢ ܪܘܚܐ. ܐܝܟܢܐ. ܚܝܠܐ
ܚܬܝܬܐ. ܣܢܝܐ. ܕܝܐ. ܘܝܒܝܫܐ ܘܪܛܝܒܐ
ܒܗ ܡܫܚܐ ܗܢܝܢ ܡܬܡܫܚܢ ܠܗ ܚܕ
ܗܘܝܐ ܘܡܢ ܡܫܡܫܢܐ. ܠܗ ܐܢܫ ܡܬܕܝܪܢܐ
ܠܗ. ܗܟܢܐ ܐܡܪ ܐܝܟ ܦܘܪܫܢܐ ܘܐܣܛܘܟܣܐ
ܘܡܫܡܫܢܐ ܠܐ ܐܝܬ ܠܗ ܒܝܢܝܐ ܘܡܫܡܫܢܐ
ܐܣܛܘܟܣܐ ܚܢܢ ܒܡܪܟܒܘܬܐ ܘܡܫܡܫܢ ܐܘܟܝܬ
ܘܕܟܝܢ ܠܗܘܢ ܡܚܕܐ ܘܡܫܡܫܢܐ
ܠܡܫܡܫܢܐ ܘܦܓܪ ܡܫܡܫܢ ܡܢ ܡܫܡܫܢܐ
ܐܦ ܡܫܡܫܢܐ. ܠܐ ܡܬܚܙܝܢ ܗܘ ܡܫܡܫܢܐ
ܡܢܗܘܢ ܠܗ ܗܘ ܕܐܬܒܪܝ. ܐܠܐ ܗܢܘܢ
ܡܫܡܫܢ ܡܫܡܫܢܐ ܡܢ ܕܐܬܒܪܝ ܡܫܡܫܢܐ
ܚܕ ܡܫܡܫܢ ܘܢܦܫܐ ܕܡܢ ܡܫܡܫܢ ܘܗܘ ܡܫܡܫܢ

ܐܬܚܡ

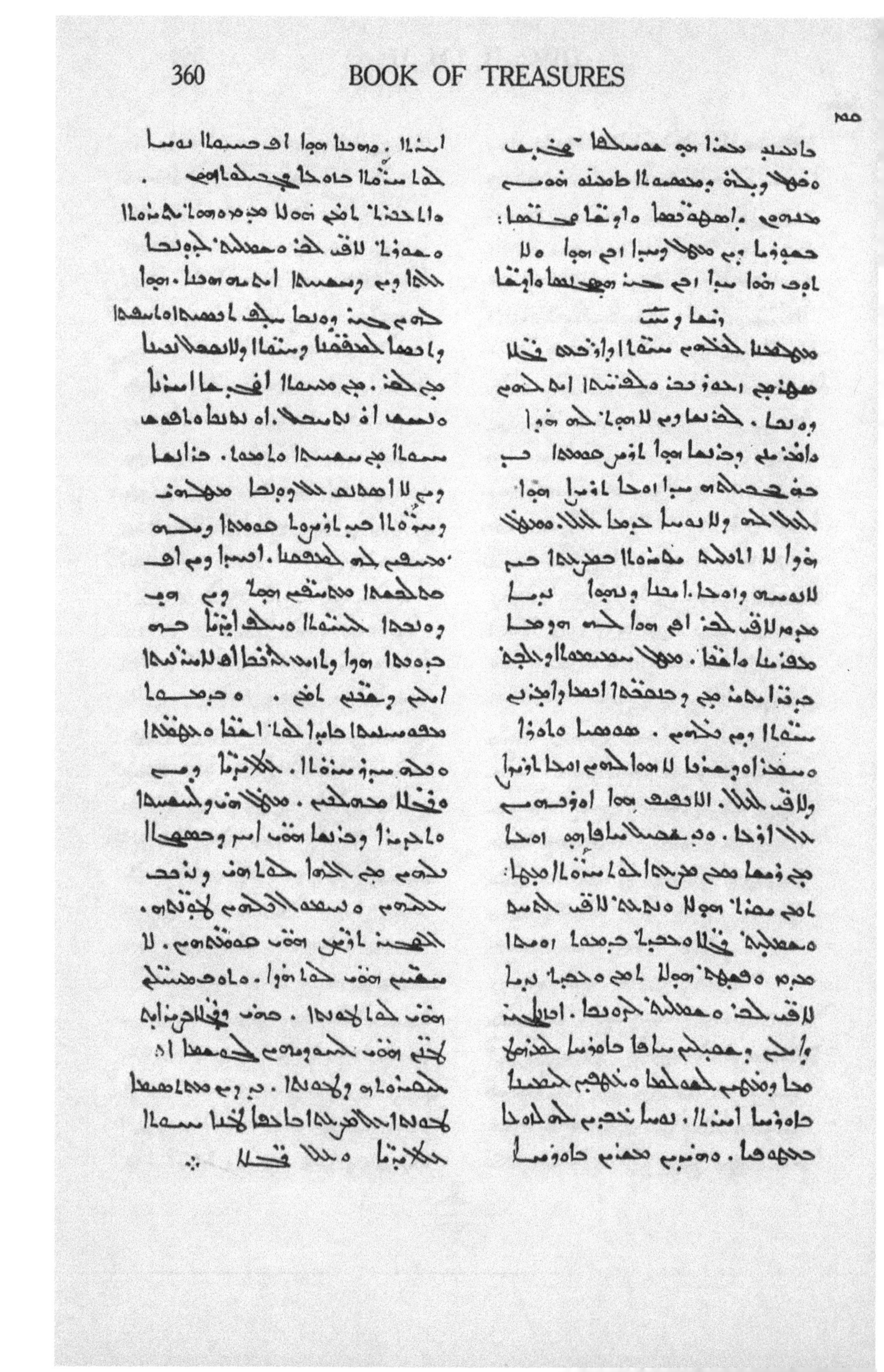

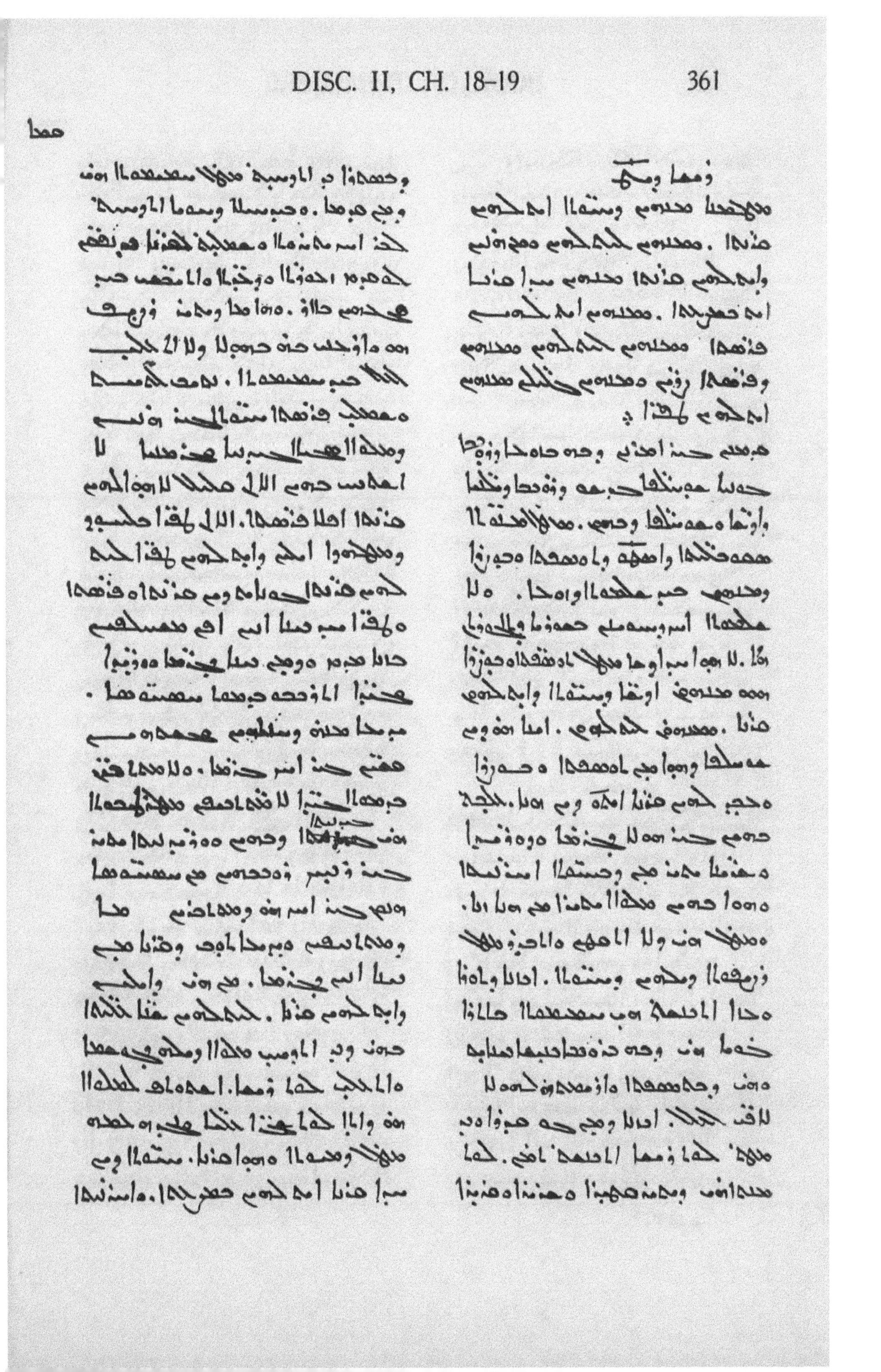

[illegible]

[illegible]

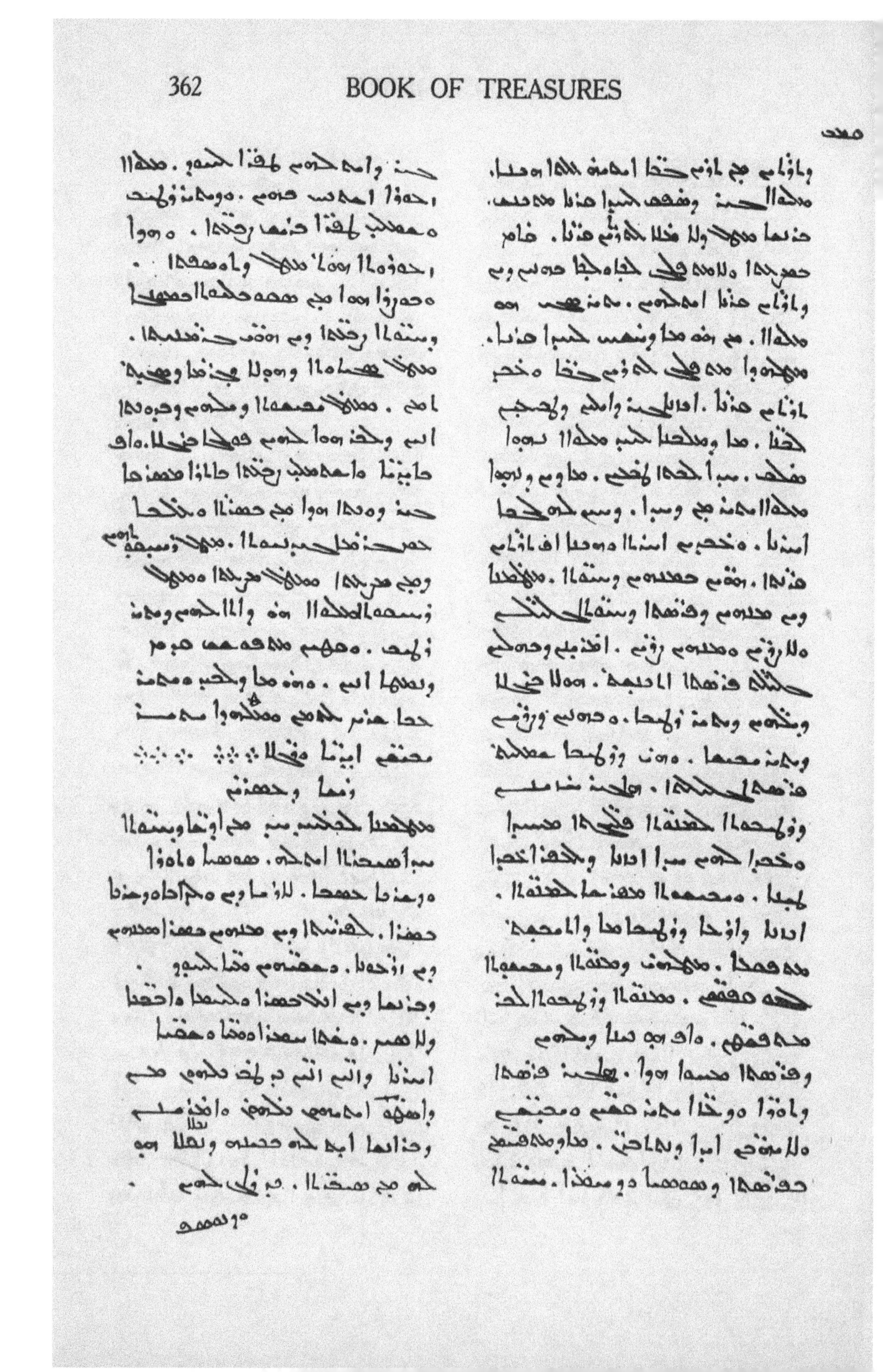

ܘܠܘܩܒܠ ܫܘܒܚܐ ܪܒܐ ܕܦܓܪܐ. ܘܗܘܐ
ܠܐ ܦܩܚ ܚܕܐ ܡܢ ܚܫ̈ܐ ܘܫܘܡܠܝܗ ܠܗܘܐ
ܡܢ ܟܠܗ̇. ܘܒܗ ܚܕܐ ܐܣܛܘܟܣܐ
ܐܝܬܘܗܝ ܗܘܐ. ܐܠܐ ܕܢܦܫܐ ܡܠܟܬܐ ܕܠܐ
ܚܣܝܟ. ܟܕ ܒܓܝܢ ܥܒܕ ܢܫܐ ܡܘܠܕܗ̇
ܚܣܝܢܘܬܗ ܕܒܪ ܟܘܡܪܐ. ܢܘܟܪܝܐ ܠܗ
ܟܢ ܠܡܘܣܟܐ ܟܠ ܥܡܟܗ ܗܘ̇ ܕܠܗ
ܠܗܡܐ ܕܚܫ̈ܐ ܘܒܗ ܒܚܣܝܠܐ ܗܘܐ ܡܬܩܢ
ܠܗ. ܐܦ ܕܟܠܗܘܢ ܡܐܢ̈ܝܗ ܢܩܒܠ.
ܗܠܟܢ ܒܝܘܠܦܢܐ ܡܬܩܒܠ ܠܡܣܒܪܬܐ
ܡܢ ܡܢ ܡܫܠܡ ܠܡܣܬܒܪ ܠܗܘܐ
ܠܡܘܣܟܐ ܕܢܩܘܡ ܒܪܡܐ ܠܡܘܠܕܐ.
ܚܫ̈ܐ. ܕܝܢ ܡܟܝܠ ܕܒܗ ܗܘܝܐ ܕܠܐ ܡܬܟܠܦ
ܡܢ ܚܕܐ ܕܐܣܛܘܟܣܐ ܟܕ ܢܦܫܐ ܡܫܘܐܫܬܐ
ܘܐܝܟ ܠܗܘܢ ܡܫܟܚ̈ܢܐ ܡܬܐܡܪ. ܕܝܢ
ܕܠܐ ܐܬܟܬܒ ܒܫܘܪܐ. ܘܐܦ ܐܡܠܐ ܕܡܬܩܒܠ
ܠܗ ܢܦܫܐ ܡܠܟܬܐ ܐܚܝܕܐ ܘܗܝܕܝܢ ܒܪܢܗ
ܗܘܦ ܕܠܐ ܐܬܟܬܒ ܒܫܘܪܐ ܠܗ ܢܬܩܒܠܘܢ
ܦܘܫ̈ܩܐ ܐܟܚܕܐ ܘܪܘܚܐ ܘܢܦ̈ܫܬܐ ܘܦܓܪܐ
ܗܘܝܐ ܡܬܩܢ ܠܗ ܐܦ ܡܫܟܚܢܐ ܕܟܠܗܘܢ
ܕܠܗ ܢܘܩܦܢ. ܘܟܠܗ̇ ܚܣܝܐ ܠܗܘܢܐ ܕܟܠܗ
ܢܦܩܢ ܡܬܩܢܐ ܠܗܘܢ ܐܦ ܕܚܣܘܟܝܐ
ܡܬܟܠܐ ܢܗܘܐ ܟܠܗܘܢ ܡܫܟܚܢܐ ܠܗ ܠܡܦܩܐ
ܘܟܦܢܐ ܘܨܗܝܐ ܘܪܘܝܐ. ܘܐܦ ܡܫܬܐ
ܠܐܝܟ ܗܘܝܐ. ܠܫܢ̈ܝܐ ܕܡܢ ܠܗ ܗܘܝܐ ܡܫܬܟܠܠ
ܘܠܗ ܒܬܘܠܘ̈ܬܐ ܡܬܟܠܐ ܗܘܐ ܠܡܫܟܚܢܐ
ܚܕܐ ܗܝ ܡܫܟܚܢܘܬܐ ܕܦܪܐ ܘܠܟܘܢ
ܘܡܫܬܠܛܐ ܘܐܦ ܚܕܐ ܘܕܟܝܘܬ ܚܕ ܗܘ ܕܝܢ:
ܘܡܠܟܘܬܐ ܚܘܫܒܢܘܗ ܐܝܟ ܟܠܗ̈ ܘܡܢ
ܐܣܛܘܟܣ̈ܐ ܢܦܫܝ̈ ܘܐܚܝܕ ܘܫܘܒܚܐ

ܐܢܫ ܕܡܫܬܟܚܐ ܡܢ ܣܘܓܐܐ ܠܐܘܡܢܬܐ
ܠܗ. ܡܛܠ ܕܠܐ ܚܣܝܢ ܚܝܠܐ ܕܪܘܚܐ ܠܗ
ܕܢܣܬܟܠ ܐܢܫ. ܕܝܢ ܗܘ ܕܪܘܚܐ ܠܗ ܡܠܐ
ܘܪܘܚܐ ܡܫܟܚܐ. ܡܛܠܗܕܐ ܒܡܫܬܘܕܥܢܘܬܐ
ܦܓܪܢܝܬܐ ܡܫܬܡܫ ܒܢܦܫܐ ܘܚܫ̈ܐ ܢܦܫܢܝ̈ܐ
ܕܝܢ ܚܕܐ. ܗܢܘܢ ܕܟܠܗܘܢ ܕܚܫ̈ܐ
ܐܢܘܢ ܒܓܘܫܡܐ ܡܫܬܡܫܝܢ ܒܗ̇. ܘܦܘܪܫܢܐ
ܠܐܦ ܒܗ ܒܓܘܫܡܐ ܗܘܐ ܡܬܝܕܥ. ܐܝܟ ܕܐܦ
ܡܢ ܟܠܗ ܕܣܘܡܟ ܕܐܝܬ ܡܢ ܐܣܛܘܟ̈ܣܐ
ܚܒܝܫܐ ܘܐܝܟ ܚܫ̈ܐ. ܘܗܢܘܢ ܒܟܠܗܝܢ
ܒܪܘܚܐ ܩܕܝܫܐ ܘܡܢ ܐܣܛܘܟܣ̈ܐ ܚܕ ܚܕ ܡܢ ܚܫ̈ܐ
ܗܢܘܢ ܐܝܟܢܐ ܒܦܓܪܐ. ܘܡܛܠܗܕܐ ܠܟܠ
ܚܕ ܚܕ ܐܝܬܘܗܝ ܐܣܛܘܟܣܐ ܟܠ ܡܐ ܕܐܝܬ ܠܗ ܘܡܢ
ܚܕ ܚܕ ܘܐܝܬܘܗܝ ܒܦܓܪܐ ܕܟܠܗܘܢ ܗܘܐ
ܗܢܘܢ ܐܝܟ ܒܦܓܪܐ. ܘܐܝܟ ܚܫ̈ܐ ܕܒܪܢܫ
ܐܚܝܕ ܕܥܡܗܘܢ ܘܠܐ ܡܬܦܠܓ. ܕܝܢ ܕܡܢ
ܚܫ̈ܐ ܕܒܗ ܒܪܢܫܐ ܐܝܟ ܒܓܘܫܡܐ
ܘܐܡܬܝ ܟܠ ܡܫܟܚܢܐ ܘܐܝܟ ܚܣܝܠܐ
ܘܐܝܬܘܗܝ ܒܦܓܪܐ ܘܡܫܟܚ ܘܐܚܝܕ. ܘܓܫܝܡ
ܐܝܟ ܡܬܬܢܝܢ. ܘܦܝܪܐ ܗܘܐ ܘܗܘܝܐ ܗܘ
ܡܢ ܗܘ ܕܐܣܛܘܟܣ̈ܐ ܚܒܝ̈ܫܐ ܐܝܟ
ܓܫܝܡܝܢ ܘܚܒܝܫܝܢ ܒܦܓܪܐ ܡܫܬܡܥܝܢ
ܠܐ ܡܫܟܚܝܢ ܒܪܢܫܐ. ܡܫܬܡܥܐ ܕܝܢ ܠܗ ܗܘܝܐ
ܠܗܘܢ ܕܝܢ ܕܓܘܫܡܐ ܐܝܬܘܗܝ ܗܘܐ. ܩܪܝܒܐ
ܡܛܠ ܕܡܢ ܟܠܗܘܢ ܚܫ̈ܐ ܘܦܘܪܫܢܐ ܘܪܘܚܐ
ܘܐܟܚܕܐ ܘܒܦܓܪܐ ܘܚܕܐ ܕܟܠܗܝܢ ܘܡܫܬܡܫܝܢ
ܒܚܝܠܗ ܠܟܠܗܝܢ. ܘܡܫܬܟܚܐ ܗܘܐ ܐܝܟ ܘܠܗ
ܗܝ ܕܡܫܬܟܚ̈ܝܢ. ܘܡܫܬܟܚ ܗܘܐ ܐܝܟ ܕܠܡܫܬܡܫܘ
ܗܘ ܘܒܗ ܗܘ ܡܣܬܒܪ ܦܓܪܐ ܘܐܦܠܐ
ܘܐܬܝܕܥ ܠܗ. ܗܝܕܝܢ ܐܦ ܠܡܫܬܡܫܘܗܝ

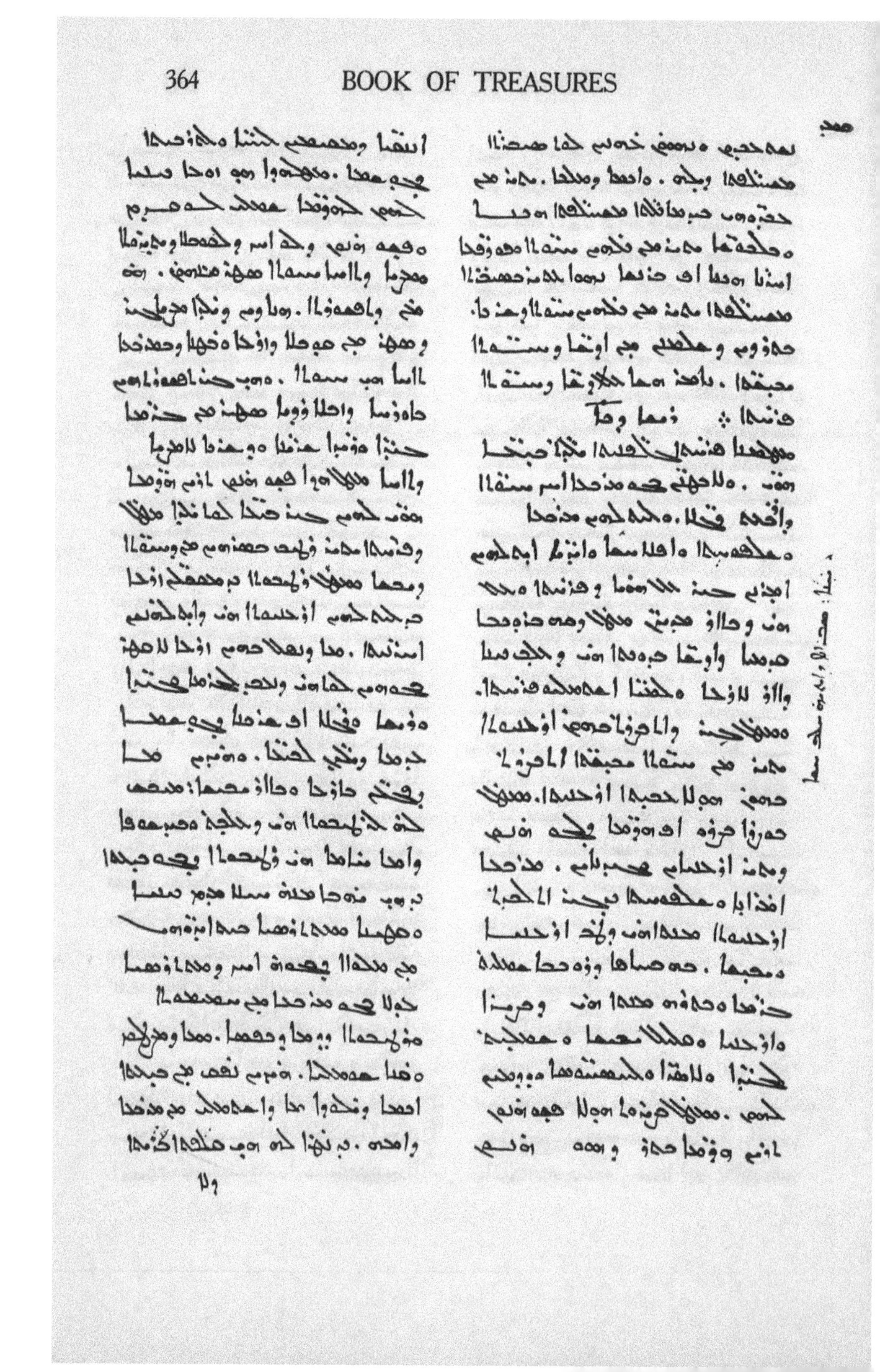

[illegible]

[illegible]

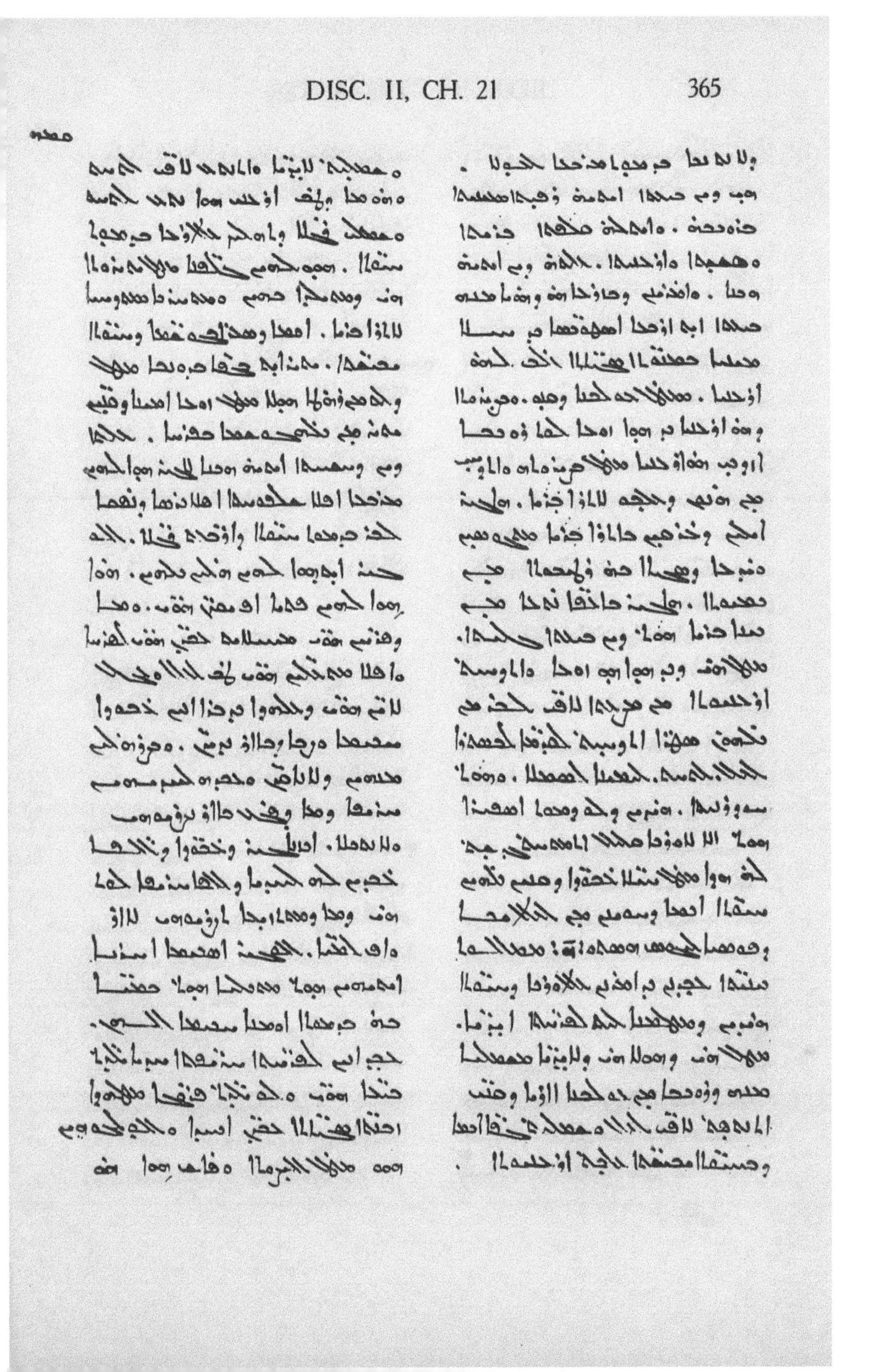

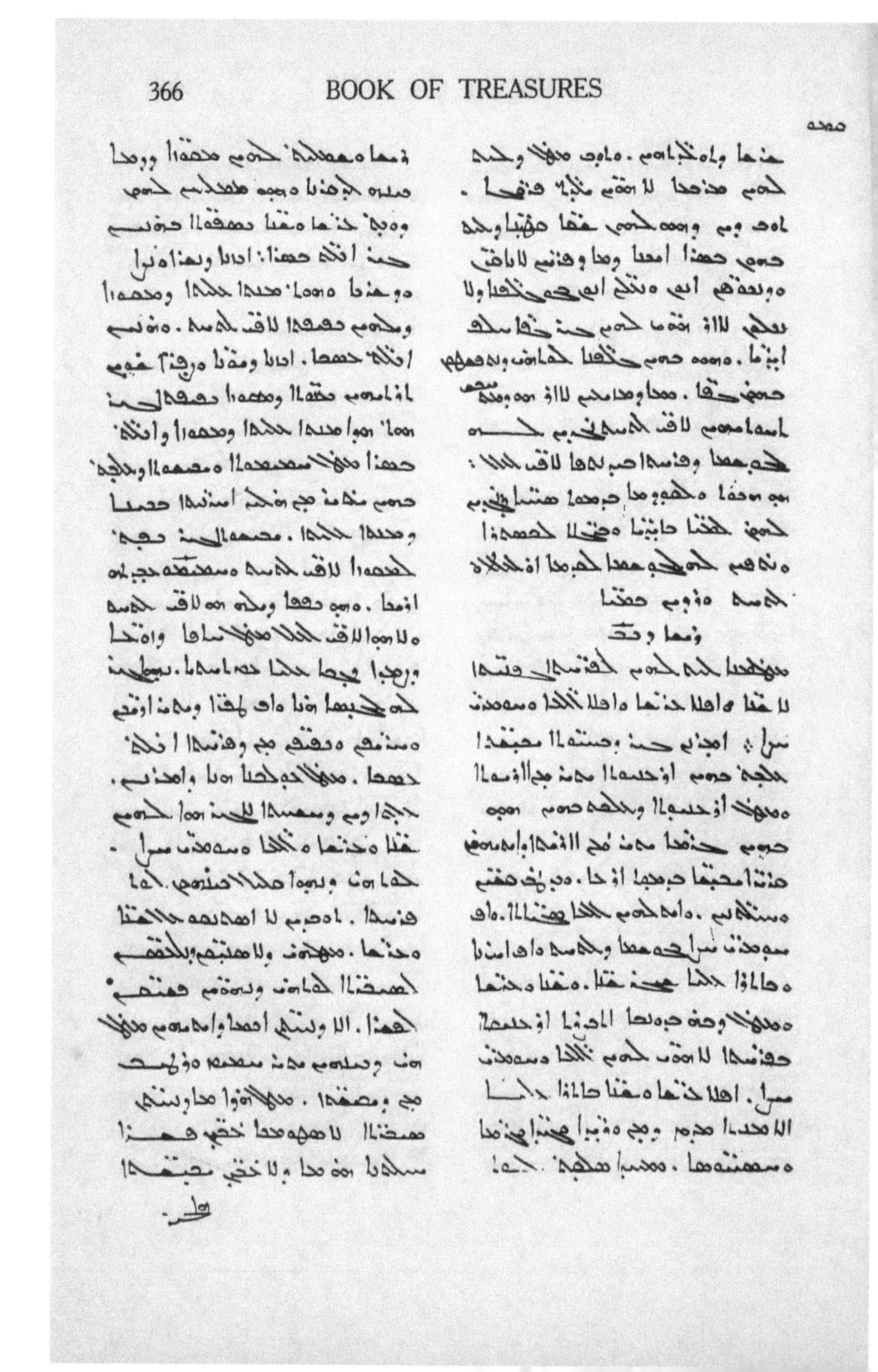

[illegible]

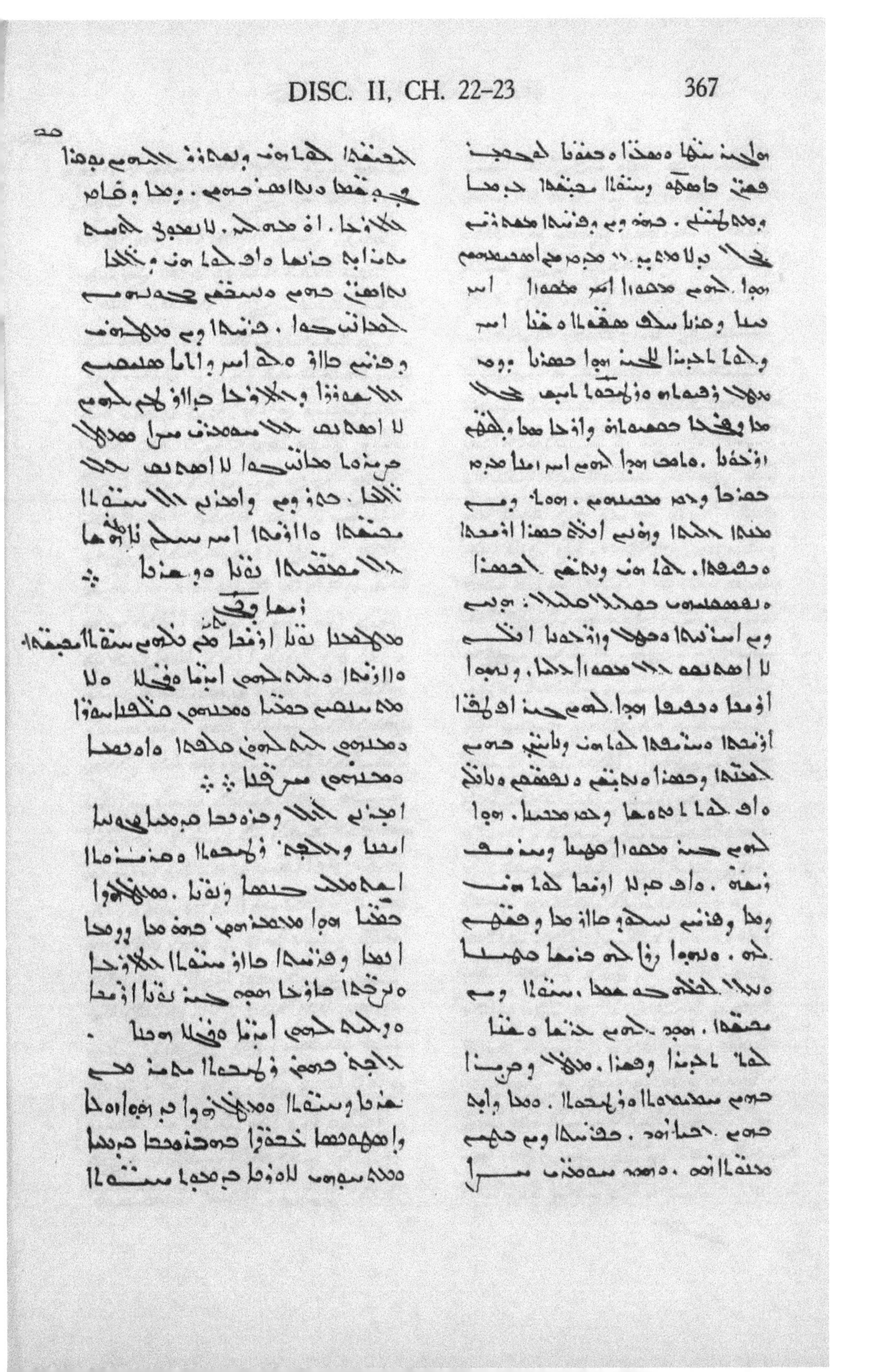

[illegible]

[illegible]

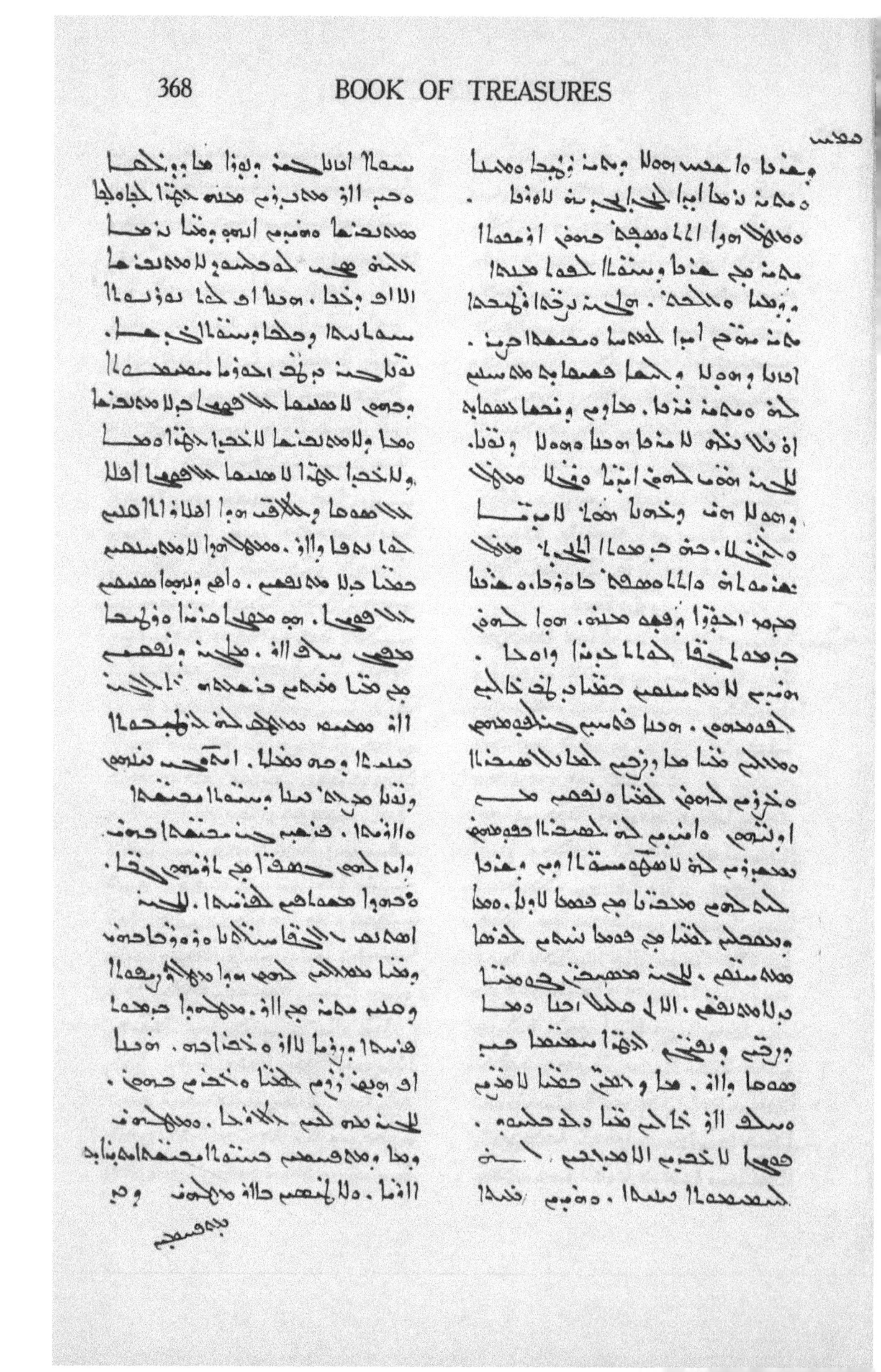

[illegible]

[illegible]

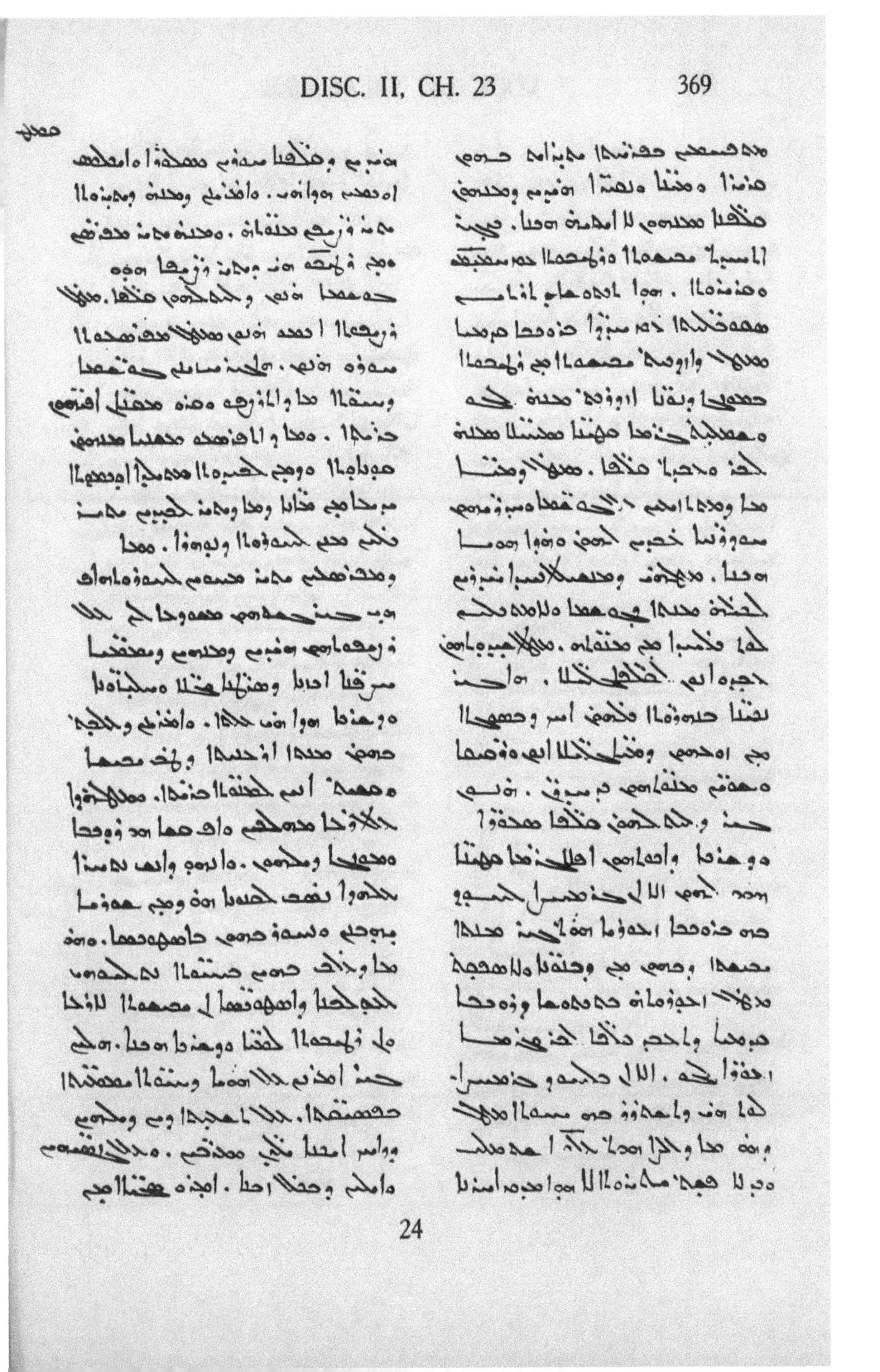

[illegible]

[illegible]

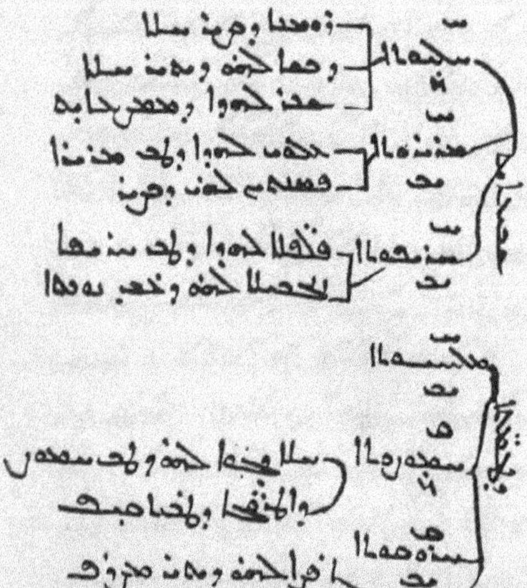

[illegible]

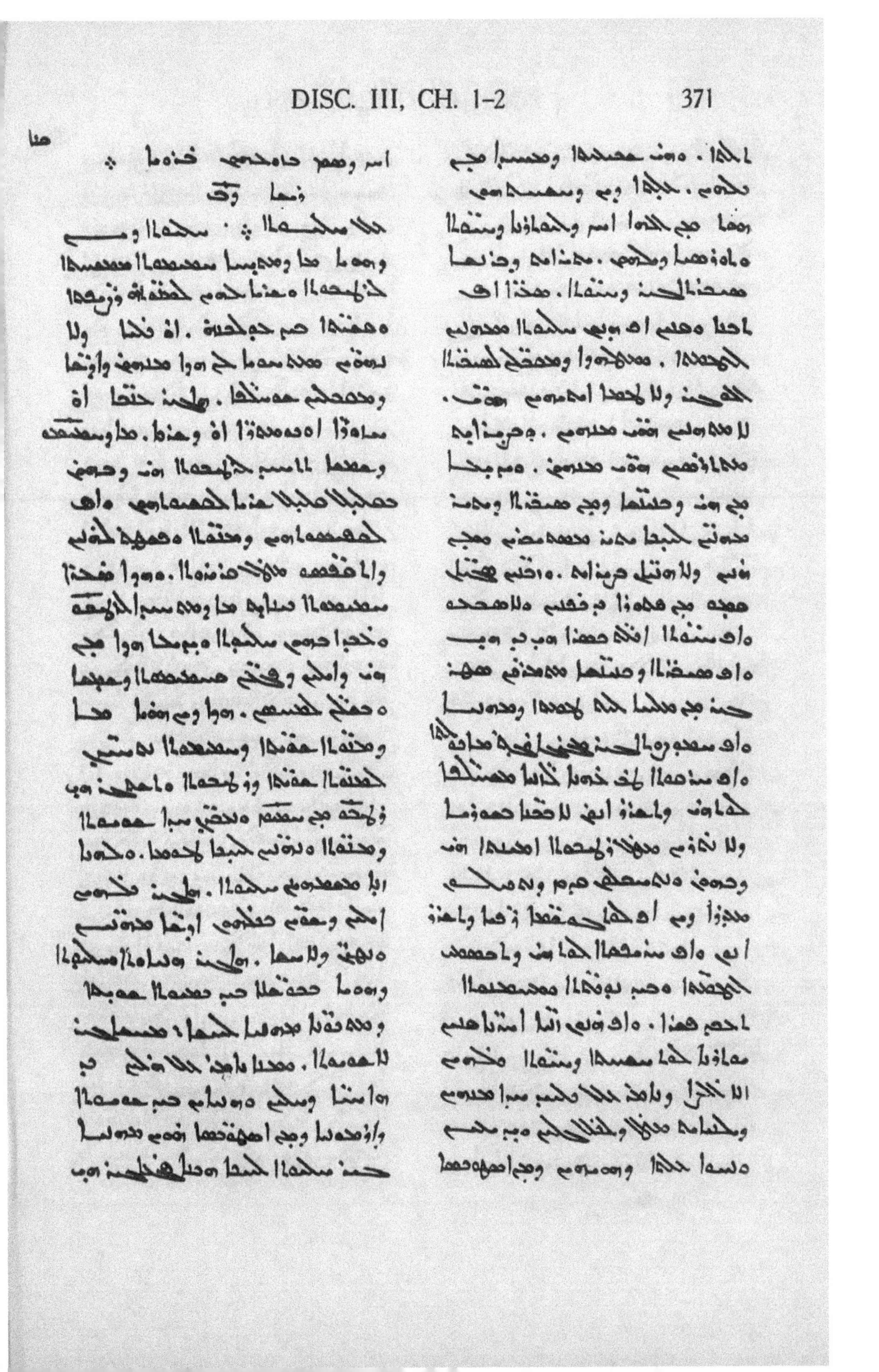

[illegible]

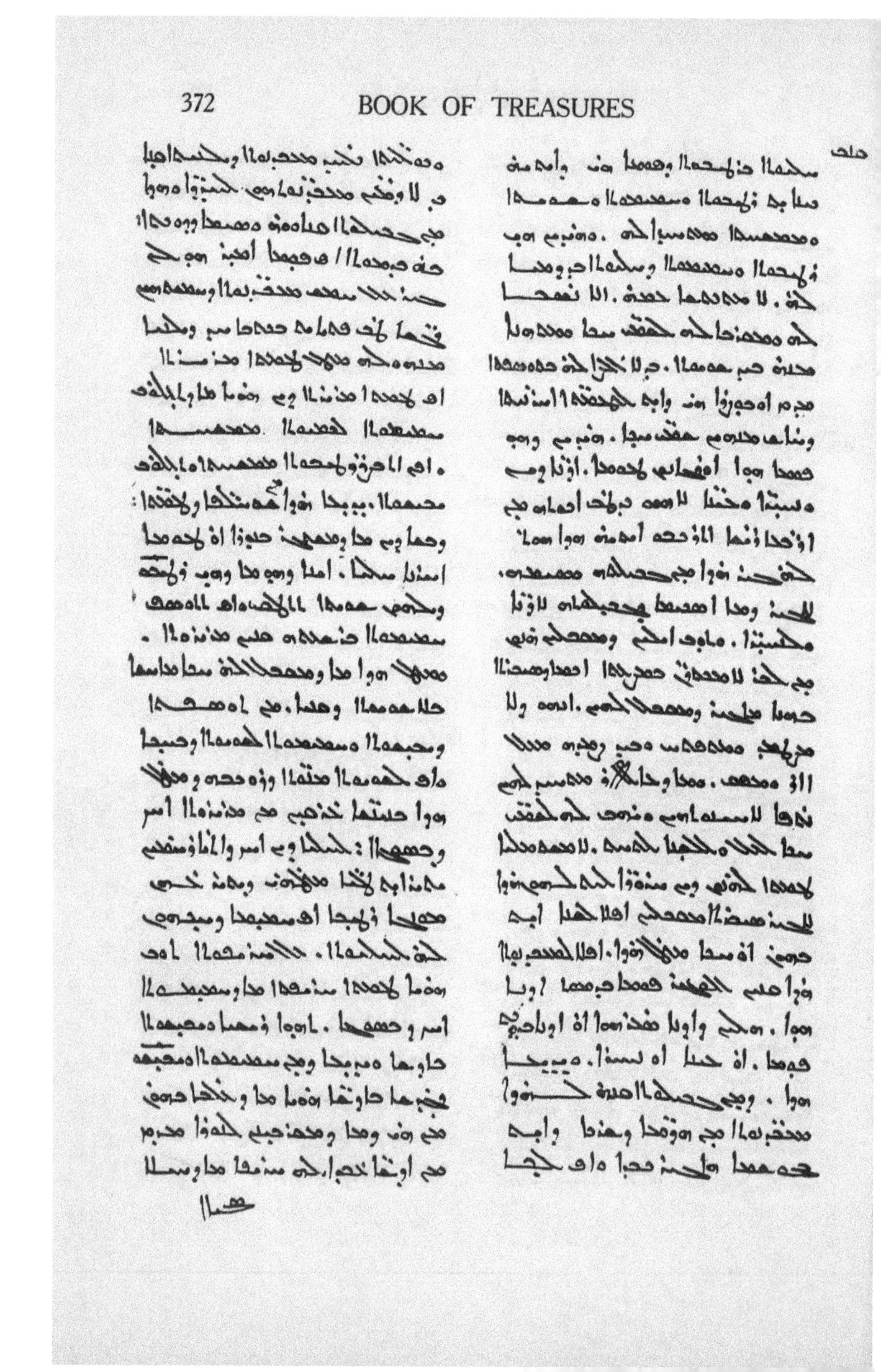

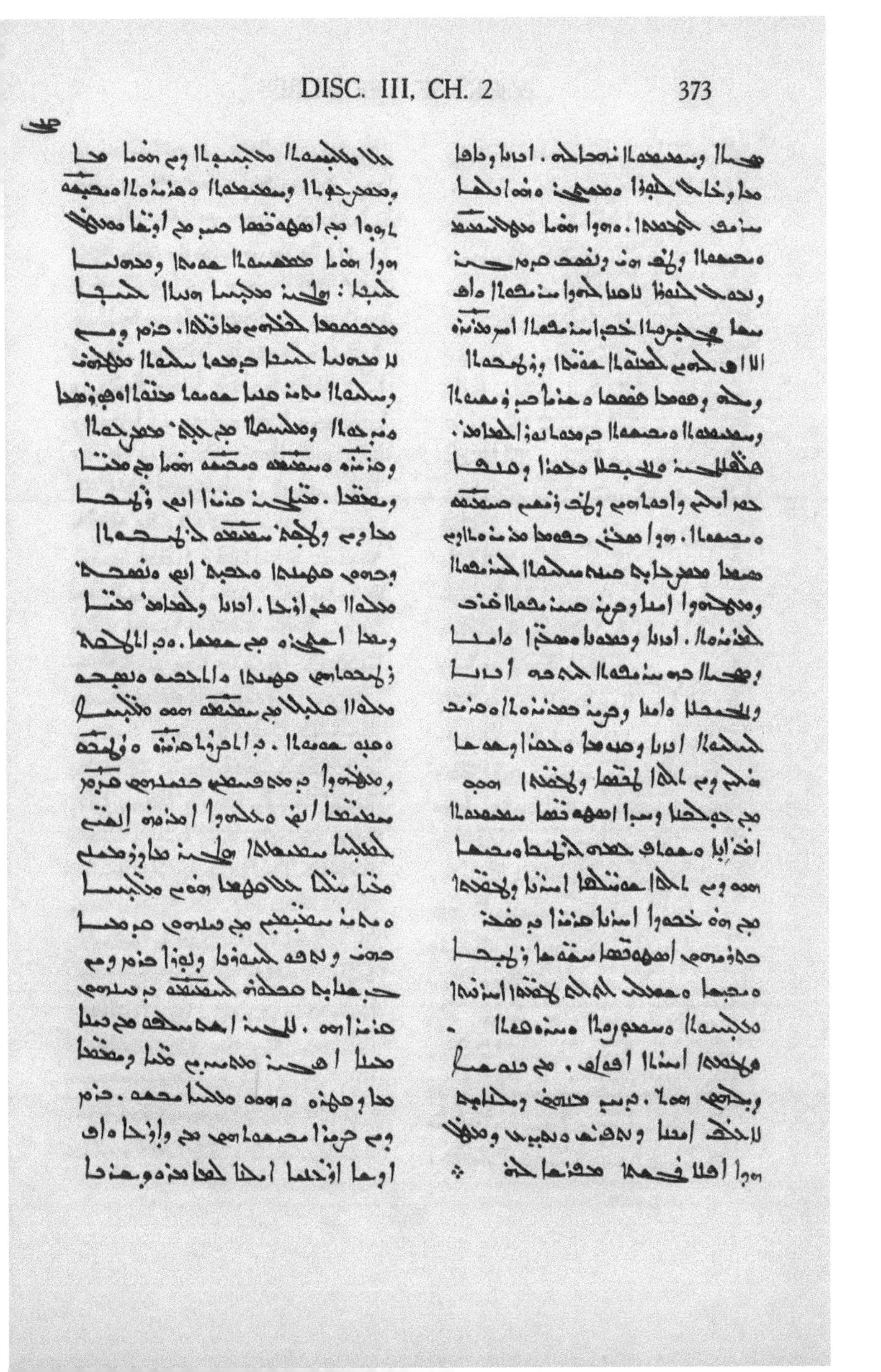

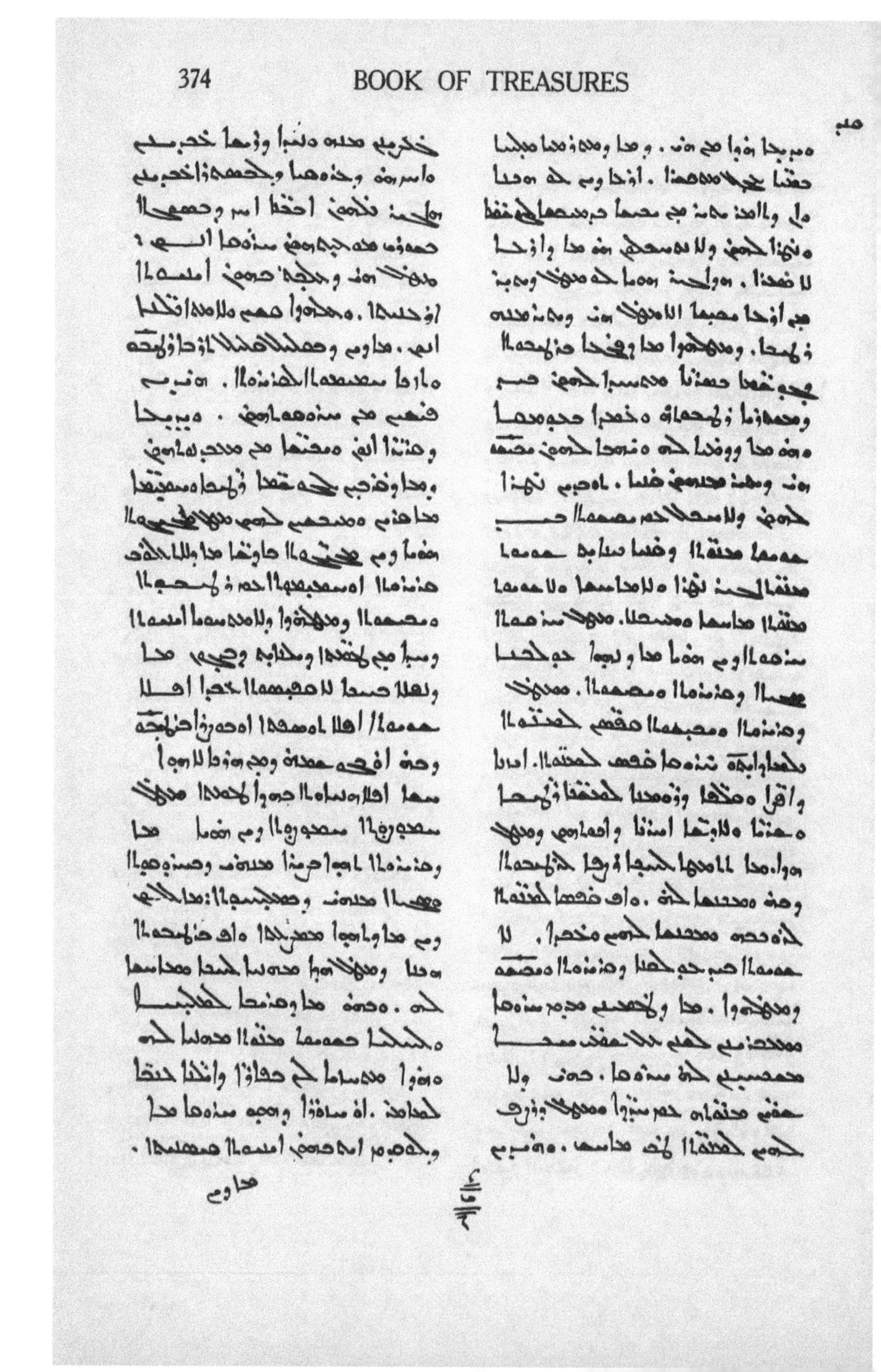

[illegible]

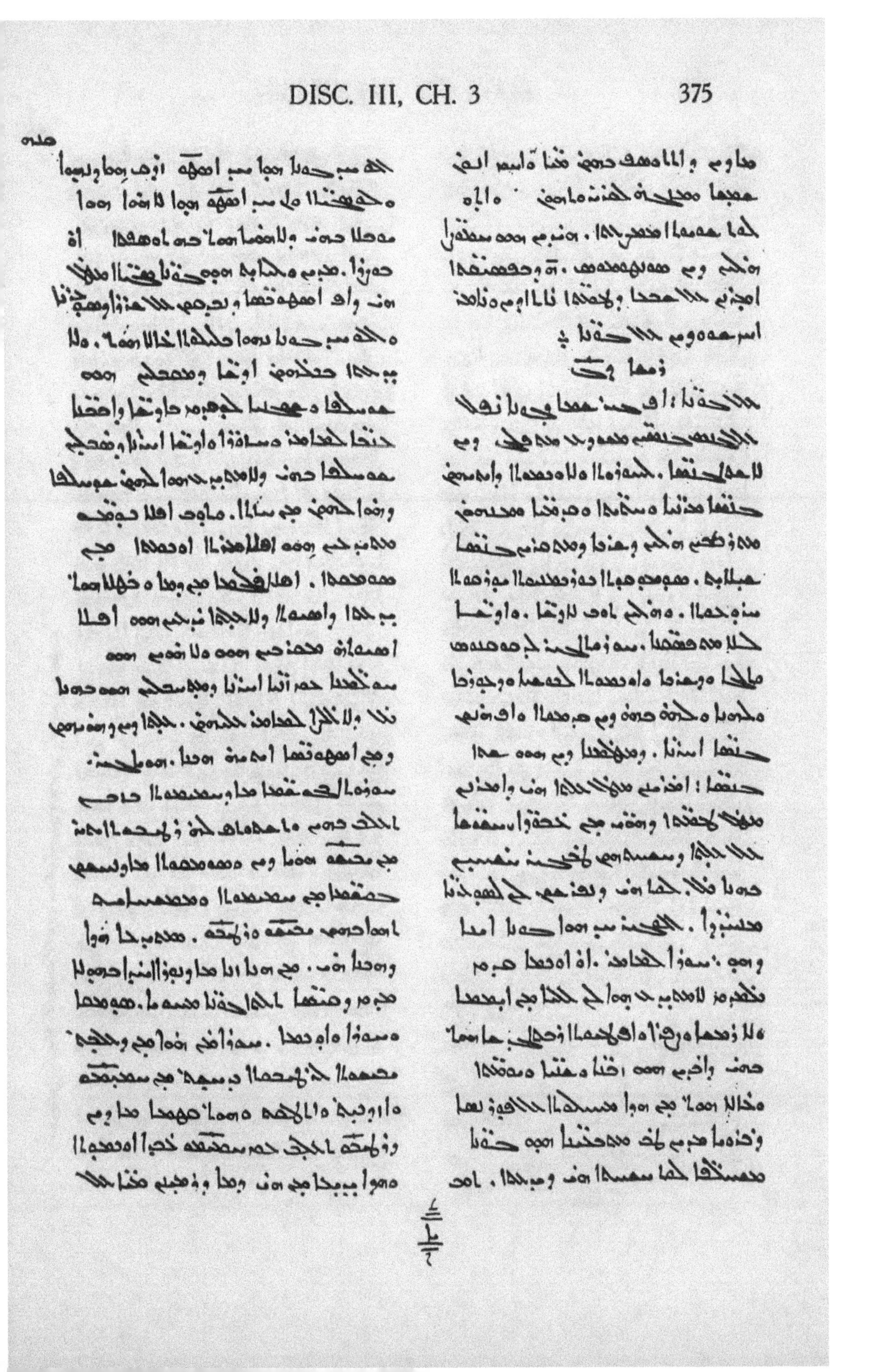

[illegible]

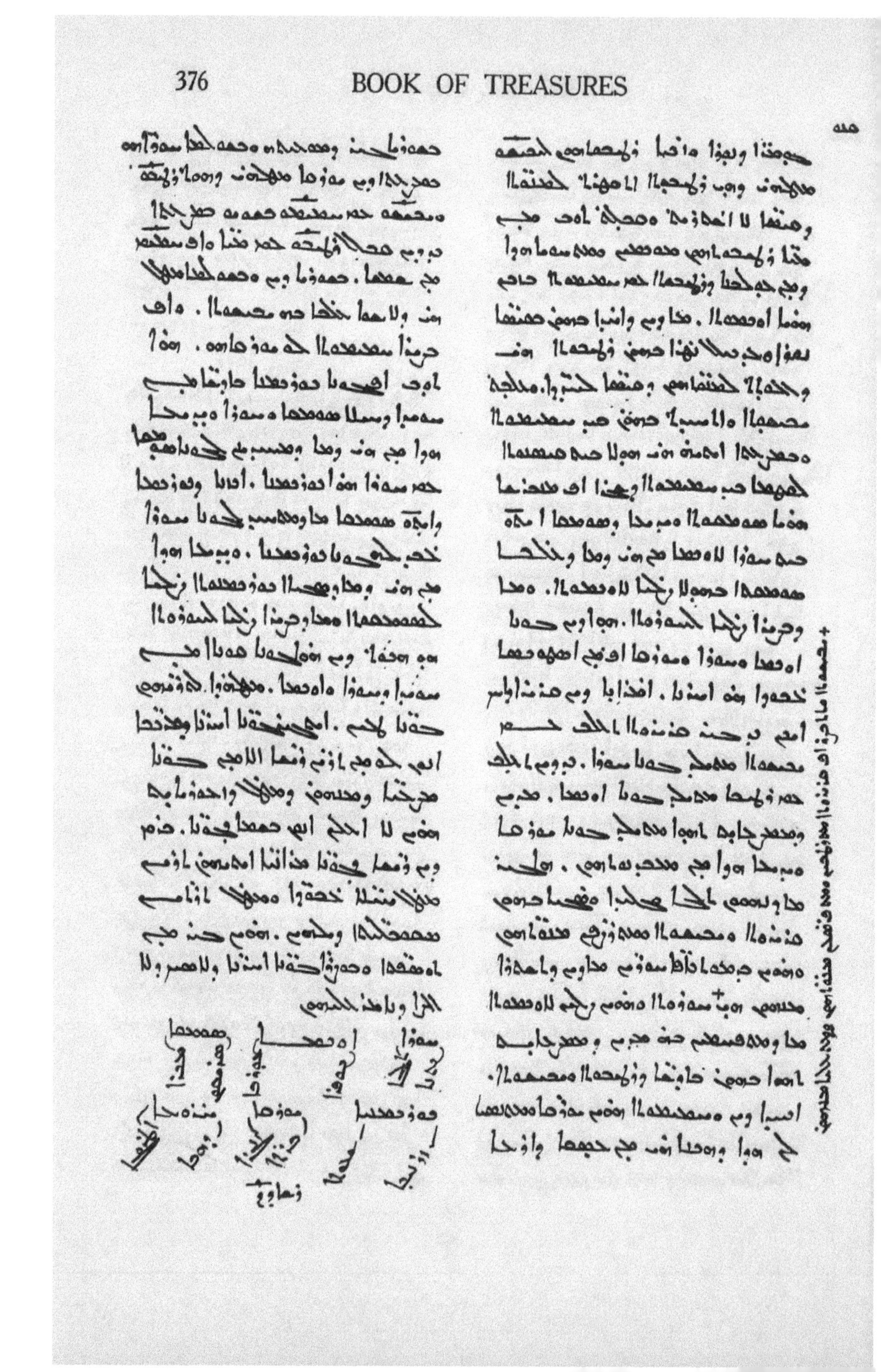

[illegible]

ܩܘ

ܪܝܫܐ ܕ ܕ̄

ܥܠ ܗܢܐ ܕܐܝܟ ܡܢܟܣܐ ܚܝܐ ܒܪ̈ܝܬܐ
ܒܗ ܕܝܢ ܕܐܬܒܪܝ ܟܠ ܗܘܝܐ ܘܡܫܬܟ̈ܚܬܐ
ܢܦܩܢ ܕܐܝܟ ܡܢܟܣܐ ܒܗܘܢ ܘܡܫܬܡܠܝܢ
ܚܝܐ ܡܢ ܪܘܚܐ ܘܐܝܟܢܐ ܕܢܦ̈ܪܐ ܐܟܡܐ ܠܗ ܐܢܘܢ .
ܒܗܘܢܐ ܕܝܢ ܠܐ ܐܬܚܙܝ ܡܢ ܪܘܚܐ ܐܘ ܚܝܘܬܐ ܡܢܗ
ܐܘ ܐܝܟ ܐܘܪܚܐ ܫܦܝܪܐ . ܐܦܢ ܕܝܢ ܗܘܐ ܗܘܐ ܠܒܪ
ܡܢ ܒܪܝܬܐ ܡܫܬܟܚܐ ܒܒܪܝܬܗ ܕܒܪ̈ܝܬܐ ܚܝܐ
ܡܫܬܡܠܝܐ ܒܝܕ ܚܝܐ ܐܘ ܡܢ ܕܢܣܒܪ ܐܘܪܚܐ .
ܐܘ ܡܬܓܠܝܐ ܐܝܟ ܒܗܘܢ ܕܚܝܐ ܡܢ ܚܝܐ
ܘܒܘܢܝܐ ܐܘ ܡܬܓܠܝܐ ܡܬܚܙܝܐ ܒܫܘܒܚܗ ܘܒܝܩܪܗ
ܘܡܬܪܥܝܢܘܬܗ . ܡܟܝܠ ܐܝܬܝܗ ܡܢ ܪܘܚܐ
ܒܪ̈ܝܬܐ ܐܦ ܠܐܬܓܠܝܬܐ ܗܘܐ ܐܝܟ ܡܫܬܡܠܝܢܘܬܗ
ܕܒܗܘܢܐ ܘܡܬܚܙܝܢܘܬܐ ܘܒܘܝܢܐ ܐܠܐ ܕܢܗܘܐ
ܒܪ ܢܗܘܐ ܘܡܬܚܙܝܐ ܥܠ ܡܫܬܡܠܝܢܘܬܐ ܘܒܚܝܠܐ
ܡܫܬܡܠܝܢܘܬܐ ܕܡܠܬܐ ܘܟܕ ܐܝܟ ܡܠܟܝܙܕܩ
ܕܐܝܬ ܕܒܪܝܬܐ ܠܗ . ܡܟܝܠ ܘܒܪ̈ܝܬܐ ܚܝܐ
ܗܘܝܐ ܡܬܓܒܠܢ ܘܐܬܚܙܝ . ܕܡܬܒܪܝܐ ܘܐܦ ܐܘܪܚܐ
ܠܐ ܡܬܚܙܝܐ ܐܝܟ ܡܠܟܝܙܕܩ ܡܢ ܢܗܘܐ ܘܒܚܝܠܐ .
ܐܠܐ ܐܠܗܐ ܕܡܠܬܐ ܕܢܚܐ ܐܦ ܕܝܠܗ ܘܡܠܬܐ ܘܕܡܘܬܗ
ܡܬܚܙܝܐ ܡܬܒܪܝܐ ܡܢ ܒܪܝܬܐ ܥܡܗܘܢ
ܘܒܚܝܠܐ ܥܠ ܐܝܟ ܡܠܟܝܙܕܩ ܕܒܟܠ ܕ ܢܗܘܐ
ܥܠ ܐܠܗܘܬܐ ܡܢ ܕܡܠܬܐ ܐܦ ܒܪ̈ܝܬܐ ܐܘ ܚܝܐ
ܗܘ ܕܢܬܚܙܐ ܥܠܝܗܘܢ ܢܗܘܐ ܘܒܚܝܠܐ ܒܗ ܚܝܐ
ܡܢ ܚܝܐ ܐܝܟ ܡܠܟܝܙܕܩ ܕܡܠܬܐ ܕܢܗܘܐ ܗܘ
ܘܒܚܝܠܐ ܥܠ ܐܠܗܘܬܐ ܕܟܠ . ܗܘ ܚܝܐ ܢܗܘܐ ܘܒܚܝܠܐ
ܗܘ ܕܢܬܚܙܐ ܥܠ ܐܠܗܐ ܗܢܐ ܐܝܟ ܡܠܟܝܙܕܩ
ܥܠ ܐܒܪܗܡ ܕܟܗܢ . ܘܥܠ ܐܠܗܘܬܗ ܐܦ ܡܫܝܚܐ
ܒܗܘ ܡܢ ܪܘܚܐ ܕܝܠܗ ܕܒܗ ܐܢܘܢ ܠܗ ܐܢܘܢ
ܫܡܝܐ ܕܡܢ ܐܒܪܗܡ ܠܗ ܒܗ ܚܝܐ ܘܐܝܟܢܐ ܕ

ܡܫܡܠܝܐ ܒܟܝ̈ܢܐ ܡܣܒܪܢܐ ܠܗܘܢ ܡܠܟܝܙܕܩ
ܐܝܟܢܐ ܕܗܢܐ ܗܟܢܐ ܠܡܫܝܚܐ . ܐܦܢ ܒܟܠ
ܡܫܡܠܝܐ ܒܚܝܠܐ ܢܗܘܪܐ ܘܫܘܒܚܐ ܕܢܗܘܐ ܡܢ
ܒܗ ܡܠܟܝܙܕܩ ܡܣܒܪܢܘܬܗ ܘܫܡܫܘܬܗ . ܘܐܝܟ
ܡܫܡܠܝܐ ܒܟܠܗܘܢ ܕܢܗܘܐ ܡܢ ܟܠ ܘܡܢ
ܐܝܟ ܡܢ ܗܘܝܐ ܚܝܠܐ ܫܘܒܚܐ ܘܢܗܘܪܐ ܕܢܗܘܐ ܠܗܘܢ
ܕܡܫܡܠܝܐ ܡܣܒܪܢܐ ܠܗ ܢܗܘܐ ܘܫܡܫܐ ܒܟܠ
ܘܡܢܗ ܒܗ ܫܡܫܐ ܡܠܟܝܙܕܩ ܘܡܣܒܪܢܘܬܗ
ܕܒܡܠܟܝܙܕܩ ܕܝܢ ܒܗܘܢ ܒܟܠܐ ܒܗܘܢ ܡܢ ܟܠ
ܘܗܘܝܐ ܒܫܡܫܐ ܡܢ ܒܗ . ܡܠܟܝܙܕܩ ܗܘܐ ܗܘܐ
ܕܟܗܢܐ ܒܟܠܐ ܕܫܡܫܗ ܘܢܗܘܐ ܒܟܗܢܘܬܐ
ܢܗܘܐ ܒܗܘܢ ܒܫܡܫܐ ܐܝܟܢܐ ܕܒܟܠܐ
ܢܗܘܐ ܕܝܢ ܐܘܪܚܐ ܕܫܡܫܐ ܐܝܟܢܐ ܠܡܫܒܚܘ
ܒܗ ܒܗܘܐ ܡܠܟܝܙܕܩ ܒܫܡܫܐ ܘܠܡܫܡܫܘܬܗ
ܐܠܗܘܬܐ ܘܡܠܐ ܕܢܗܘܐ ܟܠܐ ܒܫܡܫܐ ܕܟܗܢܘܬܐ
ܡܫܡܠܝܐ ܒܗܘܢ ܕܡܫܡܫܐ ܡܣܒܪܢܐ ܒܟܗܢܘܬܐ
ܘܟܗܢܐ ܕܒܗ ܡܫܡܠܝܐ ܒܫܡܫܐ . ܗܘ ܕܝܢ ܕܢܗܘܐ
ܠܡܫܒܚܘ ܒܗ ܒܗܘܐ ܡܠܟܝܙܕܩ ܒܫܡܫܐ ܗܘ ܒܟܠܐ
ܒܟܠܐ ܡܫܡܠܝܐ ܒܚܝܠܐ ܫܘܒܚܐ ܘܡܫܡܫܢܘܬܐ ܘܢܗܘܐ
ܡܫܡܠܝܐ ܒܟܠ ܢܗܘܐ ܘܡܫܡܫܐ ܡܫܡܠܝܐ ܚܝܠܐ
ܫܘܒܚܐ ܘܐܝܟ ܒܗܘ ܡܢ ܕܝܢ ܘܢܗܘܪܐ
ܒܗܘܢܐ ܫܡܫܐ ܘܡܫܡܫܝܢ ܕܟܠ ܟܠܗܘܢ ܡܢ
ܒܟܠܐ ܘܠܐ ܒܟܗܢܐ ܒܟܠܐ ܒܫܡܫܐ . ܘܐܝܟ
ܗܘ ܕܢܫܡܫ ܒܗܘܢ ܒܟܠ ܫܡܫܐ ܡܠܟܝܙܕܩ
ܐܡܫܡܫܢܘܬܗ . ܘܐܝܟܢܐ ܕܢܫܡܫ ܟܠܗܝܢ ܡܢ
ܒܟܗܢ̈ܘܬܐ ܗܘܝܢ ܒܝ ܗܘܝܢ ܦܩܘܕ̈ܐ

ܪܝܫܐ ܕ ܗ̄

ܥܠ ܗܢܐ ܕܡܬܦܠܓܢܐ ܡܫܡܠܝܐ ܕܢܗܘܐ ܟܠܗܘܢ ܚܝܠܐ
ܘܢܗܘܐ ܒܗ ܡܫܡܫܐ . ܒܟ̈ܗܢܐ ܘܢܬܒܥܘܢ ܡܢܗ ܘܐܝܟܢܐ
ܡܫܡܠܝܐ ܒܗܘܢ ܡܢܗ ܒܟܠܐ ܀ ܘܐܦܢ ܕܝܢ

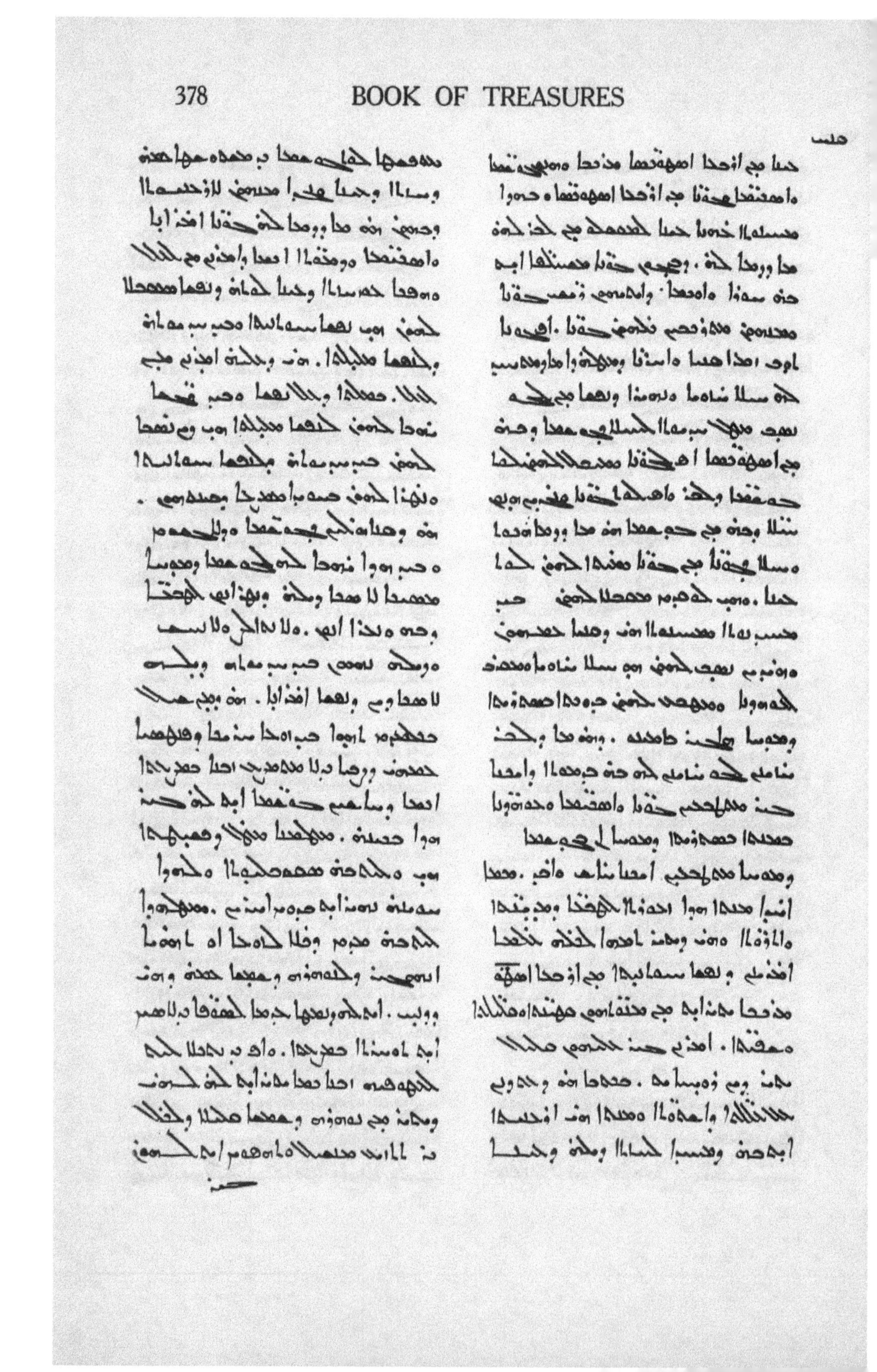

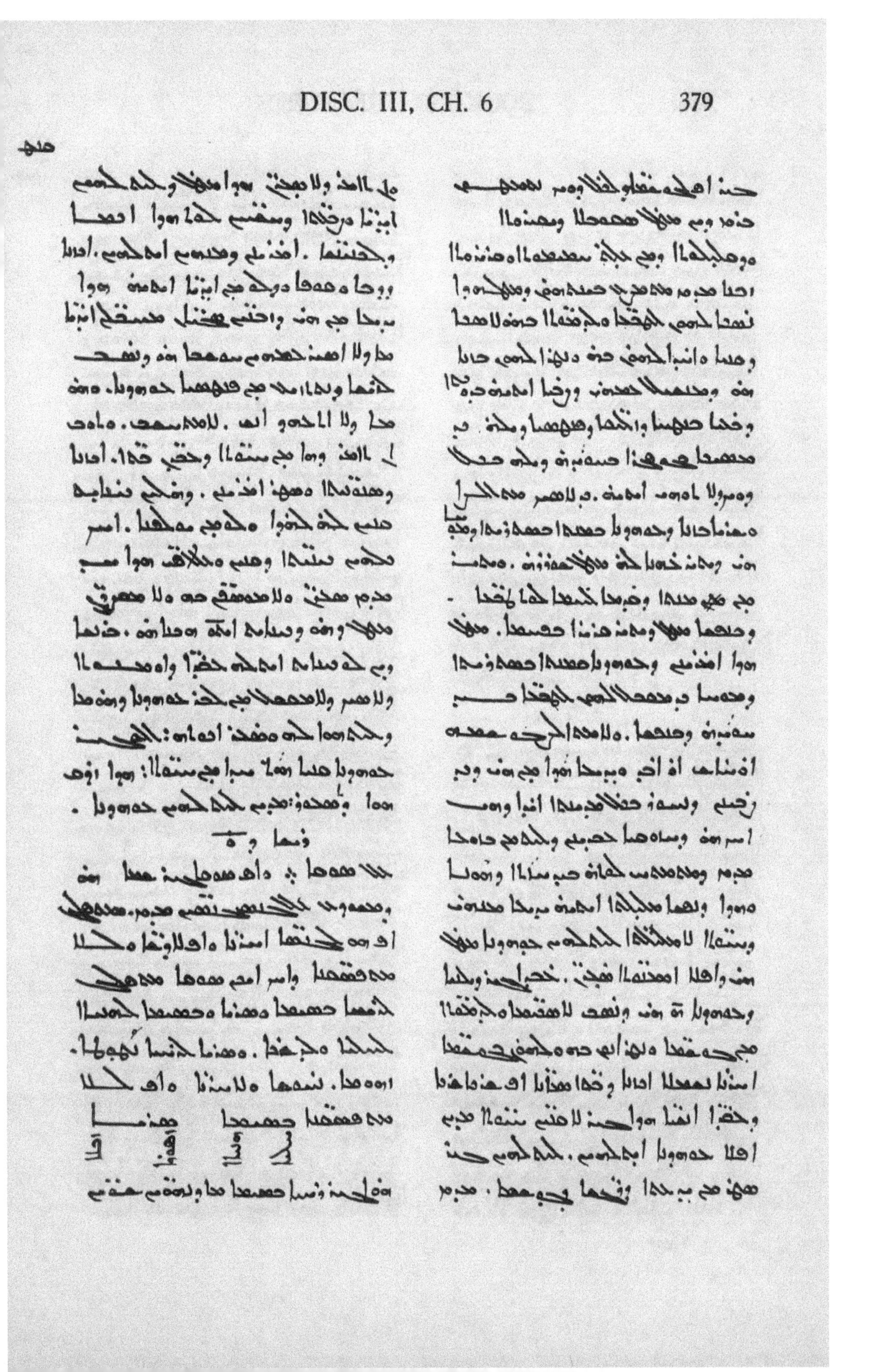

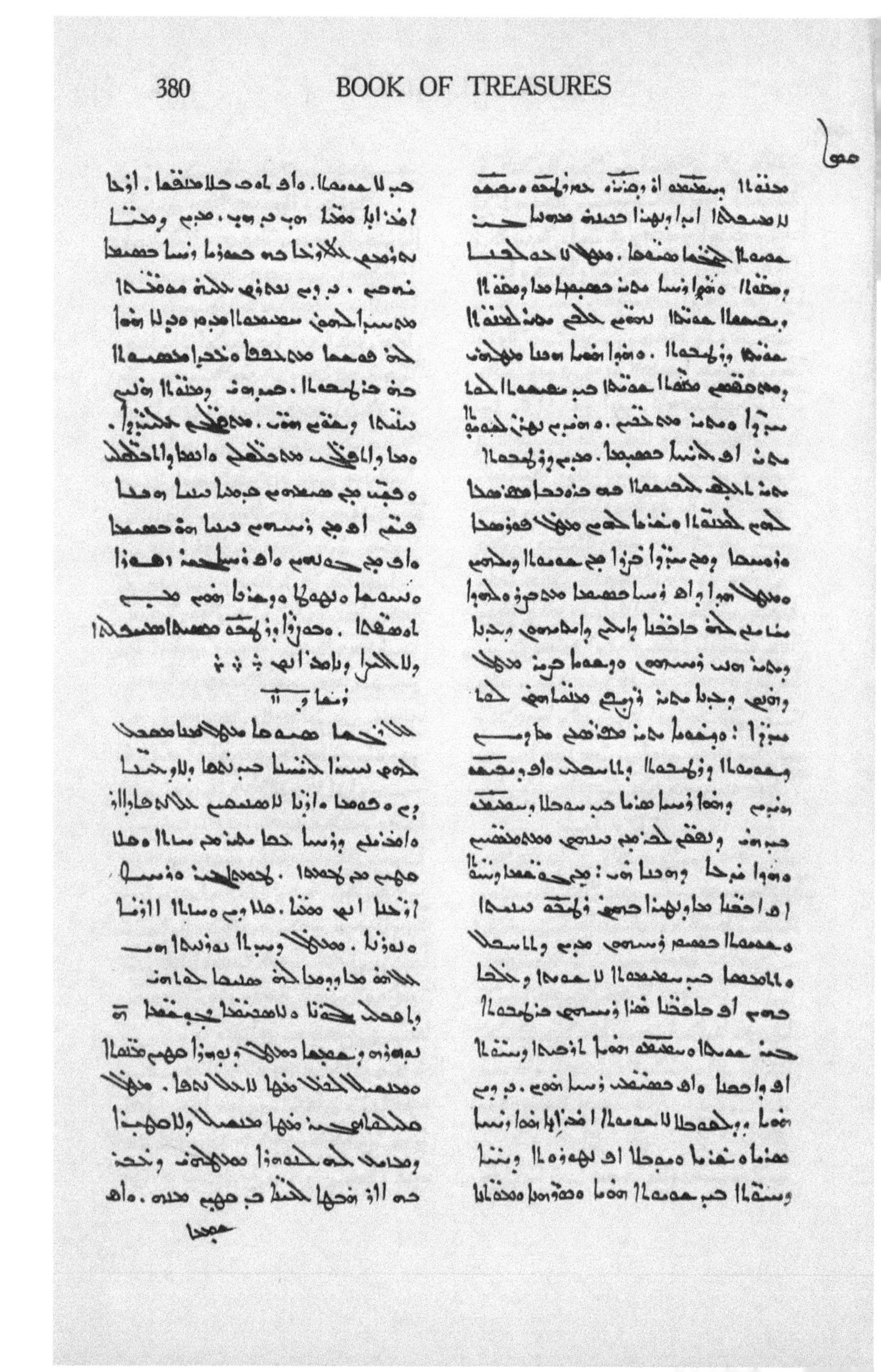

[illegible]

[illegible]

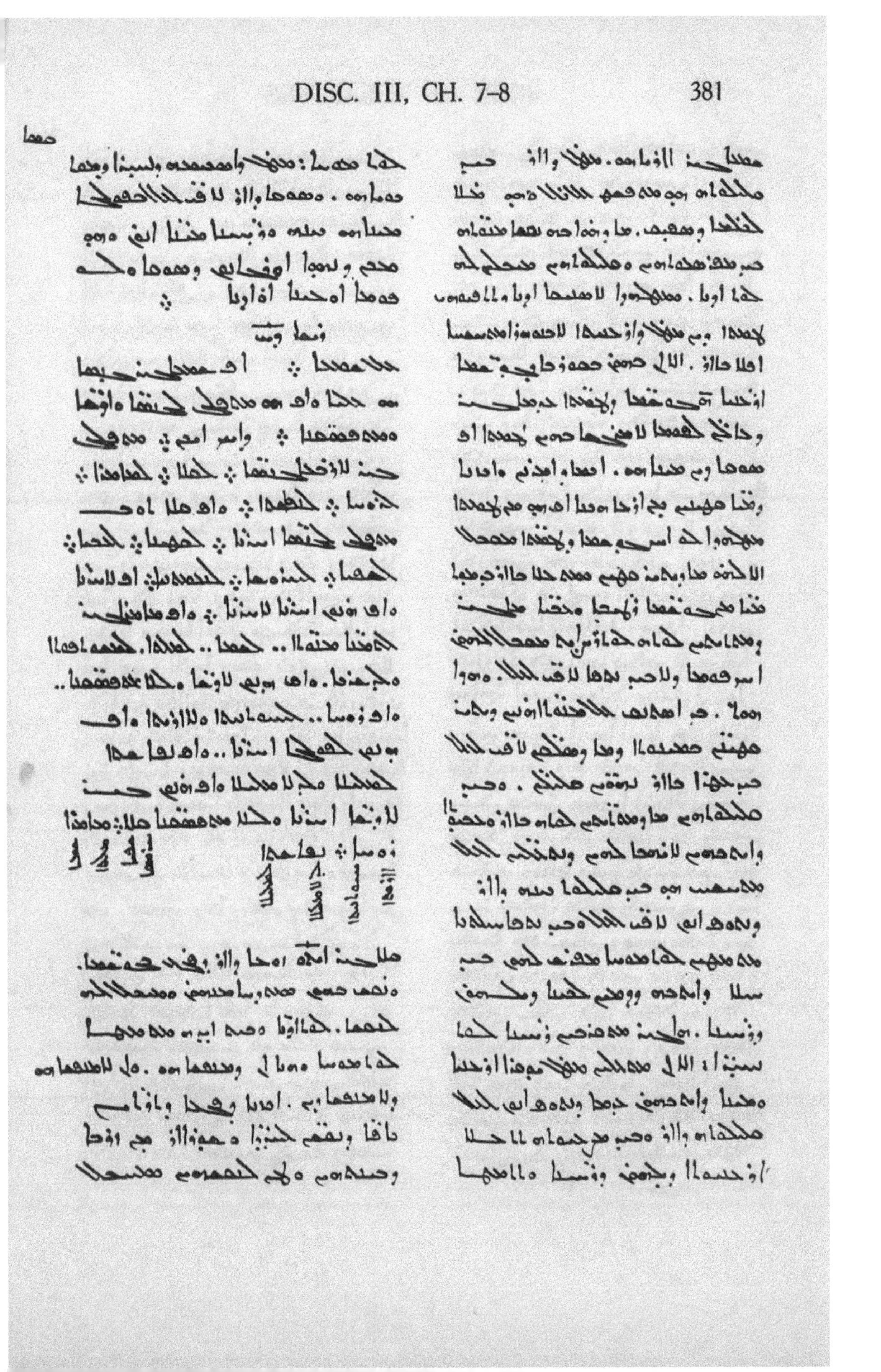

[illegible]

[illegible]

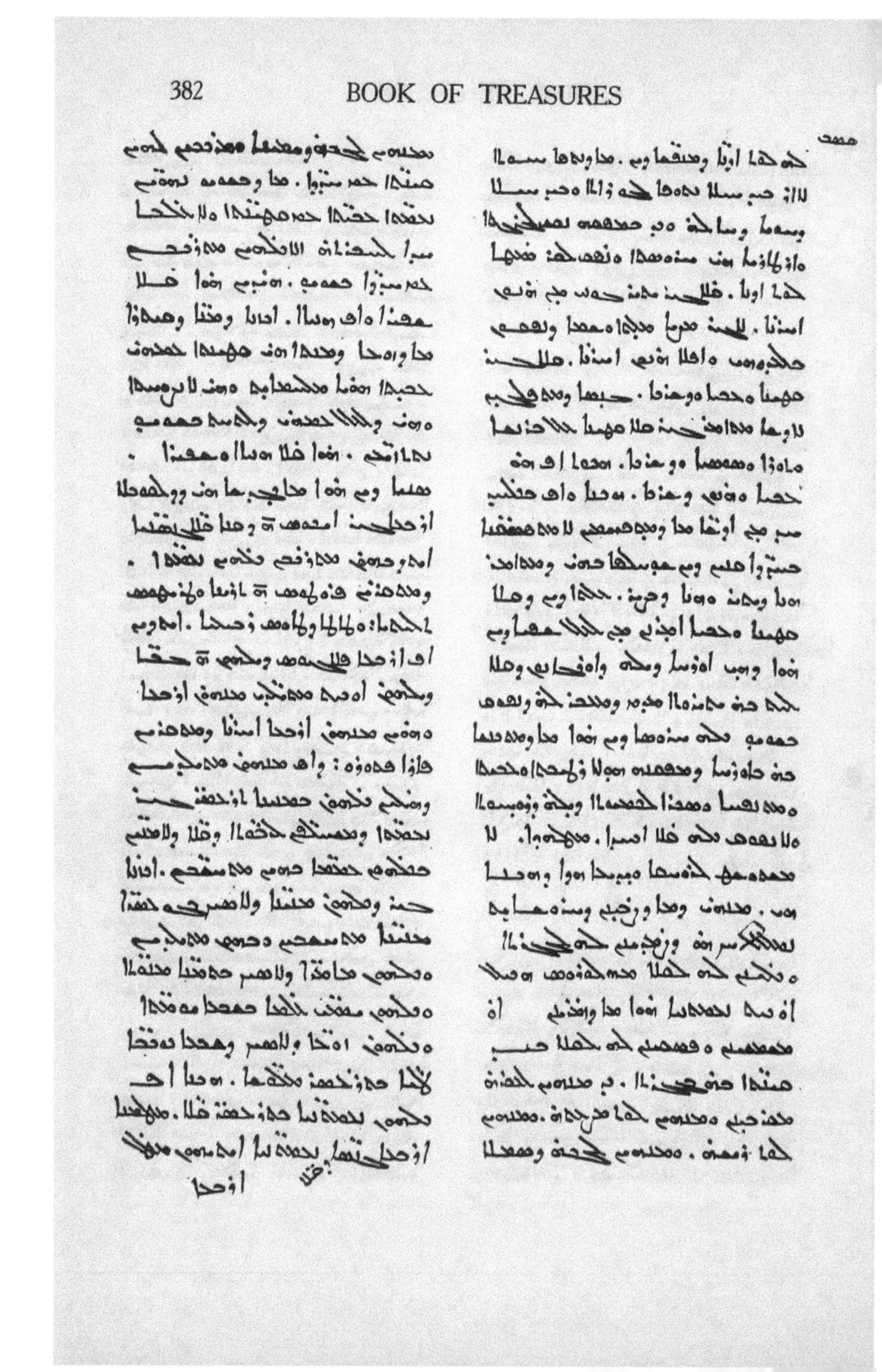

ܐܪ̈ܒܥܐ ܐܣܛܘܟܣ̈ܐ ܘܡܢܗܘܢ ܡܬܪܟܒܝܢ ܓܘܫ̈ܡܐ
ܘܟܠ ܐܝܟܢܝܢ . ܐܝܟܢܐ ܕܡܢ ܪܘܚܝܢ ܘܗܘܢܝܢ
ܐܪ̈ܒܥܐ ܟܝ̈ܢܐ ܘܟܠܐ ܠܡܬܢܝ̈ܐ ܗܘܐ ܟܕ
ܚܝܠܐ ܘܚܘܝܐ ܡܛܠ ܕܝܠܗ ܠܟܠܐ ܠܐܦ̈
ܕܢܐ ܓܒܪܢܝܐ . ܟܕ ܪܫܝܡ ܠܗ ܡܢ ܟܠܐ . ܡܛܠ
ܗܘܐ ܐܦ ܡܢ ܡܫܬܡܫ . ܟܕ ܡܬܕܢܚ ܠܗ ܠܓܒܐ
ܘܓܒܐ ܒܗ ܓܒܪܐܝܬ ܐܘ ܡܬܒܥܐ ܠܡܫܡܠܝܘ
ܥܡ ܚܙܘܐ ܘܡܟܝܠ ܡܫܬܟܚ ܡܬܘܕܥ ܘܗܦܟ
ܗܢܐ . ܘܟܠ ܠܐܪܥܐ ܗܘܐ ܡܚܐ ܘܚܝܠܐ ܕܝܗܒܐ
ܘܝܠܗ ܠܟܠܐ ܠܐܦ̈ ܠܗܘܢ ܠܗ ܐܠܐ ܐܝܟܢܐ
ܓܒܪܐܝܬ ܘܡܛܠ ܦܫܛ ܠܗ ܘܡܫܬܠܡܐ .
ܡܛܠ ܡܢ ܗܘ ܥܡܠܐ ܚܕܐ ܘܢܛܪ ܠܐܦ̈
ܠܗܘܢ ܐܝܟ ܗܘ ܕܥܒܕܐ ܢܦܫܗ ܟܠܐ ܘܡܢ
ܐܠܗܘܬܐ ܗܘܬ ܠܗܘܢ ܡܢ ܗܘ ܥܡܠܐ ܥܡܝܩܐ
ܘܡܛܠ ܡܢܗܘ ܚܕܐ . ܟܕ ܥܠܝܗ ܡܬܝܩܪܐ
ܢܦܫ ܠܡܬܪ̈ܘܢ ܡܛܠ ܠܗܘܢ ܥܡ ܚܙܘܐ
ܘܟܢ ܡܢ ܠܓܘܝܐ . ܠܐܘ ܟܠܐ ܪ̈ܘܚܢܐ
ܗܘܐ ܠܟܠܐ ܡܢ ܗܘ ܠܐܪܥܐ ܒܡܣܬܟܠܐ ܘܠܥܠ
ܠܐܠܗܐ ܘܡܫܬܠܛܢܘܬܐ ܥܠ ܡܢ ܡܢܗ ܒܚܝܠܐ
ܘܚܝܠܐ ܡܛܠ ܡܢ ܗܘ ܠܐܠܗܐ ܗܘ̈ܝܢ ܟܝ̈ܢܐ
ܟܠܗܘܢ ܕܝܠܗ ܓܒܪܐܝܬ ܕܢܗܘܐ ܟܠܗܘܢ
ܗܘ ܥܡܠܐ ܘܩܢܝܢܐ ܗܘ ܠܐܢܫܐ ܗܘ ܠܐܪܥܐ
ܘܗܝܢ ܠܐܝܟ ܐܪ̈ܥܬܐ ܒܡܣܬܟܠ ܘܠܐܪ̈ܥܢܝܐ
ܟܠܗܘܢ ܡܢ ܡܢܐ ܘܐܪ̈ܥܢܝܢ ܐܣܛܘܟܣ̈ܐ
ܠܐܝܟ ܟܕ ܡܬܢܚܝܢ ܥܠ ܟܦܪ̈ܐ ܘܠܐ
ܣܡ̈ܝܢܐ ܘܗܢܝܢ ܟܕ ܢܟܦ̈ܪܐ ܡܢ ܟܠܗܝܢ
ܗܘ ܟܠܐ ܘܡܢ ܟܠܗܝܢ ܗܘ̈ܝܢ ܐܝܟܢܐ ܗܘ
ܟܕ ܣܓܝܐܐ ܟܠܗܝܢ ܐܚܝܕܐ ܘܗܘ ܟܕ ܡܬܢܐ
ܟܠܗܝܢ ܐܝܟܢܐ . ܟܕ ܗܘ ܟܠܗܐ . ܘܗܢܘܢ
ܠܐܝܟ ܣܦ̈ܝܩܐ ܟܕ ܗܘܝܢ ܢܦܩܝܢ ܗܘܘ

ܟܕ ܡܬܡܫܚܢܐ ܠܟܠܐ ܗܘ ܥܡܠܐ ܠܗ ܠܐܣܛܘܟܣܐ
ܘܐܣܛܘܟܣܐ ܟܠܐ ܒܡܬܟܢܫܢܘܬܐ ܡܬܦܠܓ ܗܘ
ܡܢܗܘܢ ܡܬܬܢܝܚܘ ܠܐܪܥܐ ܒܡܬܒܥܘܬܐ ܘܗܘ
ܠܐܪܥܐ ܗܘܐ ܡܫܬܚܠܦܐ ܠܟܠܐ ܐܣܛܘܟܣܐ ܠܐܪܥܐ
ܘܗܘܝܢ ܠܐܝܟ ܕܡܬܪܟܒܐ ܡܬܪܟܒܝܢ ܒܟܝܢܘܬܐ
ܘܡܫܬܥܒܕܐ ܠܟܠܐ ܗܘܝܢ ܣܡ̈ܝܢܐ ܘܡܬܪܟܒܐ
ܗܘܝܢ ܘܡܢ ܠܐܦ̈ܝܐ ܐܚܪܢܐ ܠܐܪ̈ܥܐ ܡܢ ܟܝܢܗ
ܗܘܝܢ ܐܪ̈ܒܥܐ ܐܝܟ ܐܘܪ̈ܚܐ ܡܢ ܟܝܢ̈ܐ ܗܘܝܢ
ܘܡܢ ܡܬܚܠܦ ܡܢ ܐܘܪ̈ܚܐ ܘܟܠܐ ܐܝܟ ܠܐ
ܡܬܡܫܚ̈ܢܐ ܡܢ ܠܐܣܛܘܟܣܐ ܘܡܬܪܘܙܐ ܘܠܐ ܡܫܬܢܐ
ܘܠܐ ܡܬܒܥ ܒܪ̈ܟܝܗܘܢ ܒܡܬܡ̈ܫܚܢܐ ܘܒܪ̈ܟܝܗܘܢ
ܟܝ̈ܢܐ ܘܐܦ ܠܗ ܒܗ ܘܡܬܠܘܬܐ ܡܬܪܟܒܝܢ
ܐܦ ܡܬܡܫܚܝܢ ܠܡܬܪܟܒܗ ܐܝܟ ܟܠܐ ܘܗܘ
ܥܡܠܐ ܐܝܟܢܝܢ ܘܐܝܟ ܕܗܘ ܠܐܪܥܐ
ܘܐܚܪ̈ܢܐ ܘܪܘܚܐ . ܘܐܦ ܠܗܘܢ ܟܕ
ܫܠܡܘ ܒܪ̈ܟܝܗܘܢ ܒܡܬܡ̈ܫܚܢܐ ܘܢܦ̈ܩܝܢ ܟܠܐ :
ܐܦ ܠܐܝܟ ܕܗܘ ܐܝܟܢܐ ܗܘ ܘܢܬܚܠܦܘܢ
ܡܢ ܣܗܕܐ ܚܝܐ . ܟܕ ܘܡܢ ܗܘ ܘܐܬܚܠܦܬ
ܘܐܝܟܢܐ ܗܘ ܒܫܘܝܘ . ܘܡܢܐ ܘܡܟܝܠ ܡܬܦܫܛ
ܟܠܐ ܡܬܡܫܚܢ ܕܡܬܝܠܕ ܡܬܦܫܛ ܠܐ ܟܠܐ
ܘܡܬܡܫܚܢܐ ܐܝܟܢܝܢ ܠܫܒܘ . ܟܠܐ ܘܡܢ
ܘܫܘ̈ܝܐ ܐܘ ܘܟܪܝܗܐ ܟܕ ܢܦܫ ܡܫܬܒܩܐ
ܠܐ ܡܬܒܥܐ ܠܡܬܝܠܕܐ ܐܠܐ ܒܡܫܡܫ ܠܟܠܗܘܢ ܗܘ
ܘܡܢ ܕܟܕ ܢܦܫܬܐ ܡܫܬܚܠܦ ܡܬܚܠܦܐ ܟܕ ܗܢܐ
ܡܬܚܠܦܐ ܗܘܐ ܗܘ ܘܟܕ ܗܘ ܡܫܚܐ ܢܦܫܗ
ܒܟܠܗ ܢܦܫܐ ܡܫܬܚܠܦܐ ܠܗ ܠܡܬܚܠܦܐ
ܟܠܗ ܐܝܟܢܐ ܗܘܐ ܗܘ ܡܚܐ ܘܡܚܐ ܠܟܠܗ
ܡܚܐ ܘܡܚܐ ܠܗ ܡܬܚܠܦ ܘܡܫܬܠܛܢ ܒܡܫܚܐ
ܡܬܚܠܦ ܡܢ ܡܫܬܚܠܦܐ ܘܡܬܚܠܦܐ ܡܢ
ܡܫܬܚܠܦܝܢ ܘܟܕ ܚܙܘܐ ܟܕ ܡܚܠܦ ܠܐܣܛܘܟܣܐ

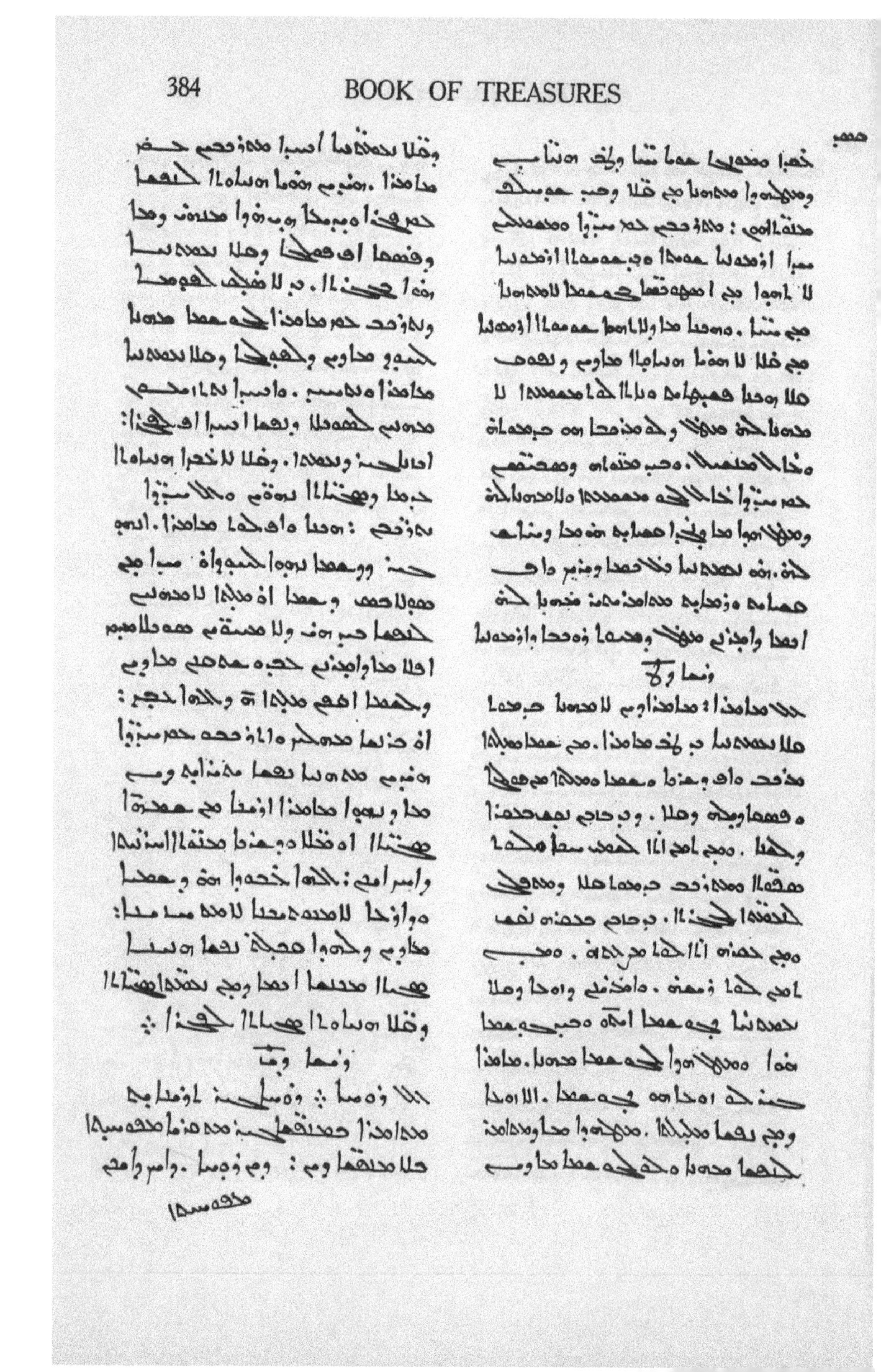

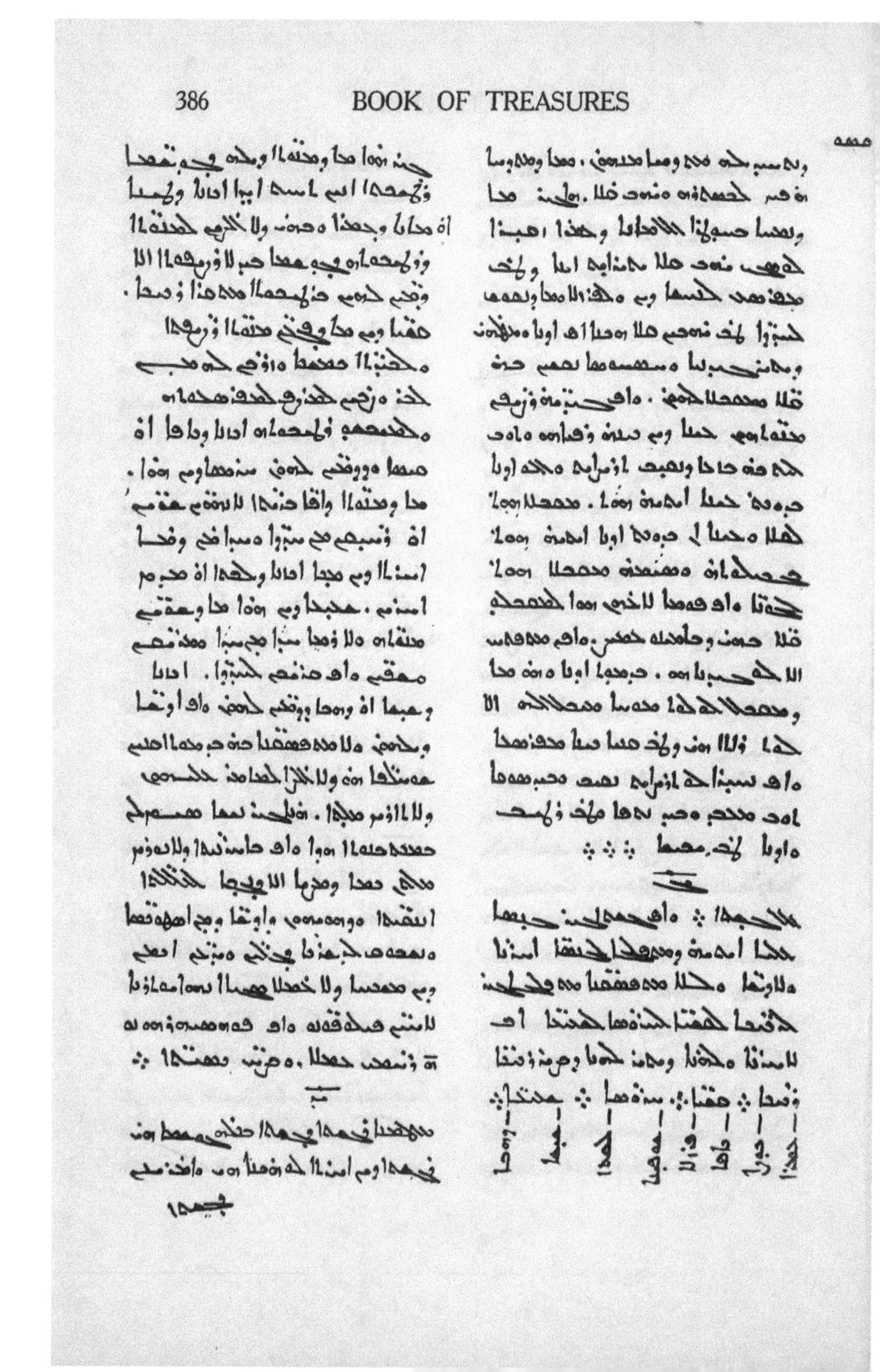

[illegible]

[illegible]

ܡܡܐ

ܒܝܘܡܐ ܕܗܘܐ ܥܡ ܟܢ̈ܫܐ ܡܗܝ̈ܡܢܐ ܘܐܒܗ̈ܬܐ
ܘܐܢܫ ܡܢ ܡܘܫܐ ܠܗܐ ܡܫܡܫܐ ܕܦܪܘܫ
ܠܠ ܢܦ̈ܫܬܐ ܗܘܢܐ . ܡܫܬܠܡܘܬܐ ܒܟܠܗ
ܒܗܘܢܐ ܗܢܐ ܦܓܪܢܝܐ . ܡܫܬܠܡܘܬܐ ܐܦ
ܗܘܝܘ ܐܘܟܝܬܐ ܐܚܪܢܐ ܕܦܓܪܐ ܒܝܬ ܠܠ
ܢܦܫܗ ܗܘܢܐ ܗܢܘܢ ܕܠܐ ܚܣܝܪܐ ܘܗܘܝܐ
ܐܝܟ ܡܫܡܠܝܢ ܠܡܚܫܒܬܗܘܢ . ܐܢܐ ܕܝܢ ܐܡܪܐ
ܠܟܠܐ ܕܗܘܝܘ ܐܝܟܐ ܕܣܘܟܢ ܡܢ ܠܠܠ :
ܠܠܐ ܒܝܘܡܐ ܡܢ ܐܘܣܝܐ ܐܝܬܝܢ .
ܒܗܘܢܗ ܡܢܗ ܡܢ ܗܘܢܐ ܢܦܫܐ
ܣܘܟܠܐ ܘܗܪܐ ܘܙܘܗܪܐ ܕܡܫܡܠܝܐ ܘܗܘܝܐ
ܘܠܐ ܗܘܝܐ ܕܣܪܝܩܐ ܘܙܥܘܪܐ . ܘܗܘ ܡܟܝܠ
ܗܘܐ ܠܗܘܢ ܡܢ ܒܗܘܢܐ . ܡܫܬܠܡܘܬܐ
ܢܦܫ ܡܢܗ ܠܗܘܢܐ ܘܠܗ ܠܗܘܢ
ܘܗܘܐ ܠܗܘܢ ܠܗ ܘܗܘܐ ܡܢ ܠܗ ܢܦܫ
ܘܡܫܬܠܡܢܐ ܘܗܘܐ ܐܝܟ ܡܫܡܠܝܐ
ܘܗܘܐ ܒܗܘܢ ܒܟܠܗܘܢ ܘܠܗܘܢ . ܡܫܬܠܡ
ܘܡܫܬܘܕܥܢܐ ܗܘܐ ܕܦܓܪܐ ܒܡܫܡܠܝܘܬܐ
ܡܫܡܠܝ ܫܡܝܢܐ ܒܗܘܢܐ . ܘܗܘܢܐ ܒܡܫܡܠܝܘܬܐ
ܘܗܘܝܘ ܕܗܘܢܐ ܘܗܘ ܡܫܡܠܝܐ ܐܝܟ ܠܗܘܢ ܡܢܐ
ܘܦܓܪܐ ܫܡܫܐ ܕܫܡܫܐ ܒܗܘܢܐ ܕܡܢ ܠܗ
ܢܦܫ ܡܢܗ ܘܡܢܐ ܗܘܝܘ ܕܗܘܢܐ . ܘܗܘ ܫܡܐ
ܡܢ ܐܦܝ̈ܢ ܒܗ . ܘܐܝܟܐ ܕܢܦܫ ܡܢ ܡܫܡܠܝܐ
ܗܘܢܐ ܡܫܡܠܝܐ ܡܢ ܐܝܟ ܕܢܦܫ ܘܐܦ ܡܢ
ܐܝܟ ܕܗܘܝܘ ܒܫܡܫܐ ܠܦܓܪܐ ܦܓܪܢܝܐ
ܗܘܐ ܗܘܢܐ ܡܢ ܠܗ ܠܚܘܕ . ܐܠܐ ܐܝܟܐ ܕܠܐ
ܒܝܬ ܦܓܪܢܝܐ ܒܗܘܢܐ ܐܦ ܒܫܡܫܐ
ܘܫܡܫܐ ܐܦ ܒܫܠܡܐ ܕܗܘܢܐ : ܐܠܐ ܠܗ
ܗܘܢܐ ܐܝܟܐ ܕܡܫܡܠܝܐ ܗܘܝܘ ܠܗܘܢ ܡܫܡܠ
ܘܒܝܬܐ ܕܠܗ ܘܗܘܐ ܕܡܢ ܦܘܫܢܐ

ܕܫܡܫܐ ܕܚܡܫܬܥܣܪ ܡܐܡܪ ܡܐܡܪܐ ܀
ܗܘ

ܡܫܬܠܡܐ ܗܢܘ ܦܓܪܐ ܫܡܝܢܐ ܡܠܗ ܟܠܗ ܐܦܐ ܕܠܗ
ܐܝܟ ܡܫܡܠܝܐ ܐܝܟܢܐ ܀
ܘܐܦ ܡܢ ܕܐܝܟܢܐ ܗܢܘ ܕܢܦܫܐ ܠܐܝܟ ܠܠܠ
ܘܐܝܟ ܠܗܘܢ . ܐܘܟܝܬ ܡܫܡܠܝܐ ܠܠܠ ܘܠܡܫܡܠܝܐ
ܘܫܡܫܐ ܠܗܘܢ ܘܐܦܝܢ ܕܡܫܬܡܠܝܢܐ ܐܝܟ
ܟܐܢܐ ܘܠܗܘܢ ܘܗܘܢܐ ܢܠܠ ܗܘܢܐ ܗܢܘ ܐܦ ܒܢܦܫܐ
ܒܝܬܐ ܘܐܝܬ ܠܠܠ ܐܝܟ ܢܘܗܪܐ ܘܐܦ ܘܡܫܝܚܐ
ܘܫܡܫܐ ܠܗܘܢ . ܐܝܟ ܐܝܟܐ ܡܫܝ̈ܚܐ ܘܐܦܝܢ
ܘܢܦܫܗ ܒܫ̈ܡܐ ܘܐܝܟܐ ܘܗܘܢܐ ܢܠܠ . ܘܗܘܢܐ ܡܫܡܠܝܐ
ܐܝܟ ܘܡܢ ܐܘܣܝ̈ܐ ܡܫܡܠܝܐ ܗܢܘ ܡܫܡܠܝܢ
ܗܘ ܘܗܘܢܐ ܡܫܡܠܝܢ ܘܡܢ ܒܗܘܢܐ ܗܘܝܘ
ܗܘ . ܐܦ ܦܓܪܐ ܫܡܝܢܐ ܡܫܡܠ ܡܢ ܢܦܫܗ
ܐܝܟܐ ܒܡܫܡܠܝܐ . ܒܗܘܢܐ ܕܡܢ ܗܘ ܘܐܘܣܝ̈ܐ
ܡܢ ܐܝܟ ܘܐܦ ܗܘܐ ܒܫܡܝܢ ܒܗ ܒܗܘܢܐ ܡܫܡܠܠ
ܘܒܗ ܐܠܐ ܐܝܟܐ ܐܘܣܝ̈ܐ ܘܗܘܢܐ
ܡܫܡܠܝܐ ܘܡܫܡܠܝܢ . ܫܡܝܢܐ ܗܢܘܢ ܢܠܠ ܘܗܘܐ
ܘܠܐܡܪ ܘܠܐ ܒܝܬ ܗܘܢܐ ܒܗܘܢܐ ܢܠܠ ܒܝܬܗܘܢ
ܒܗܘܢ ܠܗ ܐܠܐ ܫܡܝܢܐ ܦܓܪܐ ܘܠܐܡܪ ܡܢ
ܘܠܐ ܒܝܬ . ܘܐܡܪܐ ܘܗܘ ܘܗܘܢܐ ܘܡܫܡܠܝܢ
ܫܡܝܢܐ ܘܦܓܪܐ ܡܢ ܚܕ ܚܕ ܡܫܡܠܝܢ
ܗܘܢܐ ܐܦ ܦܓܪܐ ܫܡܝܢܐ ܒܝܬ ܫܡܝ̈ܢܐ ܒܗܘܢܐ
ܐܢܫܐ ܘܦ ܡܢ ܚܕ ܚܕ ܡܢ ܗܘܢܘܢ ܦܓܪ̈ܐ
ܐܝܟܢܐ . ܠܗܘܐ ܘܢܦܫܗ ܐܝܟܢܐ ܗܘܝܘܬܐ
ܐܝܟ ܒܝܬ ܐܠܗܐ ܠܗ ܗܘ ܘܒܗܘܢ ܗܘ ܡܫܡܠܝܢ
ܐܝܬܘܗܝ ܘܡܫܡܠܝܢ ܠܢܦܫܢ . ܘܡܫܬܠܡܘܬܐ
ܘܢܦܫ ܢܠܠ ܘܢܦ ܠܠ ܡܫܡܠܝܢ ܡܢܗܘܢ ܦܓܪܐ
ܘܡܫܡܠܝܐ ܗܘܐ . ܡܫܡܠ ܘܡܢ ܘܡܫ ܡܫܡܠܝܢ
ܗܘܫܠܠ ܘܡܫܡܠ ܘܡܢ ܡܫܡܠ . ܐܝܟܘ

ܒܪܡ ܕܡܫܡܫܢܘܬܐ ܢܣܒ ܦܘܠܚܢ ܡܢ ܡܕܒܪܢܘܬܐ
ܗܝ ܕܢܣܬܟܠ ܐܘܪܚܬܗ ܗܝ ܕܠܐ ܡܫܬܟܠܐ ܚܕܐ ܡܢ
ܐܡܪܢܢ ܗܝ ܕܒܕܬܐ ܠܚܝܘܬܐ ܘܡܫܝܢܐ ܗܝ
ܘܠܐ ܢܦܩܐ ܠܗ ܠܒܠܝܘܬܐ ܡܢ ܗܦܟܝܗ ܐܝܢܐ ܕܐܝܬܝܗ
ܘܐܝܟܢ ܕܠܗܕܐ ܠܗ ܗܘܐ ܒܡܠܬܐ ܦܠܚܝܢ ܗܘܘ
ܠܚܝܘܬܐ ܘܒܝܫܘܬܐ ܐܘ ܠܐܝܠܢܐ ܐܘ ܠܟܘܟܒܐ ܐܘ
ܠܦܘܪܫܐ ܐܚܪܢܐ ܘܡܕܡ ܕܦܛܝܠܐ ܠܐܠܗܘܬܐ ܗܘܬ ܠܗ ܐܦܠܐ
ܒܓܠܝܐ ܕܡܬܚܙܝܐ. ܡܛܠ ܕܡܬܚܙܝܐ ܘܡܫܬܡܥܐ ܠܐ
ܦܐܫ ܗܘܐ ܠܐܠܗܘܬܐ ܩܘܡܐ ܗܘܬ ܠܗ ܐܦܠܐ ܡܘܡܐ
ܘܣܓܕܝܢ ܒܝܫܬܐ ܗܘܘ ܚܢܦܐ. ܘܐܦ ܠܚܝܘܬܐ ܘܐܝܟ
ܘܡܩܝܡܝܢ ܦܠܚܝܢ ܗܘܘ ܚܢܦܐ. ܘܐܦ ܒܝܫܬܐ ܐܘܪܫ
ܒܗ ܒܙܒܢܐ ܐܝܟ ܗܝ ܒܪ. ܠܐ ܡܦܠܚܝܢ ܗܘܘ
ܒܡܫܒܚܘܬܐ ܘܠܗܪܐ ܘܦܠܚܝܢ ܗܘܘ ܘܐܟܝܢ ܗܘܘ
ܚܢܦܐ. ܗܟܢܐ ܕܝܢ ܘܐܦ ܒܕܘܟܬܐ ܘܒܠܚܘܕ ܠܐ
ܡܦܠܚܝܢ ܗܘܘ ܒܝܬܐ ܘܣܓܕܝܢ ܐܘ ܕܒܬܐ
ܕܒܡܫܟܢܐ ܗܘܘ ܡܫܒܚܝܢ ܗܘܘ ܠܐܘܪܚܬܐ ܐܘ ܕܫܠܝܛܐ
ܘܚܢܦܐ ܗܘܘ ܠܗܘܢ ܘܐܦܝܢ ܗܘܘ ܡܕܒܪ
ܐܝܟ ܕܝܢ ܘܐܦܠܐ ܡܬܘܕܝܢ ܗܘܘ ܡܢ ܠܒܢܝܢܫܐ
ܘܒܢܝܢ ܗܘܘ ܗܝܟܠܐ ܒܠܐ ܡܕܒܚܐ ܗܘܘ ܒܓܘܦ
ܘܒܪܝܪ ܗܘܐ ܘܦܘܩ ܗܘܘ ܦܠܚܝܢ ܗܘܘ ܐܘ ܠܐ
ܐܢܫ ܗܘܘ ܠܦܘܠܚܢܐ ܘܡܬܒܥܒܥܝܢ ܗܘܘ ܗܘܘ
ܚܕܐ ܡܕܡ. ܘܐܦ ܠܐܠܗܘܬܗ ܗܘܐ ܠܗ
ܦܠܚܐ ܦܓܪܐ ܠܐ ܡܫܬܡܥܝܢ ܗܘܘ ܒܚܕܐ ܐܘ ܐܦܠܐ
ܡܬܟܠܐ ܢܦܫܝܢ ܗܘܘ. ܗܘ ܕܡܠܟܐ ܚܝܐ ܡܠܦܐ
ܘܡܢ ܗܠܝܢ ܗܘܐ ܘܠܐ ܦܪܫܝܢ ܡܢ ܡܫܝܚܐ ܠܐ
ܡܬܚܙܝܐ. ܗܘ ܕܡܠܐ ܚܝܐ ܕܐܦܐ ܗܘܘܐ.
ܡܢ ܠܡܠܐ ܕܠܐ ܐܝܬܘܗܝ ܐܠܗܐ ܕܟܠܢܫ ܡܢܗܘܢ ܕܠܐ
ܡܫܘܠܐ ܐܢܫܐ ܕܢܦܫܝܗ ܗܘܘ ܠܗ ܡܢ ܗܘܐ.
ܘܐܦ ܗܠܝܢ ܠܐ ܦܫܝܛܝܢ ܗܘܘ ܢܦܫܐ ܒܡܘܬܐ
ܘܡܣܬܟܠܝܢ ܗܘܘ ܡܛܠܗ ܘܡܬܬܕܪܝܢ ܗܘܘ

ܒܗܘܐ ܠܗ ܡܬܐܘܪܐ ܕܚܙܐ. ܡܢ ܡܬܐܡܪ ܗܘܐ
ܗܘܐ ܪܫܐ ܒܡܫܟܢܐ. ܘܐܝܟܢ ܕܡܢ ܡܪܝܐ
ܠܐܟܪܝܢ ܗܘܘ. ܡܛܠ ܕܟܠܗ ܗܘܐ ܒܒܘܫܐ
ܘܟܪܝܢ ܗܘܐ ܚܝܠܐ ܘܡܫܡܠܝܢ ܦܓܪܐ
ܐܝܟܢܐ ܡܫܬܟܚܢܐ ܕܚܢܦܐ. ܡܛܠ ܗܢܐ
ܡܫܡܠܝܢ ܕܐܝܬ ܒܗ ܟܪܘܒܐ ܐ ܪܒܐ
ܕܢܫܡܥܘܢ ܘܢܫܡܥܘܢ ܗܘ ܘܢܒܥܐ
ܐܝܟ ܡܢ ܗܘ ܕܠܟܠ ܘܐܝܟܢܐ ܡܫܡܫܢܘܬܐ
ܕܒܗܐ. ܘܗܘܐ ܡܛܠ ܗܘ ܘܒܡܫܚܐ ܐܝܟ
ܐܝܟ ܠܗ ܘܠܐܠܗܐ ܒܪܐ ܠܗ ܘܠܗ ܡܫܡܫܐ
ܦܓܪܐ ܠܦܘܪܢܣ ܘܠܡܫܡܫ ܠܚܢܦܐ ܀

ܡܗ

ܒܠܒܘܢ ܕܒܪܬܐ ܘܟܠܐ ܘܐܬܢܒܐ ܘܪܡܫܐ
ܠܗ ܐܘܦܪܘܣ ܐܝܟ ܐܣܛܘܟܣܐ ܕܐܢܫܐ
ܒܪ ܕܝܢ ܕܡܠܟܝܢ ܡܢ ܡܠܬܐ ܕܡܠܠܬܐ
ܘܡܫܡܫܐ ܦܓܪܐ ܘܡܬܬܟܦܝܢ ܐܦ
ܢܦܫܐ ܡܠܟ ܡܢ ܥܒܘܪ ܥܠܝܠܐ ܡܬܠ
ܡܬܐܘܪܐ ܕܗܘܐ ܡܛܠ ܠܗ ܠܗ ܡܫܡܫܐ
ܗܘ ܕܠܐܡܪ ܒܠܒܘܬܐ ܘܪܡܫܐ ܘܫܡܫܐ
ܘܟܠܗܘܢ ܕܒܢܝܐ ܐܝܟܢ ܘܠܗ
ܐܘܦܪܘܣ ܡܛܠ ܕܐܣܛܘܟܣܐ ܐܝܟ
ܡܫܡܫܐ ܡܬܒܪ ܒܠܚܘܕ. ܐܦ ܒܪܝܬܐ
ܕܐܪܣܛܘ ܕܠܟܘܢ ܘܡܬܒܪܝܢ ܒܠܚܘܕ
ܗܘ ܕܒܗ ܗܘܐ ܠܒܪܘܬܐ ܟܪܝܢ ܗܘܘ
ܒܓܝܢ ܘܐܦ ܠܦܪܘܛܣܛܘ ܕܠܗ
ܡܬܐܪܝܢ. ܐܡܪ ܕܝܢ ܠܒܐܡܪ ܒܪܬܐ
ܘܐܡܫܡܫܐ ܘܟܠܐ ܘܪܡܫܐ ܒܗ ܠܗܦܟܐ
ܐܘܦܪܘܣ ܐܝܟ ܘܡܠܒܫܬܐ. ܘܐܦ
ܡܫܒܚ ܠܫܡܫܐ ܠܪܡܫܐ ܘܡܠܐ ܕܐܡܗ ܒܪܬܐ
ܘܡܦܠܚܘܢ ܘܒܢܝܗܘܢ ܐܘܦܪܘܣܐ ܘܦܪܨܘܦܐ

ܒ

ܕ݁ ܐܡܪܝܢ ܕܠܗ ܡܢ ܐܪܒܥܐ ܐܣܛܘܟܣ̈ܐ
ܓܘܫܡܗ ܕܐܝܬܘܗܝ ܒܫܘܚ̈ܠܦܐ ܕܠܐ ܫܠܝܢ ܐܣܛܘܟܣ̈ܘܗܝ
ܗ̈ܢܝܢ ܕܟܠܗܝܢ ܡܬܚ̈ܠܦܢ ܘܫܢܝܢ
ܡܫܬ̈ܚܠܦܐ ܘܠܐ ܫܪܝܪܐ ܘܐܦ ܐܝܬܝܗܘܢ ܐܣܛܘܟܣ̈ܘܗܝ
ܐܡܪ ܕܐܝܬܝܗܘܢ ܒܓܘܫܡܐ ܢܦܫܢܝܐ ܘܦܓܪܢܝܐ
ܘܫܦܝܪܐ . ܘܐܦ ܠܓܘܫܡܐ ܗܠܝܢ ܕܗܘܐ ܒܗܢܐ
ܘܠ ܡܢ̈ܝܐ ܕܟܠܗܝܢ ܠܗ ܡܢ ܡܕܡ ܕܡܢ ܟܠܗ
ܒܗ ܕܢܦܫܗ ܡܬ̈ܩܢܝܢ ܘܒܓܘܫܡܐ ܕܝܢ ܕܝܠܗ
ܘܡܢܐ ܠܓܘܫܡܝܗܘܢ ܐܝܟ ܐܝܟܢܐ . ܘܐܦ
ܘܡܬܚܙܝܢ ܡܢ ܟܠ ܚ̈ܝܐ ܘܡܕܡ ܕܗܢܐ
ܘܡܫܬܡܠܝܢ ܘܐܦ ܫܪܝܪܐ ܘܐܝܟܢܐ ܕܟܠܗܘܢ
ܡܕܡ ܕܝܠܗ ܘܗܘܐ ܒܗ ܘܒܓܘܫܡܐ ܗܘܐ
ܒܒܪܝܬܐ ܘܠܐ ܐܝܟ ܕܡܢ ܡܬ̈ܚܙܝܢ ܚ̈ܝܐ
ܐܦܠܐ ܡܢ ܗܢܐ ܘܡܬܩ̈ܢܢ ܐܠܐ ܒܓܘܫܡܐ
ܢܦܫܐ ܕܟܠܗ ܐܝܬܝܗ ܘܡܬܚܙܝܐ ܒܗ
ܐܝܟ ܡܬܩܢܐ ܕܠܗ ܘܒܓܘܫܡܐ ܘܡܢ ܕܝܠܗ
ܘܐܦ ܫܪܝܪܐ ܕܠܗ ܢܦܫܐ ܘܚܝܐ ܐܝܟ
ܓܘܫܡܗ ܒܓܘܫܡܐ ܟ̈ܠܐ ܕܢܦܫܐ ܒܗ ܪܘܚܐ
ܟܠܗ ܘܗܘ ܒܓܘܫܡܐ ܐܝܟ ܢ̈ܦܫܬܐ
ܐܘܣܝܐ ܐܝܬܝܗ ܗܘ ܡܕܡ ܕܐܝܬܘܗܝ ܘܠܐ ܦܓܪܐ
ܒܓܘܫܡܐ ܗܘ ܡܕܡ ܕܗܘܐ ܘܦܓܪܐ ܒܗܘܐ
ܡܬܩܢ . ܫܪܝܪܐ ܠܐܝܬܐ ܐܝܬܝܗ ܡܟܝܠ
ܟܠܗܝܢ ܘܡܬܚ̈ܠܦܢ ܘܫܢܝܐ ܘܠܐ ܒܓܘܫܡܐ
ܠܗ ܚܘܝܐ ܘܐܘܣܝܐ ܐܦ ܒܓܘܫܡ̈ܐ . ܐܝܟ
ܒܪ ܕܐܝܬܘܗܝ ܘܠܐ ܦܓܪܢܝܐ ܒܓܘܫ̈ܡܐ ܐܝܟ
ܢܡܪ ܠܗ ܥܡ ܟܠܗܝܢ ܒ̈ܢܝܢܐ ܢܚܙܐ ܗܘ ܠܗ
ܒܪ ܡܕܐ ܚܕ ܠܗ ܡܟܝܠ ܗܘܐ ܘܠܐ ܡܬܗܘܐ
ܒܓܘܫ̈ܡܐ ܗܘܐ . ܐܠܐ ܐܡܪܝܢ ܕܢܦܫܐ ܐܘܣܝܐ
ܐܝܬܘܗܝ ܠܗ ܡܟܝܠ ܗܘܐ ܘܐܡܪܝܢ ܕܢܦܫܐ
ܒܓܘܫܡܐ . ܘܐܦ ܡܬܩܢܐ ܡܢ ܕܝܠܗ

ܠܟܠܗܝܢ ܘܡܢܐ . ܐܝܟ ܕܐܡܪܝܢ ܠܗ ܗܢܐ
ܓܘܫܡܐ ܡܕܡ ܗܘܐ ܐܡܪܬ . ܘܐܦ
ܘܡܬܚܙܝܢ ܗܘܐ ܡܬܩܢܝܢ ܟܠܗܝܢ ܡܬܩܢܢܐ
ܘܡܢܐ ܡܬܩܢܐ ܘܡܬܚܙܝܢ ܐܘ ܟܠܗܝܢ
ܘܐܡܪ ܕܠܗ ܡܬܩܢܐ ܒܝܫ ܚ̈ܝܐ ܡܕܡ
ܘܐܝܟܢܐ ܕܗܘ ܗܘܐ ܡܬܩܢܝܢ ܠܗ ܒܓܘܫܡܐ
ܟܠܗ ܘܗܠܝܢ ܗܘܐ ܘܡܬܩܢܐ ܘܗܘܐ ܐܘܣܝܐ
ܟܠܗ ܕܝܠܗ ܓܘܫܡܐ ܡܬܩܢܐ ܐܝܬܝܗ ܘܡܕܡ
ܘܡܬܚܙܝܢ ܠܟܠܗܘܢ ܚ̈ܝܐ ܐܝܬܝܗ ܒܓܘ ܐܘܣܝܐ
ܘܡܕܡ ܕܒܓܘܫܡܐ ܐܝܟ ܗܘܐ ܡܬܩܢܝܢ ܗܘܝܢ
ܓܘܫܡܐ ܗܘܐ . ܗܝܕܝܢ ܕܝܢ ܐܡܪܝܢ ܕܗܘ
ܡܬܩܢܐ ܠܗ ܚ̈ܝܐ ܐܡܪ ܗܘܐ ܠܦܓܪ
ܟܠ . ܘܐܦ ܗܘܐ ܒܓܘܫܡܐ ܘܗܝ ܗܘܐ
ܐܡܪܝܢ ܕܟܠܗ ܒܗ ܚ̈ܝܐ ܘܡܬܩܢܝܢ
ܟܠܗ ܘܐܝܬܝܗ ܐܝܟ ܟܠܗ ܘܢܦܫܐ
ܟܠ . ܐܪܐ ܠܝ ܐܝܟ ܗܘܐ ܒܓܘܫܡܐ ܐܝܬܝܗܘܢ
ܟܠܗ ܐܚܪܐ ܡܬܩܢܐ ܠܐܝܬܝܗܘܢ ܗܘܘ
ܕ ܡܢ ܦܓܪ ܗܘܐ ܘܡܢ ܕܝܠܗ ܠܓܘܫܡܐ
ܡܬܩܢ ܘܒܢ̈ܦܫܐ ܠܗ ܗܘܘ ܠܟܠܗ
ܘܠܐ ܟܠ ܡܕܡ ܒܗܢ ܚ̈ܝܐ ܢܦܫܗ ܘܟܠܗ
ܒܓܘ ܐܝܬܝܗܘܢ ܗܘܘ ܐܠܐ ܒܪ̈ܢܫܐ ܗܘܘ ܘܡܬܚܙܝܢ
ܘܐܡܪ ܠܝ ܐܡܪ ܘܐܝܟ ܐܘܣܝܐ ܡܬܩܢܐ
ܡܬܚܙܝܢ ܗܘܘ ܦܓܪܐ ܘܒܓܘܫܡܐ ܡܢ ܕܐܝܬܝܗ
ܐܡܪܐ ܗܘܐ ܘܟܠܗ ܟܠܗ ܠܗ ܡܕܡ ܘܦܓܪ
ܡܬܚܙܝܢ . ܐܡܪܐ ܘܠܐ ܐܝܬܝܗ ܗܘ ܕܦܓܪ
ܟܠܗ ܘܐܝܟ ܘܠܐ ܡܬܩܢܝܢ ܒܓܘܫܡܐ
ܘܐܝܬܘܗܝ ܗܘܐ ܓܒܪ . ܘܐܦ ܦܓܪܐ ܘܒܓܘܫܡܐ
ܗܘܘ ܗܘܝܢ ܘܠܗ ܐܝܬܝܗܘܢ ܗܘܘ . ܘܐܦ ܐܘܣܝܐ
ܗܘ ܡܕܡ ܘܠܗ ܐܝܬܘܗܝ . ܠܐ ܗܘܐ ܐܦܠܐ ܒܪ̈ܢܫܐ
ܐܝܟ ܕܗܘܐ ܒ ܠܗ ܐܝܬܝܗܘܢ ܗܘܘ ܐܘܣܝܐ

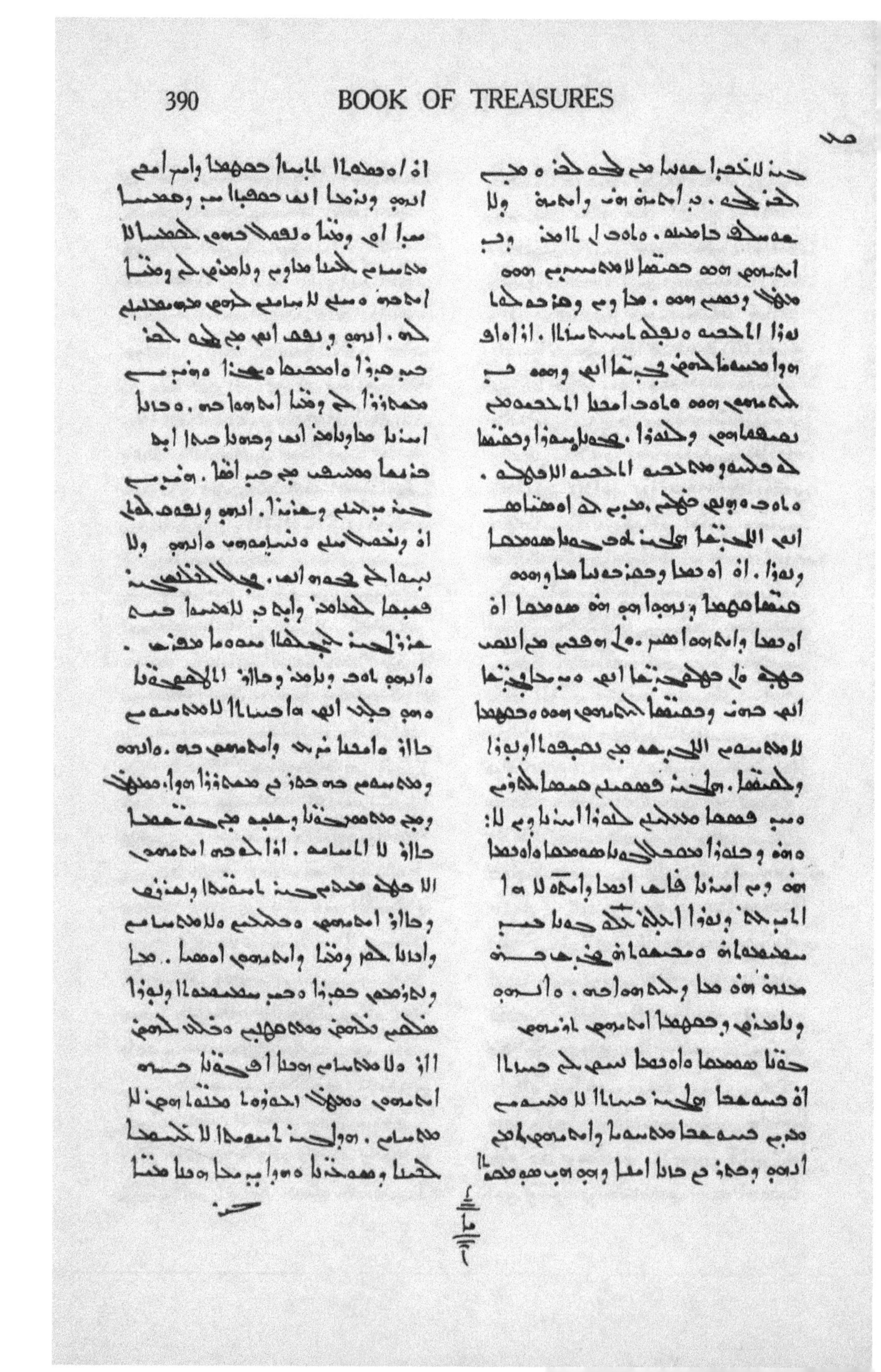

[illegible]

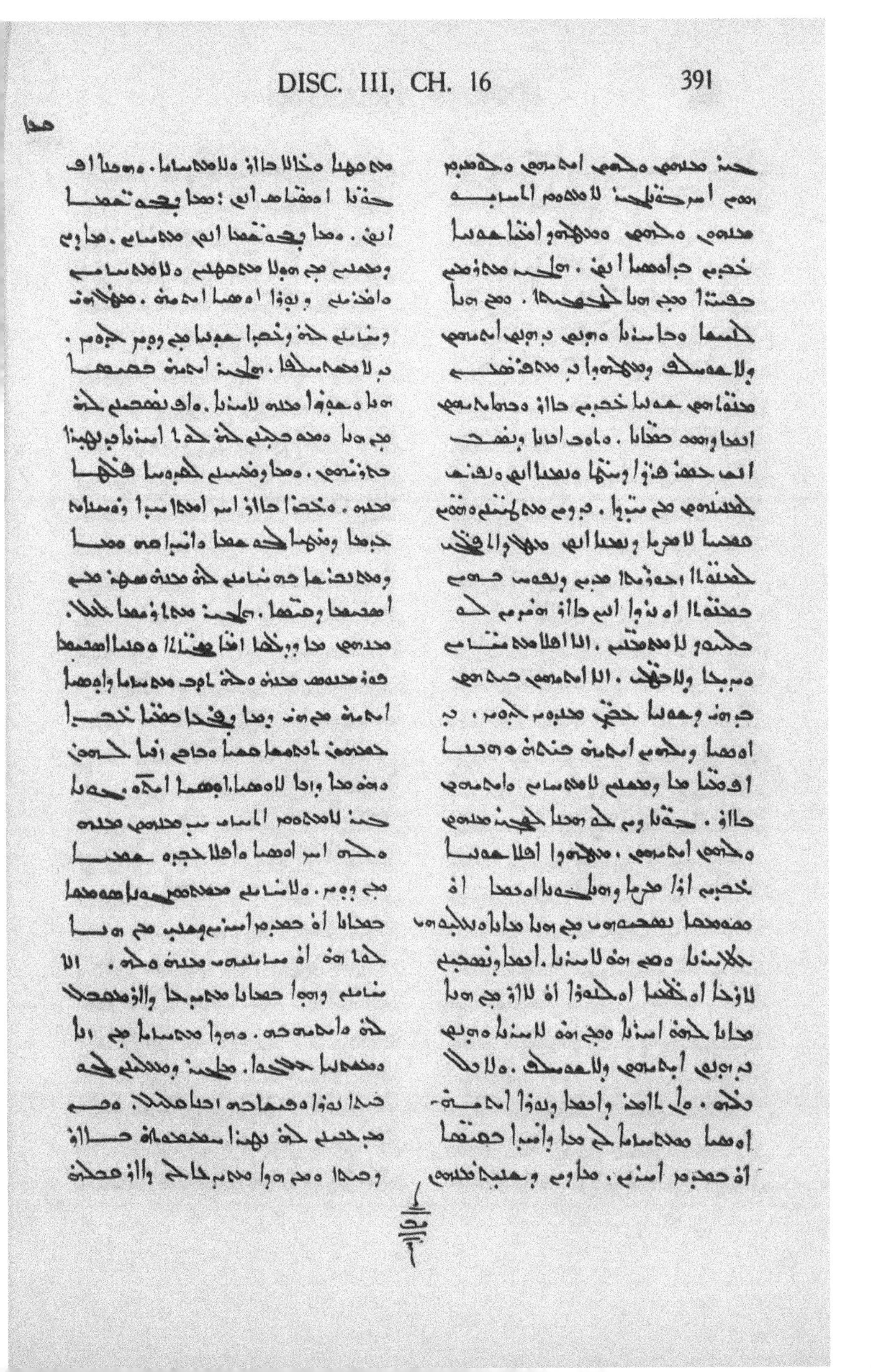

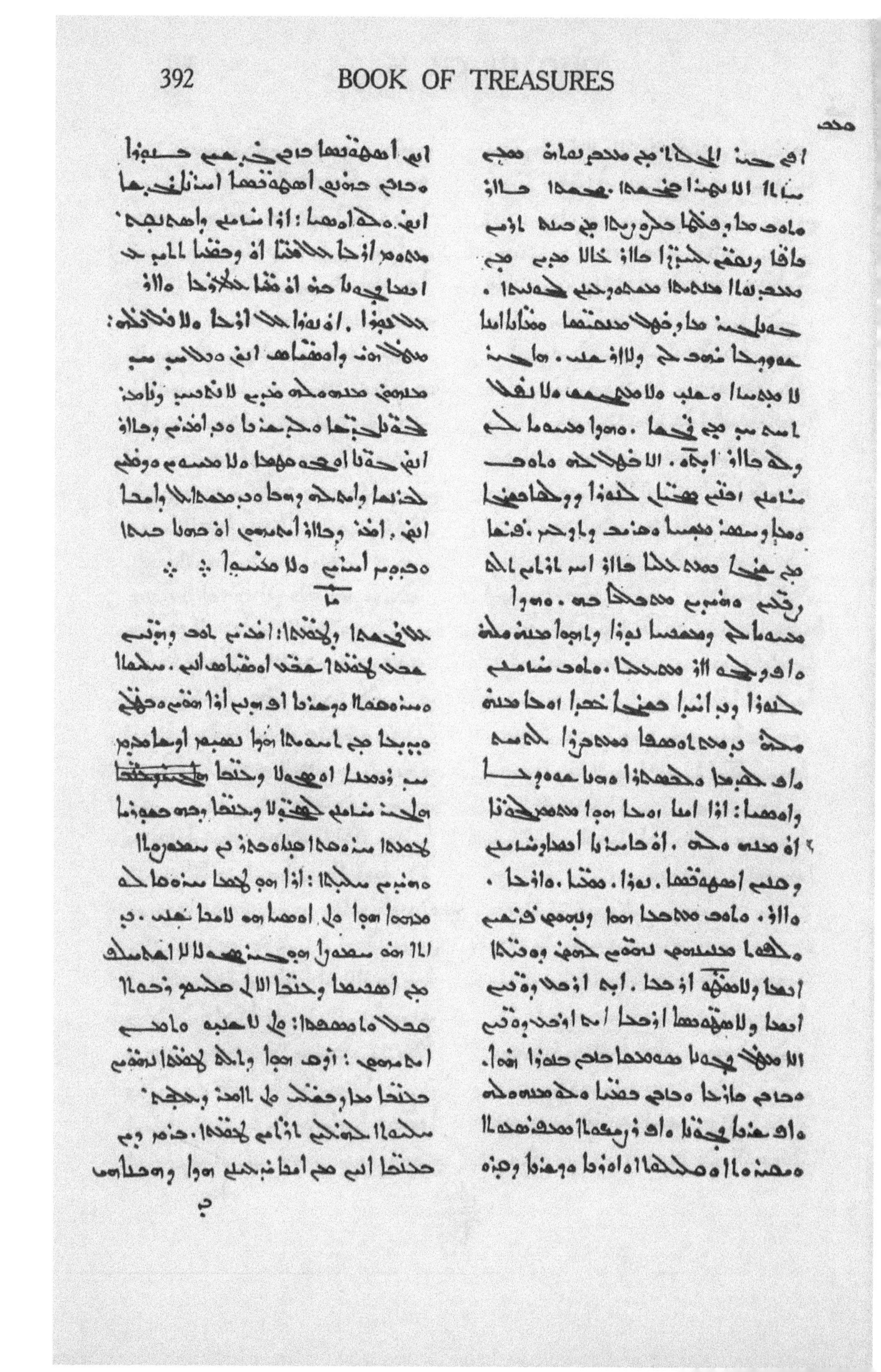

ܟܕ ܡܗܝܡܢܐ ܠܐ ܡܬܚܫܝܢ: ܠܡܗܝܡܢܐ ܠܒܪ:
ܡܢ ܚܫܘܬܐ ܕܠܗܘܢ ܘܡܢ ܦܘܪܥܢܐ ܕܠܗܘܢ
ܘܡܗܝܡܢܐ ܕܡܗܝܡܢܘܬܐ ܗܝ ܠܐ ܡܬܦܠܓܝܢ
ܘܐܝܟܢܐ ܡܬܬܥܝܩܝܢ ܕܠܐ ܢܐܡܪ. ܗܢܘ ܕܒܟܠܗ
ܥܡܐ ܕܐܝܠܝܢ ܕܒܥܝܢ ܢܚܝܘܢ ܠܗ ܕܢܚܐ
ܘܕܐܠܨܝܢ ܠܗ ܡܢ ܣܢܝܩܝܢ ܘܐܝܟ ܕܐܡܪ:
ܐܝܟ ܡܠܟܐ ܕܐܠܗܐ ܐܝܟ ܐܢܫ ܕܐܘܟܠ
ܒܥܘܕܪܢܐ ܐܘܟܝܬ ܡܬܘܩܢܐ ܐܝܟ ܗܘܐ ܒܟܪܝܗܐ:
ܐܡܪܐ ܘܠܐ ܐܘܟܝܬ ܡܬܥܕܪܢܝܬܐ ܘܠܐ ܡܬܡܢܚܐ
ܗܘܐ. ܘܠܐ ܡܬܦܪܢܣܐ ܗܘܐ ܒܟܠܗ ܒܪܢܫܐ
ܘܐܟܘܬ ܐܡܝܢܐ ܡܬܥܕܪܢܐ ܐܝܟ ܒܗܘ. ܘܐܝܟܢܐ ܕܐܡܪ
ܐܝܟ ܕܒܪܢܫ ܘܐܡܪ ܐܝܟ ܡܢܗܘܢ ܢܩܘܡܘܢ ܠܡܫܝܚܐ ܐܡܝܢܐ
ܒܥܘܕܪܢܐ ܠܐ ܐܬܝܠܕܬ ܕܟܕ ܡܬܝܠܕ ܐܝܟ ܗܢܘܢ
ܐܠܐ ܡܬܒܥܝܢ ܡܢ ܡܠܐ ܘܡܫܬܠܡ ܥܒܕܐ ܠܡܫܡܫ
ܐܠܗ ܫܠܝܚܐ ܡܢ ܡܫܝܚܐ ܦܪܘܩܐ ܐܬܠܒܫ ܐܝܟ
ܡܢ ܚܝܐ ܘܐܢ ܐܦ ܟܐܘ ܠܗ ܡܢ ܚܝܐ ܘܡܫܡܫ
ܘܡܫܡܫ ܐܬܠܒܫ ܐܝܟ ܫܠܝܚܐ. ܐܦ ܡܫܡܫܢܐ
ܡܬܝܕܥ ܐܡܪܢܢ. ܘܡܢ ܐܝܟܢܐ ܚܝܐ ܡܫܡܫ ܚܫܐ
ܣܢܝܩܝܢ ܘܡܬܥܕܪܢܝܢ ܐܘ ܣܢܝܩܝܢ ܠܥܘܕܪܢܐ
ܚܝܐ ܘܠܡܫܡܫܢܘܬܐ ܡܫܡܫ ܡܢ ܡܠܐ ܘܡܢ ܚܝܐ
ܘܡܫܡܫ ܦܪܘܩܐ ܘܐܝܟܢܐ ܕܐܦ ܗܢܘܢ .
ܡܫܬܡܫܝܢ ܗܘܘ ܒܗ ܘܐܢ ܡܫܡܫ ܐܝܟ ܒܫܠܝܚܐ
ܠܡܫܠܡܘ ܡܢ ܐܝܟ ܚܝܐ ܫܡܫܘܗ. ܘܡܫܡܫ ܒܗ
ܐܚܪܢ ܠܗ ܡܫܡܫ ܡܢ ܦܘܩ ܘܐܝܟ ܕܡܫܡܫ ܐܢܫ
ܠܗܘܢ ܒܟܠ ܪܘܚܐ ܦܪܘܩܐ ܐܝܟ ܫܠܝܚܐ ܘܠܗܘܢ
ܗܢܘܢ ܘܡܫܡܫ ܝܛ

ܠܐ ܦܪܘܩܐ ܕܪܘܚܐ: ܐܦ ܠܐ ܪܘܚܐ
ܕܡܢ ܗܢܐ ܟܕ ܗܘܐ ܐܦܪܫ ܘܐܬܦܪܫܘ ܐܝܟ:
ܕܐܝܠܝܢ ܕܠܐ ܡܬܦܠܓܝܢ ܘܢܦܫ ܡܫܝܚܐ
ܟܕ ܠܐ ܡܫܬܠܚ: ܢܦܫܐ ܐܦ ܒܪܢ ܕܠܗ ܘܦܠܚܝܢ

ܟܕ ܦܫܝܛܝܢ ܠܡܣܟܢܐ ܣܓܝܐܐ ܠܒܪ: ܒܥܘܕܪܢܐ
ܠܐ ܗܘܐ ܗܘܐ ܪܘܚܐ ܩܕܝܫܐ ܘܢܦܫܗ. ܒܥܘܕܪܢ ܘܦܠܓ
ܗܘܐ ܠܗ ܪܘܚܐ ܐܦ ܐܡܪ ܐܢܐ ܐܢܐ ܠܗ ܪܘܚܐ:
ܐܡܪ ܗܘܐ ܡܫܝܚܐ ܘܠܗ ܐܝܟ ܐܘܟܝܬ ܘܢܦܫ
ܠܥܘܕܪܢܐ. ܘܐܡܪ ܐܡܪ ܗܘܐ ܘܐܦ ܡܢ ܕܒܟܠܗܘܢ
ܠܥܘܕܪܢܐ ܠܒܢܝܢܫܐ ܘܠܡܬܒܥܝܢܗܘܢ ܘܠܒܪ ܟܠܗܘܢ
ܘܢܒܐ ܪܘܚܐ ܩܕܝܫܐ. ܐܦ ܠܐ ܡܬܒܥܝܢ ܗܘ ܪܘܚܐ
ܒܫܘܡܠܝܐ. ܘܡܢ ܐܡܪ ܟܠ ܗܘ ܫܘܪܝܐ: ܘܡܢܗܘܢ
ܡܢܐ ܠܝ ܐܘܟܝܬ ܐܡܪ ܗܘ ܪܘܚܐ ܠܡܫܝܚܐ ܒܥܘܕܪܢܐ
ܟܕ ܡܬܦܠܓܝܢ ܗܘܘ ܘܥܘܕܪܢܐ. ܥܠ ܡܫܝܚܐ ܗܘ
ܡܬܐܡܪ ܐܘܪܚܐ: ܡܫܡܫܝܢ ܠܐ ܡܬܦܠܓܝܢ ܒܥܘܕܪܢ
ܡܠܐ ܡܫܝܚܐ ܘܥܘܕܪܢܐ. ܠܒܪ: ܡܠܐ ܘܡܢܗ
ܟܕ ܢܦܩ ܡܢܝܢܐ ܘܐܬܠܒܫ ܟܕ ܦܫܝܛܝܢ ܟܠܗܘܢ
ܒܡܫܡܫܢܐ ܥܠ ܐܡܪ ܘܡܢ ܡܢ ܡܠܐ ܡܫܡܫ
ܡܢ ܡܫܡܫܢ ܒܠܐ ܐܡܪܐ ܘܗܘ ܢܦܩ ܐܝܟ ܫܠܝܚܐ
ܘܡܫܡܫ ܘܫܡܫܗ ܠܡܫܝܚܐ ܘܐܡܪ: ܐܦ ܠܐ ܟܬܒ
ܐܠܐ ܡܫܡܫ ܥܒܕ ܠܗ ܝܚ

ܟܠܐ ܩܠ

ܟܠ ܩܠ ܟܕ ܡܢ ܫܡܝܐ ܐܝܟ ܐܠܐ ܐܘ ܣܓܝܐܐ ܐܠܐ
ܗܘܝܢ ܘܡܫܡܫܝܢ ܘܡܪܕܝܐ ܗܘܐ ܗܘܐ ܡܢ ܒܪܢܫܐ
ܒܪ ܐܢܫ ܒܢ ܫܠܝܚܐ ܠܗ ܐܘܪܚܐ ܘܡܢ ܐܢܫ
ܒܗܘܢ. ܟܕ ܗܠܐ ܘܡܫܡܫ ܠܐ ܡܫܡܫܝܢ ܡܢ ܟܠ
ܘܢܦܫܗ ܒܐܘܪܚܐ ܡܫܡܫܝܢ ܠܡܫܡܫ ܐܦ ܐܡܪ
ܗܘܐ ܗܘ ܗܠܐ ܗܘ ܘܢܦܫܗ ܒܐܘܪܚܐ ܒܗܘܢ
ܒܠܚܘܕ ܐܘ ܒܗ ܟܐܘ. ܥܠ ܒܠܚܘܕ ܐܡܪ ܗܘܐ
ܠܐ ܡܫܬܡܫܝܢ ܗܘܘ ܟܠ ܢܦܫܐ. ܥܠ ܟܐܘ ܗܘܐ
ܘܢܦܫܗ ܐܘܪܚ ܗܘܐ ܘܢܦܫܗ ܡܫܡܫ ܗܘܐ
ܘܡܫܡܫ ܒܐܘܪܚܐ: ܐܝܟ ܡܠܟ ܘܡܫܡܫ ܗܘ
ܗܠܐ: ܐܡܪ: ܒܠܚܘܕ ܐܘ ܟܐܘ. ܥܠ ܒܠܚܘܕܝܗܘܢ
ܐܡܪ ܗܘܐ. ܘܡܢܐ ܠܡܫܡܫܘ: ܐܦ ܟܐܘ

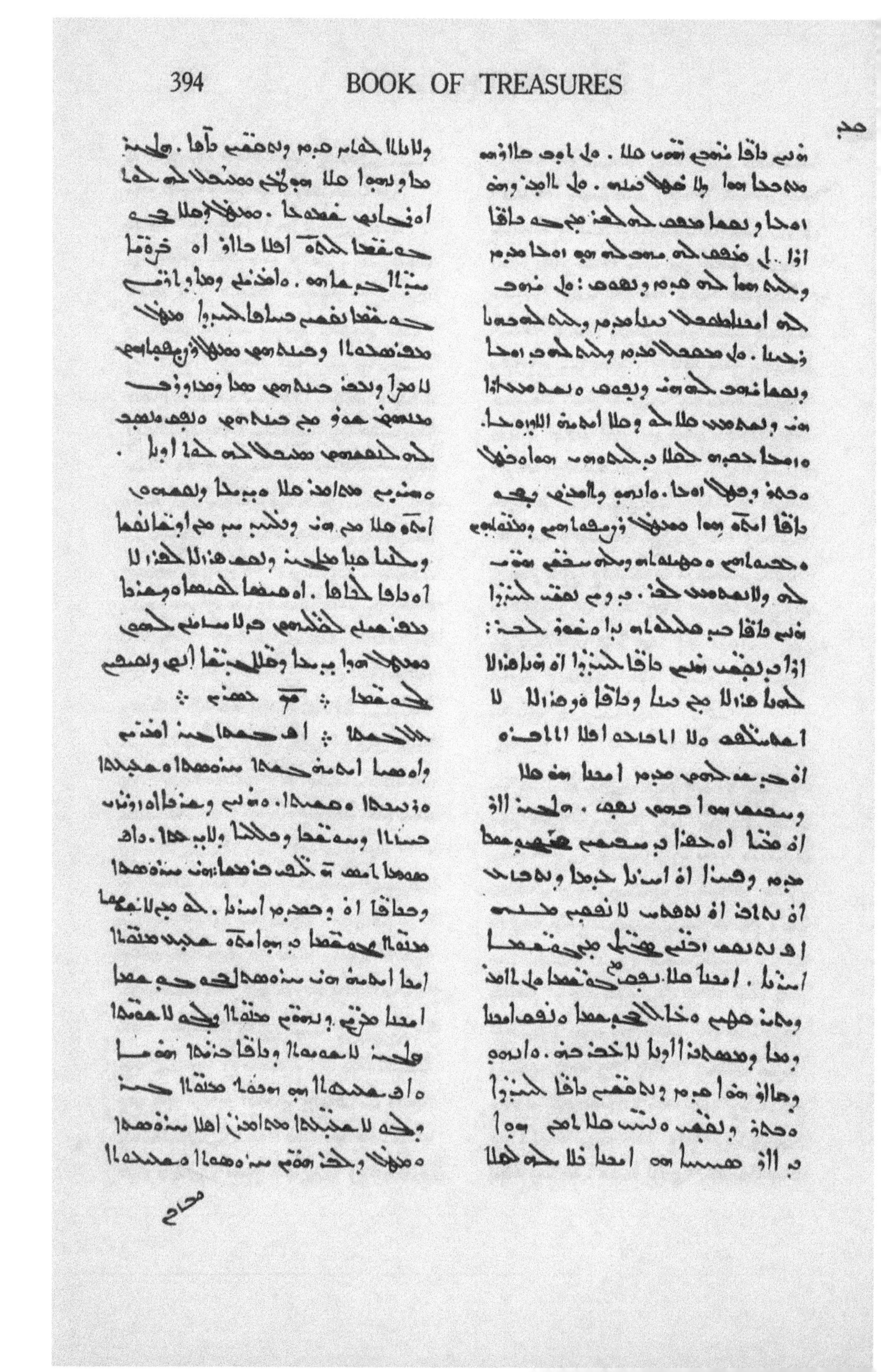

[illegible]

ܘܟܠܗ ܐܣܛܘ̈ܟܣܐ ܡܟܠܗ ܠܐܦ̈ܝܢ ܒܟܝ̈ܢܐ
ܐܝܟ ܕܡܬܩܪ̈ܝܢ ܘܡܬܗܦܟܝܢ ܒܗܘܢ ܒܚܕܕܐ. ܘܡܛܠ
ܐܢܝܢ ܐܚܪ̈ܢܐ ܕܫܘܚܠܦܐ ܗܘܦܟܐ ܕܝܢ . ܘܙܘܥܬܐ
ܘܡܫܘܚܬܐ ܘܡܢܝܢܐ ܠܟܝ̈ܢܐ ܐܝܟ ܕ ܗܘܝܢ
ܘܟܠܗ ܗܢܘܢ ܘܟܠܗ ܠܐ ܡܬܚܠܦܢ
ܠܓܘܫܡܐ ܕܢܦܠܘܢ ܐܝܟ ܕܐܣܟܡ ܣܡܝܡܐ ܩ̈ܕܡܐ
ܢܦܠܝܢ ܒܟܝ̈ܢܐ ܐܝܟ ܡܬܚܙܝܢ ܗܘܝܐ ܕ ܒܐܦܠܘܬ
ܐܝܟܢܐ ܘܠܡܫܡܠܝܘ ܡܟܝܢܘܬܐ ܐܝܟ ܠܗ ܗܘ
ܘܐܝܟܢܐ ܐܘܣܝܐ ܗܘ . ܘܐܝܟ ܐܝܟ . ܐܠܐ ܕܗܘܝܐ
ܟܝܢܐ ܗܘܝܐ ܡܘܬܐ ܟܝܢܐ ܕܡܢ ܐܠܗܐ
ܘܡܬܢܨܝܢ ܗܘܝܐ ܡܚܝܠܐ ܟܝܢܐ ܐܘܣܝܐ
ܣܘܟܠܐ ܡܟܝܢܐ ܘܡܣܬܟܠܝܢ ܠܗܘܢ ܕܚܝܐ
ܘܡܬܦܠܓܐ ܕܡܢ ܡܫܡܫܝܢ ܡܬܒܥܝܢ ܐܘ ܡܬܬܘܣܦ
ܡܬܝܕܥ ܟܠ ܡܢ ܡܫܡܫܝܢ ܡܫܡܠܝܐ ܢܦܫܐ ܡܢ ܦܓܪܐ :
ܡܬܦܠܓ ܕܝܠܗ ܡܫܡ̈ܫܢܘܬܐ ܕܝܠܗ ܕܗܘܝܘ
ܐܬܚܙܝܘ ܐܝܟܢܐ ܕܡܢ ܡܬܦܪܫܝܢ ܚܘܕܬܐ
ܟܝܢܐ ܕܝܠܗ ܚܘܕܬܐ ܡܫܬܟܚܝܢ ܟܠܐ ܘܐܝܬܝܗ
ܘܡܫܬܡܫܢܘܬܐ ܕܟܠܗ ܐܝܬܝܗܝܢ . ܘܡܛܠ
ܗܘܐ ܕܡܢ ܡܬܦܪܫܝܢ ܐܘ ܫܪ̈ܝܢ ܐܘ ܐܦܢ ܠܐ
ܡܬܚܠܦ̈ܢ ܠܗܘܢ ܘܡܫܬܡܫ ܕܝܠܗܘܢ ܐܝܟ
ܡܢ ܫܡܫܐ ܠܡܫܬܡܫܘ ܕܢܚܬ . ܘܕܝܢ ܚܕܘܐ
ܐܪܒܥܐ ܟܝܢܐ ܘܟܠܡܕܡ ܠܐ ܡܫܬܚܠܦ ܐܠܐ
ܕ ܢܦܫܐ ܡܫܡܫܐ ܘܕܝܠܗ ܡܢܗܘܢ ܟܝܢܐ
ܡܫܡܫܢܘܬܐ ܐܘ ܒܠܚܘܕܝܗ ܕ ܢܦܫܐ ܕ ܚܝܝܢ
ܐܠܐ ܕ ܦܓܪܝܢ ܘܒܟܝܢܐ ܒܟܝܢܐ ܐܘ ܕ
ܢܦܫܐ ܟܝܢܐ ܘܟܠ ܚܝܐ . ܘܐܝܟ ܫܪܝܪ
ܒܓܫܡܐ ܘܫܘܚܠܦ ܘܡܛܠ ܗܢܐ ܐܘ ܡܢ
ܦܪܘܫܘܬܐ ܐܘ ܡܢ ܡܫܡܫܢܘܬܐ ܐܘ ܡܢ
ܡܫܡܫܢܘܬܐ ܐܘ ܡܢ ܓܘܫܡܐ ܘܠܡܫܡܫ ܐܝܟ
ܘܗܢܘܢ ܠܗ ܐܝܬܝܐ ܡܢ ܟܠܗܘܢ . ܐܦܢ ܕܝܢ

ܒܝܕܐ ܐܘ ܡܢ ܟܠܗ̇. ܐܘܪܚܐ ܐܘ ܡܫܘܚܬܐ ܘܫܘܥܝܐ
ܡܬܝܠܕܐ ܡܛܠ ܕܡܬܢܣܒܐ. ܐܘ ܫܘܚܠܦܐ ܡܬܦܪܫܝܢ
ܠܗ ܐܘ ܦܪܫܬܐ ܐܘ ܡܥܒܕܐ ܐܘ ܙܘܥܐ ܡܫܬܠܡܝܢ
ܡܬܝܕܥܐ . ܠܘܬ ܡܬܦܪܫܝܢ ܠܗܘܢ ܐܘ ܫܪ̈ܝܢ
ܐܝܟ ܕܡܬܦܪܫܝܢ ܠܗ ܠܐ ܡܬܗܘܝܐ ܘܫܘܚܠܦܐ
ܗܘܝܢ ܐܝܟ ܕܐܝܬ ܐܢܫܐ ܡܬܦܪܫܝܢ ܐܘ ܡܢ
ܣܘܟܠܐ ܐܘ ܡܢ ܫܘܪܝܐ ܘܡܫܡ . ܗܘܐ ܡܬܠܝܕܢ
ܕܡܬܐ ܐܘ ܣܟܠܐ ܕܗܘ ܡܛܠ ܕܟܠܗ ܐܝܟܢܐ
ܡܬܚܠܦ ܒܗܘ ܡܬܠܝܕܢܐ ܘܟܠܗ ܒܠܚܘܕ ܗܘ ܐܦܠܐ
ܗܘ ܕܐܝܟܢܐ: ܟܠ ܐܢܫܐ ܐܝܬܘܗܝ ܕܗܘ ܘܐܝܟܢܐ
ܠܗ ܕܝܠܗ ܐܝܬܘܗܝ . ܐܝܟܢܐ ܕܝܢ ܐܝܬܘܗܝ ܟܠ
ܐܘ ܣܘܟܠܐ ܠܗ ܕܡܢ ܐܘܪ ܘܠܒܝܫ ܢܦܫܐ ܘܦܓܪܐ
ܢܦܫܐ ܟܠ ܐܢܫܐ ܘܒܚܕܘܬܐ ܘܙܘܥ
ܠܗܘܢ ܘܠܐ ܙܒܢܐ ܘܠܟܠ ܟܝܢܐ ܡܬܚܫܒ
ܘܠܗܘܐ ܘܗܟܢܐ ܒܟܠ ܟ̈ܝܢܐ ܐܝܟ ܕܗܘܝܢ
ܕܡܫܬ̈ܡܫܐ ܟܝܢ ܗܘܝܢ . ܘܐܝܟ ܕܠܐ ܡܫܬܡܫܢܘܬܐ
ܢܦܫܐ ܕܟܝ̈ܢܐ ܡܢ ܘܡܫܡܠܝܢ ܙܘܥܬܐ
ܘܡܫܡܫܢܘܬܐ ܘܦܪܫܘܬܐ ܘܗܠܟܘܬܐ . ܕܐܪܒܥܐ
ܟܝܢ ܕ ܡܫܡܫܝܢ ܠܗ ܟܠܗܘܢ ܚܝ̈ܠܐ ܡܫܡܫܢܐ
ܠܗ ܟܠ ܡܫܡܫܢܘܬܐ ܡܢ ܘܡܫܡܫܝܢ ܠܗ
ܠܟܝ̈ܢܐ ܡܢ ܗܢܐ ܐܘ ܡܢ ܐܘ ܐܦ ܡܢ ܙܘܥܐ
ܡܢ ܡܢܐ . ܘܕܝܢ ܐܝܟܢܐ ܘܐܦ ܡܫܡܫ
ܡܢ ܗܘ ܕܐܝܬܘܗܝ ܠܐ ܐܝܬܘܗܝ ܡܫܡܫܢܐ . ܡܢܐ
ܘܐܢܫܝܢ ܟܠܗܘܢ ܕܡܫܡܫ ܡܫܡܫܢܘܬܐ
ܗܘܐ ܘܐܝܟ ܫܪܝܪ ܒܓܫܡܐ ܐܝܟܢܐ
ܘܠܐ ܕܪܫܝܢ ܠܗ ܡܫܡܫ . ܘܐܝܟ ܕܠܐ
ܡܫܡܫܝܢ ܠܗ ܟܠܟܝܢ ܘܐܝܟ ܕܡܫܡܫܢܐ
ܢܦܫܐ ܐܝܟ ܦܓܪܐ ܚܝܐ ܐܘ ܡܫܡܫܢܐ ܘܚܝܐ
ܡܫܪܝܐ ܚܝܐ ܡܫܡܫܢܘܬܐ ܡܢ ܟܠ ܘܦܓܪܝܢ
ܠܗܘܢ ܕܗܘ ܐܘܣܝܐ ܡܢܝܢܐ. ܘܗܘܝܢ ܗܘܐ

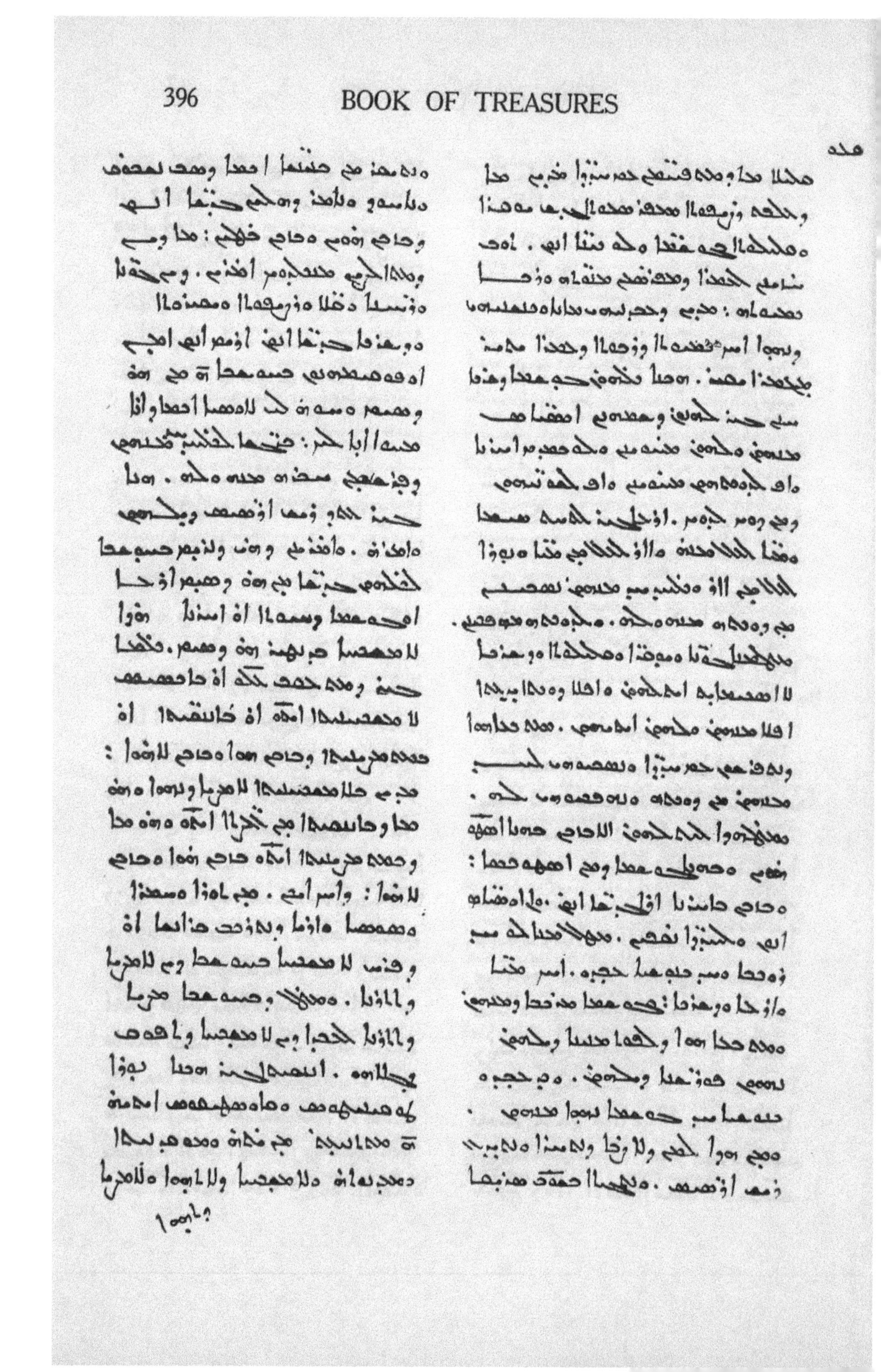

[illegible]

[illegible]

ܕܐܝܬܘܗܝ ܘܠܐ ܐܝܬܘܗܝ. ܡܫܡܫܢܘܬܐ ܕܝܢ ܗܕܐ
ܡܫܦܪܐ ܠܗܘܢ ܠܗܠܝܢ ܡܫܡܠܝܐ. ܡܫܡܫܢܐ
ܕܐܝܬܘܗܝ ܘܠܐ ܐܝܬܘܗܝ ܡܫܡܫܢܘܬܐ ܕܒܪܢܫܐ ܠܟܠܗ
ܥܒܕܐ ܕܢܦܫܗ ܕܐܢܫܐ ܗܘ ܗܘ ܦܘܠܚܐ
ܘܠܗ ܡܢ ܐܢܫܐ ܗܘܝܐ ܡܫܡܫܢܘܬܐ. ܘܠܗ
ܠܐ ܡܫܡܫܐ ܕܐܝܬܘܗܝ ܘܒܢܝܢܐ ܘܡܪܝܐ ܕܐܝܬܘܗܝ
ܗܢܘܢ ܕܟܕ ܐܡܝܢ ܘܟܕ ܠܐ. ܟܕ
ܓܝܪ ܡܫܐ ܡܢ ܘܡܫܡܫ ܐܘ ܡܫܬܡܠܐ. ܡܪܝܐ
ܕܐܝܬܘܗܝ ܘܠܐ ܐܝܬܘܗܝ ܡܫܬܡܠܐ ܘܗܘܐ ܡܫܬܪܪܐ
ܐܘ ܡܬܬܟܠܐ. ܘܐܢܫܐ ܕܠܐ ܡܫܡܫܢܘܬܐ ܠܐ
ܡܫܬܪܪ ܟܕ. ܡܫܬܟܚܝܢ ܟܕ ܟܠ
ܡܫܡܠܝܢܐ ܗܝ ܟܠܗ: ܡܫܬܡܠܐ ܘܟܕ
ܗܘܝܐ ܘܟܕ ܠܐ. ܦܪܝܫ ܘܐܬܦܪܫ ܕܐܝܬܘܗܝ
ܡܫܬܘܫ ܕܒܗ ܟܕ ܗܘܝܐ ܘܟܕ ܠܐ ܗܘܝܐ
ܥܠܡܐ ܠܐܗܘܐ ܡܫܡܫ ܟܠܐ ܫܒܩ ܒܒܢܝܢܐ:
ܡܢ ܐܦܠܐ ܡܫܡܠܐ ܐܢܫ ܕܢܗܘܐ ܠܟܠܗ.
ܕܢܦܩ ܠܐܡܪ ܒܫܘܠܡܐ ܘܕܢܡܘܬ ܡܢܗ
ܠܥܠܡܘܢ ܒܢܝܢܐ. ܘܡܢ ܙܒܢ ܗܘ ܟܕ
ܘܠܐ ܡܫܡܫܢܐ ܐܝܟ ܗܘ ܕܢܬܠ ܠܟܠ ܐܢܫܐ
ܘܗܘܐ ܡܢ ܐܘܪܚܐ ܐܘܪܚܐ ܘܦܪܝܫܐ: ܡܢ ܕܝܢ
ܕܗܘܐ ܡܫܬܘܕܥ ܗܘ ܟܕ ܘܠܐ ܡܫܡܫܢܐ ܡܫܬܘܕܥ
ܡܢ ܕܝܢ ܕܒܟܠܗ ܡܫܬܘܕܥ: ܗܘܐ ܚܕ ܘܠܐ
ܠܐܘܪܚ ܕܢܫܬܠܐ ܡܢܗ ܐܝܟܢ ܕܦܠܓ
ܥܕܪܐ. ܘܐܦܠ ܥܠܝܢ ܕܢܫܬܘܐ ܒܫܘܠܡܐ ܡܢܗ
ܚܘܒܐ ܐܘ ܛܘܪܦ ܕܫܘܡܗܘܢ ܠܟܠܗ
ܕܗܢܘܢ ܡܫܪܝܐ ܟܕ ܡܬܟܠ ܘܡܫܚܠܦܐ
ܡܫܡܫܢܐ ܕܐܝܬ ܡܪܝܐ ܠܟܢ ܕܢܫܬܘܐ ܗܘܢ
ܠܐܝܟ ܫܘܪܐ ܘܐܦ ܐܘܪܚܐ ܘܐܘܪܚܐ ܘ ܦܪܫܐ.
ܒܡܫܬܘܠܐ ܠܟܠܗ ܘܫܘܡܗ. ܘܟܕ ܐܡܪܗ
ܐܘܪܚܐ ܐܚܪܢ ܘܫܘܪܐ ܘܦܪܘܫܐ

ܠܟܠܗܘܢ ܒܢܝ̈ܢܫܐ ܚܕܐ ܠܐ ܡܫܡܫܢܐ
ܟܕ ܕܝܢ ܐܫܬܪܝ ܐܘܟ ܕܐܫܬܡܠܝ ܗܘܐ ܐܡܝܢ
ܒܫܪܪܐ ܕܙܘܕܩܐ ܘܡܫܬܡܠܝܢܘܬܐ ܡܪܝܐ ܒܟܢ
ܕܐܬܪܐ ܡܫܐ ܘܡܫܬܘܬܐ ܠܟܠܗܘܢ ܒܢ̈ܝܐ
ܘܕܥܝܢ. ܡܫܠܝܢ ܕܢܦܫܐ ܠܟܠܒܐ ܕܡ̈ܠܐ
ܒܪܝ ܘܐܬܪܘܬܐ ܠܟܠܗܘܢ ܒܢ̈ܝܐ ܕܐܬܘܪ
ܗܘ ܕܡܫܡ ܘܡܫܡܫܢܐ ܗܘܐ ܠܟܠܗܘܢ ܡܢ ܙܘܒܚܐ
ܕܒܟܠܗ ܐܘܡܢܐ ܡܫܐ ܘܐܬܕܪܫ ܠܟܠܗܘܢ
ܕܒܗ ܠܐܡܝܢ ܘܒܫܠܗ. ܘܡܫܠܝܢ ܗܘ ܕܡܢܝܢ
ܟܠܐ ܒܪܝܬܐ ܘܐܘܡܢܘܬܐ ܡܫܬܡܫܢܘܬܐ ܐܪܝܬܐ
ܙܢܝܬܐ ܡܫܡܫܢܐ: ܕܐܡܝܢ ܗܘܐ ܡܬܦܠܓܐ
ܕܝܢ ܗܘܐ ܡܫܦܪܢܘܬ ܘܡܫܡܠܝܐ ܗܘܐ
ܠܒܢ̈ܝܐ ܘܐܡܝܢ ܗܘܐ ܕܗܘܐ ܒܫܘܪܝܐ
ܘܡܠܗ ܕܙܘܒܚܐ ܘܠܐ ܐܬܪܐ ܘܡܢ ܐܡܬܐ
ܡܫܬܘܬܐ ܘܐܡܝܢ ܕܝܢ ܗܘܐ ܠܒܪܐ ܗܘ
ܘܠܗ ܙܘܒܚܐ ܕܒܫܘܠ ܒܢ̈ܝܐ ܐܡܝܢܘܬܗ
ܘܐܬܪܡܝ ܕܐܡܫܠܗ ܐܝܟ ܒܪܢܫܐ ܡܢ
ܠܡܫܬܐ ܕܒܗܐ ܠܡܫܡܫܘܬܐ ܡܢ ܐܢܫܐ
ܐܡܬܗ ܫܡܫܐ ܗܪܟܐ ܙܕܝܩܐ ܡܫܡܫܐ
ܟܕ ܡܫܡܫ ܘܟܕ ܡܫܠܡܢܐ. ܠܐ
ܒܪܝ ܡܪܝܐ ܕܐܬܙܕܩ ܚܝ ܡܫܠܡܢ ܗܘ
ܡܫܠܗ ܘܐܢܫܐ ܡܢ ܠܡܫܬܐ ܕܒܗܐ
ܠܡܫܡܫܘܬܐ. ܘܕܐܘܪܚܐ ܐܝܟ ܕܝܢ
ܕܐܡܝܢ ܦܠܓ ܡܢ ܫܦܝܪܘܬܗ ܘܐܘ ܫܡܝܗ
ܐܘ ܒܪܗ ܐܘ ܡܫܡܗ ܐܘ ܐܬܪܗ ܘܐܫܬܡܫܘܬܐ
ܗܘܝ ܘܙܘܒܚܐ ܡܫܬܪܝܢ ܘܡܫܠܡܢܘܬܗ
ܟܐ ܘܡܠܠܐ ܦܫܘܛ ܡܢ ܗܘ ܕܝܢ ܡܢ
ܡܫܬܡܫ ܡܫܡܫܐ ܡܫܠܡ ܡܫܡܫ ܘܙܘܒܚܐ ܐܘ
ܠܟܠܗ ܒܫܪܪܐ ܐܝܟ ܘܙܘܒܚܐ. ܘܐܦ
ܕܐܪܝܐ ܐܡܫܘܢ ܕܝܢܐ ܡܢ ܐܢ̈ܫܐ ܕܫܡܝܐ

ܡܣ

ܕܐܘܪܚܐ ܐܝܟ. ܘܡܬܬܡܢ ܡܢ ܐܘܪܚܐ ܗܘܘ ܀
ܐܝܟ ܗܘܐ ܕܐܟܢܐ ܡܬܐܡܪ ܕܗܘܝܢ ܒܐܬܪܐ
ܒܪ ܚܠܦ ܣܓܝܐܐ ܘܙܒܢܐ ܀ ܘܒܫܢܐ
ܣܓܝܐܐ ܘܡܝܬܐ. ܘܒܗ ܒܐܬܪܐ ܐܬܪܐ
ܘܡܝܬܐ. ܘܒܗܘܢ ܐܬܪܐ ܐܬܪܐ ܘܙܒܢܐ.
ܘܟܠܗ ܗܝ ܡܢ ܒܪ̈ܝܬܐ ܐܘܣܝܐ ܗܘܐ ܀
ܐܝܟ ܗܘܐ ܕܟܠܗܝܢ ܒܗܝܢ ܐܣܪ̈ܬܐ ܡܢ
ܠܥܠ ܡܢ ܠܫܡܝܐ ܕܐܝܟ ܗܕܡ
ܒܗ ܒܪ̈ܝܬܐ ܘܠܐ ܐܝܟ ܦܓܪܐ ܡܬܕܡܐ ܘܠܐ
ܢܦܩ ܒܟܝܢܗ ܒܪܝܬܐ ܡܬܕܡܝܢ ܡܬܡܨܐ
ܡܘܬܐ ܀ ܡܫܘܕܥ ܗܘ ܡܢ ܫܡܝܐ ܒܪܝܬܐ
ܠܫܡܝܐ. ܒܪ ܠܐ ܐܝܬ ܗܘܐ ܐܘܣܝܐ. ܘܐܦ
ܒܗܘܢ ܡܢ ܫܡܝܐ ܒܪ ܗܘ ܠܐ ܡܬܚܙܝܢ
ܘܒܟܝܢܗ. ܒܪ ܠܐ ܡܫܟܚܐ ܗܘ ܐܘܣܝܐ
ܒܢܦܫܗ. ܘܐܦ ܗܝ ܕܡܢ ܫܡܝܐ ܗܝ
ܘܒܪܝܬܐ ܠܫܡܝܐ ܗܝ ܐܘܣܝܐ ܗܝ ܘܠܐ
ܡܬܚܙܝܐ ܒܟܝܢܗ. ܘܐܠܐ ܗܝ ܕܐܝܬܝܗ
ܕܝܢ ܒܗ ܒܗܝ ܀ ܘܡܝܢܐ. ܘܠܐ ܗܘܐ ܕܐܝܬܝܗ
ܡܬܚܙܝܢܐ ܒܗ ܒܗ ܕܐܝܬ ܡܬܚܙܝܢܐ ܘܠܐ ܗܘܐ
ܘܐܝܬܗ ܟܝܢܐ ܘܐܠܐ ܗܘ ܕܗܘܐ ܐܪܥܐ
ܘܐܠܐ ܡܫܘܚܬܐ ܠܐ ܐܘ ܒܪܐ ܐܠܐ ܐܘ ܠܐ ܐܠܐ
ܐܠܐ ܕܝܢ ܘܐܠܐ ܡܫܘܚܬܐ ܐܦ ܟܠܗ ܒܪ
ܡܢ ܗܕܐ ܐܝܬܝܗ ܗܝ ܗܝ ܟܠܗܘܢ ܡܢ
ܒܪ̈ܝܬܐ ܢܬܦܣܩ ܒܗܘܢ ܡܢ ܫܡܝܐ ܒܪܝܬܐ
ܠܫܡܝܐ ܘܐܝܬܝܗܘܢ ܡܫܟܚܢ ܘܡܬܕܪܝܢ
ܘܠܐ ܡܬܚܙܝܢ ܠܗ. ܠܥܠ ܡܢ ܐܝܟ
ܡܘܬܐ ܗܘܝܢ ܘܒܗܘܢ ܣܓܝܐܐ
ܐܝܬܐ ܘܒܗܘܢ ܐܘܣܝܐ. ܐܠܐ ܗܘܝܢ
ܘܒܗܘܢ ܡܫܚܬܐ ܒܪ ܒܗܘܢ ܡܘܬܐ
ܐܘ ܡܝܬܐ. ܐܘ ܙܒܢܐ ܐܠܐ ܕܫܡܝܐ

ܒܪ ܡܢ ܐܪܥܐ ܐܘܣܝ̈ܐ ܐܝܬܐ. ܐܝܬܐ ܕܒܪ
ܡܢ ܒܪ̈ܝܬܐ ܘܠܐ ܡܬܚܙܝܢ ܒܗܘܢ ܒܪ̈ܝܬܐ
ܘܙܘܥܐ ܘܗܘܝܬܐ ܘܒܗܘܢ ܒܪ̈ܝܬܐ
ܘܒܟܠܗܘܢ ܐܝܬܐ ܕܡܫܬܐ ܘܒܪܝܬܐ ܘܐܪܥܐ
ܡܢ ܡܫܝܚܘܬܐ ܒܪܝܬܐ ܠܫܡܝܐ. ܒܗܘܢ ܕܣܓܝܐܐ
ܘܒܪܝܬܐ ܘܙܒܢܐ ܘܡܝܬܐ ܘܠܐ ܡܬܕܪܝܢ ܘܒܗܠܝܢ
ܡܬܚܙܝܢ. ܐܠܐ ܕܟܠܐ. ܘܐܘܣܝܐ ܐܝܬܘܗܝ
ܘܒܙܘܥܐ ܗܘ ܕܐܝܬ ܠܗ ܠܐܘܣܝܐ. ܘܠܐ ܐܝܬ
ܐܝܟ ܒܪܝܬ ܕܐܝܬ ܕܐܝܬ ܠܗ ܒܪ̈ܝܬܐ
ܘܕܐܝܬ ܠܗ ܢܦܫܐ. ܐܝܟ ܗܠܝܢ
ܐܘܣܝܐ ܘܡܘܬܐ ܐܘܣܝܐ ܠܗ. ܡܛܠ ܗܢܐ
ܒܪ ܐܝܬܝܗ ܡܢ ܟܠܗ ܡܕܡ ܢܦܫܐ ܒܪܐ
ܐܘܣܝܐ ܘܠܐ ܡܬܚܙܝܢ ܒܗܘܢ ܘܡܬܕܪܝܢ
ܡܬܕܪܝܢ ܘܒܗ ܒܪܝܬܐ ܢܦܫܐ ܗܘ ܒܪ
ܗܘ ܡܬܕܪܝܢ. ܘܐܝܟ ܐܝܟ ܒܪܝܬܐ ܠܫܡܝܐ
ܘܒܪܝܬ ܗܘ ܘܐܠܐ ܘܐܘܣܝܐ ܡܢ ܡܫܝܚܘܬܐ
ܒܪܝܬܐ ܠܫܡܝܐ ܘܐܝܬܐ ܗܘ ܒܪܝܬܐ ܘܡܝܬܐ
ܘܠܫܡܝܐ ܡܢ ܡܫܟܚܐ ܘܒܪܝܬܐ ܟܠܐ
ܡܫܟܚܐ ܡܢ ܐܪܥܐ ܡܝܬܐ ܘܐܪܥܐ ܘܙܒܢܐ.
ܘܐܘܣܝܐ ܠܐ ܐܬܡܫܟܚ ܡܢ ܗܘ ܕܐܝܬܐ
ܒܪܝܬܐ ܡܢܗܘܢ ܘܟܠܗ. ܗܟܢܐ ܐܪܥܐ
ܘܐܪܥܐ ܘܗܘ ܡܫܟܚܢ ܒܗ ܕܟܠܗܘܢ
ܘܠܐ ܢܬܒ ܡܢ ܡܬܚܙܝܢ ܡܢ ܕܒܗܘܢ ܘܠܐ
ܐܬܡܫܟܚ. ܡܢ ܕܐܬܡܨܝܬ ܗܘ
ܒܗܘܢ ܐܝܟ ܡܫܝܚܘ ܘܠܐ ܐܪܥܐ ܕܡܬܚܙܝܢ
ܐܝܬܘܗܝ ܡܘܬܐ. ܒܪܝܬܐ ܒܪ̈ܝܬܐ
ܡܬܕܪܝܢ ܘܡܬܚܙܝܢ ܗܘ ܘܒܗܘܢ
ܐܝܬܐ ܐܝܬܘܗܝ ܒܪܝܬܐ ܘܠܐ ܐܝܬ. ܐܝܟ ܠܫܡܝܐ
ܘܒܪܝܬܐ ܘܙܘܥܐ ܘܡܝܬܐ ܘܐܘܣܝܐ
ܠܫܡܝܐ ܘܡܠܬܐ ܘܒܪ̈ܝܬܐ ܘܒܪ̈ܝܬܐ

ܐܦ

ܐܘ ܡܫܡܫܢܘܬܗ ܕܒܪܢܫܐ ܘܐܣܝܪܘܬܗ ܕܦܓܪܐ
ܘܡܛܠ ܗܕܐ ܐܬܐܠܨܘ ܕܢܫܬܡܥܘܢ ܗܘܘ
ܡܢ ܦܩܘܕ ܦܘܠܚܢܗ. ܡܐܪܝܢ ܘܐܚܪܢ
ܒܗܕܐ ܐܟܙ ܐܫܬܡܥܘ ܠܦܬܓܡܗ ܘܫܡܥܘܗ
ܡܫܡܫܢܘܬܗ ܘܠܡܥܒܕ ܒܝ̈ܬܐ ܐܚܪܢܐ ܘܢܦܫܗ
ܠܗ ܡܢ ܦܘܠܚܢܐ ܘܢܛܪ ܡܛܠ ܗܢܐ ܘܐܦ ܦܘܩܕܢܗ
ܒܗܘܢ. ܘܐܫܟܚܐ ܐܢܫ ܗܘ ܕܠܐ ܡܫܬܡܥ ܠܗ
ܗܘ ܡܛܠ ܕܫܡܫܐ ܘܢܗܪܐ ܐܢܘܢ ܐܘܡܢܘܬܐ ܕܝܠܗ
ܕܦܘܠܚܢܐ ܘܡܫܡܫܢܘܬܐ ܘܥܡܠܐ ܦܘܠܚܢܐ ܘܠܐ
ܡܫܡܫܢܝܬܐ ܡܫܡܫܢܘܬܐ ܐܠܐ ܕܐܦ ܟܕ ܡܫܠܡ ܡܫܡܫ
ܢܟܘܦ ܒܝ̈ܬܐ ܘܢܫܬܥܒܕ ܡܢ ܦܘܠܚܢܗ ܘܢܛܪܘܢ
ܠܐ ܡܫܠܡ. ܡܛܠ ܐܬܬܣܝܡܬ ܡܫܡܫܢܘܬܐ ܐܦ
ܡܫܟܚܐ ܘܕܝܢ ܐܘܡܢܐ ܒܝ̈ܬܐ ܢܗܘܐ ܐܘܡܢܐ
ܡܢܗ ܘܠܗ. ܕܦܘܠܚܢܐ ܘܥܡܠ ܦܘܠܚܢܐ ܒܢܝ̈ܢܐ
ܘܥܡܠ ܡܥܠܬܐ: ܐܘ ܡܫܬܥܒ̈ܕܐ ܠܐ ܐܬܘܪܝܗ ܒܗܘܐ
ܡܫܟܚܐ ܘܦܓܪܐ ܗܘܐ ܘܡܫܡܫܢܘܬܐ ܘܡܫ̈ܡܫܐ
ܡܫܡܫ ܫܡܫܢܘܬܗ ܐܝܟ ܐܢܫ ܕܡܫܬܡܫ ܠܗܘܢ
ܐܡܬܝ ܕܗܘܐ ܡܫܠܡ ܡܢ ܐܘܪܚܐ ܡܫܬܕܪ
ܕܐܡܪܝܢ ܘܡܫܝܚܐ ܘܐܡܪܝܢ. ܘܐܝܟ ܗܢܐ
ܘܡܫܡܫܢܐ ܘܫܪܝܪܐ. ܘܕܝܢ ܡܫܡܫ ܠܟܠ ܡܠܦܢܘܬ
ܙܕܩ ܫܡܫܘܬܐ ܘܢܦܫܗ ܕܒܫܡܫܐ ܡܫܡܫܐ ܐܢܫ
ܐܠܐ ܐܦ ܠܗ. ܘܡܢ ܐܝܬܘܗܝ ܐܘܪܚܐ ܕܡܫܬܥܒܕ
ܠܡܫܡܫ ܘܫܡܫܢܘܬܐ ܘܡܫܡܫ ܡܫܡܫܢܘܬܗ
ܘܐܡܪܝܢ ܕܢܟܦܐ ܘܡܫܡܫܢܐ ܘܢܦܫܐ ܠܟܠܗ
ܡܫܝܚܐ ܘܐܡܪܝܢ ܐܡܬܝ ܡܢ ܕܒܥܐ ܘܐܒܕ ܘܠܗܘܢ
ܟܠ ܘܫܡܫܐ ܡܫܝܚܐ ܕܡܫܠܡ ܠܟܠܢ ܡܫܟܚ
ܙܕܝܩܐ. ܠܗܢܐ ܗܘ ܕܟܠܗ ܗܘܦܟܗ ܒܢܝܐ ܗܘܐ.
ܘܡܢܐ ܕܝܢ ܠܒܢܝܐ ܗܘܐ. ܢܡܠܠ ܕܢܒܥܐ
ܡܫܟܚܐ ܘܒܠܚܘܕ ܐܝܟ ܠܗܢܐ ܘܠܐ ܐܫܟܚܗ ܐܝܟ
ܠܡܫܠܡܐ ܘܢܫܒܘܩ ܡܠܠ ܘܠܒܘܫܐ ܘܐܠܐ ܐܡܪ

ܫܨܗ

ܡܢ ܟܠܗܝܢ ܘܡܛܠ ܦܓܪܐ ܒܝܬܐ ܘܢܦܫ ܗܘܐ
ܡܫܝܚܝ. ܐܡܪ ܐܦܪܝܡ ܠܟܠܝ ܘܠܝ ܒܒܝ̈ܬܐ ܡܫܬܡܫܢ̈ܐ
ܘܒܗܕܐ ܒܝ̈ܬܐ ܐܝܟ ܕܗܘܘ ܕܡܫܬܡܫܝܢ ܗܘܘ
ܘܐܘܦ ܡܫܠܡ. ܐܘܪܐ ܡܢ ܡܕܡ ܗܘܐ ܡܢ ܡܕܡ ܘܐܝܬܘ̈ܗܝ
ܐܘܠܐ ܡܕܡ. ܘܡܛܠ ܕܡܫܠܡ ܟܠ ܡܕܡ ܚܝܠܗ
ܠܗ ܐܡܬܝ ܡܢ ܠܐ ܡܕܡ ܗܘܐ ܡܕܡ ܘܡܢܗ ܟܠ
ܡܕܡ. ܘܐܦܪܝܡ ܘܡܢ ܡܕܡ ܘܐܝܬܘ̈ܗܝ ܗܘܘ ܘܡܢ
ܡܕܡ ܘܐܝܬܘܗܝ ܗܘܐ ܕܡܫܬܡܫܝܢ ܗܘܘ ܡܕܡ
ܘܢܗܘܘܢ. ܘܠܗ ܡܢ ܠܐ ܡܕܡ. ܠܡܫܬܡܫܢܘ ܐܬܡܫܚ
ܗܘܘ ܠܡܫܡܫܢܘ ܗܘܘ ܘܢܗܘܘܢ. ܗܘ ܡܫܠܡܢ
ܘܐܡܬܝ ܠܐ ܗܘܐ ܥܠ ܗܘܐ ܗܘ ܕܗܘܐ ܘܗܘ ܘܐܡܬܝ
ܗܘ ܕ ܗܘܐ ܐܡܬܝ. ܥܠ ܗܘܐ ܐܡܬܝ. ܠܐ ܡܫܒܚܐ
ܘܡܢ ܐܘܪܚܐ ܘܕܐܝܬܐ ܢܗܘܐ ܙܕܩܐ ܘܐܦ ܠܐ ܡܢ
ܙܕܩܘ ܫܡܫܐ ܘܢܘܗܪܐ. ܘܐܦܠܐ ܡܢ ܡܕܡ
ܐܘ ܡܫܬܡܠܐ ܘܫܡܝܐ ܠܟܠܗܘܢ. ܘܫܡܫܐ
ܗܘ ܐܘܪܚ ܘܫܡܝܐ ܡܫܠܡܢ ܘܢܗܘܐ ܡܕܡ ܘܗܘܝܐ
ܐܡܬܝ ܗܘܐ. ܡܫܠܡ ܕܗܘܐ ܠܡܫܡܫܢܐ ܐܘܪܚ
ܗܘ ܘܐܡܬܝ ܗܘ ܘܗܘܐ ܐܝܬ ܐܠܐ ܗܘܐ ܠܡܫܬܐ
ܒܟܠܗ ܡܫܠܡܢ ܠܡܫܬܡܫܢܐ ܘܗܘܐ ܐܠܐ ܕܐܡܬܝ
ܐܡܬܝ ܒܪܝܬܐ ܚܕܐ ܡܫܟܚܐ. ܠܡܫܡܫܢܐ
ܘܗܘܐ ܚܕܐ ܡܫܟܚܐ. ܐܠܐ ܘܗܘܐ ܠܟܠܐ ܐܘܪܚܐ
ܡܫܘܪܐ ܐܘܪܚܐ ܡܫܢܝܐ ܗܘܬ ܡܫܡܫܐ ܕܗܘ ܙܒܢܐ
ܐܝܟ ܕܒܪܐ. ܗܘ ܡܫܡܫ ܘܫܡܫܐ ܡܫܟܚ ܠܟܠܗܘܢ
ܠܡܫܠܡ ܗܘ ܕܗܘ ܐܡܬܝ. ܘܡܢ ܟܠܗ
ܡܫܝܚܐ ܠܟܠܗ ܘܡܢ ܠܐ. ܐܝܟܢܐ ܘܐܘܡܢܐ ܡܫܘܕܥܐ
ܘܒܗܕܐ ܘܐܦ ܠܐ ܐܡܪܝܢ ܘܟܕ ܗܘܐ ܐܘ
ܡܢܐ ܐܘ ܐܢܐ ܐܘ ܐܢܬ. ܐܠܐ ܘܐܡܫܡܫ ܡܢ
ܒܦܘܠܚܢܐ ܗܘܐ ܒܫܘܝܐ ܟܠ ܐܘܪܚܐ ܘܡܫܬܡܫܝܢ
ܗܘܘ. ܘܡܫܡܫܝܢ ܒܟܠ ܐܝܟ ܠܗ ܘܡܢ ܠܗܕܐ
ܡܫܡܫ ܐܘ ܐܘܪܚܐ. ܕܒܗ ܗܘܘ ܡܢ

ܩܦ

ܗܘ ܡܠܐ ܕܐܡܪܗ ܠܗܘܢ ܠܐ ܡܕܡ ܒܠܚܘܕ ܗܘ ܡܠܐ
ܕܐܬܡܣܟܢ ܒܗܘܢ ܘܗܘܘ ܠܒܢܘܗܝ ܗܘܐ ܠܒܢܝܢ: ܒܢܝܐ.
ܒܢܝܐ ܒܪܡ ܕܢܦܠ. ܠܐܠܗܐ: ܐܡܬܘܐ ܐܡܪܐ
ܗܘܐ ܗܘܐ. ܘܠܗ ܠܒܢܘܗܝ ܗܘܐ ܐܡܪܐ ܕܠܐ ܗܘܐ
ܘܠܒܢܝ: ܐܝܟܐ ܐܬܬܗ. ܘܠܐ ܡܬܬܐܡܪܐ ܕܡܦܩܐ ܐܘܟܐ
ܡܬܗܦܟ ܘܠܐ ܡܦܩܐ ܠܐ ܡܬܬܐܡܪܐ ܕܡܠܐ ܒܟܠܗ
ܒܗ ܐ ܘܒܥܝܬܐ ܐܦ ܐܘܟܐ ܐܚܪܢܐ. ܘܐܡܪ ܐܚܪܢܘ
ܕܢܘܗܝ ܡܢ ܡܠܐ ܕܐܬܡܣܟܢܘ ܘܐܬܡܣܟܢ ܒܗܘܢ
ܕܢܘܗܝ ܗܘܐ ܡܠܐ ܕܟܠ ܒܗܘܢ ܕܠܐܡܣܟܢ ܒܝ
ܡܬܚܫܒܝܢ ܠܒܢܘܗܝ ܗܘ ܕܐܡܪܗ. ܕܐܠܗܐ: ܕܟܠܗ ܗܘܐ
ܡܢ ܕܗܘܐ ܕܐܡܪܗ ܒܝ ܒܪܡ ܕܢܦܠ ܕܟܠܗ. ܗܘ
ܕܟܠܗ ܠܐ ܡܬܐܡܪ ܗܘܐ ܘܐܒܪܗ ܗܘܐ ܘܐܦ
ܡܬܩܪܐ ܘܡܠܐ ܘܟܠܗ ܕܟܠܗ ܗܘܐ ܕܗܘܐ ܒܝ
ܗܘ ܕܟܠ ܒܥܘܬܐ. ܒܗܘܢ ܕܢܦܠ ܕܗܘܐ
ܐܡܪܗ ܘܒܫܡܝܐ ܒܟܠܗܡ. ܘܡܛܠ ܕܐܝܟ
ܗܘܐ ܘܡܬܚܫܒ ܗܘܐ ܕܐܡܪܝܢ ܐܘܩܝܐ ܘܟܠܗ
ܒܗܘ ܢܦܠ ܕܐܬܩܝܡ ܗܘ ܕܢܦܠ ܘܐܦܠܐ
ܒܟܠ ܢܦܠ ܘܐܘܩܝܐ ܢܦܝܫ ܒܝ ܟܠܗ ܐܠܐ ܗܘܐ ܒܝ
ܗܘܐ ܘܡܬܬܪܝܡ. ܘܒܗ ܡܬܩܢܐ ܘܟܠܗ ܡܠܐ
ܘܟܠܗ: ܡܬܚܫܒ ܗܘܐ ܘܡܛܠ ܒܝܬܐ ܘܒܗܘܢ
ܡܕܡ ܗܘܐ ܘܐܦܠܐ ܒܗ ܡܕܡ ܢܦܝܫ. ܒܗܘ
ܦܠܓ ܕܐܡܪ ܐܚܝܐ ܘܐܚܝ ܡܬܝܕܥܐ ܕܡܬܚܫܒܐ
ܟܠ ܡܠܐ ܐܠܗܝܐ ܗܘ ܡܬܩܪܝܐ ⟨……⟩

ܫܠܡ ܡܐܡܪܐ ܐ ܠܒܝܬܐ

ܘܐܦ ܒܝ ܟܠܗ ܐ ܚܕܒ ܐ ܠܗܘܢ ܡܬܚܫܒ
ܒܡܐܡܪܐ ܕܒܝܬܐ.
ܟܠ ܡܐܡܪܐ ܐܚܪܢܐ : ܒ ܐ ܒ
ܦܬܪܘܢ ܕܡܬܚܫܒ ܡܢ ܒܢܝܢܐ ܘܐܘܢܝܐ
ܘܫܘܢܝܐ ܘܡܬܬܚܝܢ ܕܪܝܫܐ ܘܕܪܝܫܐ ܘܦܩܕܐ
ܘܡܬܚܫܒ ܡܢ ܕܟܠܗ ܕܐܠܗܐ ܐܘܟܐ

ܘܐܬܐܡܪ ܒܗܘܢ ܕܐܡܪܗܘܢ ܐܠܗܐ ܟܠܗܘܢ
ܡܬܩܪܝܐ. ܘܟܠܗܘܢ ܢܦܫ ܘܐܡܪܐ
ܒ ܐܘܟܐ ܗܢܐ ܐܬ ܗܘ ܘܡܬܚܫܒܐ. ܢܦܫ
ܡܬܚܫܒ ܐܘܢܝܐ ܘܠܐ ܡܬܚܫܒ ܠܗ ܘܗܘܐ ܘܡܠܐ ܐܝܟܐ
ܘܫܡܝܐ ܕܦܩܕ ܠܐ ܘܐܚܪܢܐ ܘܐܝܟܐ ܘܐܡܪܝܢ
ܘܐܘ ܢܦܠ ܘܦܩܕܐ ܘܡܬܚܫܒܐ ܘܡܫܝܚܐ ܒܗ
ܐܘܢܝܐ ܐܡܪܐ ܘܒܥܝܬܐ. ܘܐܡܪܝܢ ܒܝ ܐܠܐܘܟܗ
ܢܦܫܗܘܢ ܐܘܢܝܐ ܘܡܢ ܐܬܩܝܡܘܢ ܒܫܠܝܐ ܘܦܪܘܩܐ
ܘܒܝ ܐܘܟܐ ܡܬܩܪܝܐ ܘܡܬܚܫܒ ܒܗܘܢ ܗܘܐ
ܘܟܠܗ ܘܟܠܗ ܘܠܗ ܕܐܡܪܐ ܘܡܬܚܫܒ ܒܗܘܢ
ܢܦܫܗܘܢ ܐܘܢܝܐ ܘܗܘܐ ܟܠ. ܡܬܩܪܝܢ ܡܢ ܐܘܢܝܐ
ܘܒܗ ܐܘܟܐ ܘܐܘܟܐ ܐܬܩܝܡܘܢ: ܒܗܘܢ ܕܡܬܒܥܝܐ
ܒܫܡܝܐ ܐܘܢܝܐ ܐܬܡܣܟܢܐ ܡܢ ܘܐܡܪܝܢ ܘܐܡܪܝܢ
ܐܡܪܐ ܐܦ ܡܬܚܫܒܐ. ܘܡܬܚܫܒ ܗܘܐ ܐܘܟܐ
ܐܬܡܣܟܢ: ܘܗܘܐܠܒܝܢ ܘܡܐܡܪܐ ܘܫܡܝܐ
ܘܦܪܘܩ ܐܘܟܐ ܐܢܫܐ ܘܐܡܪܝܢ. ܡܢ ܡܬܚܫܒܝܢ
ܘܡܬܚܫܒ. ܒܪܝܡ ܐܘܢܝܗܘܢ ܡܬܚܫܒܝܢ
ܘܡܬܚܫܒܝܢ: ܡܬܚܫܒ ܗܘܐ. ܘܐܝܬܝܐ. ܡܬܚܫܒ
ܒܗܘܢ ܒܢܝܢܐ ܘܐܘܟܐ ܘܡܬܚܫܒ. ܘܡܢ ܒܝܬܐ
ܗܘܐ ܘܗܘܢܐ ܗܘ. ܡܢ ܗܘ ܘܒܝ ܡܬܚܫܒܝܢ
ܠܗܘܢ ܠܐܘܟܐ. ܘܡܠܐ ܡܬܚܫܒ ܠܐܢܫܐ ܐܘܟܐ
ܐܘܟܐ ܡܬܚܫܒܝܢ ܐܡܪ ܡܠܐ ܡܬܐܡܪ ܒܝܬܐ
ܐܢܫܐ: ܘܡܠܐ ܕܟܠܗ ܡܬܚܫܒ ܡܠܐ ܘܗܘܐ
ܒܢܝܐ ܘܡܠܐ ܦܪܘܩ ܘܦܩܕ ܒܝܬܐ. ܡܠܐ
ܘܡܢ ܘܡܬܚܫܒ. ܠܐܘܟܐ ܡܬܚܫܒܝܢ ܘܦܩܕ
ܠܒܢܝܗܘܢ ܒܪܡܐ. ܗܘܢܐ ܐܦ ܘܗܘܐ ܡܬܐܡܪܐ
ܘܡܠܐ. ܒܪܝܡ ܡܬܚܫܒܝܢ ܐܘܢܝܗܘܢ
ܠܗܘܢ ܒܒܝܬܐ ܘܐܘܟܐ ܘܐܝܟ ܒܝܬ
ܡܢ ܡܬܚܫܒܝܢ ܡܠܐ ܘܡܬܚܫܒܝܢ ܠܒܢܝܐ:
ܡܪܢܐ ܡܬܚܫܒܢܐ ܢܦܫܗ ܡܢ ܒܗܘܢܐ
ܕܗܦܟ

ܘܢܬܚܘܦܛ ܠܡܐܬܐ ܡܢ ܚܛܗܘ ܀
ܩ̄ ܥܠ ܕܗܘܐ: ܫܘܪܝܗ ܕܕܘܒܪܐ
ܫܪܝܐ ܡܢ ܗܝܡܢܘܬܐ. ܘܡܫܬܡܠܝܐ ܒܬܪ ܕܢܛܪ
ܘܒܬܪ ܡܬܠܐ ܕܡܫܡܠܐ ܡܢ ܗܝܡܢܘܬܐ. ܡܫܬܠܡ
ܒܬܪ ܕܒܬܪ ܡܛܠ ܡܠܘܗܝ ܕܡܫܝܚܐ ܢܦܩܐ
ܗܘ ܕܗܝܡܢܘܬܐ ܫܘܪܝܗ. ܡܛܠ ܕܒܬܪ
ܗܢܢ ܕܗܘܐ ܒܬܪ ܫܡܥܗ ܘܫܡܥܐ ܠܐܢ
ܒܬܪܡܝܟ. ܡܢ ܕܗܘܐ ܡܢ ܗܝܡܢܐ ܘܐܦ
ܒܬܪ ܡܡܠܠܘܬܗ ܠܗ ܡܛܠ ܕܒܬܪ ܫܡܥܢ
ܘܐܦ ܐܢܐ ܙܕܩ ܡܢ ܕܗܘܐ ܘܗܝܡܢܐ. ܘܐܦ
ܐܚܪܐ ܘܐܝܟ ܒܬܪ ܡܢ ܢܛܘܪ ܡܛܠܗܘܢ
ܠܟܠ ܦܩܕ ܠܗܘܢ. ܗܪܟܐ ܕܝܢ ܒܬܪ ܐܘܟܝܬ
ܡܢ ܢܛܘܪ ܘܡܢܗ ܘܡܬܚܙܝܐ ܠܗܝܡܢܘܬܐ
ܡܫܡܠܝܐ ܠܗܘܢ ܡܛܠ ܕܪܡܙܐ ܠܡܣܗ
ܡܢ ܦܘܩܕܢܐ ܕܡܠܐ ܢܝܫܐ ܐܝܬܘܗܝ. ܐܠܐ ܠ
ܕܠܐ ܢܬܦܠܓ ܠܗ ܫܒܝܠܐ ܗܢܐ. ܡܛܠܗܢܐ
ܢܛܘܪ ܡܢ ܣܦܪܐ ܦܘܩܕܢ ܡܢ ܢܡܘܣܐ ܘܦܩܕ
ܐܝܠܝܢ ܕܠܐ ܡܒܥܐ ܘܠܐ ܡܛܠܗܘܢ. ܐܝܟܢܐ
ܕܩܒܠܘ ܦܘܩܕܢܐ ܗܢܐ ܘܥܒܕ ܒܗ ܗܢܐ
ܘܠܐ ܡܬܢܛܪ. ܒܗܝ ܕܝܢ ܐܝܬܝܗ ܗܘ ܡܢ
ܐܝܟ ܐܚܪܢܐ ܐܘܪܚܐ ܡܛܠܗܘܢ ܘܡܬܝܕܥ
ܒܗܘܢ ܐܝܟ ܡܕܡ ܕܚܝܐ. ܐܝܟ ܕܒܪ
ܡܛܠܗܢ ܕܪܘܚ ܦܪܘܫܘܬܐ ܠܟܠܗ ܪܘܚܐ
ܗܘ ܐܝܬܘܗܝ ܡܛܠ ܐܝܟܢܐ ܠܟܠܗ ܟܠܗ. ܠܐ
ܡܫܡܗ ܟܠܗ ܘܡܛܠ ܐܝܬܘܗܝ ܐܠܐ ܠܟܠ
ܡܛܠܗ ܡܢ ܗܢܐ. ܐܝܟ ܕܝܢ ܕܠܗܘܢ ܘܡܢܐ
ܘܡܛܠ ܡܬܚܙܝܢ ܠܟܠ ܗܢܐ ܡܛܠ ܦܘܩܕܢܘܗܝ
ܡܢ ܟܠܗ ܠܐܪ. ܒܠܐ ܡܬܚܙܝܢ ܠܗܘܢ ܡܦܩܕ ܦܩܕܐ
ܠܒܪܐ ܕܝܢ ܡܛܠ ܕܡܠܟܗ ܠܗ ܐܘܪܚܐ ܡܫܬܡܥ
ܘܢܦܩܢ. ܘܗܢܝܢ ܢܬܟܠ ܥܠܘܗܝ ܗܘܐ ܐܦ

ܡܚܐ
ܗܢܘܢ ܐܢܫܐ ܡܛܠ ܕܡܠܟܐ ܕܥܡܗ ܐܘܪܚܐ
ܡܒܝܢ ܡܫܪ ܡܢ ܐܝܟ ܀ ܒ̄ ܀
ܡܛܠܗܢܐ ܕܗܘܐ ܠܐ ܡܫܬܡܥܝܢ. ܘܐܠܐ ܕܒܗ
ܘܡܛܠܗܢܐ ܕܢܦܩܘܢ ܡܢ ܗܘ ܕܗܘܐ ܠܐ
ܡܫܬܠܛܝܢ ܘܠܐ ܫܠܛ ܠܐܢܫ ܐܘܪܚܐ ܘܠܐ ܡܢ
ܡܢܐ ܘܠܐ ܡܢ ܗܢܐ ܘܐܦ ܡܢ ܕܗܘ ܗܝܡܢܐ. ܗܢܘܢ
ܕܝܢ ܕܗܢܐ ܡܫܬܟܚ ܡܦܩܐ ܪܘܚܢܐ
ܡܒܝܢܐ ܠܡܫܬܡܥܘ ܡܬܡܫܚܝܢ ܒ ܠܗܘܢ
ܫܡܥܝܢ ܢܦܫܗ ܡܫܡܠܝܐ. ܘܗܘܐ ܡܢ ܐܡܝܢܐܝܬ
ܘܐܘܪܚܐ. ܐܝܟ ܕܒܗ ܡܫܡܠܝܐ ܗܘܐ ܕܐܪܥܐ ܘܐܦܪܫ
ܘܐܦ ܒܗ ܡܫܡܥ ܡܫܡܠܝܐ ܘܡܛܠܗ. ܡܢ
ܐܡܝܢܘܬܐ ܘܐܘܪܚܐ. ܒܬܪ ܕܝܢ ܦܪܫ ܐܘܪܚܗܘܢ
ܡܢ ܫܒܝܠܐ ܕܝܢ ܕܡܫܡܠܝܐ ܕܡܦܩܢܘܬܐ ܕܡܢܗ
ܗܘܝܐ. ܕܠܟܠܗ ܡܢܐ ܘܐܝܟ ܕܗܘ ܡܢܗ ܐܝܬܘܗܝ ܒܬܪܗ
ܘܫܡܫܐ ܡܢ ܐܡܗܬܗܘܢ. ܘܡܬܠܝܢ ܡܢ
ܟܠܗܝܢ ܒܡܦܩ ܒܬܪܐ. ܘܡܬܠܡܝܢ ܫܡܫܐ
ܡܛܠܗ ܕܗܘ ܒܗ ܡܢܐ ܒܬܪܐ. ܒ ܦܫܝܛ ܡܛܠ ܐܝܬ
ܒܬܪ ܐܝܬܘܗܝ ܘܫܡܫܐ ܒܫܪܪܐ ܗܘ ܡܢܐ ܕܒܗ
ܠܗܘ ܡܢܐ ܕܗܘܐ ܐܝܟܢ ܘܐܠܝܨܘ ܠܗܘܢ
ܒܬܪ ܐܝܬ ܠܗ ܗܘ ܡܢܐ ܕܒܗ. ܘܫܡܫܐ
ܗܘܐ ܘܒ ܐܢܫܐ ܒܫܡܝܐ ܗܘ ܕܡܬܚܙܐ
ܗܘܘ ܐܘܪܚܐ ܕܗܘܐ ܘܗܝܡܢܘܬܐ ܘܫܪܪܐ:
ܐܡܪܐ ܕܦܫܝܛ ܗܘ ܐܝܟ ܒܬܪܐ ܕܡܫܡܠܝܐ
ܡܫܬܠܡ ܠܗܝܡܢܐ ܘܒܗܘܐ ܘܐܝܬܐ ܕܡܫܬܠܡ
ܫܡܫܐ ܢܒܥܐ ܠܚܝܐ ܘܐܝܬܐ ܘܫܪܪܐ
ܡܢܗ ܐܝܬ ܠܗ ܡܫܡܠܝܢ ܡܫܬܠܡ ܒܗܘܐ
ܘܒܬܪ ܐܝܬ ܫܪܪܐ ܡܫܬܠܡ ܠܗܝܡܢܐ. ܡܢܗ ܐܝܬ
ܐܦܪܐܝܬ ܒܬܪܐ ܕܒܢܝܢ ܘܒܬܪ ܘܗܘܐ
ܡܢܗ ܐܝܬ ܒܬܪܗ ܘܫܡܫܐ ܠܟܠܗ ܐܝܬ
ܐܝܬܘܗܝ ܠܗܘܢ ܟܠܗ ܡܫܪܪܐ ܘܐܝܬܘܗܝ

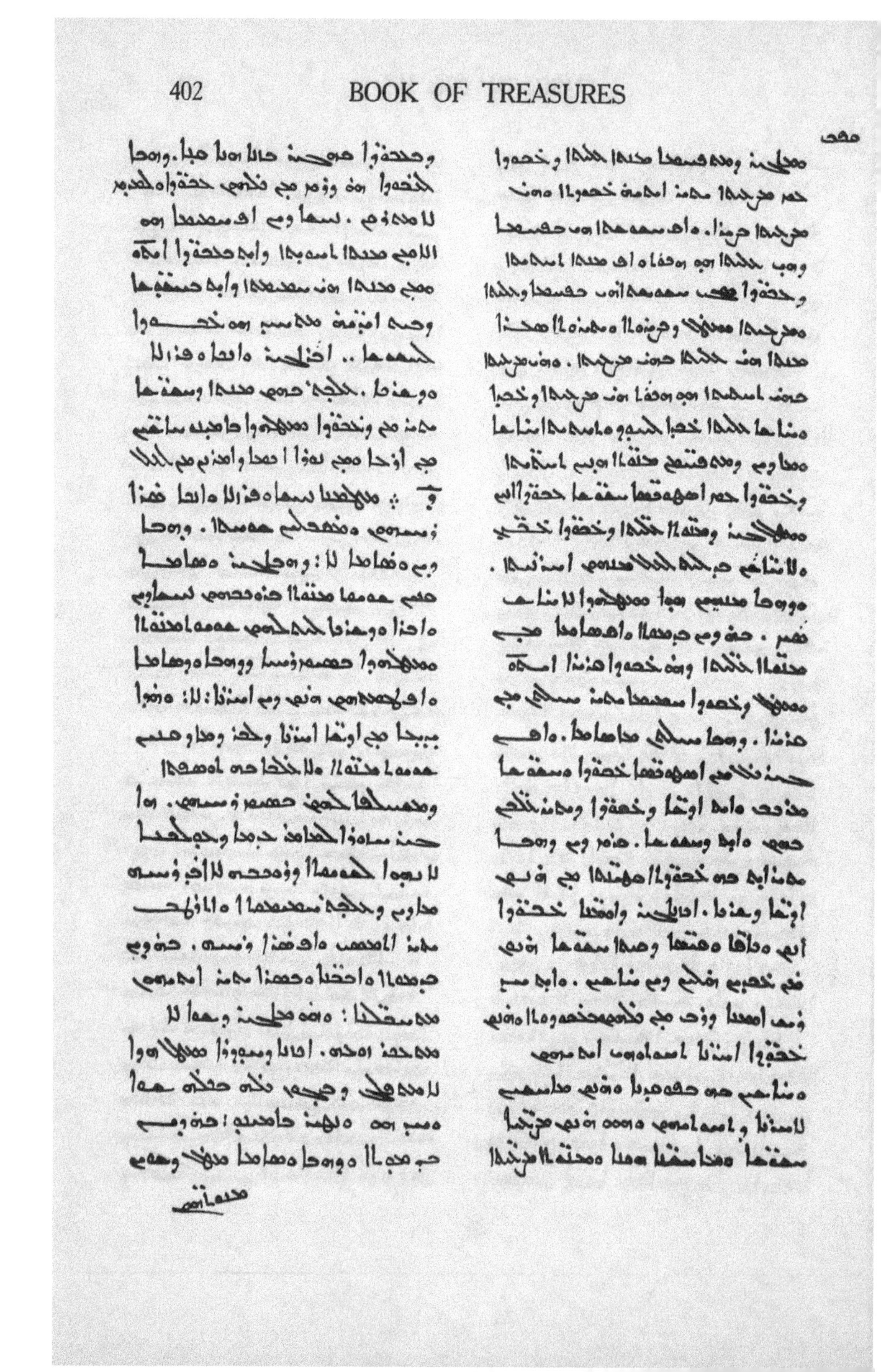

ܩܗ

[illegible]

[illegible]

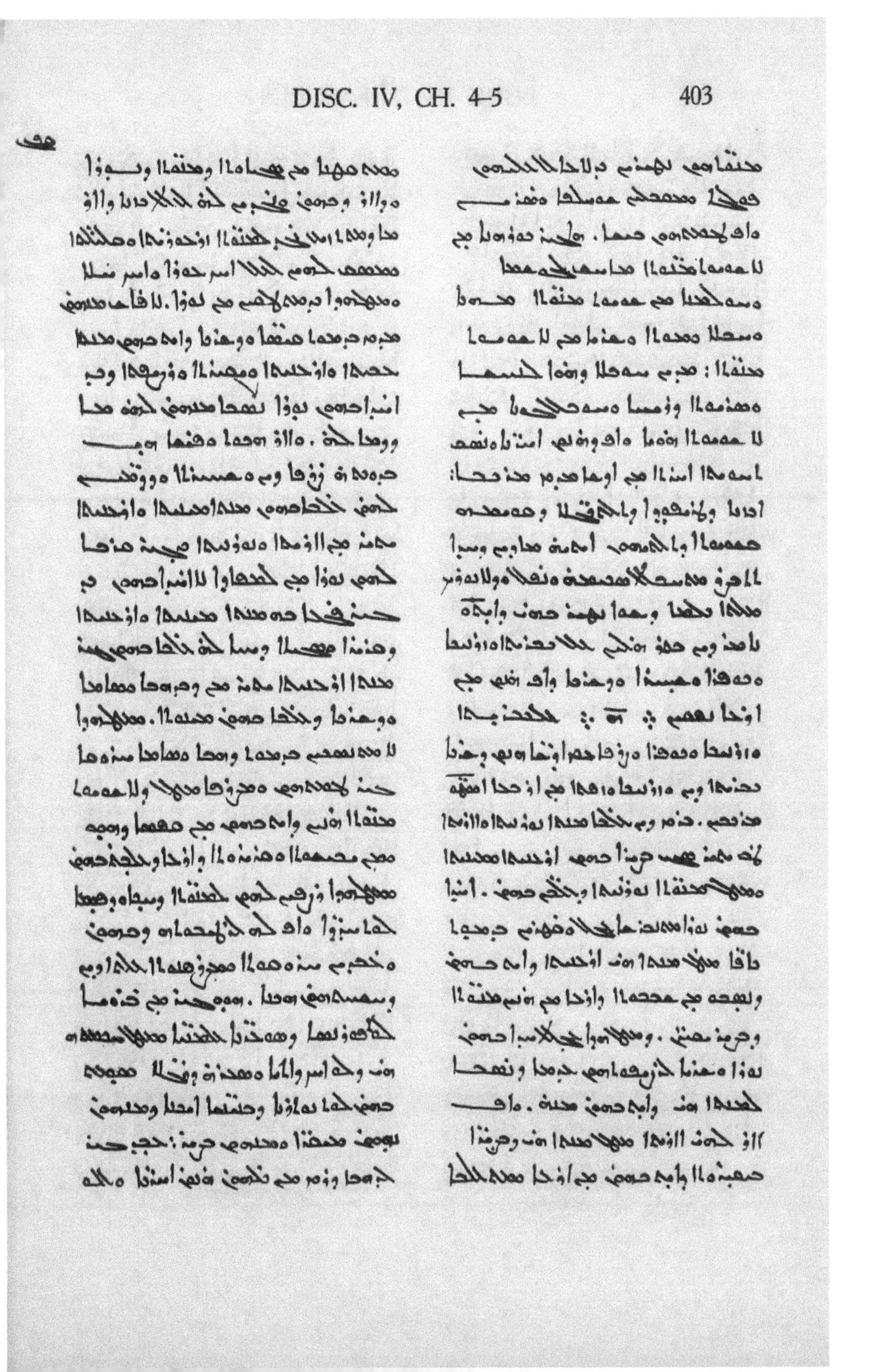

[illegible]

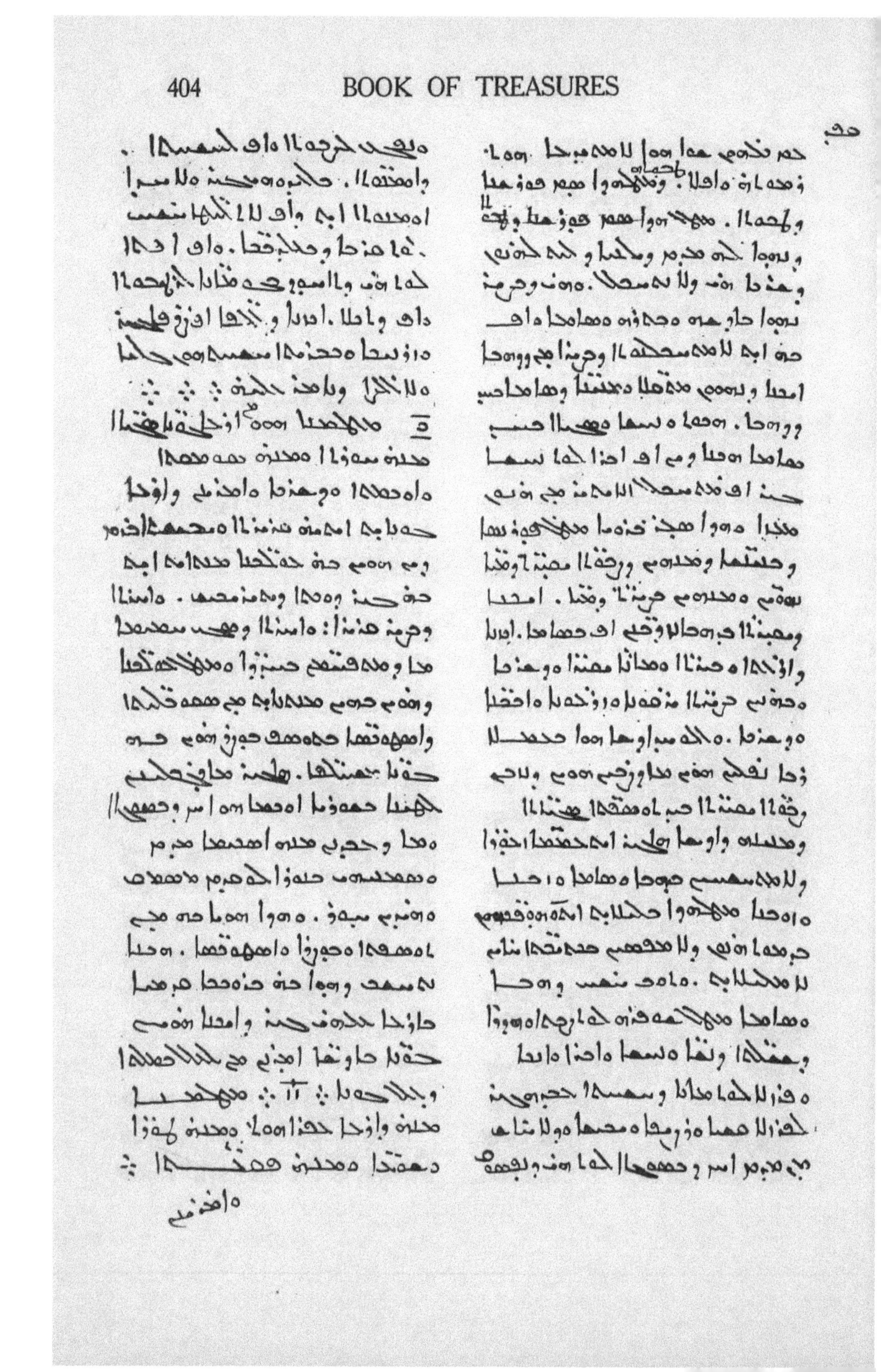

[illegible]

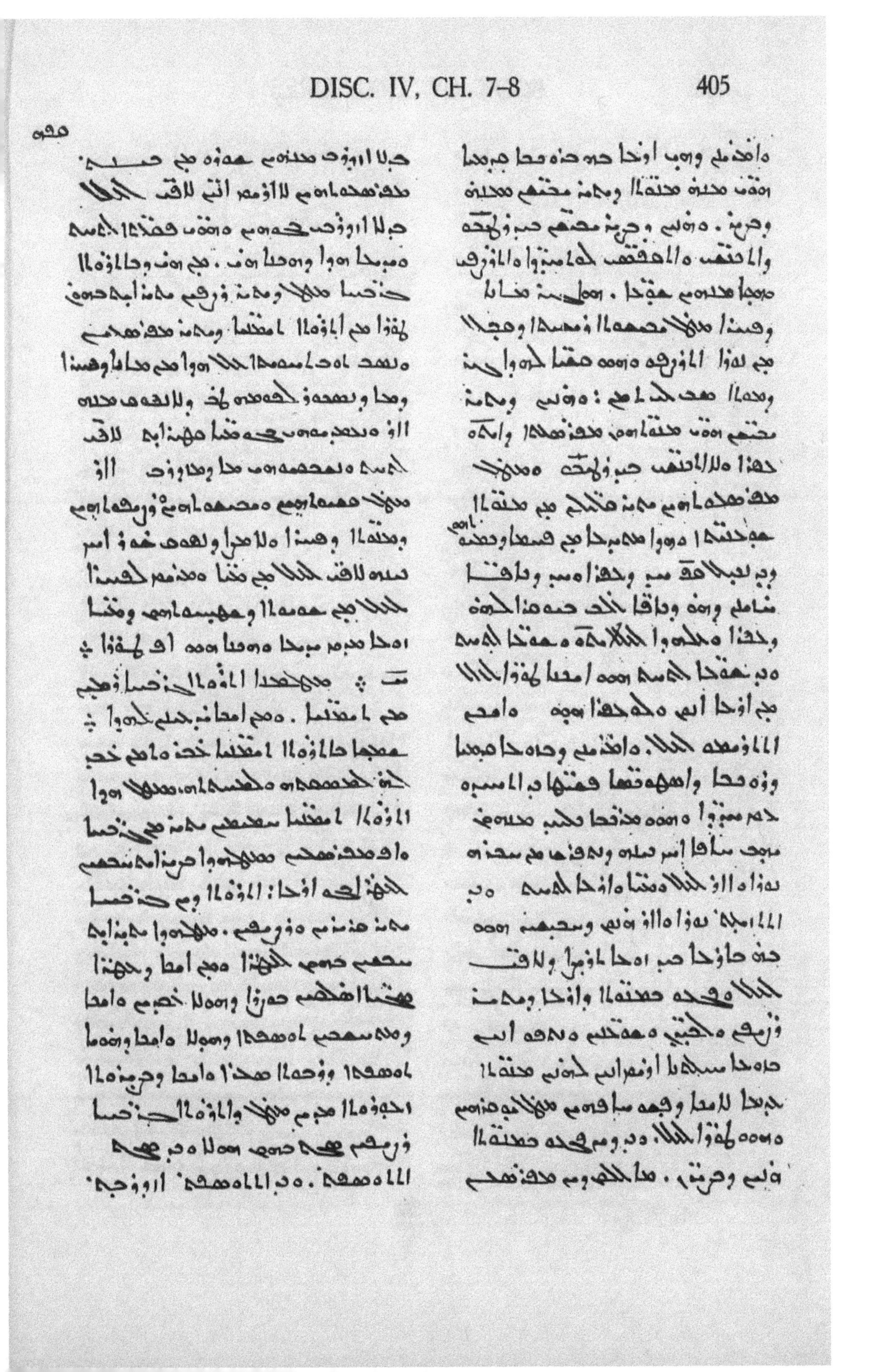

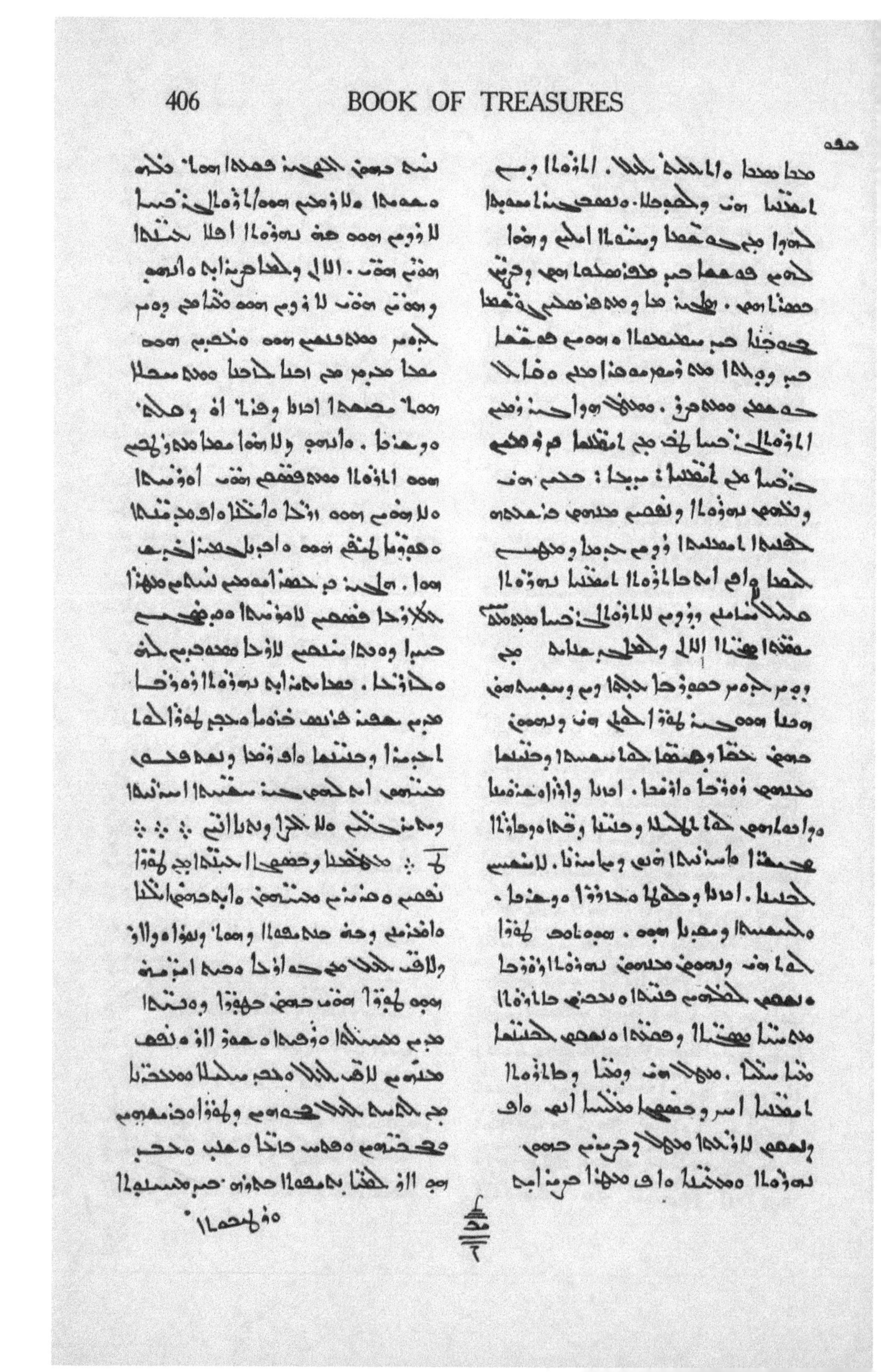

[illegible]

[illegible]

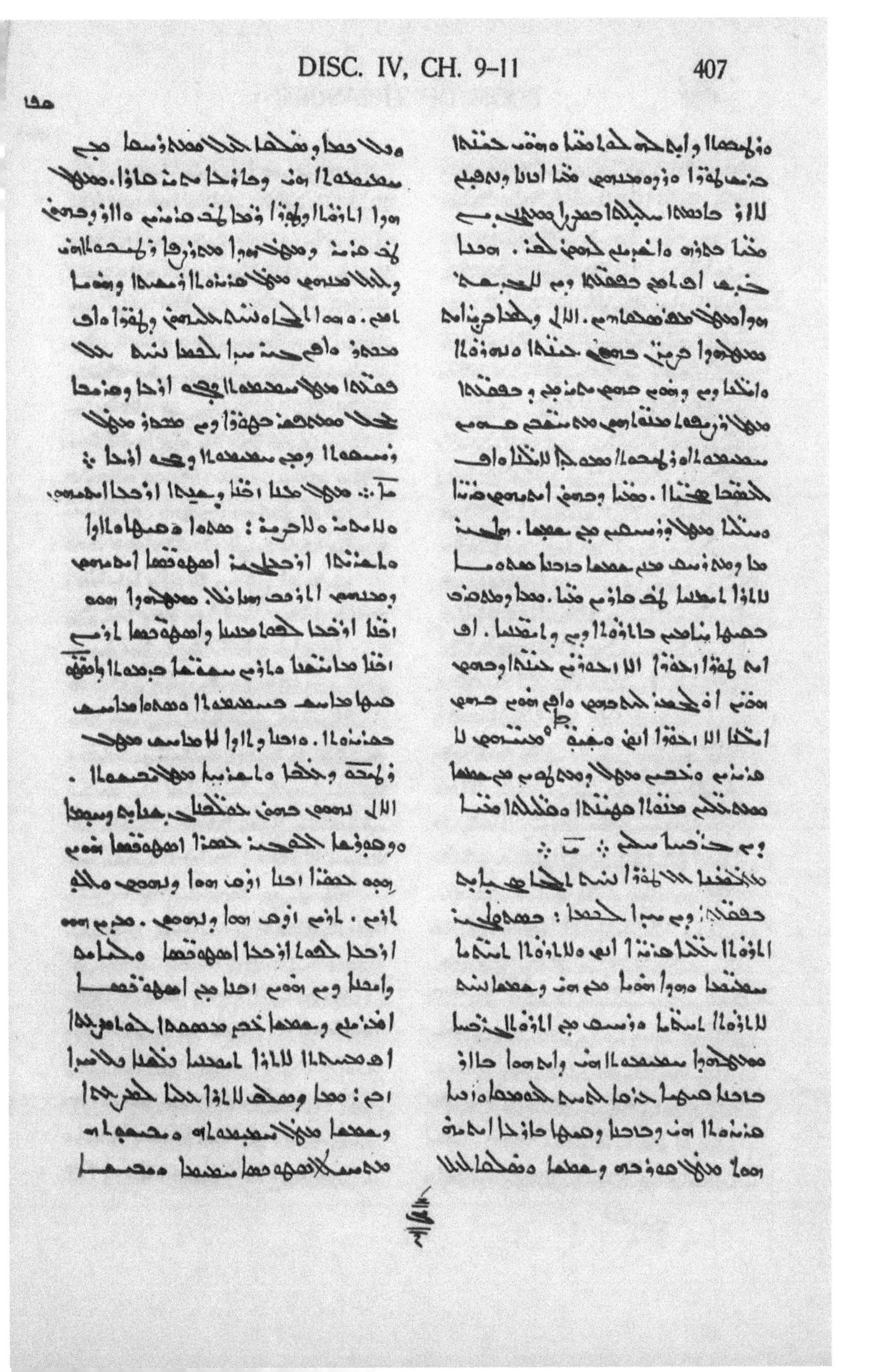

ܩܦܐ

[illegible]

[illegible]

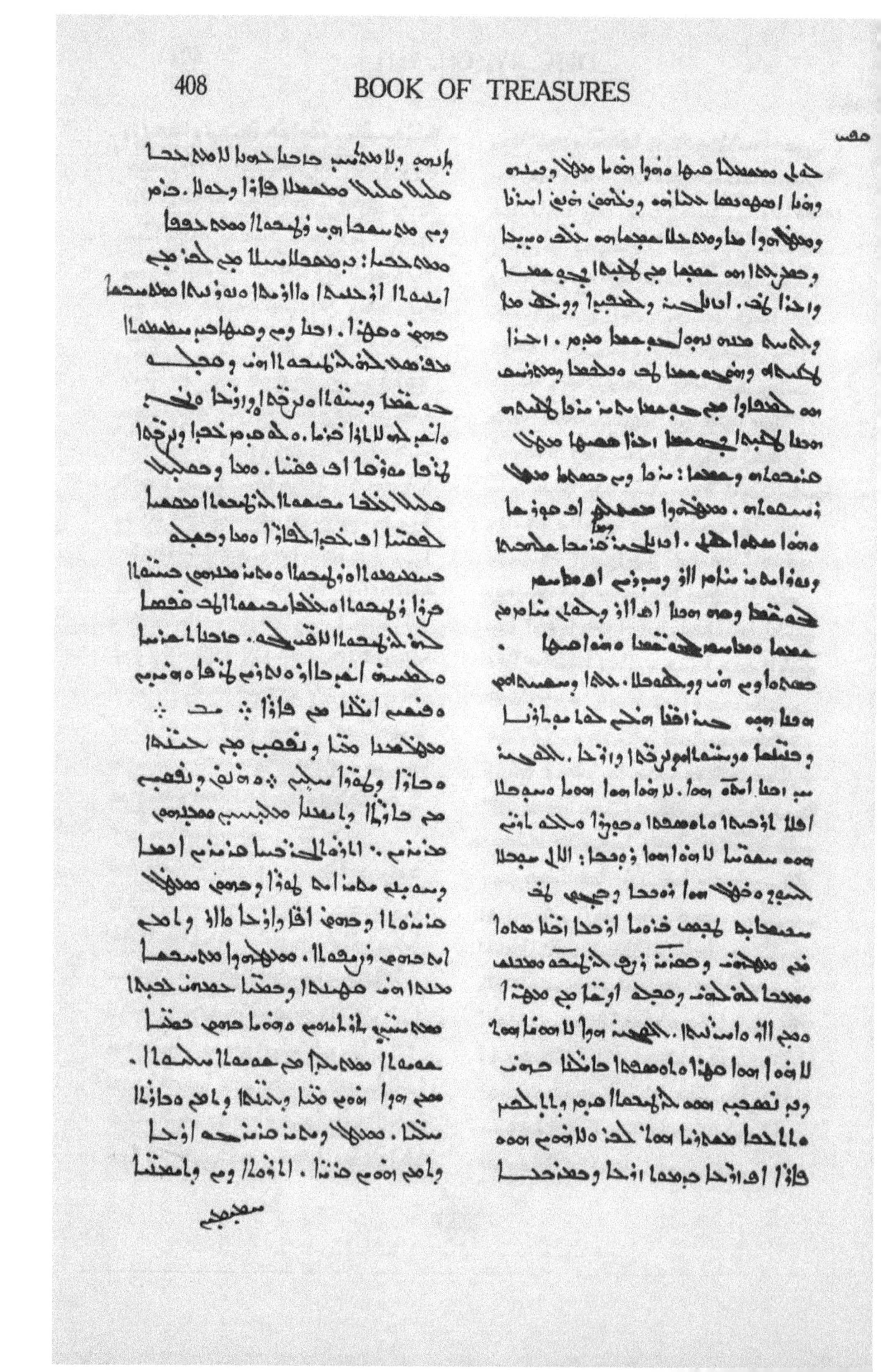

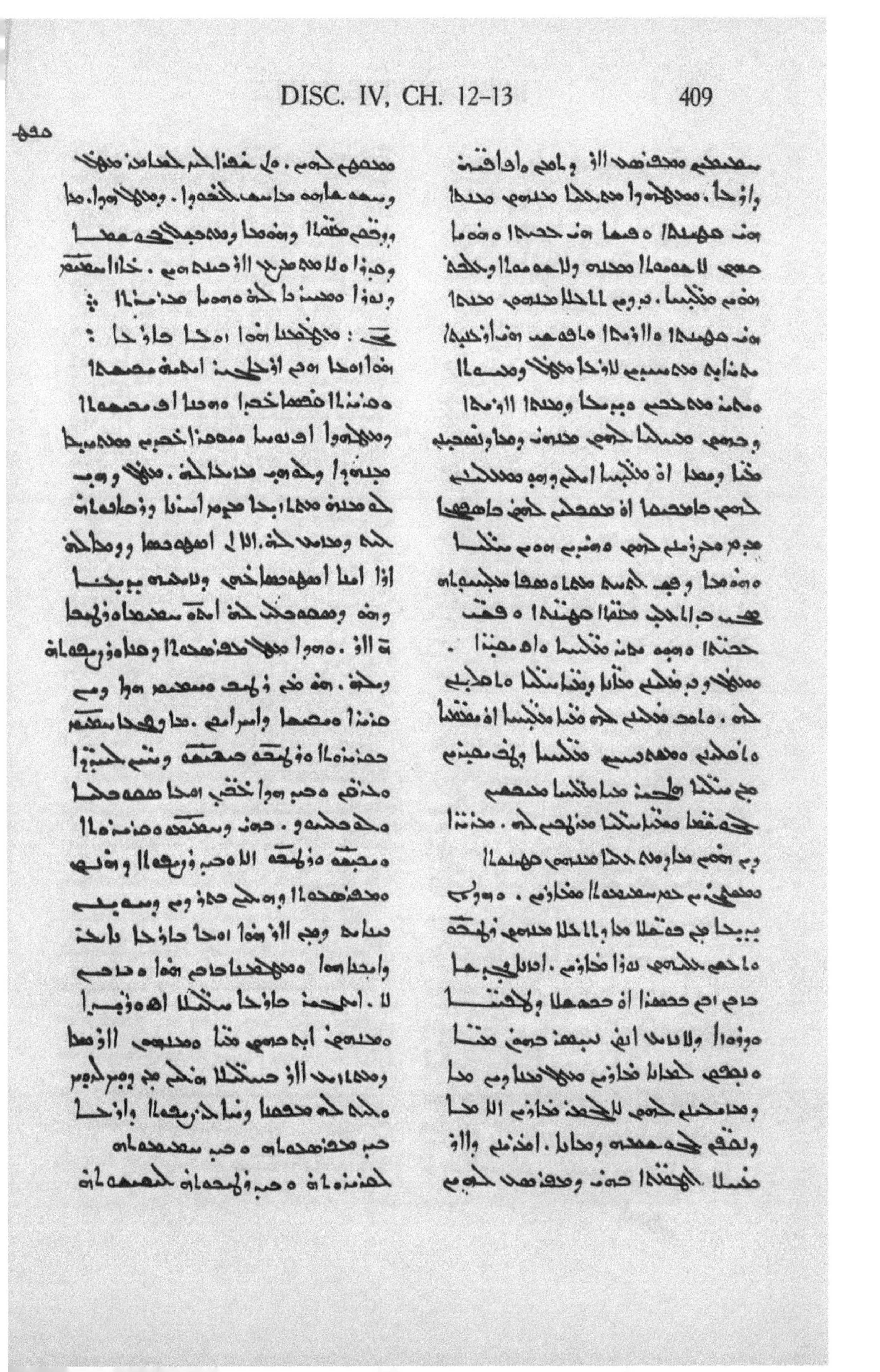

[illegible]

ܩ

ܘܕܠܐ ܢܦܫ ܡܢܗܘܢ. ܘܫܟܠܠ ܗܘܝ ܘܥܒܪ ܐܝܟܢܐ
ܘܡܝܐ ܘܗܘܐ ܢܘܪܐ ܘܐܪܥܐ. ܡܛܠ ܗܕܐ ܕܝܢ
ܗܕܐ ܕܠܒܝܫ ܡܫܡܫܢܘܬܐ ܡܕܡ ܕܗܘ ܥܠܬ
ܘܠܒܪ ܐܣܛܘܟܣܐ ܥܡ ܦܪܘܫܘܬܐ ܘܢܗܘܐ ܐܝܟ
ܐܬܪ ܡܢ ܚܝܠܗܘܢ. ܘܡܕܡ ܗܘܐ ܘܗܘ ܚܝܐ
ܗܘ ܡܢ ܗܢܝܢ ܘܟܠܗ ܡܢ ܐܪܥܐ ܘܡܝܐ ܘܗܘܐ
ܐܣܛܘܟܣܐ ܥܡ ܦܪܘܫܘܬܐ ܡܟܐ ܕܟܝܠ ܚܢܢ ܘܐܡܪ
ܣܥܪܝܗܝ. ܗܠܝܢ ܗܕܐ ܕܡܬܚܙܝܢ ܒܦܓܪ̈ܢܝܬܐ
ܟܠ ܗܢܐ ܗܕܐ ܘܡܢܝܬܗ ܕܐܪܥܐ ܘܐܣܟܝܡ ܚܙܝܢ.
ܚܘܪܝ ܡܢܗܘܢ ܘܡܢܗ̈ܝܢ ܠܐܦ ܠܥܠ ܘܕܢ̈ܚܝ
ܚܡܝܡܝܢ ܠܐܣܛܘܟܣܐ ܘܟܠܗܘܢ ܠܗ̇ ܚܫܝܐ
ܡܕܡ ܕܡܕܐ ܘܗܘܐ ܠܐܪܥܐ ܕܡܠܐ ܘܗܘܐ
ܐܢܫܐ ܚܘܝܢ ܘܕܝܠܗ ܐܦ ܚܘܝܢ ܠܗ ܘܠܐ
ܡܣܬܟܠ ܠܚܘܪܐ ܡܛܠ ܕܢܣܒܐ ܗܘܝ ܢܘܪ
ܗܘܐ ܗܢܐ ܘܡܢܗ ܕܡܬܚܙܝܐ ܠܗ
ܘܣܝܡܐ ܠܗ. ܗܟܢܐ ܗܢܐ ܐܦ ܐܪܥܐ
ܗܕܐ ܘܡܝܐ. ܘܐܦ ܘܚܕܐ ܠܐܘܪܐ ܘܢܦܫ
ܐܝܟ ܕܐܬ ܘܢܣܝܒ ܠܗ ܗܘܐ ܐܦ ܥܒܪܐ
ܘܒܗܢܐ ܘܡܕܡ ܕܢܣܒ ܚܘܝܢ ܒܢܦܫܐ ܚܠܝܠܬܐ
ܕ ܘܡܢ ܚܘܝܢ ܦܪ̈ܐ ܘܥܡܘܕܝܢ ܐܢܫܐ ܗܘܘ
ܢܗܘܐ ܘܣܠܩ ܐܝܟ ܒܟܢܝܐ. ܘܟܝܪ ܐܝܟ ܟܘܢܝܐ
ܘܡܘܬ ܦܠܐ ܒܢܝܐ ܒܡܕܡܗ ܗܘ ܕܝܠܝܬ ܕܝܢ ܕܝܢܐ
ܐܦ ܕܐܪܥܐ ܐܘܟ ܕܝܢ ܫܐܠܝܢ ܠܗ ܠܗܘܢ ܐܦ
ܒܗ ܦܝܚܐ ܘܣܦܝܬܐ. ܘܡܛܠ ܕܡܢܗܝܢ ܐܬܐ ܐܪܥ ܘܢܦܫ
ܕܐܦ ܡܢ ܚܕܐ ܦܢ ܘܕܦܝܢ ܘܡܫܚܚ ܕܐܝܬܝܗܘܢ
ܡܕܡ ܨܡܚܐ ܚܕ ܘܦܩܐ ܘܡܕܐ ܚܕ ܕܒܗܝ
ܡܢ ܠܗ ܚܕܪ. ܐܟܚܕܐ ܒܦܓܪܐ ܕܡܕܐ ܘܡܢ
ܗܒ ܘܗܕܐ ܘܡܢ ܗܢܝܢ ܐܘܟ ܕܝܢ ܡܬܡܙܓܢ
ܠܗ ܒܟܢܝܢ ܗܢܐ ܐܡܪܐ ܐܠܗܝܢ ܐܝܢ. ܡܛܠ
ܗܢ ܕܠܗ ܒܟܢܝܢ ܟܠܐ ܡܬܚܙܝܢ ܕܐܪܥܐ

ܘܐܢܛܒܝ ܗܘܐ: ܐܦ ܒܪܢܫ ܡܬܗܦܟܢܐ
ܐܝܟܢ ܐܠܐ ܐܝܬܘ ܒܪ ܚܘܒܐ ܢܘܣܐ ܘܠܗ ܒܟܢܝܢ
ܟܕܒܝܢ ܐܣܛܘܟܣܐ ܘܗܘܐ. ܐܝܟ ܕܡܠܝܢ ܕܐܪܥܐ
ܟܠ ܡܢ ܐܪܥܐ ܠܗ ܒܟܢܝܢ ܟܕܒܝܢ
ܐܣܛܘܟܣܐ. ܡܫܡܫܢܘܬܐ ܥܡ ܦܪܘܫܘܬܐ ܡܟܐ ܡܫ
ܠܗ ܐܠܐ ܐܝܟܢ ܐܝܟ. ܗܟܢܐ ܐܦ ܒܗ ܐܪܥܐ
ܡܠܝܢ ܕܐܢܫ ܡܫܬܡܫ ܕܐܪ ܡܫܡܫܢܐ
ܥܡ ܦܪܘܫܘܬܐ ܘܗܘ ܘܗܘܐ ܐܘܪܐ ܚܢܢ ܡܫܡܫܢ
ܠܦܪܘܫܘܬܐ ܒܢ ܢܣܒܘܬܐ ܡܫܡܫܢܘܬܐ ܗܘ
ܡܢ ܟܠܠ ܘܓܒܐ ܓܒܐ. ܗܘܐ ܕܝܢ ܠܗ ܠܣܢ
ܘܡܓܒܐ ܘܡܥܒܕ ܐܚܪ̈ܢܝܬܐ ܘܒܡܕܝܢ̈ܬܐ
ܡܬܐ ܐܢܫ ܐܪ ܡܢ ܚܝܠܗܘܢ ܡܢ ܕܡܕܡ ܠܒܝܫ
ܘܡܛܠ ܕܡܬܚܫܚ ܡܢ ܡܫ̈ܟܐ ܢܦܫ. ܘܒܝܕ
ܗܦܘܟܗ ܟܕܒ ܕܟܠܐ ܘܠܗ ܗܘܐ ܠܒܘܫ:
ܒ ܟܠ ܢܘܚܐ ܚܡܪ ܗܘܠܝܢ ܢܐܡܪ ܟܠ
ܣܠܩ ܕܢܘܚܐ ܕܡܠܟ ܘܕܐܝܬ ܐܡܬܝ ܗܘܐ
ܐܝܬܝ ܒܪܢܫ ܗܘܕܐ ܐܡܪܘ ܕܐܝܬܘ ܒܟܢܐ
ܗܘ ܡܫܡܫܢܐ ܘܗܢܘܢ ܠܗܘܐ ܒܟܠܐ ܗܘ
ܘܟܠ ܦܪ̈ܐ ܘܡܠܟܘܬܐ ܐܡܬܝ ܘܒܟܕܝܢ
ܡܠܐ ܟܠ ܢܘܚܐ ܚܝܘܬܐ ܘܟܝܢܐ ܘܢܦܫܐ:
ܐܦ ܒܪܢܫ ܐܪܡܣܛܝܛܘܣ ܒܟܬܒܘܗܝ ܘܟܠ
ܦܘܢܝܐ ܐܡܪ ܘܐܝܬܘܗܝ ܡܢ ܐܣܛܘ ܡܫܬܘܚܐ
ܘܡܠܐ ܡܢ ܐܪܥܐ ܐܡܬܝܚܢܝܢ ܗܘܝܐ ܘܐܣܛܘ
ܗܘ ܡܫܡܫܐ ܒܗܘܪܐ ܗܘ ܘܢܟܝܢ ܢܘܚܐ
ܘܒܐܝܠܢ ܢܦܩܢ ܡܢ ܐܪܥܐ ܡܫܬܡܫܐ ܗܘܐ
ܢܘܚܐ ܢܦܩܝܢ ܒܐܝܠܢ ܘܠܐ ܡܬܚܙܐ ܡܫܬܠܝܢ ܒܟܠ
ܐܬܘܗܝ ܢܘܚܐ ܘܒܐܝܢ ܠܗܘܢ ܡܫܡ.
ܗܘ ܒܐܝܠܢ ܢܦܩܢ ܡܢ ܐܪܥܐ ܐܦ ܒܟܝܢܗ
ܘܒܕܝܢܗ. ܐܢܐ ܠܐ ܗܘܐ ܘܡܣܬܒܪܐ ܗܘ ܘܢܘܚܐ
ܐܘ ܒܠܐ ܘܡܠܗܘܢ. ܘܐܝܟ ܗܘ ܡܫܬܡܫܢܐ

ܐܡܬܝ

ܟܕ ܦܘܩܕܢܐ ܕܡܫܡܫܝܢ ܠܗ ܡܫܐܠܝܢ ܘܪܘܚ
ܡܠܐ ܘܟܕ ܪܘܚܐ ܟܕ ܡܫܡܫ ܕܡܢ ܟܝܢܗ
ܡܫܩܠܐ. ܐܝܟܢܐ ܕܡܢܐ ܕܢܫܬܟܚ ܕܚܝܐ
ܡܢܐ ܕܚܠܘ ܡܫܠܡܝܢ ܕܡܫܡܗ ܡܬܘ
ܟܕ ܗܘ ܐܪ̈ ܟܠܗܝܢ ܚܫܝܫ ܘܟܠܗܝܢ ܡܪܡܢ
ܘܟܠܗܝܢ ܪܓܫ. ܘܟܠܗܝܢ ܡܫܝܫ. ܐܪܐܡ ܟܗܢܐ
ܡܫܝܚܐ ܕܡܬܠܫ ܡܢ ܐܪܥܐ ܠܐܝܠܐ ܟܗ ܟܕ ܡܬܗ ܠܐ
ܐܦ ܐܦܐ ܕܡܫܝܚܐ ܘܡܢܐ ܕܡܘܡܢܝܢ ܐܪ ܡܢܐ
ܡܢܐ ܕܡܢܐ ܕܪܘܚ ܢܘܗ ܐܘܚܐ ܕܡܠܐ ܡܡܢܐ ܒܗ
ܠܗ. ܐܝܟܢܐ ܕܒܐ ܠܐܦܠܐ ܠܗ ܚܝܐ ܡܫܝܚܐ
ܘܠ ܚܫܝܫ ܐܪ ܟܫܝܢܐ ܘܠ ܡܫܝܫ. ܘܠ
ܦܪܢ ܟܫܡܬܐ ܘܠ ܪܓܫ ܠܟܘܠܗܝܢ
ܡܫܡܠܐ ܘܐܘܚܐ ܕܐܝܕܐ ܗܘܢܐ ܡܫܡܫܐ
ܗܘ ܐܪ ܘܟܕ ܪܘܚܐ ܡܢܐ ܕܐܬܪܚܡ
ܐܝܕܐ ܡܢ ܙܢܐ ܢܦܫ ܪܘܚܐ. ܟܡܐ ܕܡܢ
ܐܝܠܐ ܐܝܬܝܗܝ ܗܘܝܐ ܗܘܐ ܠܡܠܟܘܬܐ ܕܫܡܝܐ
ܡܫܬܡܫ ܠܐܪ ܟܠܗܝܢ ܡܫܡܠܡܢܐ ܟܡ ܦܪܝܫ
ܘܟܠܗ ܐܦ ܦܪܝܫܬܐ ܟܡ ܡܫܬܡܫ ܟܕ ܡܢ
ܕܝܢ ܐܘܟܝܬ ܟܡ ܡܫܡܫܬܐ ܟܕ ܪܘܚܐ
ܘܗܘܐ ܟܝܢܐ ܡܟܝܠ ܕܡܫܡܫܬܐ ܠܐ
ܡܫܬܟܚܐ ܐܪ ܡܢ ܟܝܢܗ ܕܟܕ ܐܝܟܢܐ
ܡܢ ܕܢܝܢ ܟܝܢܢ ܕܡܫܝܚܐ ܗܘܐ ܕܗܘܢܐ
ܗܘ. ܡܢ ܗܘܝ ܕܡܢܐ ܘܡܦܢ ܟܝܢ ܡܫܝܚܐ
ܪܓܫܐ ܠܗܘܐ ܐܬܪܡܝ ܠܪܘܚܐ ܘܢܦܫܐ
ܡܢ ܪܘܚܐ ܐܦ ܢܦܫܐ ܘܦܓܪܐ
ܟܠܐ ܗܘܐ ܕܡܢ ܗܢܐ ܡܢܐ ܕܡܫܬܡܫ
ܐܠܐ ܟܕ ܟܡ ܦܪܝܫܐܝܬ ܘܦܫܝܛܐ ܘܡܫܡܫܐ
ܟܡ ܪܓܫܬܐ ܡܬܘܪܝܢ ܐܪ ܡܫܬܡܠܝܐ
ܘܦܫܝܛܐ ܘܡܫܝܚܐ ܡܢ ܠܗ ܟܠܗ ܘܟܕ
ܪܘܚܐ. ܐܝܟ ܡܢܐ ܕܡܢܐ ܘܟܕ ܪܘܚܐ

ܐܝܟܢܐ ܕܟܕ ܐܘܚܐ ܡܫܬܘܐ ܐܝܟ ܕܠܒܝܫ: ܦܓܪܗ
ܡܩܦܐ ܕܡܫܬܡܗܐ ܠܟܘ ܡܢ ܡܩܦܝܢ ܢܗܝܐ
ܟܕ ܪܘܚܐ ܐܘܚܐ ܕܠܐܦ ܠܟܠܢ:
ܐܝܟܢܐ ܗܘܐ ܡܫܬܪܝܐ ܕܟܝܢܗ ܡܢ ܟܡܢܐ ܐܝܟܢܐ
ܘܠ ܦܫܝܛܐ ܕܠܐܦ ܠܟܠܢ. ܐܪܟ ܗܘܐ
ܕܟܕ ܟܡܢܐ ܪܓܫܐ ܐܦ ܡܫܝܚܐ ܕܝܢ
ܕܟܕ ܐܘܚܐ ܟܕ ܦܫܝܛܐ ܕܠܐܦ ܠܟܠܢ
ܘܐܝܢܘܢ ܕܐܡܪ ܕܗܘܐ ܟܡܫܝܚܐ ܘܡܬܝܕܥܐ
ܘܐܬܡܫܝܚܐ ܟܡ ܡܫܝܚܐ ܦܫܝܛܐ ܕܟܕ ܒܢ
ܘܐܬܡܫܝܚ ܕܡܢܐ ܕܠܡܪܢ ܠܟܘܪܐ ܟܡܪܐ
ܐܦ ܟܕ ܐܦ ܡܢ ܐܝܕܐ ܪܓܫܐ ܦܫܝܛܐ
ܟܕܐ ܟܡܢ ܪܓܫܐ ܕܝܢ ܘܡܫܝܚܐ ܠܗ
ܡܫܝܚ ܡܫܝܚܐ ܠܐ ܟܕܝܢ ܐܠܐ ܕܡܫܡܫܝܢ
ܠܟܡܢ ܪܓܫܗ ܠܐ ܡܫܬܡܫܝܢ ܠܟܡܢ. ܐܝܟܢܐ ܕܝܢ
ܗܘܐ ܦܘܪܫܢܐ ܡܢ ܚܝܐ ܡܫܝܚܐ ܘܡܬܝܕܥܐ
ܡܫܡܫܠܐ ܪܘܚܐ ܕܠܐ ܠܗ ܠܗ ܠܐ ܐܦܫܪ
ܡܫܬܠܐ ܠܟܢ ܕܚܝܠܐ ܕܟܡܢ ܚܝܐ ܘܟܠܗܝܢ
ܡܫܬܐܠܐ ܘܗܘܐ ܘܦܗܠܐ. ܐܪܐ ܠܗ ܡܢ ܚܝܐ
ܡܫܝܚܐ ܗܘܐ ܟܡܪܝܗ. ܐܠܐ ܡܢ ܡܢ ܐܝܟ ܐܝܕܐ
ܗܘ ܡܫܬܠܐ ܘܡܫܡܫܢܐ ܐܪܝܢ ܐܝܟ ܡܫܡܫܢ
ܡܫܡܫܢܐ ܘܐܪܝܢ ܟܕܐ ܘܡܫܝܚܐ ܡܫܡܫܐ
ܗܘ ܐܪܐ ܐܝܕܐ ܗܘ ܟܕܟܗ. ܐܝܬ ܠܚܢ
ܪܓܫܢ ܕܠܐ ܪܘܚܐ ܘܡܢܗ ܗܘ ܡܐܪܓܐ
ܠܓܒܐ ܠܗ ܪܘܚܢܢ ܐܝܟ ܐܘܚܐ ܐܪܝܢ
ܕܝܢ ܟܕܝܢ ܕܦܘܩ ܟܠܐ ܘܗܘܐ ܠܐܠܗܐ ܕܠܗ
ܡܫܬܠܠܐ ܐܦ ܟܕܢ ܗܘܐ ܡܢ ܟܕ ܦܘܩܕܢܐ
ܐܠܗܐ ܕܝܢ ܟܕ ܡܫܡܫ ܐܝܕܐ ܐܦ ܡܢ
ܠܒܗ ܐܝܕܐ ܕܡܫܝܚܐ ܘܦܪܝܫܐ ܐܘܪܚܐ
ܘܡܫܝܚܐ ܟܕ ܟܘܡܪܝܢ ܕܪܘܚܢ
ܠܗ. ܗܘܐ ܗܘܐ ܠܐܠܗܢ ܢܗܘܘܢ ܠܐܪ

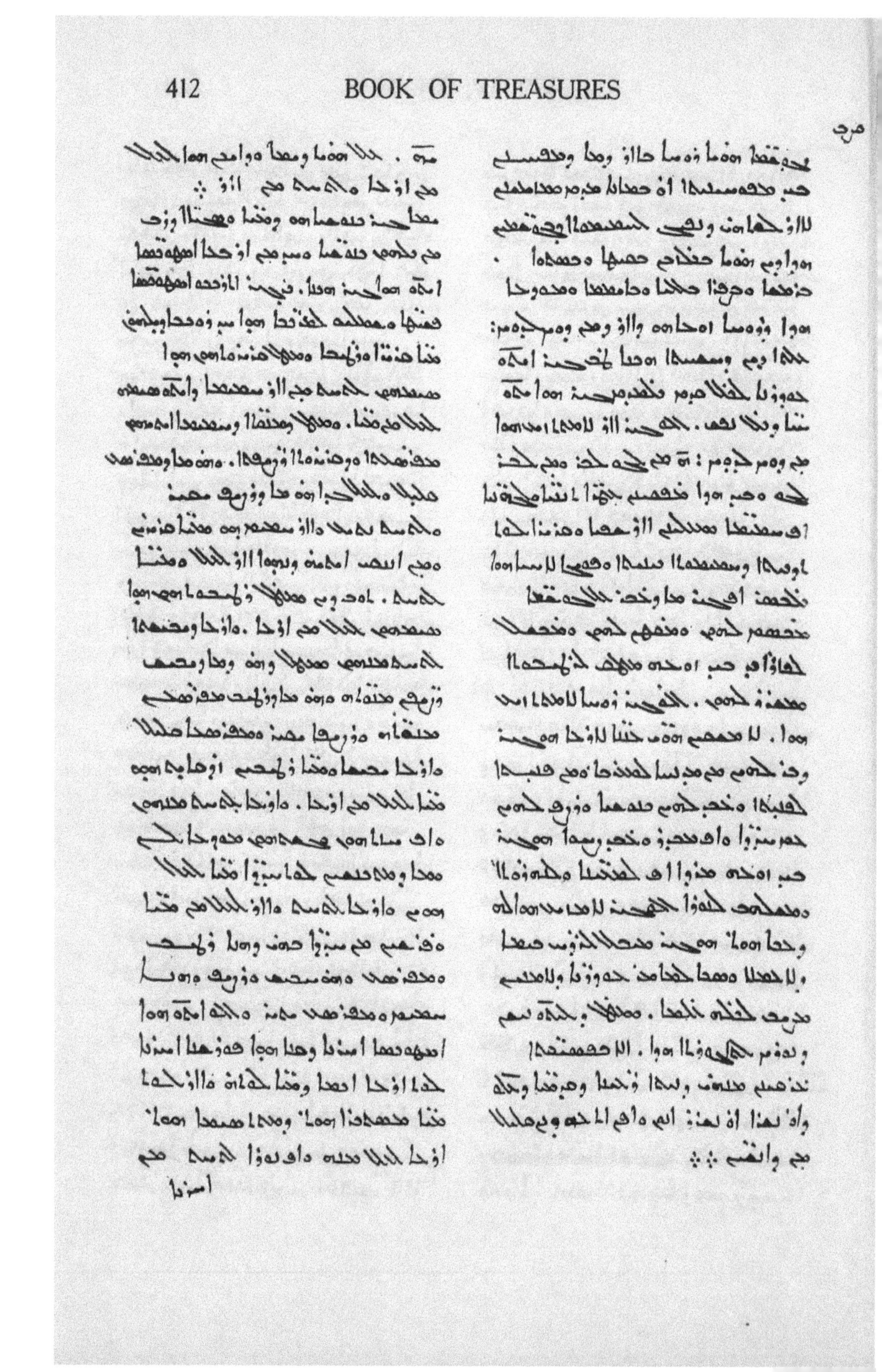

[illegible]

ܐܝܟܢܐ ܒܗܘܢܐ ܗܢܐ ܘܡܫܬܡܠܝܘܬܐ ܕܐܬܐܡܪ ܒܗ
ܐܣܛܘܟܣܐ ܡܢ ܚܙܘܐ ܩܕܡܝܐ ܕܐܡܪ ܠܝ ܦܘܠܘܣ
ܡܟܝܠ ܕܗܝܕܐ ܠܟܘܢ ܗܘ ܡܠܐ ܕܗܝܕܝܢ ܗܘܐ ܡܠܝܐ
ܟܠܗ ܡܢ ܐܘܪܚܐ ܕܝܘܠܦܢܐ ܘܡܦܩܢܘܬܐ ܀
ܝܗ: ܘܡܫܬܡܠܝܢܘܬܐ ܐܝܟ ܕܗܘܝܐ ܗܘܐ ܟܠܗ ܘܡܢܐ
ܠܗ ܠܟܠܗ ܡܢ ܢܦܫܗ ܐܘܪܚܐ ܗܘܐ ܡܛܠ ܐܠܐ ܠܟܠ
ܡܢ ܡܚܫܒܬܐ ܡܕܡ ܘܦܪܫ ܘܠܗ ܒܫܘܠܡܐ
ܒܡܕܡ ܕܗܘܐ ܠܐܣܛܘܟܣܐ ܒܟܠ ܚܙܘܐ ܡܢܗܘܢ
ܘܡܢܐ ܕܒܗ ܒܫܢܝܐ ܕܐܘܪܚܐ. ܐܦ ܐܬܚܙܝ ܘܡܟܝܠܐ
ܘܐܘܪܚܐ ܕܡܫܬܡܠܝܢܘܬܐ ܘܕܪܘܚܢܝܘܬܐ ܕܒܠܚܘܕܝܗܘܢ
ܘܒܗ ܒܗ ܐܘܟܝܬ ܡܫܬܠܡ ܘܠܗ ܡܫܟܚܘܬܐ
ܘܐܘܪܚܐ ܡܫܟܚܐ ܗܘܬ. ܐܠܐ ܐܝܟ ܒܗ ܘܦܓܪܐ ܘܪܘܚ
ܘܢܦܫܐ ܘܡܐܢܐ ܡܫܬܟܚܐ ܗܘܐ ܘܐܡܪ ܕܝ
ܡܛܠ ܗܘܐ ܠܗܘܢ ܠܟܢܐ ܒܫܡܝܐ ܒܝܬܐ
ܡܫܬܠܡܢܘܬܐ. ܐܝܟ ܕܝܕܥܝܢ ܘܡܢܐ ܗܘ ܕܡܫܬܡܠܐ
ܠܐܘܪܚܐ ܗܕܐ ܘܠܡܬܒ ܡܬܝܕܥ ܡܫܬܐ ܘܐܝܟܢ
ܕܢ ܠܟܠܗ ܗܢܐ ܗܘ ܡܕܡ ܘܡܫܟܚ ܠܗ ܘܡܠܐ
ܘܦܐܫ ܡܬܬܣܝܡ ܠܗܝ ܘܗܝܕܝܢ ܘܡܬܓܠܐ
ܡܢ ܐܘܪܚܐ ܘܩܢܝܢ ܘܗܢܐ ܘܠܟܠܗ ܟܠܗ
ܡܬܠܘܝܢ ܡܢܐ ܡܫܡܠܝܐ ܗܘܐ ܐܦ ܡܫܬܡܠܝܢ
ܘܡܢ ܐܝܬܝܗ ܐܣܛܘܟܣܐ ܡܢܝܐ ܠܗܘܢ ܘܥܘܡܪܐ
ܝܘ: ܘܡܫܬܡܠܝܢ ܗܘܘ ܒܢܝܐ ܘܡܠܐ ܡܢܝܢܐ
ܡܢܐ ܘܡܠܐ ܡܬܚܙܝܐ ܡܢ ܫܡܝܐ ܡܢ
ܥܘܠܡܐ ܘܥܠܡܐ ܕܒܝܬܐ ܠܚܝܐ. ܡܫܬܠܡ ܗܘܐ
ܗܝܕܝܢ ܒܗܘܢ ܢܦܫܐ ܒܗ ܐܦ ܠܟܠܗ ܗܘ
ܡܫܝܚܐ ܘܪܘܚܐ ܘܡܫܬܡܠܝܐ ܘܐܝܬܘܗܝ
ܘܡܐ ܠܗ ܘܡܫܬܡܠܝܐ ܘܡܢܐ ܠܒܢܝܐ
ܘܒܗ ܘܐܬܚܙܝ ܠܗ ܚܙܘܐ. ܒܕ ܫܡܝܐ
ܫܡܝܐ ܡܢܗ ܘܡܠܟܐ ܕܫܡܝܐ ܘܫܠܝܡ
ܠܡܫܬܡܠܝܐ. ܐܝܟܢܐ ܕܝܢ ܐܦ ܡܫܬܡܠܝܢ ܡܢ

ܐܡܠܝܘܬܐ ܕܐܝܟܢܐ ܕܐܘܪܚܐ ܡܫܬܡܠܝܐ ܡܢ ܩܕܡ ܕܡܫܬܡܠܝܘܬܐ
ܘܒܝܬܐ ܢܦܫܐ ܘܡܢ ܥܘܠܡܐ ܘܡܫܬܡܠܐ ܘܡܫܬܡܠܝܘܬܐ
ܒܪ ܒܗ ܡܢ ܗܘ ܕܡܠܐ ܘܪܘܚܢܝ ܡܢܐ ܠܟܠܗ
ܡܫܬܡܠܐ ܕܢܦܫ ܒܗܘܢ ܘܡܫܬܡܠܝܢ ܡܢܗ ܡܢܝܢܐ
ܠܥܘܡܪܐ ܗܘ ܕܡܟܝܠ ܡܢ ܗܘܐ ܗܟܢ ܡܫܬܡܠܝܐ
ܘܐܦ ܡܫܬܡܠܝܢ ܕܒܫܘܠܡܐ ܠܡܫܬܡܠܝܘܬܐ ܐܝܟ ܕܒܫܘܠܡܐ
ܡܫܬܡܠܝܐ ܐܝܟ ܡܫܬܡܠܝܢܘܬܐ ܕܫܡܝܐ ܗܘܐ
ܗܟܢܐ ܗܘܐ ܒܗ ܘܫܡܝܐ ܗܘܐ ܡܫܬܡܠܝܢ
ܠܫܡܝܐ ܗܘ ܫܡܝܐ ܘܫܡܝܐܘܬܐ. ܗܘܝܢ ܕܝܢ
ܒܪܢܫܐ ܕܢܦܫܗ ܠܗ ܒܗܘܢ ܠܫܡܝܐ ܗܘܐ
ܘܗܘܝܢ ܡܛܠܐ ܘܪܘܚܐ ܘܡܫܝܚܐ. ܘܒܪܢܫܐ
ܘܠܐ ܗܘܐ ܗܘܝܢ ܠܟܠܗܘܢ. ܟܠܗܘܢ ܕܝ ܡܫܬܡܠܝܢ
ܐܡܪ ܗܘܐ ܡܫܬܡܠܝܢ ܗܘܐ. ܡܢܝ ܘܡܢ
ܠܒܪܢܫܐ ܕܡܢ ܕܟܠܗ ܢܦܫܐ ܠܝ ܠܡܫܬܡܠܝܢ
ܒܪܢܫܐ ܕܟܠܗܘܢ ܢܦܫܐ ܘܗܘܐ ܡܠܐ ܘܫܡܝܐ
ܠܗܘܢܐ: ܝܙ: ܡܫܬܡܠܝܐ ܗܘܝܢ ܡܫܬܡܠܝܢ
ܠܐܘܪܚܐ ܕܡܢܐ ܕܟܠܗܘܢ ܒܪܢܫܐ ܘܐܦ ܡܫܬܡܠܝܐ
ܐܡܪܝܢ ܗܘܘ ܚܝܠܐ ܠܐܘܪܚܐ ܘܢܦܫܐ
ܡܬܚܙܝܢ ܡܢܐ ܘܡܫܬܡܠܝܐ ܡܬܚܙܝܢ ܒܫܡܝܐ
ܐܡܠܝܘܬܐ ܕܐܘܪܚܐ ܗܘ ܘܒܗ ܐܡܪܝܢ ܟܠܗܘܢ
ܒܫܡܝܐ ܒܫܡܝܐ ܠܝ ܡܫܬܡܠܝܐ ܗܘ ܡܫܬܡܠܝܐ.
ܘܠܝ ܪܘܚܢܝܐ ܗܘ ܪܘܚܢܝܐ. ܘܠܝ ܕܫܡܝܐ ܗܘ
ܪܘܚܢܝܐ. ܘܠܝ ܒܫܡܝܢܝܐ ܗܘ ܒܫܡܝܢܝܐ
ܘܕܫܡܝܐ ܡܢ ܐܣܛܘܟܣܐ ܫܡܝܐ ܡܢ ܟܠܗ
ܘܡܫܬܡܠܝܐ ܘܡܢ ܒܫܡܝܢܝܐ ܡܫܬܡܠܝܢ ܪܘܚܢܝܐ
ܘܡܢ ܫܡܝܐ ܘܫܡܝܐ ܡܫܬܡܠܝܢ ܗܘܝܢܐ
ܘܡܫܬܡܠܝܢ ܗܘܝܐ. ܗܘܝܢ ܟܠܗ ܒܫܡܝܢܝܐ
ܐܡܪܝܢ ܡܫܬܡܠܝܐ. ܡܫܬܡܠܐ ܘܫܡܝܐ
ܘܒܫܡܝܐ ܡܫܬܡܠܝܢܘܬܐ. ܡܠܐ ܘܡܠܬܐ ܐܡܪܝܢ
ܠܟܠܗ ܘܠܫܡܝܢ ܡܢ ܐܘܪܚܐ ܘܡܠܐ ܡܫܬܡܠܝܐ

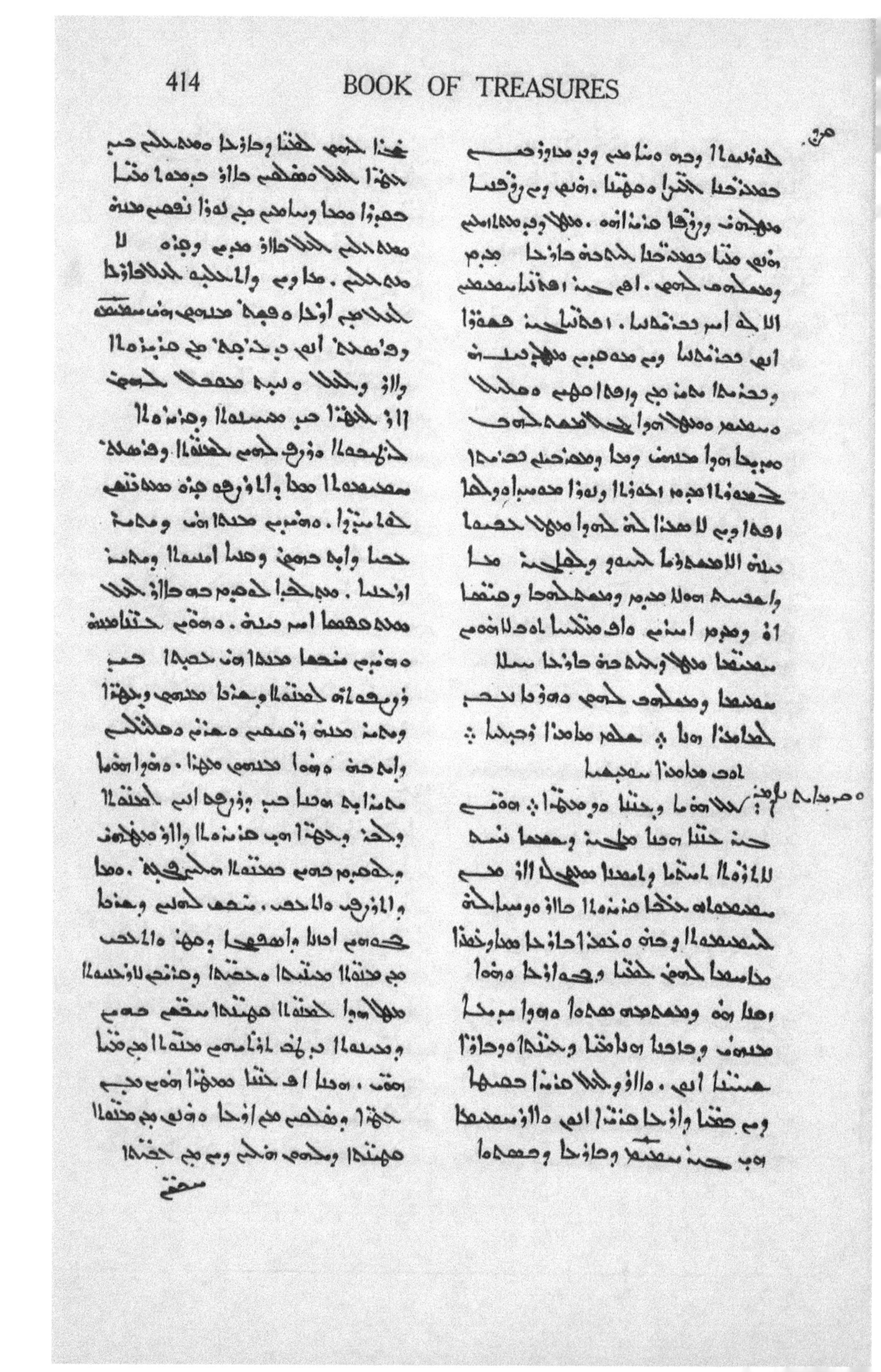

[illegible]

[illegible]

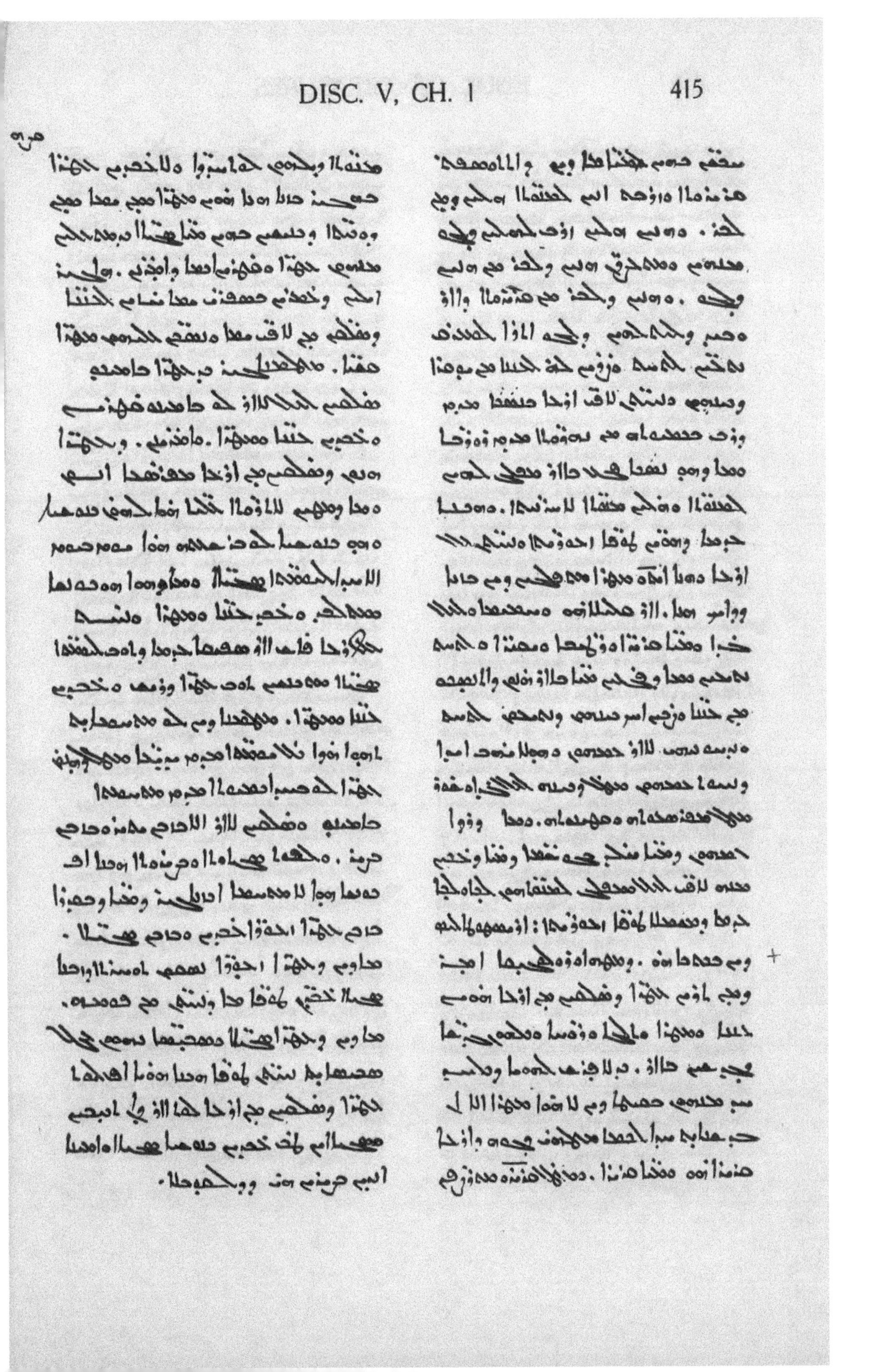

ܟ : ܡܫܠܡܢܐ ܒܪܝܫ ܒܢ̈ܝܐ ܣܓܝ ܘܒܪܝܫ ܡܘ
ܡܢܗܘܢ ܘܒܪܝܫ ܡܫܡ̈ܫܢܐ : ܐܦܣܩܘܦܐ ܕܗܢܘܢ
ܣܘܓܐܐ ܡܢܐ ܕܡܫܡܫܝܢ ܡܢ ܡܢܝܢܐ ܗܢܘܢ ܕܠܐ ܒܥܠܬܐ
ܐܠܗܘܬܐ ܒܗܘܢ ܡܬܠܐܡܢ ܡܫܡܠܝܘܬܗ ܕܚܝܐ ܡܬܚܙܝܐ
ܗܘܐ ܓܝܪ ܒܗܘܢ ܢܗܘܪܗ ܕܝܕܥܬܐ ܡܫܬܡܠܝܢ
ܠܗ ܣܘܓܐܐ ܒܡܬܠܐ ܢܗܘܪܗ ܡܢܐ ܕܡܬܝܕܥܢ
ܡܢܝܢܐ ܕܡܠܟܘܬܐ ܘܕܒܢ̈ܝܐ ܡܫܡ̈ܫܢܐ ܒܗܘܢ ܡܫܡܠܝܐ
ܡܬܡܠܐ ܢܗܘܪܗ ܕܚܟܡܬܐ ܡܢ ܕܢܚܬ ܒܗܘܢ ܡܢ
ܒܟܠ ܚܕ ܡܬܦܪܫܢܘܬܐ ܢܗܝܪܐ ܕܡܬܝܕܥܝܢ
ܗܘܝܢ ܠܗ ܕܪܓܐ ܡܫܡܠܝܢ ܡܢܝܢܗܘܢ ܒܗܐ
ܚܕܐ ܘܠܐܓܪ ܒܗܘܢ ܢܗܘܪܗ ܕܝܕܥܬܐ ܡܫܬܡܠܝܢ
ܠܗ ܐܘܣ̈ܝܐ ܘܢܗܝܪ ܐܝܣܘܡܐ ܡܢ ܐܣܟܝܡܐ
ܘܗܐ ܕܗܠܝܢ ܡܢܐ ܚܢܝܢ ܠܗ ܕܪܓܐ ܒܗܐ
ܐܘܣܝܐ ܘܠܗ ܒܟܠܗܘܢ ܐܣܟܝܡܐ . ܐܠܐ ܡܢܐ
ܐܣܪܐ ܕܡܢ ܕܪܓܝܢ ܪܚܡ ܠܐܘܣܝܐ ܘܡܢ
ܕܡܫܝܢ ܪܚܡ ܒܡܘܪܐ ܘܠܘܬ ܡܟܘܢ
ܢܟܘܢ ܡܢܐ ܕܪܓܐ ܒܪܝܬܐ ܓܒܪܐ ܒܗܘܢ
ܫܡܝܐ ܒܕܡܘܬ ܐܘܣܝܐ . ܡܢܐ ܕܡܫܡܫܝܢ
ܓܒܪܐ ܒܗܘܢ ܘܡܫܡܫܝܢ ܗܢܘܢ ܕܡܢ ܒܢ̈ܝܐ ܗܘܐ ܡܫܡ̈ܫܢܐ
ܡܢܐ ܕܐܝܬܘܗܝ ܫܡܝܐ ܒܡܬܝܕܥܢܘܬܐ ܒܡܫܪܝܐ ܡܢܐ
ܒܪ ܘܒܡܫܡܫ ܫܡܝܐ ܐܝܟ ܠܐܠܗܐ ܕܡܫܡܫܝܢ
ܗܘܐ ܕܡܢ ܒܝܬܐ ܡܢ ܐܠܗܐ ܡܫܡܫܢ ܕܫܡܝܐ
ܡܢܐ ܕܡܬܠ ܓܒܪܐ ܒܗܘܢ ܒܗܘ ܠܐܠܗܘܬܐ
ܘܡܢܐ ܕܓܒܪ ܡܫܡܫܢܐ ܕܫܡܝܐ ܗܘܐ ܡܫܡܫܢܐ
ܗܘ ܓܒܪ ܐܠܗܐ ܡܫܡܫܢ ܘܡܫܡܫܢܐ ܕܢܗܘܢ
ܒܗܘܢ ܒܗܘܐ ܕܡܠܟܐ . ܡܠܐܟܝܢ ܕܗܘܐ ܡܬܒܪܝ
ܫܡܝܐ ܒܠܗ ܠܗ ܡܫܡܫܝܢ ܠܐ ܠܟܠ ܡܫܡܫܝܢ
ܘܠܓܒܪ ܐܠܗܐ ܡܫܡܫܢ ܠܗ . ܘܒܝܕ
ܠܐ ܡܬܡܠܝܢ ܡܫܡܫܢܘܬܐ ܘܒܝܬܐ ܗܘܐ ܕܗܘܐ
ܗܘ : ܡܫܡܫܢ ܕܠܡܫܡܫ ܒܗܐ ܠܡܫܡܫܢܐ

ܡܫܡܫܢܐ ܡܢܐ ܕܡܫܟܚ ܫܡܝܐ ܠܟܠ ܒܗ
ܡܢ ܡܕܢܚܐ ܠܡܥܪܒܐ ܘܓܪܒܝܐ ܡܬܦܪܫܝܢ
ܠܐܘܣܝܐ ܕܫܡܝܐ ܓܒܪܐ ܐܝܟ ܡܫܝܚܐ ܕܡܢܗ
ܓܘܢܐ ܕܠܫܡܝܐ ܕܠܐܘܣܝܐ ܘܗܕܐ ܡܢ
ܕܝܠܗ ܕܫܡܝܐ ܠܐܘܣܝܐ ܗܘܐ . ܒܗ ܕܡܢ
ܒܡܫܚܐ ܕܠܒܫ ܫܡܝܐ ܕܟܠ ܒܢ̈ܝܐ ܡܦܪܫ ܒܐܝܕ
ܘܡܢ ܒܓܘܢܐ ܘܗܝ ܐܝܟ ܡܫܟܠܠܢܐ ܓܘܢܐ
ܒܗ ܒܢ̈ܝܐ ܘܡܢ ܡܫܡ̈ܫܢܐ . ܡܫܠܡܢܐ
ܠܗ ܟܠܗܘܢ ܕܝܠܗ ܫܡܝܐ ܒܪ ܘܐܝܟ
ܠܡܫܟܠܠ ܒܢ̈ܝܐ ܠܐܘܣܝܐ ܗܘܐ . ܘܐܦ ܡܢ
ܘܡܢܐ ܕܐܝܟ ܒܐܝܕ ܕܬܚܬܝܐ ܡܬܦܪܫܐ ܗܘܝܐ
ܗܘܐ . ܘܗܘܝܢ ܒܢ̈ܝܐ ܡܫܡ̈ܫܢܐ ܠܗ ܒܟܠܗ
ܫܡܝܐ ܗܘܝܢ ܡܫܡܫܝܢ ܠܐܠܗܐ ܕܐܝܬܘܗܝ
ܡܢܐ ܗܘ ܕܬܚܬܝܐ ܕܐܝܟ ܗܘܢ ܟܠܗ ܡܫܡ̈ܫܢܐ
ܕܒܢ̈ܝܐ ܫܡܝܐ ܒܫܡܝܐ ܡܫܡܫܝܢ ܗܘ . ܘܡܫܡܫܢܘܬܗ ܗܘܐ
ܐܝܟ ܕܗܘܝܢ ܡܫܡܫܝܢ ܒܪ̈ܝܬܐ ܘܬܚܬܝܐ ܘܡܢܐ ܕܡܬܚ
ܡܫܡܫܢܘܬܐ ܕܫܡܝܐ ܡܬܦܪܫܐ ܘܐܝܟ ܡܫܡܫܢܘܬܗ
ܒܪܢܫܐ . ܒܗܘ ܬܚܬܝܐ ܫܡܝܐ ܗܘܝܢ
ܘܗܘܝܐ ܡܬܡܠܐ ܠܥܠܡܐ ܘܫܡܝܐ ܘܡܢ
ܐܝܟ ܡܫܡܫܢܐ ܡܢܐ ܠܥܠܡܐ ܕܥܠܡܝܢ ܘܫܡܝܐ
ܕܒܢ̈ܝܐ . ܘܐܝܟ ܡܬܠܐ ܘܕܝܢ ܒܐܝܕ ܬܚܬܝܐ
ܘܡܢ ܕܝܢ ܒܝܕ ܒܗ ܫܡܝܐ ܘܡܢ ܒܢܝܢܐ
ܐܝܬܘܗܝ . ܘܡܬܠܐ ܗܘܐ ܠܐܠܗܐ ܡܢ
ܫܡܝܐ ܡܢܐ ܘܡܢ ܡܫܡܫܢܘܬܐ ܕܟܝܢܐ
ܡܢ ܡܫܡܫܢܘܬܐ ܕܟܠ ܒܢ̈ܝܐ . ܐܝܟ ܕܐܡܪ ܘܡܢ
ܘܕܡܢ ܡܢܐ ܒܡܫܡܫܢܘܬܐ ܣܘܓܐܐ ܕܬܚܬܝܐ
ܘܡܫܡܫܢܘܬܐ ܠܗ ܡܢ ܟܠ ܫܡܝܐ ܓܒܪܐ ܓܘܢܗ
ܐܠܗܘܬܐ ܘܡܫܡܫܝܢ ܒܡܫܡܫܢܐ ܕܒܪ
ܒܗ ܐܠܗܐ ܡܢܐ ܡܢ ܫܡܝܐ ܠܗ ܓܘܢܗ
ܡܫܡܫܢܐ ܕܗܘ ܕܐܝܬ ܕܢܚܬ ܠܥܠܡܐ
ܐܠܗܐ

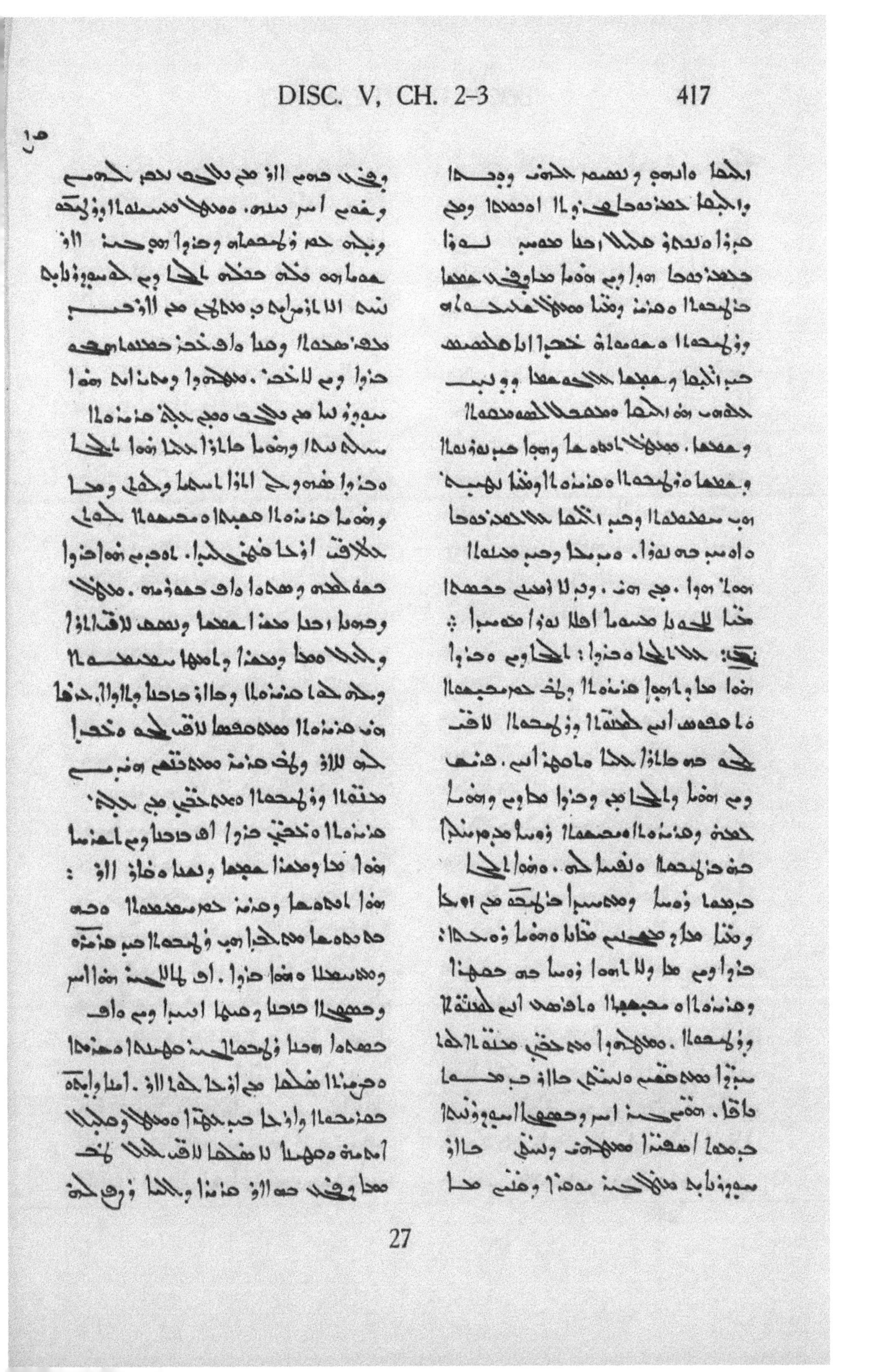

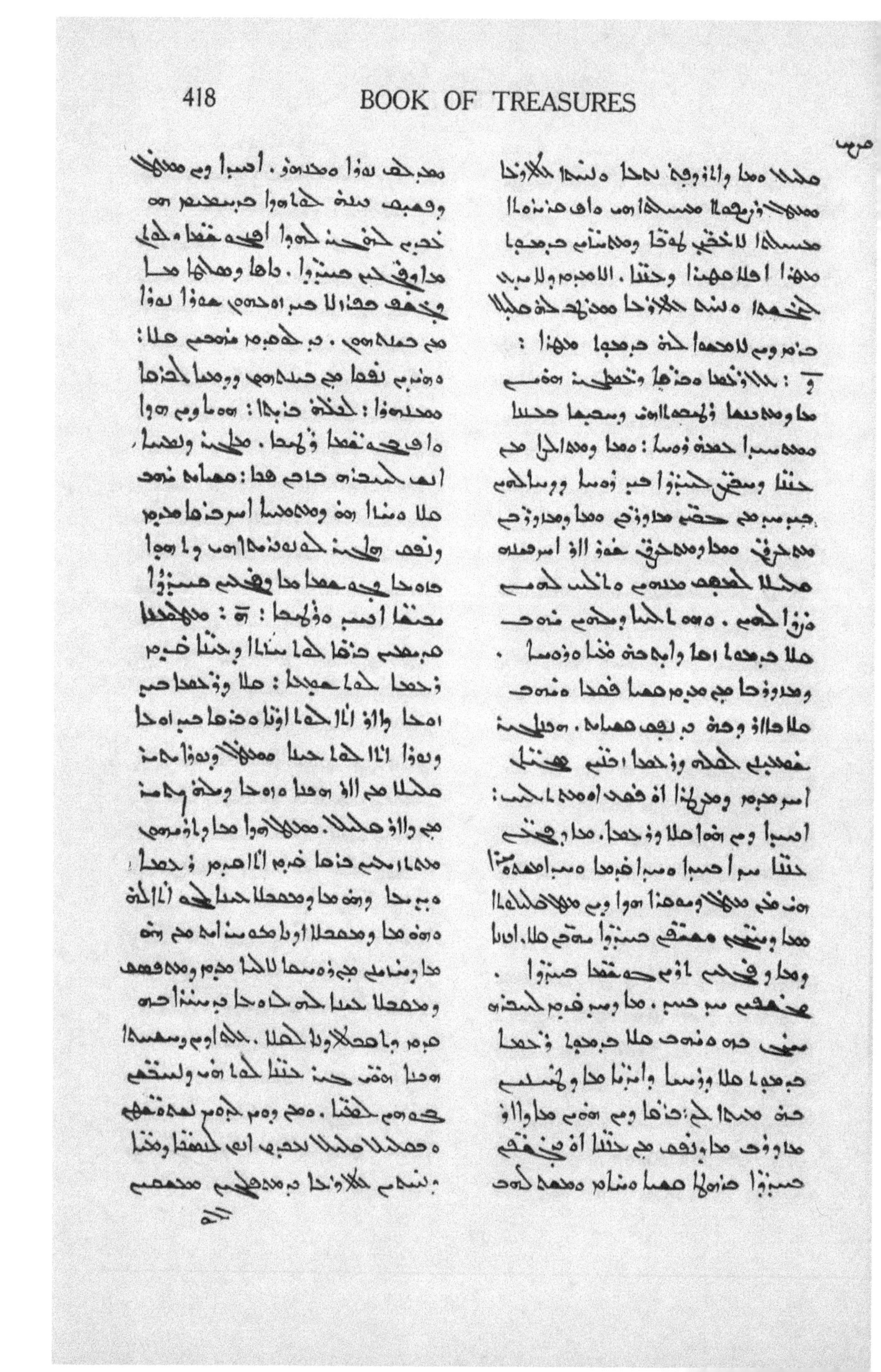

[illegible]

[illegible]

ܐܠܐ ܒܪ ܟܠ ܚܛܝܢ ܗܘܘ ܫܒܝܩܝܢ ܗܘܘ
ܠܦܬܓܡܗ ܕܢܝܐ ܘܠܐ ܡܫܬܘܠܝܢ ܗܘܘ
ܡܢܗܘܢ. ܘܦܘܪܥܢܐ ܕܒܥܝܢ ܗܘܘ ܡܢܗ ܥܠܝܗܝܢ
ܡܐ ܕܚܘܐ ܡܫܝܚܐ ܢܣܒܗ ܡܢܗ ܘܡܣܒܪܢܐ
ܠܗ ܐܘܪܚܐ ܘܡܚܘܐ ܒܗ ܠܗ. ܘܐܦ ܒܦܘܩ̈ܕܢܐ
ܗܠܝܢ ܕܦܩܕܝܢ ܗܘܘ ܡܢܗܘܢ ܒܚܛܗܐ ܫܒܩ.
ܐܝܟ ܕܡܚܘܐ ܗܘܐ ܡܫܝܚܐ ܕܠܐ ܐܝܬܘܗܝ ܗܘܐ ܠܢܡܘܣܐ ܐܣܝܪܐ
ܗܘܝܘ ܕܒܪܝܗ. ܡܕܝܢ ܠܗ ܫܠܝܛܘܬܐ ܥܠܝܗ
ܐܠܗܝ ܠܡܫܪܐ ܚܛ̈ܗܐ ܘܠܡܥܒܪ ܕܢܣܬܟܠ ܡܢܐ
ܕܣܠܩ ܥܠ ܟܠ ܡܛܠ ܗܘ ܕܥܠܝܢ ܘܥܠ ܟܠ
ܡܛܠ ܗܘܐ ܐܠܗܐ ܠܗܘ ܕܡܫܪܐ ܚܛܗܐ. ܘܗܘ ܡܢܐ
ܘܡܢ ܗܝܟܠ ܡܫܟܚܝܢܢ ܠܡܐܡܪ ܣܘܟܠܐ ܢܣܒ
ܘܒܗ ܚܝܠܐ ܕܒܪ ܐܢܫ ܕܐܬܐ ܕܢܚܐ ܘܢܚܙܘܢ ܠܡܒܗܢܬܗ
ܘܐܘܪܚܐ ܐܚܪܝܢ ܕܗܘ ܘܢܫܩܘܠ ܠܚܛ̈ܗܐ ܡܫܠܡܢܘܬܐ ܘܣܘܓܐܐ
ܘܐܝܟ ܒܗ ܢܬܒܪܘܢ ܘܢܚܒܫ ܠܦܩ̈ܕܐ ܘܚܘܒܗ
ܢܡܘܣܐ ܠܐܘܪܚܐ ܗܕܐ ܘܢܣܒ ܕܝܠܗ. ܐܠܐ ܡܩܒܠ
ܒܡܘܗܒܬܐ ܐܝܬܘܗܝ. ܗܠܝܢ ܚܛ̈ܗܐ ܡܢܐ ܕܢܫܩܘܠ
ܗܠܝܢ ܡܢ̈ܐ ܠܐ ܡܬܒܪܝܢ ܗܘܘ ܒܐܘܪܚܐ ܘܒܟܝܢ
ܗܘܘ ܢܡܘܣܐ ܚܝܪܐ ܘܦܠܐ ܘܡܬܗܦܟܝܢ ܗܘܘ
ܠܦܘܩ̈ܕܢܐ ܘܠܥܒ̈ܕܐ ܕܢܫܡܫ ܗܘܘ ܠܐܠܗܘܬܗ.
ܘܐܦ ܢܦܩܝܢ ܗܘܘ ܠܚܝܘܬܗ ܕܐܘܪܚܐ. ܘܐܠܗܘܬܗ
ܠܐ ܡܫܡܫܐ ܗܘܐ ܘܗܘ ܐܝܬܘܗܝ ܠܥܒܕܐ ܕܦܘ̈ܩܕܢܐ
ܒܒܪ ܢܫܐ ܘܐܦ ܠܐ ܡܫܡܫ ܗܘܐ ܠܢܡܘܣܐ
ܘܦܘܩ̈ܕܢܘܗܝ ܕܝܠܗ. ܘܐܦ ܒܗ ܐܠܗܐ ܠܡܫܝܚܐ
ܘܢܚܐ ܡܢ ܐܠܐ ܦܘܩܕܢܗ ܕܠܒܝܫܘܬܗ
ܠܒܪܢܫܐ ܡܢ ܟܠܗ ܘܟܠܗ ܘܢܬܕܡܐ ܒܗܘܢ
ܐܢܫܐ ܘܡܬܦܪܫܝܢ. ܐܢܫܘܬܐ ܕܐܝܬܝܗ ܐܣܝܪܐ
ܒܗܘܢ ܘܫܒܩܬ ܢܦܫܗ ܒܗ ܘܚܠܡܝܢ ܒܗ
ܘܢܦܠܐ ܒܗܘܢ ܡܢܬܐ ܘܢܫܩܘܠ ܒܗ ܐܝܟ ܐܣܝܪܐ
ܘܠܚܘܒ. ܐܠܐ ܕܒ̈ܢܝܢܫܐ ܡܫܬܟܚܐ

ܪܟ

ܘܡܬܚܠܦ ܡܫܬܢܐ ܘܐܚܪܢܝܬܐ. ܐܠܗ ܒܪܢ
ܠܐ ܬܘܕܝܬܐ ܕܦܩ̈ܕܢܐ ܘܫܘ̈ܪܝܐ ܕܡܫܝܚܐ ܘܡܫܡܫܢܐ
ܕܝܠܗ ܗܘܐ ܘܢܦܫܐ ܐܚܪܝܬܐ ܚܝܐ ܗܘܐ. ܘܐܦ
ܠܗ ܠܚܛܗܐ ܐܘܪܚܐ ܡܪܝܐ ܗܘܐ ܘܢܦܫܐ. ܐܠܐ ܠܒܪ ܦܘܩ̈ܕܢܐ
ܗܘܝܢܐ ܠܦܘ̈ܪܫܢܐ ܘܦܘܩܕܢܐ ܠܐ ܡܪܝܐ ܗܘܐ ܠܡܫܡܫܘ
ܡܫܡܫܢܐ ܡܟܝܠ ܠܗ ܘܡܛܠ ܐܘܪܚܐ ܢܡܘܣܐ
ܘܢܐܡܪ ܡܢ ܗܝܟܠ ܢܚܘܐ ܕܗܘܝܘ ܘܐܠܗܐ
ܕܐܘܪܚ ܡܫܝܚܐ ܠܗ ܚܝܠܐ ܘܫܘܠܛܢܐ ܐܘܪܚܐ:
ܐܣܝܪܐ ܘܡܢ ܐܝܬܘܗܝ ܡܫܝܚܐ ܘܡܫܬܥܒܕܢܘܬܐ
ܘܡܛܠ ܕܗܘܝܘ ܟܠܗ ܡܫܝܚܐ ܐܝܟܢܐ ܕܐܚܪܢܐ ܘܫܠܡܬ
ܘܗܘܝܐ ܕܟܢ̈ܫܐ ܡܫܬܟܚ. ܘܡܛܠ ܗܘܐ ܦܪܝܫ
ܕܐܝܬܘܗܝ ܒܟܠܗ ܦܓܪܐ ܘܡܛܠ ܗܢܐ ܠܚܘܪܢܐ
ܡܫܟܚܝܢܢ ܠܟܝܢܗ ܡܫܬܟܠܐ. ܕܒܗ ܐܝܬ
ܘܡܪܐ ܘܢܫܡܥ ܐܘܪܚܗ ܘܐܦ ܘܡܢܐ
ܘܡܛܠ ܗܘ ܣܝܡ. ܕܒܟܠܗܘܢ ܐܝܟ ܐܣܝܪܐ
ܡܪܢ ܟܢܫܝܐ ܒܡܐ ܕܗܘܐ ܘܡܫܬܥܒܕ ܠܟܠ
ܐܘܪܚܐ ܘܢܦܫܐ ܐܣܝܪܐ ܡܫܬܟܚܐ. ܘܡܛܠ
ܗܘܐ ܕܐܘܪܚܐ ܘܗܘܝܘ ܡܢ ܡܛܠ ܕܝܠܗ ܘܡܢ ܗܘܐ
ܘܐܝܬܘܗܝ ܠܗ ܒܗܘܢ ܠܗ ܡܫܬܥܒܕܐ ܡܢ
ܡܢ ܗܘܝܘ ܘܢܦܫܐ ܡܫܬܥܒܕܐ ܗܘܐ ܘܗܘܝܐ
ܗܝ ܡܢ ܗܘ ܘܡܛܠ ܐܝܟܢܐ ܐܘܪܚܐ ܘܢܦܫܗ
ܡܢ ܡܫܝܚܐ ܕܟܢ ܕܡܢ ܒܡܫܝܚ ܠܟܠܗܘܢ
ܐܦ ܘܒܗܘܢ ܘܦܘܩܕܢܐ ܕܡܫܝܚܐ ܐܚܪܝܢ. ܐܦ
ܐܠܐ ܒܪܢ. ܢܦܫ ܠܗ. ܕܒܪ ܒܪܢ ܢܫܗ
ܕܠܐ ܐܘܪܚܐ ܡܢܗܘܢ ܐܝܟ ܚܕܐ ܚܕܐ ܘܡܫܝܚܐ
ܟܐܘܪܚܐ ܠܐܚܪܢܐ ܘܡܫܬܠܛܐ ܒܢܦܫܗ ܘܐܘܪܚܐ
ܠܗܐ ܗܘ ܘܠܡܬܪܘܪ ܒܗ ܡܢ ܡܫܬܟܚܐ ܘܚܦܝܐ
ܘܟܠܐ ܗܘ ܘܠܡܩܡ ܒܢܦܫ ܠܗܘܪܐ. ܘܒܟܠܐ
ܘܡܛܠ ܡܢ ܕܒܢܦܫ ܢܦܫܐ ܠܐ ܐܘܪܚܐ ܐܣܝܪܐ
ܢܦܫ ܘܢܦܫܐ ܘܡܬܩܢ ܡܢܐ ܒܗܘܢ ܐܘܪܚܐ :

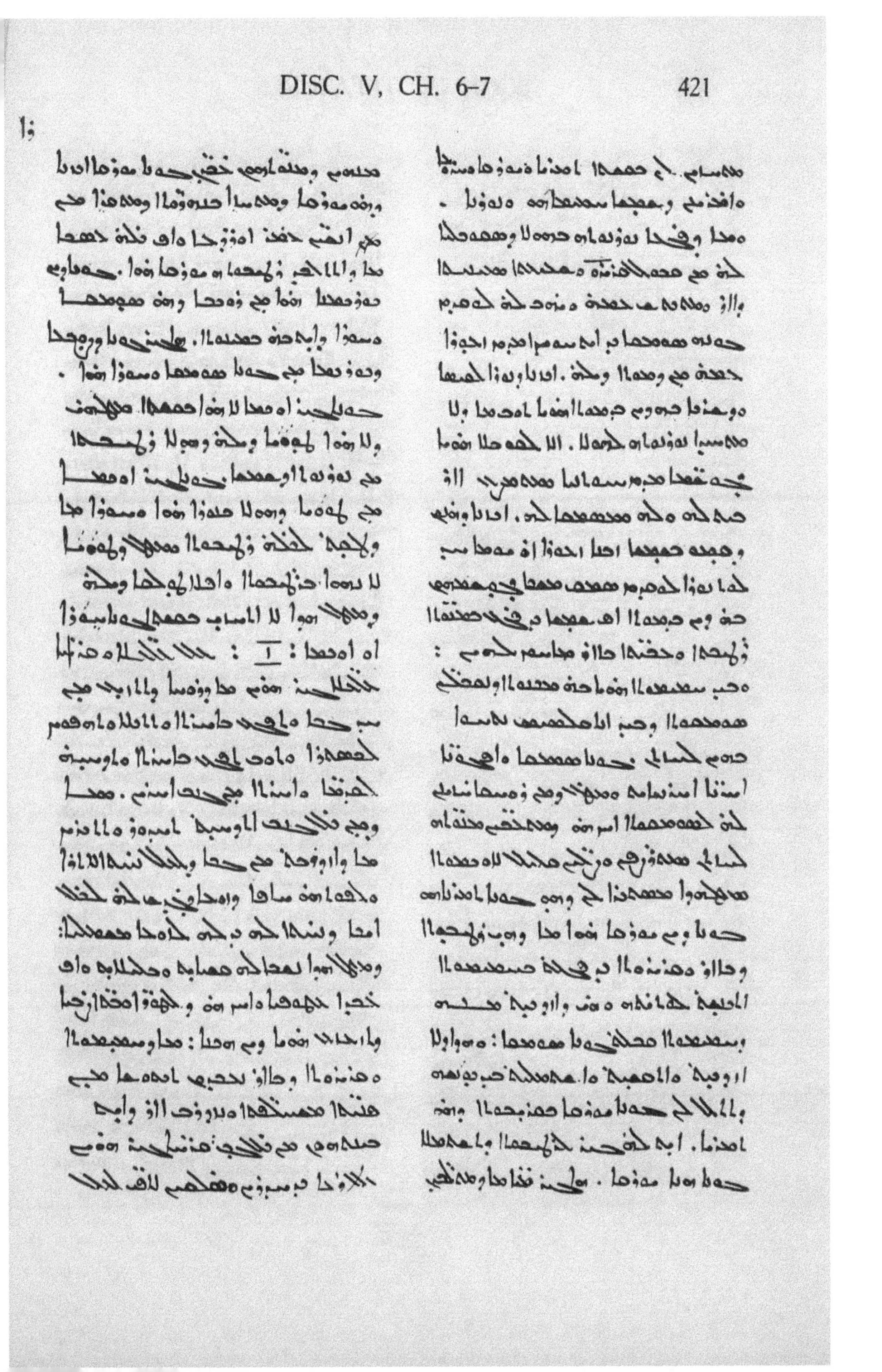

[illegible]

[illegible]

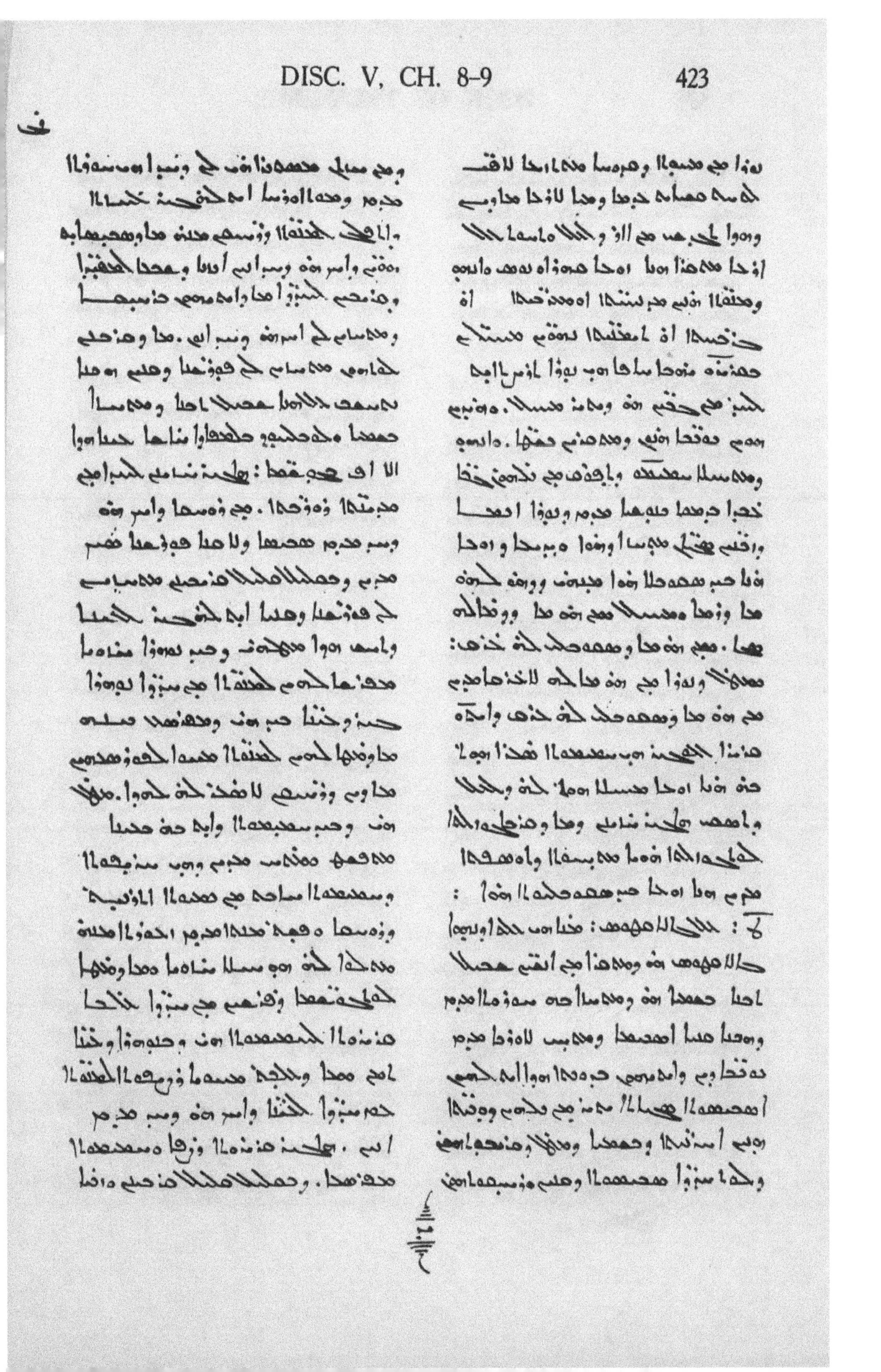

[illegible]

ܢܘ

ܗܘ ܡܫܡܫܢܘܬܐ ܗܢܝܢ ܡܥܕܢܐ ܘܐܦܢܝ ܐܝܟ
ܠܥܢ̈ܝܐ ܘܢܒܥ ܐܣܝܘܬܐ ܡܢ ܥܩ̈ܪܐ ܘܠܥܠ̈ܥܐ
ܘܒܝܫ ܡܛܠ ܕܫܟܝܚܘܬܐ ܘܫܪܝܪܘܬܐ ܠܐ ܩܢܝܢ
ܠܒܪ̈ܢܫܐ ܗܢܘܢ ܕܡܢ ܡܛܠܬܗ ܟܠܗܝܢ ܘܫܡܫܘܬܐ
ܩܢܝܢ ܠܒܪܢܫܐ ܘܡܬܕܢܝܢ ܠܥܢ̈ܝܐ ܐܝܟ ܕܙܕܩܐ
ܩ̄ : ܥܠ ܐܣܝܘܬܐ ܕܥܝܢ̈ܐ ܘܕܫܘܪܐ :
ܡܬܚܫܒܐ ܗܘܬ ܐܣܝܘܬܐ ܠܡܫܘܪܐ ܐܦ ܠܥܝܢܐ
ܐܣܝܘܬܐ ܕܫܘܪܐ ܗܘܬ ܡܢܐ ܕܢܗܘܐ ܐܠܐ ܕܩܝܡܬܗ
ܡܫܡܫܐ ܕܟܠܝ ܘܡܢ ܟܕ ܐܦܪܘܩ ܘܡܫܡܫܘ
ܡܬܚܫܒܐ ܗܘܬ ܕܒܥܝܢ ܗܘ ܒܝܘܡܐ ܘܫܘܪܐ ܠܐ ܩܦ̈
ܠܗܠܝܢ ܣܘܓܐ ܐܢܫܐ ܩܢܝܢ ܠܗ ܒܥܒܕܝܢ ܠܗܘܢ
ܥܠܘܗܝ ܘܐܘܪܚܗ ܕܡܠܟܗ ܠܡܫܪܝܘܬܐ ܗܘ
ܥܒܝܕܐ ܡܢ ܩܕܡ ܫܢ̈ܝܐ ܣܓܝܐܬܐ. ܘܡܠܐ
ܘܐܠܘܡܝܬ ܡܬܚܙܝܐ ܡܢܝܢܬܗ ܣܝܕܪܘܬܗ
ܡܫܡܫܐ ܠܗ ܐܣܝܘܬܐ ܘܐܝܟ ܗܘ ܕܡܚܪܙܩ
ܐܬܐܡܪ ܩܢܝܢ ܠܗ ܐܝܟ ܠܒܪܐ ܕܗܘܐ
ܗܘ ܥܒܕܐ ܡܢ ܐܪܥܐ ܘܡܠܐ ܐܝܠܐ ܘܡܠܝܢ ܢܒܥܝܢ
ܚܝ̈ܠܐ ܕܪܡܐ ܘܩܒܠܐ ܗܘ ܐܣܝܘܬܐ. ܘܡܠܝ ܠܥܡܡ
ܕܪܡܐ ܗܘ ܐܣܝܘܬܐ ܩܢܝܢ ܠܗ ܡܫܡܠܝ ܠܗܘܢ
ܟܕ ܐܠܐ ܕܒܟܠ ܘܡܠܐ ܐܣܝܘܬܐ ܡܢ ܩܕܡ. ܡܛܠ
ܘܠܗ ܟܕ ܐܠܐ ܠܐ ܗܘܐ ܗܘܐ ܘܒܝܕܗ ܗܘܐ
ܘܡܛܠ ܟܕ ܐܘܬܐ ܘܐܦ ܗܘܐ ܡܬܚܫܒ ܘܠܗ
ܟܕ ܐܘܬܐ ܥܠ ܪܝܫܐ ܐܝܟܢܐ ܘܟܕ ܐܘܬܐ. ܐܘ
ܟܢ ܡܢܐ ܘܢܗܘܐ ܐܠܐ ܠܟܒܪܐ ܘܟܕ ܡܢܐ
ܘܫܒܝܩ ܐܢܫ ܘܟܕ ܣܝܡܐ ܘܢܣܒ ܘܝܕܥ ܐܦ ܡܫܡܫ ܡܛܠ
ܗܘ ܘܡܫܡܫܢܐ ܒܪܢܫܐ ܟܕ ܐܝܬ ܘܡܫܐ ܐܣܝܘܬܐ
ܒܥܐ ܘܢܬܠ ܗܘ ܒܝܘܡܐ ܘܒܬܪܗ ܘܒܫܘܡܠܝ
ܥܒܕܐ ܩܢܝܢ ܠܗ ܣܝܕܪܘܬܗ. ܡܛܠ
ܘܣܝܕܪܘܬܗ ܘܡܥܒܕ ܐܢܫܐ ܡܫܡܫܐ ܠܗ
ܒܝܘܡܐ ܘܥܒܕܐ ܠܡܫܡܠܝܘ ܘܡܛܠ ܗܘܐ

ܡܢܐ ܬܒܥ: ܟܕ ܥܒܘܕܐ ܡܫܡܫ ܠܗ .
ܣܝܕܪܘܬܗ ܩܢܝܢ ܠܗ ܘܡܠܐ ܐܣܝܘܬܐ
ܥܒܕܐ ܕܝܢ ܘܡܕܒܪ ܡܫܡܫܐ ܠܗ ܘܡܕܒܪܐ
ܘܐܣܝܘܬܐ ܘܥܒܕܐ ܡܢ ܟܕ ܥܒܘܕܐ ܡܢ ܩܕܡ
ܘܐܝܟ ܕܐܦ : ܩܐ : ܥܠ ܗܘܐ ܕܥܡܠܐ
ܘܕܘܟܝܐ ܕܡܢ ܐܣܛܘܟܣܐ ܘܥܠܠܬܐ
ܕܟܠܗܘܢ: ܐܦܢܝ ܒܪ ܥܒܘܕܐ ܘܡܕܠܝܢ
ܘܐܦܢܝ ܗܘܝܐ ܣܝܡܬܐ ܐܝܟ ܠܥܡܡ
ܠܐܣܛܘܟܣܐ ܘܕܒܝܢ ܡܢ ܚܕܕܐ ܘܐܦܢܝ
ܗܘܝܐ ܘܢܦܠܝܢ ܠܚܕܕܐ ܣܝܡܬܐ ܕܝܢ
ܡܬܩܝܡܐ ܠܗܪܟܐ ܘܗܠܝܢ ܠܥܠܡܐ ܘܢܦܠܝܢ
ܕܝܢ ܠܚܕܕܐ: ܡܬܩܝܡܐ ܘܠܗܠܝܢ ܐܠܗܝܐ
ܘܗܪܟܐ ܠܗܠܝܢ ܘܠܥܠܡܐ: ܘܡܢܝܢ ܐܝܟܢܐ
ܗܘܐ ܡܬܚܙܐ ܗܘܐ ܕܠܘ ܒܠܚܘܕ ܟܕ ܒܪܢܫܐ
ܘܗܘܝܐ ܣܝܡܬܐ ܘܐܣܛܘ. ܘܡܛܠ ܗܘܐ ܗܘ
ܟܕ ܗܘܐ ܘܡܬܠܐ. ܘܐܝܬܝܗ ܐܝܟܢܐ
ܘܕܝܢ ܕܒܕܒܪ ܐܠܐ ܐܝܟ ܐܣܛܘܟܣܐ
ܥܠܡܐ ܟܠ ܗܘ ܘܠܗ ܕܟܠܗܘܢ ܘܗܘܐ ܡܕܡ
ܗܢܘܢ ܐܪܒܥ ܗܘܝܐ ܣܝܡܬܐ ܘܢܦܠܝܢ
ܐܠܝܢ ܘܗܢܘܢ ܘܣܝܡܬܐ ܡܛܠ ܕܡܢܗܘܢ
ܘܠܚܕܕܐ ܘܗܘ ܐܦܢܝܢ ܡܬܚܫܠ ܐܝܬ ܐܦ ܗܢܘܢ
ܐܘܦ ܘܢܦܠܝܢ ܐܠܘܢ ܥܠ ܚܕܕܐ. ܡܛܠ
ܘܡܠܬܐ ܡܬܚܫܒ ܒܥܒܕ ܕܡܢܐ ܡܢ ܩܕܡ
ܐܒܪܘܬܐ ܡܢ ܗܘܝܐ ܘܗܢܘܢ ܣܝܡܬܐ
ܕܪܘܚܐ ܘܗܘܝܐ ܗܢܘܢ ܘܢܦܠܝܢ ܗܘܐ
ܘܚܝܘܬܐ ܡܢ ܡܢ ܗܢܘܢ ܗܘܝܐ ܘܢܦܠܝܢ
ܕܪܘܚܐ ܘܗܘܝܐ ܗܢܘܢ ܘܣܝܡܬܐ ܠܗܘܢ
ܘܗܘܐ ܐܝܟ ܪܘܚܐ ܠܚܘܝܐ. ܡܢ ܗܢܘܢ
ܣܝܡܬܐ. ܘܐܝܟ ܪܘܚܐ ܘܡܠܝܐ ܡܢ
ܗܢܘܢ ܘܢܦܠܝܢ ܣܝܡܬܐ ܕܪܘܚܐ
ܘܕܝܢ

ܕܒܥܠܡܐ ܡܢ ܗܢܘܢ ܕܡܫܬܡܠܝܢ ܡܫܡܫܢܘܬܐ
ܡܢ ܗܕܐ ܕܥܠܡܐ: ܡܢ ܗܢܘܢ ܕܝܢ ܕܡܫܬܡܠܝܢ
ܘ ܫܡܫ ܗܘܐ ܡܫܝܚܢ ܫܪܝܐ ܘܡܫܠܡܢܐ ܕܟܠܐ
ܡܕܡ ܕܒܥܠܡܐ ܗ̄ܘ ܡܢ ܗܢܘܢ ܫܘܡܠܝܐ ܘܐܦ
ܗܘܐ ܚܒܝܒܘܬܐ: ܘܠܐ ܡܬܦܪܫ ܐܝܟ
ܗܘܢ. ܡܛܠ ܕܡܢ ܦܢܝܢ ܠܒܪܢܫܐ ܘܫܡܫܘܬܐ
ܕܗܘܐ ܟܠܗ ܥܠܡܐ ܘܪܚܡܐ ܪܒܪܒܐ ܡܫܬܡܥ ܗܘܐ
ܫܡܫܘܬܐ ܘܡܫܡܫܢܘܬܐ ܡܫܬܐܠ ܕܠܗ ܡܫܝܚܐ
ܐܝܟ ܡܫܬܐܠ ܫܡܫܘܬܐ ܕܗܕܐ ܡܫܡܫܢܐ
ܕܫܡܫ: ܕܟܠܗܘܢ ܡܫܬܐܠ ܠܗ ܫܡܫܐ
ܗܘܝܐ ܐܦ ܒܗܝܠܐ ܕܫܡܫܐ ܗܘܬ ܗܘܐ
ܡܫܡܫܢܐ ܕܡܢ ܫܡܫܘܬܐ ܘܡܫܡܫܢܘܬܐ
ܘܫܡܫܗ ܘܡܫܡܫܗ. ܟܕ ܕܝܢ ܡܫܬܡܠܝܢ
ܡܢ ܫܡܫܘܬܐ ܘܗܕܐ ܕܡܫܡܫܝܢ ܡܫܡܫܢܝܢ
ܡܥܒܕܝܢ ܘܡܫܬܐܠܝܢ ܠܐ ܟܕ ܚܫܒܝܢ ܚܫܒܐ
ܘܡܫܡܫܢܐ ܕܐܘܫܦܢܐ ܘܡܫܝܚܐ ܘܫܡܫܐ
ܘܫܡܫܐ ܗܘܐ ܘܗܘܝܐ ܗܘ ܡܫܡܫ ܘܡܫܡܫܢ
ܡܫܡܫܢܐ ܐܝܟ ܐܚܝܐ ܕܫܡܫܐ ܘܐܘܡܢܐ
ܡܫܡܫܢ ܫܡܫܐ ܐܝܟ ܐܚܝܐ ܘܫܡܫܢܐ
ܘܐܦ ܐܚܝܐ ܦܠܚܐ ܘܐܚܝܐ ܡܫܡܫܢܐ ܘܐܦ ܥܡܠܐ
ܘܟܠܗ ܫܡܫ ܒܫܡܫܘܬܐ. ܡܫܡܫܢ ܫܡܫܢܐ ܡܫܡܫܢ
ܫܡܫܐ ܘܫܡܫܐ: ܡܫܬܐܠ ܘܗܢܘܢ ܕܠܐ ܫܡܫ
ܘܗܢܘܢ ܕܠܫܡܫ ܐܝܟ ܫܡܫܢ ܐܫܡܫܢܐ ܘܗܢܘܢ
ܡܫܡܫܢ ܗܘܬ ܘܫܡܫܐ ܘܐܫܡܫܘܬܐ ܘܫܡܫܐ
ܗܢܘܢ ܕܡܢ ܠܐ ܡܫܡܫ ܫܡܫܐ ܡܫܡܫ ܟܕ ܫܡܫ
ܘܡܫܡܫܢ ܫܡܫܐ ܘܡܫܡܫ ܗܘ ܕܫܡܫ ܫܡܫܢܐ
ܫܡܫ ܫܡܫܐ ܠܫܡܫܘܬܐ ܘܐܝܬ ܗܘܬ ܡܫܡܫܢܘܬܐ ܘܐܘܡܢܘܬܐ
ܘܡܫܡܫ ܕܐܫܡܫܢܐ ܕܫܡ ܫܡܫܐ ܘܫܡܫܢܘܬܗܘܢ
ܫܡܫܢܐ ܠܗ ܠܫܡܫܐ ܫܡܫ ܫܡܫܐ ܘܫܡܫܐ
ܐܠܗ̄ܐ ܗܘ ܗܘ ܡܫܝܚܐ ܘܐܠܗ̄ܐ ܘܠܐ ܡܫܬܡܠܝܢ

ܡܫܡܫܢܘܬܐ ܫܪܝܪܐ ܘܡܫܒܚܬܐ ܘܡܫܡܫܢܘܬܐ
ܘܒܛܝܠܬܐ ܕܗܘ ܡܫܝܚܐ ܘܐܠܗ̄ܐ: ܡܢܗ ܐܝܟ
ܘ ܫܡܫܘܬܐ ܘܡܫܡܫܢܘܬܐ ܘܐܫܡ̄ܫܗ ܠܫܡܫܘܬܐ
ܘܗܘ ܡܫܝܚܐ ܘܐܠܗ̄ܐ ܘܠܐ ܡܫܬܡܠܝܢ ܡܫܡܫܢܘܬܐ ܫܪܝܪܐ
ܘܒܛܝܠܐ . ܘܗܘ ܡܫܝܚܐ ܘܐܠܗ̄ܐ : ܘܫܡܫܐ ܘܫܡܫܐ
ܘܫܡܫܐ ܕܫܡܫܐ ܠܫܡܫܢ ܫܡܫܐ ܘܐܫܡܫܗܘܢ
ܘܠܐ ܡܫܡܫܢ ܫܡܫܐ . ܐܦ ܐܚܝܕ ܫܡܫܘܬܐ ܘܫܡܫܘܬܐ
ܘܐܫܡܫܘܬܐ ܗܘܬ ܠܫܡܫܢ ܗܘܐ . ܘܐܚܝܕܝܢ
ܘܗܘܪܐ ܗܘܐ ܠܫܡܫܐ ܘܗܘܘܫܡܫ ܘܐܠܗ̄ܐ ܘܠܐ
ܠܫܡܫܐ ܫܡܫܐ . ܟܕ ܗܘ ܕܝܢ ܠܫܡܫܐ ܠܗ
ܫܡܫܐ ܗܘ ܡܫܬܐܠ ܡܫܡܫܢܐ ܘܡܫܡܫܢܘܬܐ
ܫܡܫܐ ܡܢ ܒܫܡܫܐ ܘܗܢܘܢ ܐܦ ܫܡܫܐ ܐܘܡܢܐ
ܐܚܝܐ ܕܝܢ ܒܫܡܫܐ ܗܘܐ . ܡܫܬܐܠ ܗܘܝܐ
ܐܦ ܐܝܢ . ܐܦ ܐܝܬ ܡܫܡܫܢܘܬܐ ܐܝܟ ܡܢ ܘܠܐ
ܕܫܡܫܐ ܐܦ ܫܡܫܐ ܗܘܐ ܐܠܐ ܡܫܡܫܢܘܬܐ
ܫܡܫܐ ܕܡܫܡܫܢܐ ܡܫܡܫܢ ܫܡܫܘܬܐ
ܘܫܡܫܘܬܐ . ܐܚܝܕܝܢ ܘܗܘܐ ܠܗܘܐ ܡܫܡܫܢܐ
ܡܫܡܫܐ ܕܝܢ ܐܡܫܡܫܐ ܐܚܝܐ ܘܡܫܡܫܢ ܫܡ
ܫܡܫܐ ܡܢ ܫܡܫܐ ܘܗܘ ܘܗܘܐ ܠܐ ܡܫܡܫܢܐ
ܡܫܡܫܢܗ ܘܫܡܫܐ ܘܐܚܝܐ ܘܠܐ ܡܫܡܫܢܐ
ܕܗܘܐ ܘܫܡܫܐ ܠܗ ܗܘܝܐ ܐܝܟ ܗܘ
ܫܡܫܐ ܘܫܡܫܢ ܡܫܡܫ ܠܗ ܠܫܡܫ ܫܡܫܐ
ܘܟܕ ܫܡܫܐ ܠܗܘܝܐ ܟܠܠܗ ܘܐܦ ܐܚܝܐ
ܘܟܠܐ ܡܫܡܫܢܘܬܐ ܘܐܚܝܐ ܘܫܡܫ ܘܗܘܐ ܗܘ
ܘܡܫܡܫܢ ܠܗ ܫܡܫܐ ܟܠܗܘܢ ܗܘܢ ܘܐܚܝܐ
ܠܗܘܢ ܘܗܘܐ ܘܠܐ ܡܫܡܫ ܐܠܐ ܫܡܫܐ
ܗܘܝܐ ܘܗܘܝܐ ܘܟܠܗ ܡܫܬܐܠ ܘܐܚܪܢ
ܠܗ ܡܫܡܫܢܘܬܐ ܫܡܫܐ ܡܫܡܫܢܘܬܐ ܘܫܡܫܢ
ܠܗ ܗܘ ܫܡܫܗ ܫܡܫܢܘܬܐ ܘܫܡܫܢܐ
ܘܫܡܫܘܬܐ ܘܫܡܫܐ ܘܫܡܫܐ ܀ ܀ ܀ ܀

[illegible]

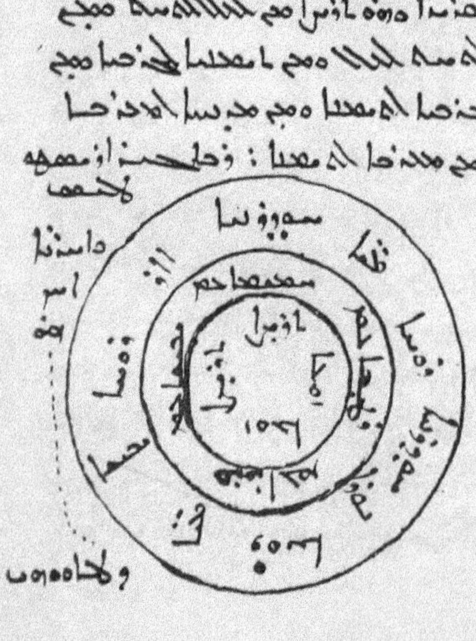

ܘܙ

ܘܡܢ ܚܘܝܕ̈ܝܐ ܟܘܟܒܐ ܗܘܝܐ ܘܠܟܠܗܘܢ
ܘܐܘܪܚܐ ܒܗܘܢ ܘܦܫܝܛܘܬ ܐܣܟܡ̈ܐ
ܒܘܢܐ . ܚܘܒܗ ܠܐܘܡܪܐ ܘܒܟܐ ܐܣܟܡ̈ܐ
ܒܠܐ ܘܟܠܗܝܢ ܒܗܘ ܕܝܢ ܘܠܟܠܠܐ ܘܒܐ
ܚܘܝܕܐ ܡܫܠܡܢܐ ܐܦ ܡܣܬܟܠܢܘܬܐ
ܒܟܒܐ ܗܘ ܘܠܟܠܣܡܝܐ ܘܐܝܟܐ ܠܐܘܡܪܐ
ܗܢܐ ܒܠܐ ܘܟܠܗܝܢ ܘܡܫܟܚܘ ܘܐܠܗܐ ܠܡܫܒܚܘܬܐ
ܘܫܒܚܐ ܐܝܟܘ ܘܐܘܣܝܐ ܘܟܝܢܐ ܠܟܠܗܘܢ
ܘܠܟܠܗܝܢ ܘܟܕ ܐܘܟܐ ܗܘ ܚܘܝܕܐ ܡܫܟܚܐ
ܘܠܗ ܗܘܝܐ ܗܘ . ܡܫܬܟܠܢܘܬܐ ܠܐ ܐܝܬܝܗ ܐܠܐ
ܡܢ ܡܫܟܚܐ ܗܘ ܡܫܬܟܠܐ ܘܐܠܗܐ : ܘܠܡܫܟܚܘ
ܐܣܟܡܐ ܘܟܝܢܐ ܘܢܟܒ ܡܫܠܡܢܘܬ ܡܫܟܚܘ
ܗܘܦܟܐ ܘܚܘܝܕܐ ܚܕ ܐܘܟܐ ܡܠܐ ܘܡܢ
ܟܡܫܘܬܐ ܘܗܘܢ ܗܘ ܘܠܗ ܟܗ ܡܫܪܝܐ
ܐܘ ܡܘܠܕܐ ܐܘ ܐܣܟܡܐ ܘܟܝܢܐ . ܘܐܦ
ܡܢ ܡܫܟܚܐ ܗܘ ܘܟܕ ܡܬܡܟܟ ܘܠܟܠܗ . ܠܐ
ܠܡܫܟܚܘ ܢܗܘܐ ܡܢ ܡܫܬܟܢ ܘܡܬܒܪܝܢ
ܡܘܫܚܬܐ . ܘܡܢ ܗܘܐ ܠܐ ܡܬܚܙܐ ܘܠܐ
ܘܠܐ ܡܘܫܚܬܐ ܗܘ ܐܦܠܐ ܗܘܝܐ . ܡܢ ܗܘܐ
ܡܬܝܕܥ ܘܐܣܟܡ̈ܐ ܗܘ ܡܫܟܚܐ . ܐܡܪܬ
ܒܗ ܠܗܘܐ ܘܠܐ ܡܬܚܫܒܐ ܗܘܐ ܠܐܘܟܐ
ܟܡܘܬܐ ܘܡܠܝܐ ܘܕܝܠܗ ܐܠܗܘܬܐ ܗܘ
ܘܐܘܡܪܐ ܠܟܠܗܝܢ ܡܬܝܕܥܢܘܬܐ ܡܢ ܡܫܟܚܐ
ܘܠܐ ܡܬܒܥܐ ܘܗܘܝܐ ܘܡܢ ܟܠܐ ܘܒܗ . ܒܗܘ
ܘܐܡܪ ܡܫܟܚܐ ܘܟܡܘܬܐ ܕܒܟܠܗ ܐܠܗܐ ܫܡܝܐ
ܘܐܘܟܐ ܘܠܗ ܘܠܐ ܟܡܘܬܐ ܒܟܒ ܐܝܟ ܡܢ
ܘܡܢ ܐܠܐ ܟܡܘܬܐ ܗܘܝܘ ܡܢ ܐܠܗܐ ܘܡܢ
ܠܐ ܡܬܝܕܥ ܗܘܝܘ ܐܝܟܐ ܘܚܘܝܕܐ ܡܢ ܚܢܢ
ܘܡܫܬܟܢܢ ܟܡܘܬܐ ܘܡܠܝܢ : ܐܡܝܢܐܝܬ
ܘܐܣܟܡܗ ܗܘ ܡܫܟܚܐ ܟܡܐ ܘܐܦ

ܒܗ ܡܫܟܚܐ ܘܠܗܘܢ ܐܘܟ ܘܟܠܗܝܢ ܐܣܟܡ̈ܐ
ܡܫܟܚ ܐܝܟ ܐܘܟܐ ܘܡܫܬܟܚܐ ܡܢܗ
ܠܟܠܗܘܢ ܘܟܠܗܝܢ ܘܡܫܬܟܢܢ ܘܠܟܠܠܐ ܒܗܢ
ܘܐܣܟܡ̈ܐ ܘܒܗܘܢ ܒܡܫܟܚܐ . ܒܡܪ ܕܝܢ
ܒܗ ܠܗ ܘܟܠܗܝܢ ܦܫܛ ܐܝܟ ܡܫܬܟܚܐ . ܒ
ܗܢܐ ܘܠܐܢܫܐ ܘܡܠܐܟܐ ܘܠܐܢܫܐ ܡܫܟܚܐ . ܘܐܦ
ܠܗ ܟܒܐ ܘܡܬܚܙܐ ܡܫܟܚ ܘܦܫܛ ܠܟܠܗܘܢ
ܘܠܟܠܠܐ ܡܫܟܚܘ ܗܝ ܠܐ ܦܫܛܘ ܐܘ ܐܡܪ ܠܟܠܗܘܢ
ܘܡܬܟܒܐ ܗܘ ܐܘ ܡܫܟܚܐ ܐܠܐ ܐܡܪ ܐܣܟܡ̈ܐ
ܫܡܝܢܐ ܡܫܟܚܐ ܒܠܐܢܫܐ ܐܘ ܦܫܛܘ ܘܡܐ
ܒܗܘܐ ܠܐܢܫܐ ܗܝܕܝܢ ܘܡܬܚܘܝܢ ܒܫܡܝܐ ܐܣܟܡ̈ܐ
ܒܗܢܐ ܒܠܐ ܘܟܠܗܝܢ ܐܘܟܐ ܗܝܢ ܘܫܢܝ ܡܬܚܘܝܢ
ܒܗܘܢ ܘܫܡܝܢܐ ܘܐܪܥܐ ܘܡܐ ܘܐܠܗܐ ܠܟܘܢ
ܘܦܫܛܘ ܢܘܗܪܐ ܒܡܫܟܚܐ ܘܡܠܝܐ ܡܫܬܡܫ ܒܡܫܟܚܐ
ܘܐܘܟܐ ܘܫܡܝܢܐ ܘܦܫܛܐ ܘܠܐܟܐ ܘܡܫܟܚܐ .
ܒܟܠܗ ܠܗܘܢ ܡܫܟܚܐ ܘܗܘܝܢ ܠܗ ܐܣܟܡ̈ܐ
ܘܐܡܪܝܢ ܘܗܘܝܐ ܗܘ ܘܡܬܚܫܒ ܘܡܫܬ
ܡܫܬܟܚ ܡܫܟܚܐ ܦܫܝܛܐ ܘܐܦ ܗܘ ܗܘܐ ܐܡܪ
ܒܗܢܐ ܐܣܟܡ̈ܐ ܫܡܝܢܐ ܘܡܠܗ . ܠܡܫܟܚ
ܡܫܟܚܐ ܘܒܟܝ̈ܐ ܐܣܟܡ̈ܐ ܗܘ ܫܡܝܢܐ ܡܫܬܡܫܐ
ܗܘܐ ܘܡܫܬܡܫ ܡܢ ܐܘܟܐ ܐܣܟܡ̈ܐ ܘܟܠܗܝܢ
ܡܕܡ ܦܫܛ ܠܗ ܐܟܡܐ ܘܦܫܛ ܠܡܫܟܚܐ ܡܢ
ܗܢܐ ܡܠܐܟܐ ܡܢ ܡܫܟܚܐ ܒܗܢܐ ܘܟܠܗܝܢ :
ܐܡܝܢܐܝܬ ܘܡܫܟܚܐ ܐܘܟܐ ܚܘܝܕܐ ܡܠܐ ܘܦܫܛ
ܘܟܠܗܝܢ ܠܐܘܡܪܐ ܐܣܟܡ̈ܐ ܗܘ ܫܡܝܢܐ :
ܡܫܬܟܚܢ ܠܡܫܟܚܐ ܗܘܝܐ ܘܟܡܐ ܐܝܟܘ
ܐܣܟܡܐ ܘܡܫܟܚܐ ܡܫܬܟܚܐ ܚܘܝܕܐ ܘܐܢܫܐ
ܘܟܠܐ ܘܗܘ ܐܡܪܝܢ ܘܐܣܟܡܐ ܘܡܫܟܚܐ ܡܫܬܟܚܐ
ܚܘܝܕܐ ܠܗ ܒܠܚܘܕ ܡܫܟܚܐ ܐܝܟܘ ܐܠܐ ܐܦ
ܡܫܬܦܐ ܒܗܘܢܐ ܘܟܠܗܝܢ . ܐܡܠܟܘܢ ܒܡܢ

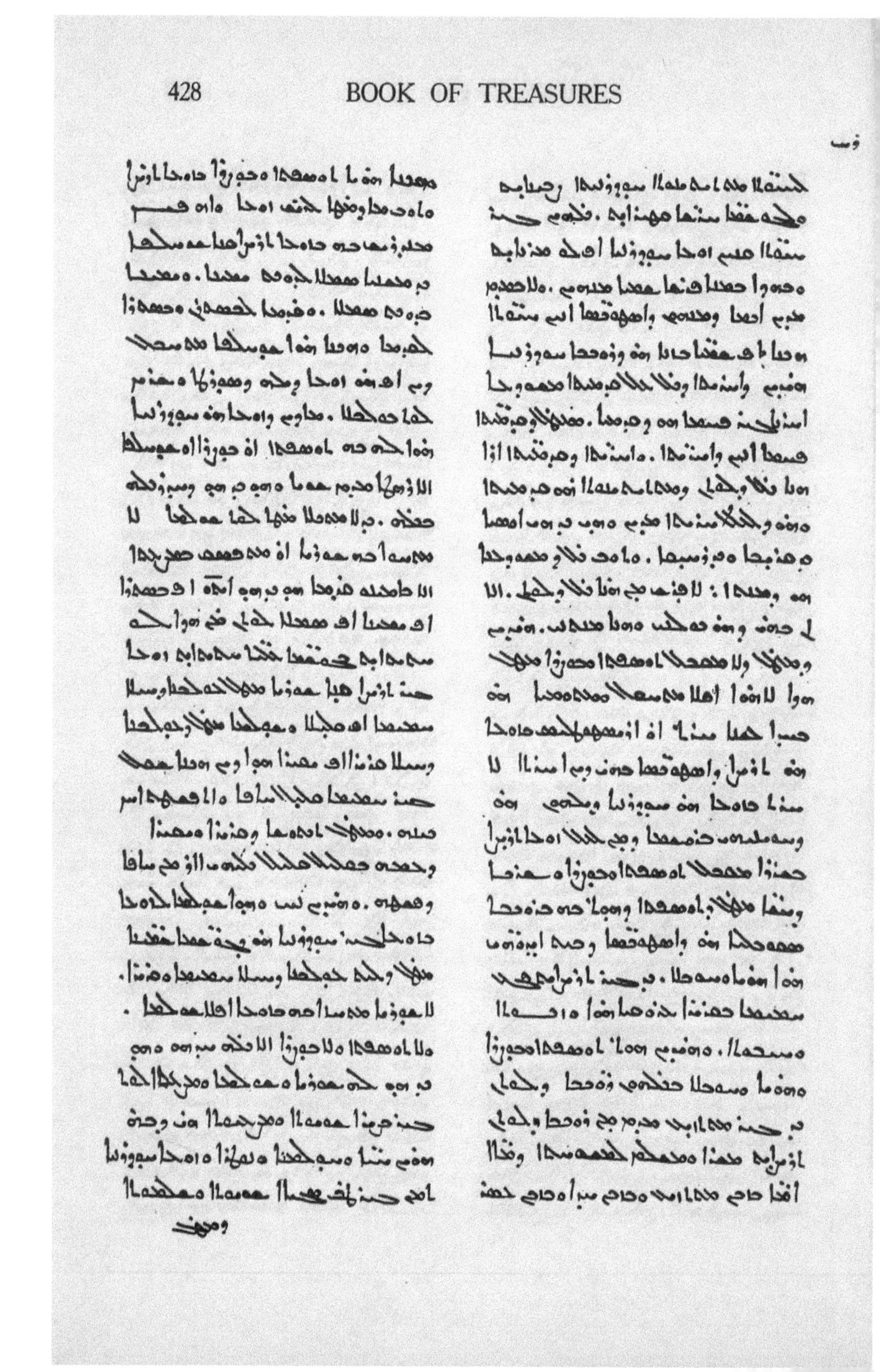

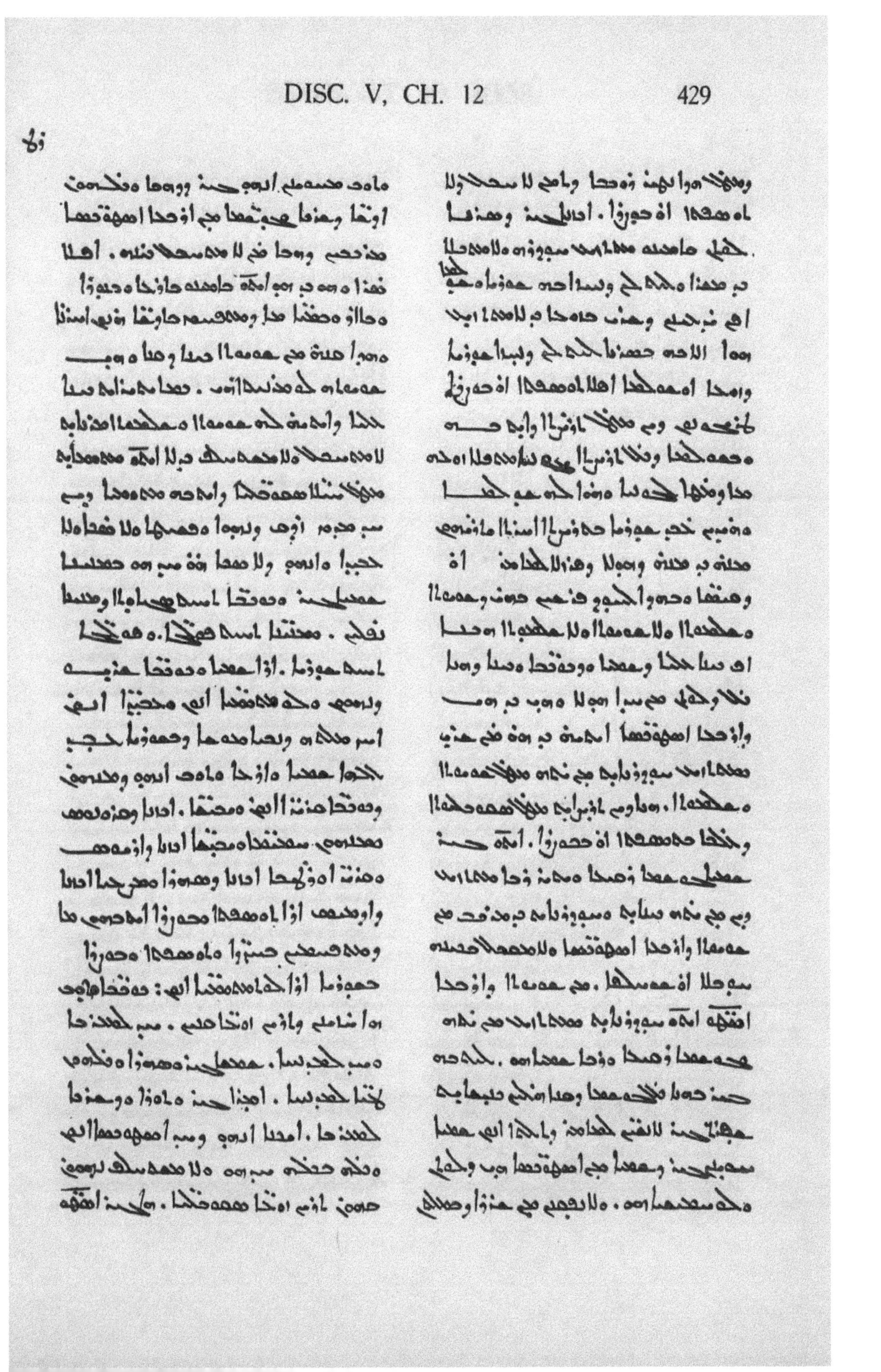

ܩܠ

[illegible]

ܘܡ

ܘܟܠܗܝܢ ܢܟܝܢ ܡܬܝܠܕܢ ܚܕ ܐܡܬܝ ܐܝܬ ܠܗ ܢܘܗܪܐ
ܠܟܠܠ ܘܐܦ ܐܐܪ ܘܠܐܪܥܐ ܘܫܡܝܐ ܠܟܠܗܘܢ. ܒ
ܝܕ ܐܝܟ ܡܐܡܪ ܐܣܛܘܟ̈ܣܐ ܡܕܒ̈ܪܐ ܡܕܡ ܡܢܗܝܢ
ܘܐܪܡ ܐܢ̈ܫܐ ܡܫܬܘܕ̈ܥܐ ܐܝܟ ܠܗܘܢ ܠܗܘܢ̈ܝܐ.
ܠ ܪܓܐ ܐܝܟ ܓܫܡܐ ܦܫܝܛܐ ܕܫܡܫ ܫܡܝܐ ܐܦܝܐ
ܡܫܘܚ ܘܐܘܪ̈ܚܐ ܡܫܬܘܕ̈ܥܐ ܕܐܪܥܐ ܡܢܘܢ
ܐܣܛܘܟ̈ܣܐ ܐܝܟ ܡܢܘܢ. ܐܪܐ ܡܢ ܐܪܥܐ ܐܣܛܘܟ̈ܣܐ
ܡܕܝܢܝܢ. ܦܪܝܢ ܠܗ ܡܫܡ̈ܫܢܐ ܐܝܟ ܐܠܐ ܐܪܥܐ ܘܡܢ
ܠܐ ܡܕܡ ܐܝܬܐ ܘܣܘܡܟ ܡܢ ܟܠܠ ܘܐܦ ܐܝܟ
ܘܫܡܝܐ ܠܗ ܪ̈ܚܡܐ ܪܘ̈ܚܐ ܡܢܗܠܟ ܣܘܓܐܗ .
ܘܡܣܘܪܐ ܠܗ ܠܢܝܢ ܡܬܦ̈ܢܝܢ ܘܡܪܘܚܝܢ ܠܗ ܠܢܝܢ
ܗܢ̈ܝܢ ܐܡܝܐ ܚܝܢ ܐܣܛܘܟ̈ܣܐ ܢܫܡܬܗ ܘܪ̈ܚܐ
ܡܫܬܟ̈ܚܐ ܘܪܚܐ ܡܫܬܘܕܥܐ ܘܪܚܐ ܟܠܠ ܘܘܪܚܐ
ܡܪܚܐ. ܘܦܬܝܟܝܢ ܗܘ ܕܫܠܐ ܫܡܝܐ ܕܫܡܝܢ
ܫܘ̈ܦܪܐ ܘܪܘܚܢܐ ܐܝܟ ܡܠܟܘܬܗ ܕܗܘܐ ܕܟܠܢ
ܡܫܬܘܚܕܢܐ ܐܦ ܕܗܘ ܕܟܠܗ. ܘܐܝܟܢ ܕܗܘܐ
ܘܟܠܢ ܐܠܗܐ ܗܘ ܕܟܠܗ ܠܗܠܝܢ ܪܘܚܐ. ܘܗܘ
ܘܟܠܢ ܡܢ ܟܠܗ ܕܟܠܗ. ܘܐܝܟ ܕܗܘ ܘܗܘ
ܒܟܠܗܘܢ. ܗܘ ܕܟܠܢ ܡܫܬܟܚܐ ܗܘ ܡܢܗܝܢ
ܗܘܠܟ ܘܪܚܐ ܘܠܗܝܢ ܗܘ ܐܣܛܘܟ̈ܣܐ ܡܫܬܡܠܝܢ ܠܗ ܠܐ
ܐܝܟ ܗܘ. ܡܕܡ ܡܢ ܐܣܛܘܟ̈ܣܐ ܗܘ ܗܘܝܐ ܘܐܝܟ ܕܫܡܫ
ܡܫܬܟܚܢܐ. ܘܐܦ ܐܝܟ ܘܪܚܐ ܡܫܬܘܕܥܐ
ܘܟܠܠ ܕܫܠܐ ܪܘܚܢܐ ܘܫܡܝܢ ܫܘܦܪܐ
ܐܝܬܘ ܘܡܫܢܐ ܘܟܠܠ ܐܣܛܘܟ̈ܣܐ ܡܫܬܟܚܐ
ܘܟܠܝ ܐܝܬܘ. ܐܪܐ ܫܪܝܪܐܝܬ ܘܐܣܛܘܟ̈ܣܐ ܗܢܘܢ
ܘܟܠܝ. ܘܐܦ ܐܝܟ ܡܢܗܝܢ ܘܡܬܬܐ ܗܘܝܐ.
ܗܘܐ ܘܡܫܬܡܠܝܐ ܐܦ ܘܦܪܝܫܐ. ܡܬܝܚܕܢܐ
ܠܗ ܚܕ ܡܕܡ ܢܦܫ ܐܦ ܫܡܫܐ ܐܘ ܦܪܝܫܐ
ܘܠܟܢ ܗܘ ܘܡܫܬܘܕܥ ܠܗܘܢ ܡܢ ܢܟܝܢ ܘܗܘ
ܡܫܬܘܕܥܐ ܗܘ ܠܗܘܢ ܡܢ ܢܟܝܢ ܦܪ̈ܝܫܐ ܦܪ̈ܝܫܐ

ܡܟܝܠ ܕܡܫܘܚܬܐ ܦܪܝܫܬܐ ܗܝ ܟܠܗ
ܟܠ ܐܪ̈ܝܐ ܘܫܡܝܐ ܘܪܘܚܢܐ. ܘܐܦ ܫܟܝܢܐ
ܘܡܬܚܠܠ: ܐܝܢܐ ܕܡܫܬܡܠܐ ܘܠܗܘܢܐ ܪܘܚܢܐ
ܐܝܟ ܘܐܝܟ ܕܐܝܟܢ ܘܗܘ ܟܕ ܗܘ ܘܠܗܘܢ
ܐܘܣܝܐ ܘܟܘܪܐ ܘܫܘܚܠܦܐ ܠܐܢܫ
ܟܠܗܘܢ ܐܪ̈ܥܐ ܡܫܬܟܚܐ ܡܫܬܘܕ̈ܥܢ ܐܝܟ ܕ
ܐܪ̈ ܠܗ ܚܝܘܗܝ ܐܣܛܘܟ̈ܣܐ ܘܠܗ ܡܫܬܘܕܥܐ: ܗܘ
ܟܪ ܕܐܦܪܝܢ ܐܝܟܢ ܘܫܡܫܐ ܘܦܪܝܫܐ ܘܪܘܚܐ
ܘܡܫܡܫܐ ܠܗ ܕܫܠܐ ܒܟܠܗ ܫܡܝܐ ܘܐܣܛܘܟ̈ܣܐ
ܫܡܫܐ ܐܝܟ ܐܠܐ ܠܗ ܫܡܝܐ ܪܘܚܐ ܕܫܡܝܢ
ܚܡܫܝܢ ܪܘܚܐ ܡܫܡܫܢܝܢ. ܗܘ ܕܝܢ ܫܠܐ ܐܦܪܝܢܐ
ܫܠܐ ܕܟܠܗ ܠܗ ܚܕ ܐܟܚܕ ܐܪܥܐ ܐܢܫܘܬܐ .
ܐܪܐ ܡܢ ܕܡܬܩܪܝܢ ܠܐܪܥܐ ܐܢܫܘܬܐ ܕܟܠܗܘܢ
ܠܗܘܢ ܫܡܫܐ ܡܫܬܐ ܢܦܫܬܗܘܢ :
ܐܝܟ ܠܗ ܫܡܫܐ ܡܕܡ ܫܠܐ ܘܟܠܗܘܢ
ܘܡܬܚܠܦ ܠ ܐܝܟ ܠܗ ܡܢ ܐܝܟ ܕܡܫܬܡܫܐ
ܫܡܫܐ ܐܠܗܐ. ܘܠܟܢ ܐܘܪܚܐ ܫܡܫܐ ܗܘ
ܘܐܣܛܘܟ̈ܣܐ ܘܐܝܟ ܠܗ ܫܡܫܐ ܘܟܠܗܘܢ ܫܡܫܐ
ܫܠܐ ܘܡܠܐܟܐ ܡܫܬܘܕܥܐ ܘܟܠܗܘܢ ܚܕ ܡܢ
ܗܠܝܢ ܐܠܐ ܐܪ̈ܥܐ. ܘܐܦ ܠܗܘܢ ܐܪ̈ܥܐ ܟܕ
ܕܡܘܬܐ ܘܡܫܠܡ ܕܟܠܗ ܠܗ ܠܟܠܢ ܕܢܐܡܪ
ܘܐܠܗܐ ܫܡܫܐ. ܘܠܐ ܐܝܟ ܟܕ ܢܦܫܐ ܘܪ̈ܘܚܐ
ܐܠܗܐ ܘܡܢ ܐܣܛܘܟ̈ܣܐ ܡܕܡ: ܕܟܠܗܘܢ
ܐܣܛܘܟ̈ܣܐ ܐܠܗܐ ܟܕ ܘܪܘܚܐ ܡܢ ܐܝܟܢ
ܐܪܥ ܐܢܫܘܬܐ ܫܠܐ ܕܡܘܬ ܡܬܚܙܝܐ ܘܚܝ̈ܐ
ܘܠܐ ܠܗ ܗܘܝܐ ܗܘ ܫܡܫܐ ܡܬܚܙܝܢ ܐܝܟ ܐܘܡܢܘܬܗ
ܡܬܚܙܝܐ ܘܐܣܛܘܟ̈ܣܐ: ܕ ܠܐ ܡܬܚܙܝܢ ܐܝܟܢ
ܠܐ ܗܘܝܐ ܕܟܠܗ ܘܐܢܫܘܬܐ ܗܘ ܫܡܝܢ ܡܢ
ܘܐܣܛܘܟ̈ܣܐ : ܘܠܐ ܐܠܗܝܢ ܟܕ ܕܫܠܐ
ܗܘ ܫܡܝܐ ܡܫܬܟܠܠ ܘܪܘܚܢܐ ܐܦ ܡܢܝܢܐ

ܘܗܘ

ܗܘܝܘ ܕܡܫܒܪܐ ܐܦ ܠܟܠܗ ܡܫܡܠܝܘܬܐ: ܐܝܟܢܐ
ܕܢܫܐ ܫܦܝܪܐ ܗܘܝܢܘܗܝ ܒܗܘܢ ܒܥܒ̈ܕܐ :
ܘܐܝܠܝܢ ܕܗܘ ܡܕܒܪ ܠܫܦܝܪܐ ܡܫܠܡܘ ܗܘܝܐ
ܘܗܘ ܡܕܒܪ: ܠܗܘ̈ܦܟܐ ܒܥܡܪܐ ܕܢܫܐ ܣܓܝܐ
ܡܗܝܡܢܐ ܐܡܬܝ ܡܫܟܚܐ ܕܘܡܝܐ ܘܐܦ ܪܘܚܐ
ܘܠܐܚܘܢ ܘܠܐ ܐܫܬܘ ܒܡܣܒܪܢܘܬܐ: ܗܘܝܢ ܗܟܢ
ܒܫܦܝܪܐ ܐܝܟ ܘܐܫܬܘܕܝܐ ܠܗ ܕܡܛܠ ܢܫܐ
ܐܝܟ ܕܐܡܪ: ܐܡܬܝ ܡܫܟܠܐ ܘܪܘ̈ܚܐ
ܒܡܛܠ ܢܫܐ ܕܡܕܒܪ ܫܦܝܪܐ ܠܡܫܡܠܝܘ ܠܡܫܡܠܝܘܬܐ
ܕܢܫܐ ܡܫܡܠܝܐ ܘܠܐ ܐܫܬܘ ܒܣܦ̈ܪܐ ܠܗܘܢ ܡܢ ܩܕܡ
ܐܠܗܐ: ܠܐ ܐܦܪܫ ܡܕܒܪܢܘܬܐ ܕܝܠܢ ܐܘ ܕܐܝܟ ܗܘ
ܠܟܠܗܘܢ ܐܝܟܢܐ ܕܡܫܡܠܝܘܬܐ. ܗܘܐ ܘܠܐ ܗܫܐ
ܡܦܢܐ ܠܗܘܢ ܕܡܗܝܡܢܐ ܠܗܘܢ ܟܠܗܘܢ
ܘܡܬܒܪܐ ܘܐܦ ܕܪܘܚܢܝܘܬܐ ܗܘܐ ܡܢܗܘܢ ܐܘ ܕܡܢܗ
ܡܗܝܡܢܘܬܐ ܡܢܗ ܕܐܠܗܐ ܪܒܐ ܡܫܡܠܝܐ
ܠܗܘܢ ܘܐܠܗܐ ܪܒܐ ܕܐ ܡܫܡܠܝ ܘܢܛܪ ܐܢܘܢ ܠܗ:
ܘܐܠܗܐ ܡܥܒܕ ܘܦܪܩ ܠܗ ܫܦܝܪܐ ܡܫܡܠܝܐ
ܘܡܕܒܪ ܒܥܒ̈ܕܘܗܝ ܘܡܫܬܡܠܝܐ ܒܗܘܢ ܠܟܠܗ
ܘܢܬܠ ܗܘ ܐܝܟܢܐ ܡܫܒܪ ܕܐܦ ܕܪܘܚܐ ܐܫܬܘ̈ܕܝܐ
ܕܡܫܬܡܠܝܐ ܗܘ ܡܗܝܡܢܘܬܐ ܘܠܗ ܣܦܪܐ ܐܝܟܢܐ
ܡܢ ܕܪ̈ܝܐ ܘܡܫܡܠܝܐ ܘܟܠ ܕܗܘ ܡܫܒܪܐ :
ܐܫܬܠܡܬ ܐܠܐ ܘܡܫܬܡܠܝܐ ܘܐܝܟ ܠܗ ܠܦ̈ܠܓܐ
ܘܪܘ̈ܚܐ ܐܝܟ ܡܫܟܠܐ ܐܘ ܕܐܦ ܐܡܪ ܡܢ ܚܝܠܐ
ܐܢܫ ܠܡܫܡܠܝܘ ܡܗܝܡܢܐ ܡܢ ܕܡܫܡܠܝܐ
ܡܫܒܪ ܡܢ ܗܘ̈ܦܟܐ ܘܐܫܬܘܕܝ ܒܡܛܠܗ. ܗܘܐ
ܒܪ ܠܗ ܠܗܘܢ ܡܫܡܠܝܘܬܐ ܐܢܫܐ ܡܫܡܠܝ
ܘܠܗ ܗܘܐ ܘܗܘ ܡܫܡܠܝܘܬܐ ܕܐܦ ܡܫܡܠ ܟܠܗ
ܡܫܡܠܝ ܡܢ ܕܝܢ ܐܠܐ ܕܒܗ ܕܡܐ ܠܡܫܡܠܝܐ.
ܐܠܐ ܐܝܟܢܐ ܡܫܡܠܝܘܬܐ ܡܗܝܡܢܘܬܐ ܒܥܒܕܘܗܝ
ܘܗܘ ܗܘܐ ܕܝܢ ܕܗܘ: ܡܫܡܠ ܡܢ ܡܫܬܡܠܝܐ

ܪ̈ܘܐ

ܘܠܗܘܢ ܡܫܡܠܝܘܬܐ ܘܠܗܘܢ ܐܫܬܘ̈ܕܝ ܘܠܐܫܬܡܠܝ
ܕܗܘ̈ܦܟܐ ܐܘ ܕܫܦܝܪ̈ܐ ܒܗܘܢ ܕܠܐ ܗܘܐ ܠܗܘܢ
ܡܗܝܡܢܘܬܐ ܡܫܡܠܝܐ ܘܡܫܡܠܝܐ ܠܟܠܗܘܢ ܐܝܟܢܐ ܘܡܫܡܠܝܐ
ܘܐܝܟ ܠܗܘܢ ܡܢ ܡܗܝܡܢܐ ܗܘܘ ܘܠܗ ܡܫܡܠܝܐ
ܒܫܦܝܪܐ ܐܝܟ ܡܕܝܢ ܠܗ ܗܘܐ ܠܗܘܢ ܡܫܡܠܝܘܬܐ
ܗܘܐ ܠܗܘܢ ܠܟܠܗܘܢ ܡܫܡܠܝܘܬܐ ܡܫܡܠܝܐ ܡܢ ܢܫܐ
ܡܫܟܚܐ ܐܝܟ ܡܫܡܠܝܐ ܗܘܬ ܠܡܫܡܠܝܐ ܘܡܫܡܠܝܐ
ܠܐ ܡܫܡܠܝܘܬܗܘܢ ܠܐ ܗܘܐ ܡܫܡܠܝ ܘܠܗܘܢ ܒܡܗܝܡܢܐ
ܐܝܟ ܐܡܬܝ ܪܘܚܐ ܩܕܝܫ ܘܡܢ ܡܫܡܠܝܘܬܗ ܡܫܡܠܝܐ
ܠܗܘܢ ܡܢܗ ܠܡܫܡܠܝܘܬܗ ܐܦܪܘܫܐ ܘܡܫܡܠܝܢ
ܐܦ ܡܫܡܠܝܐ. ܠܗܘܢ ܡܫܡܠܝܐ ܐܘܡܢ ܡܢ ܦ̈ܠܓܐ
ܕܡܢܗ ܠܡܫܡܠܝܘ ܡܫܡܠܝܐ ܩܕܡ ܘܠܐ ܠܡܫܡܠܝܘ
ܐܘ ܡܫܡܠܝ ܒܗ ܐܘ ܡܫܡܠ ܐܝܟܢܐ ܘܡܫܡܠܝܐ
ܡܫܡܠܝܘܬܐ ܘܪܘ̈ܚܐ ܘܣܓܝܐ ܐܝܟܢܐ ܘܡܫܡܠܝܐ
ܒܡܫܡܠܝܘ ܗܘܐ ܐܘ ܡܫܡܠܝܐ ܒܗ ܡܫܡܠܝܘܬܐ ܐܠܐ
ܐܝܟܢܐ ܕܡܫܡܠܝܐ ܐܫܬܘ̈ܕܝܐ ܘܗܘ ܗܘܐ ܘܐܝܟ ܠܗܘܢ
ܡܗܝܡܢܘܬܗ ܘܐ ܡܫܡܠܝܘܬܗ. ܡܫܡܠܝܐ ܘܠܟܠ
ܐܠܐ ܡܢ ܐܢܫ ܘܠܐܫܬܘܕܝ ܠܡܫܡܠ ܘܐܦܪܫ
ܠܗ ܘܠܐܫܬܘܕܝ ܠܟܠܗܘܢ ܠܐ ܐܫܬܘܕܝܐ
ܡܫܡܠܝ ܘܐܝܟ ܐܫܬܡܠܝ ܒܥܒܪܐ ܕܒܠܗ ܠܗܘܢ ܠܡܫܡܠܝܘ
ܡܫܡܠܝܘܬܐ ܐܘ ܪܘܚܐ ܡܢ ܡܫܡܠܝܐ ܘܦܘ̈ܠܓܐ
ܘܡܫܡܠܝܘܬܐ ܐܝܟ ܐܡܝܢܐ ܒܗܘܢ ܘܡܫܡܠ
ܡܕܝܢ ܘܐܝܟ ܠܗܘܢ ܡܗܝܡܢܘܬܗ ܐܦܪܐ ܘܫܦܝܪܐ
ܐܫܬܘ̈ܕܝܐ ܐܝܟ ܘܡܫܡܠܝܐ ܡܫܡܠܝܐ ܐܡܝܢܐ :
ܕܢܫܐ ܕܝܢ ܒܡܠܬܗ ܐܫܬܘ̈ܕܝ ܐܝܟܢܐ ܐܝܟ ܠܗ
ܡܫܡܠܝܐ ܐܡܝܢܐ ܡܫܡܠܝܐ ܘܪܘܚܐ ܘܠܐܚܘܢ
ܘܦܘܠܓܐ ܡܫܡܠܝܐ ܐܡܝܢܐ ܕܡܫܡܠܝܘܬܐ
ܡܫܡܠ: ܡܢ : ܒܟܠܗ ܕܡܫܡܠܝܐ ܘܐܝܟ̈ܢܐ
ܘܡܢ ܐܫܬܘ̈ܕܝܐ : ܐܠܐ ܒܡܫܡܠܝ ܐܝܟܢܐ
ܘܗܘ ܡܫܡܠ ܐܡܝܢ ܡܢ ܡܫܡܠܝܘܬܐ ܘܡܫܡܠܝܘܬܐ

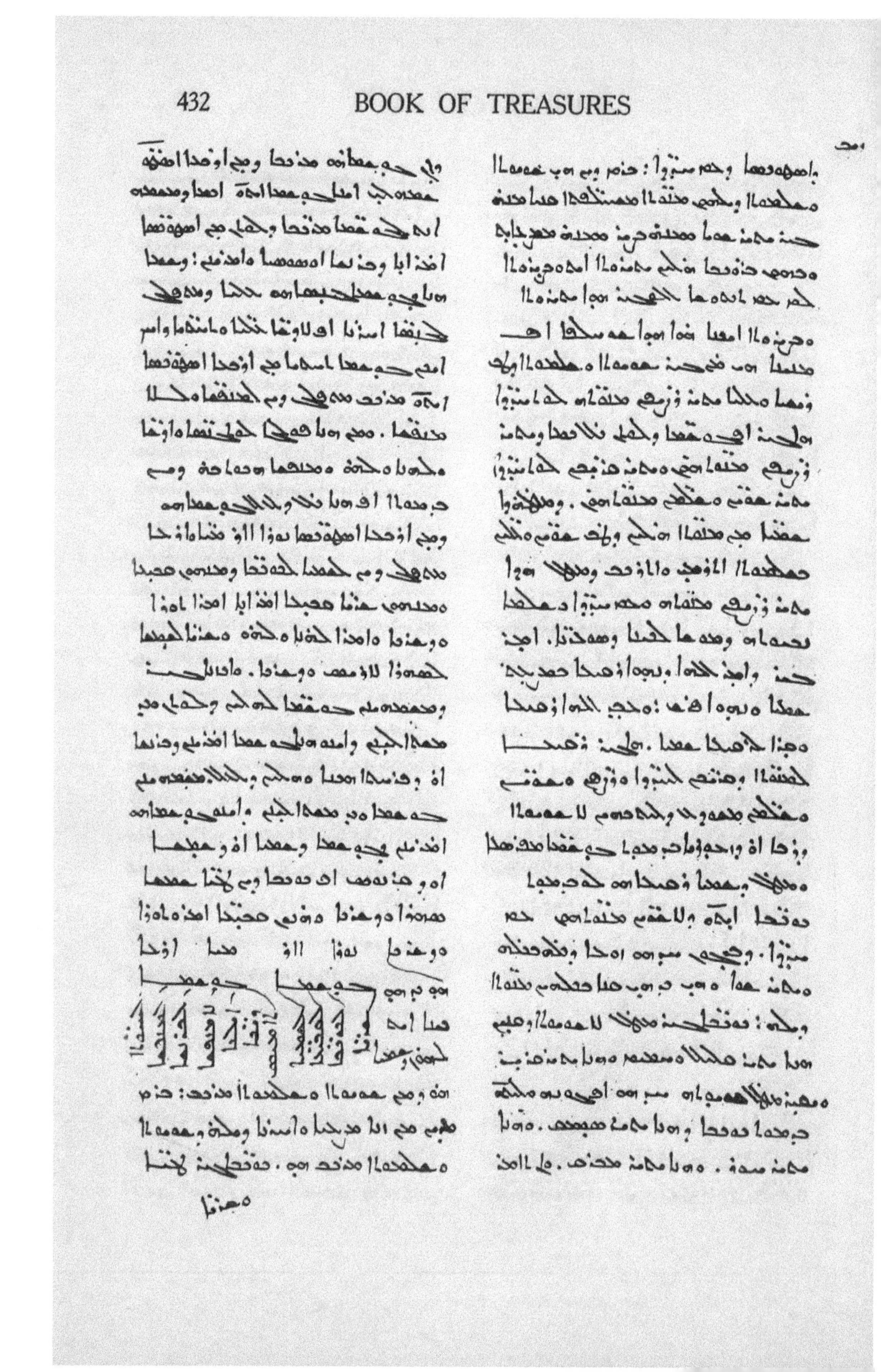

[illegible]

[illegible]

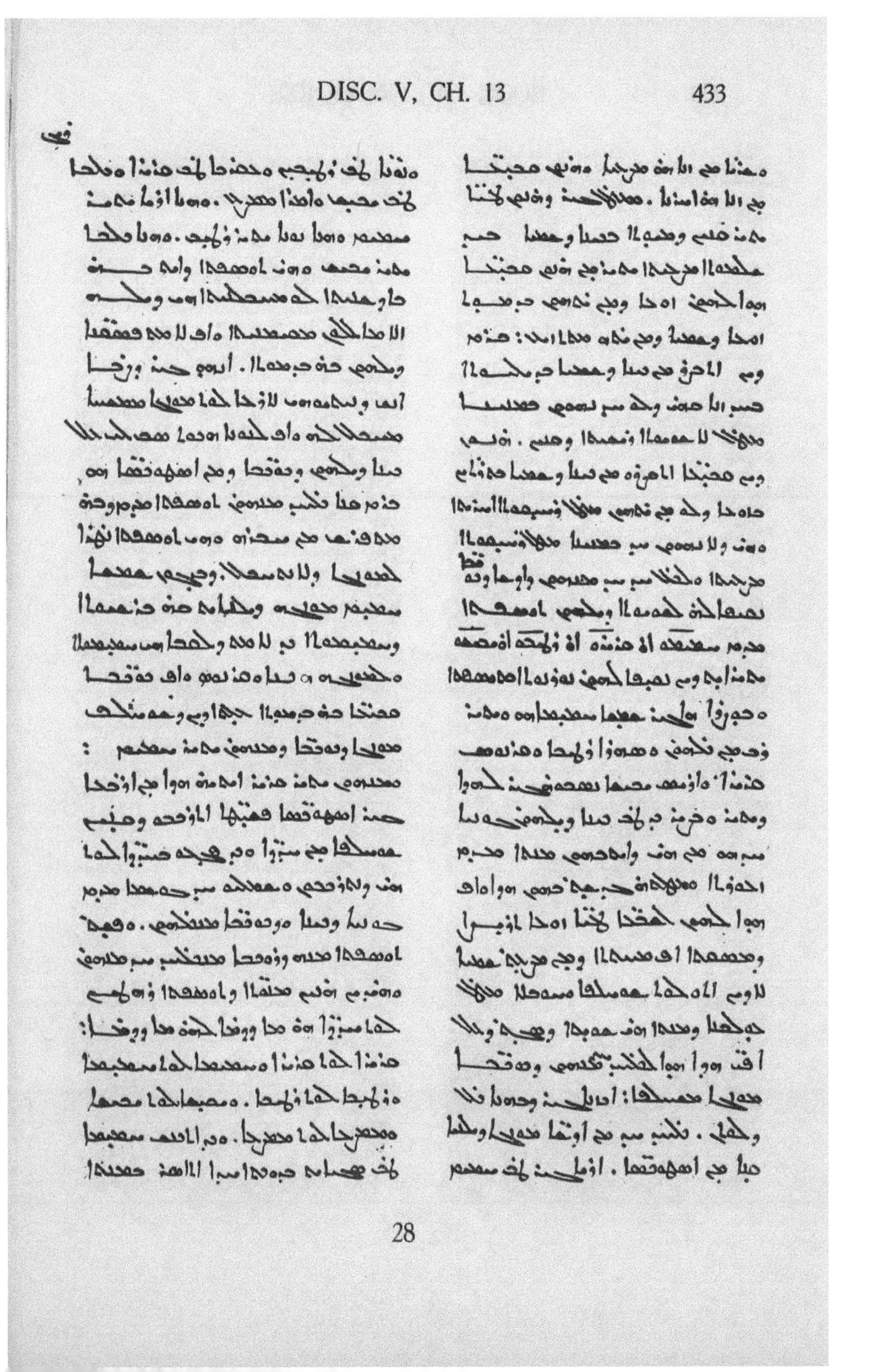

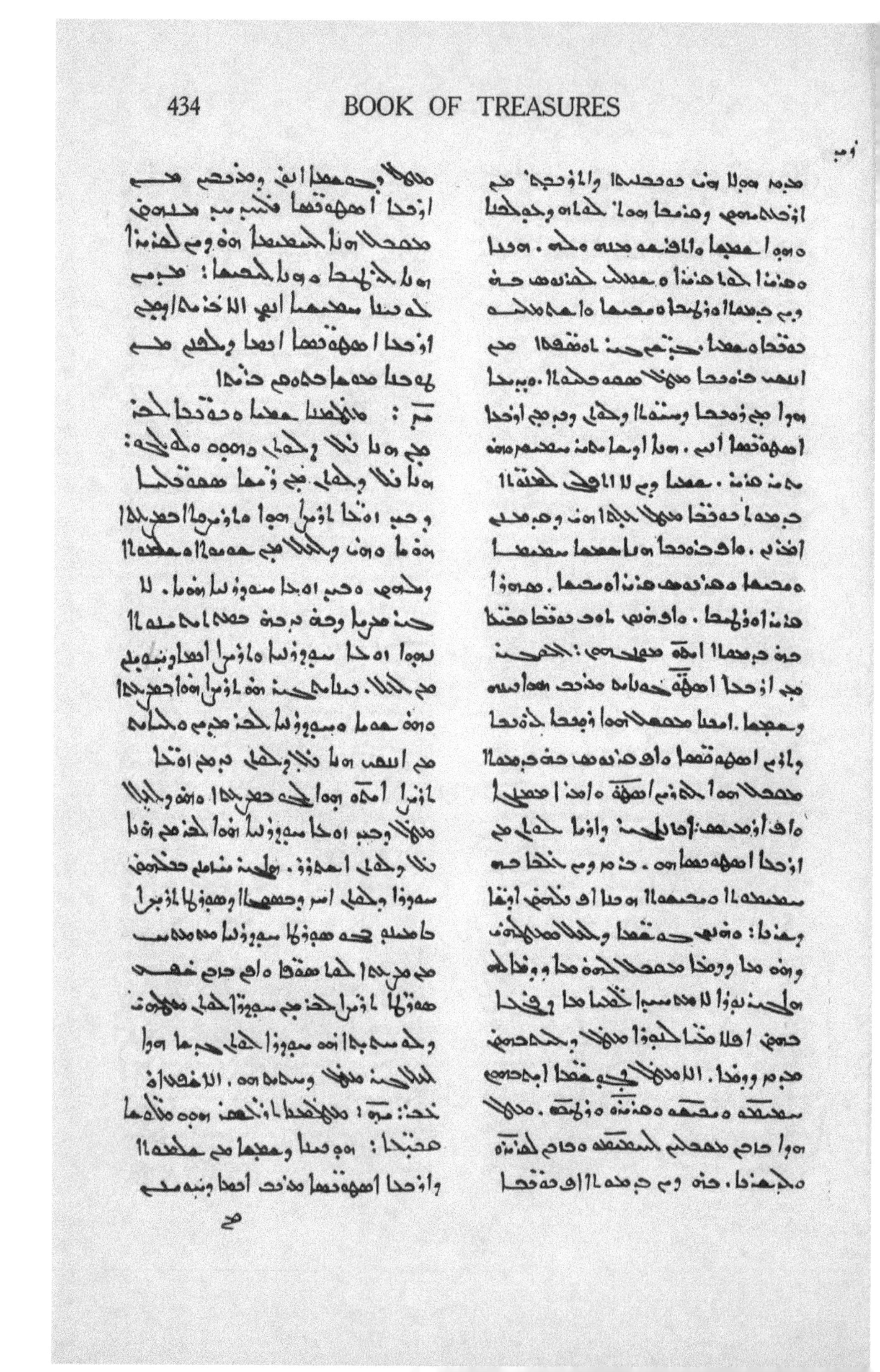

[illegible]

[illegible]

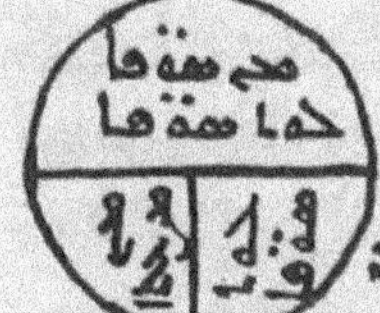

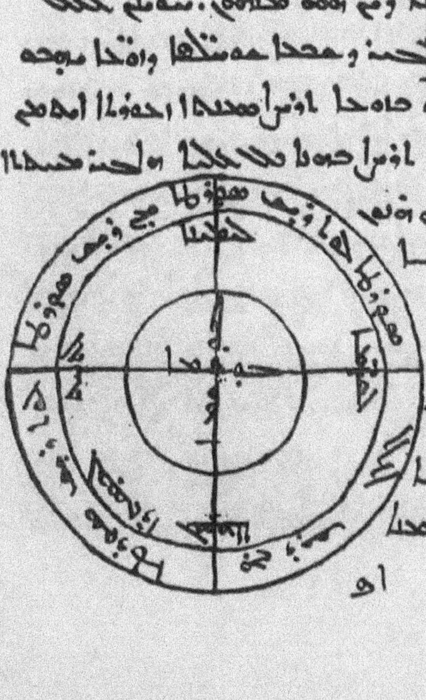

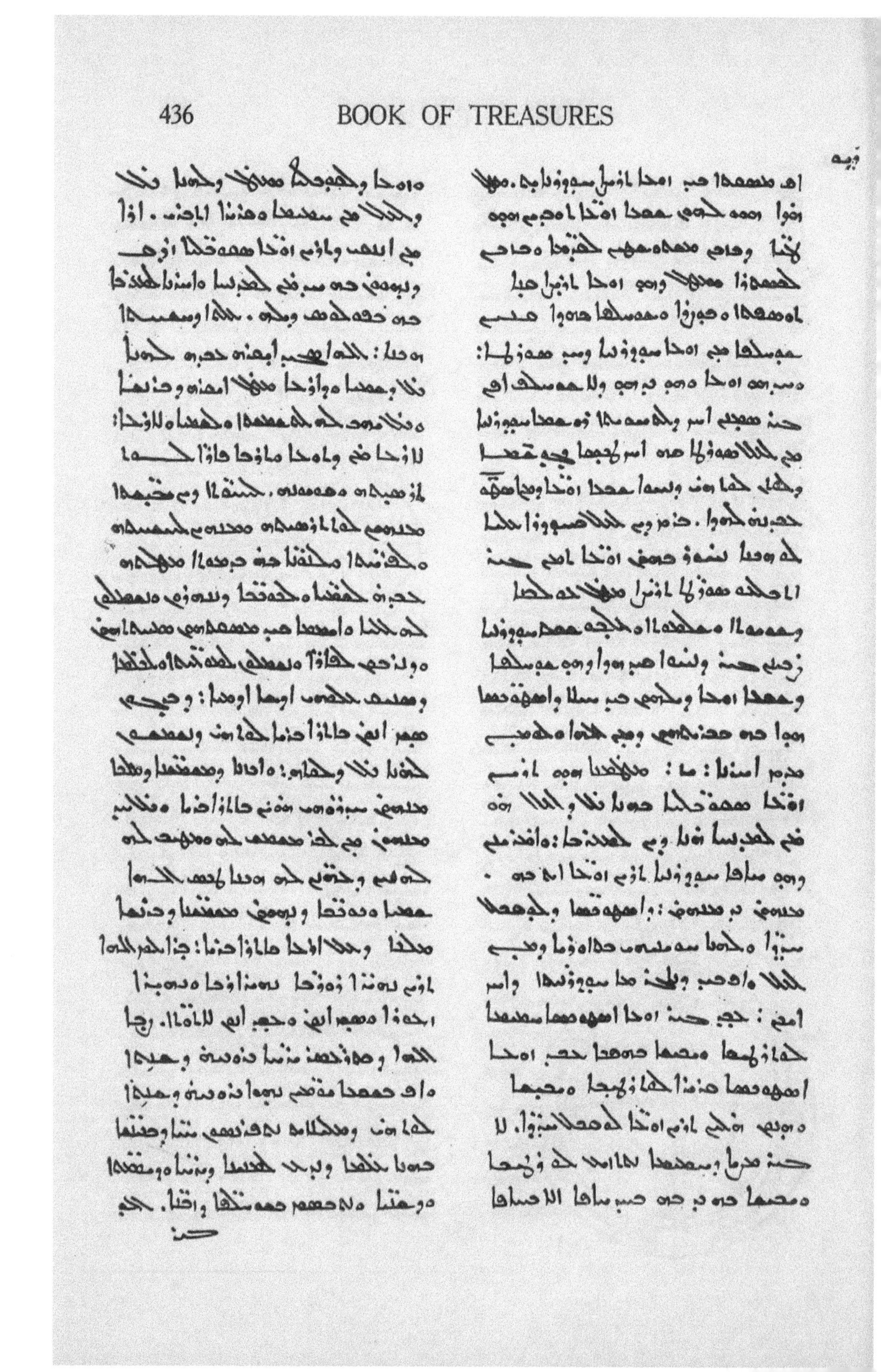

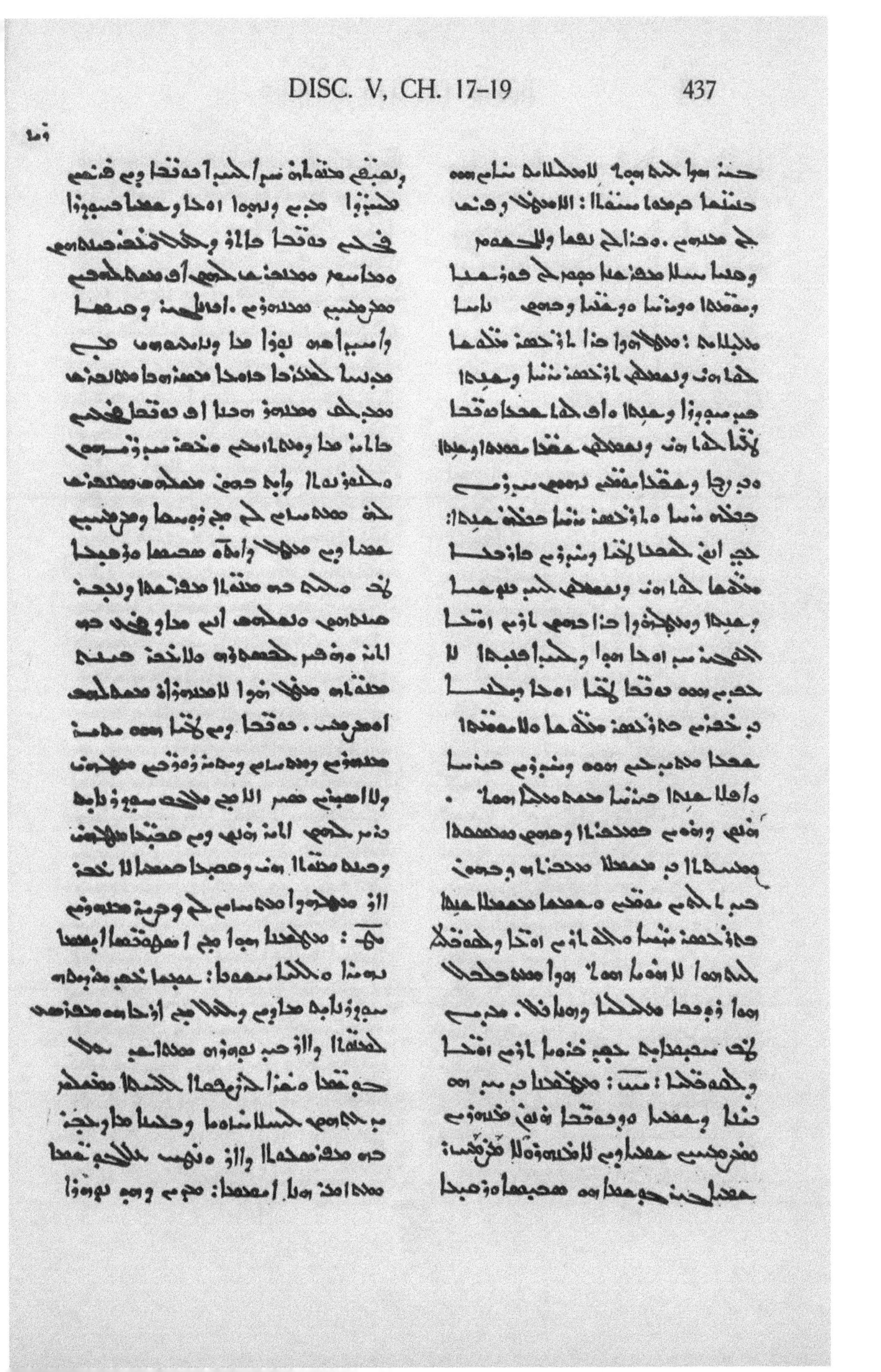

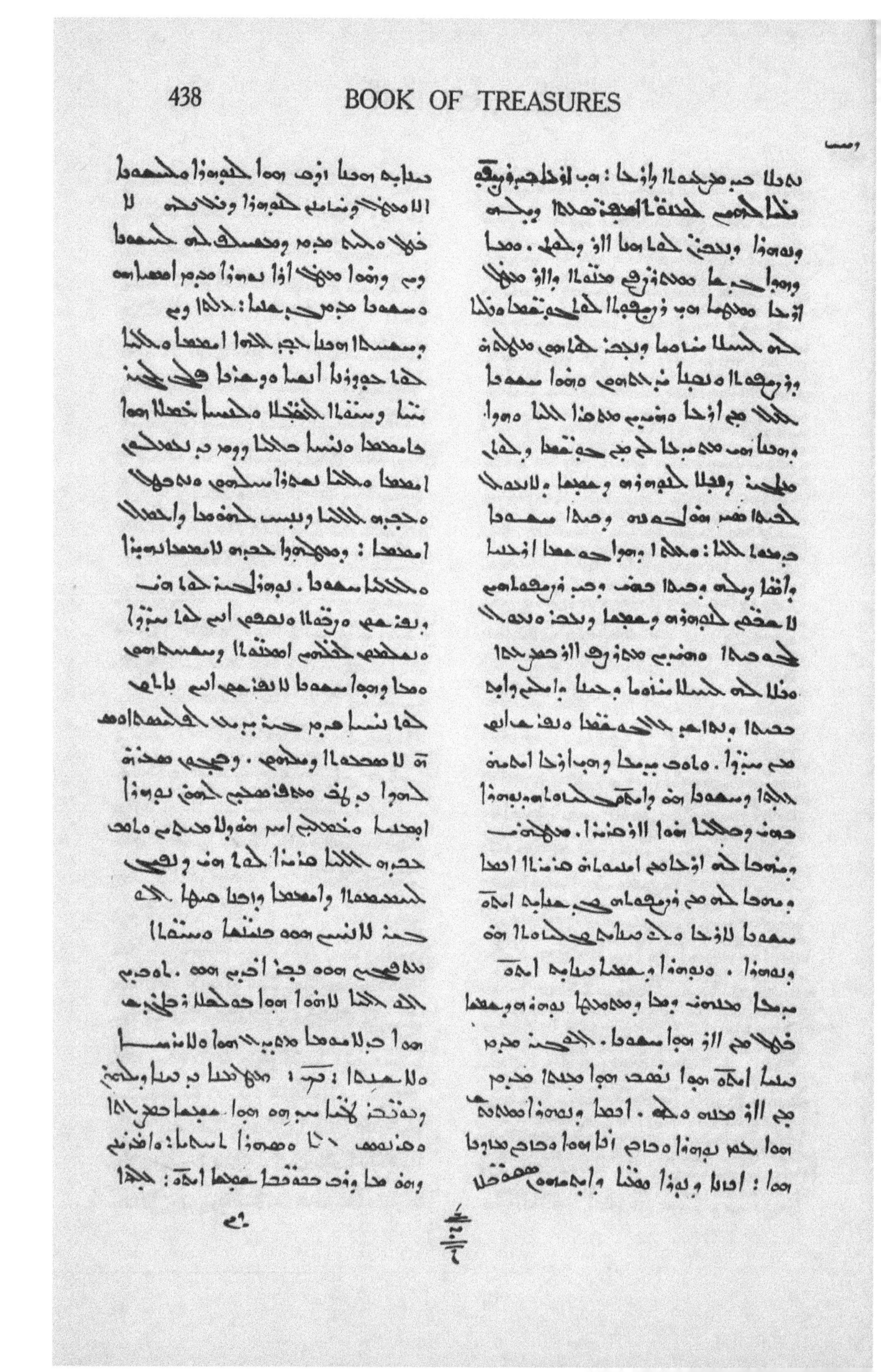

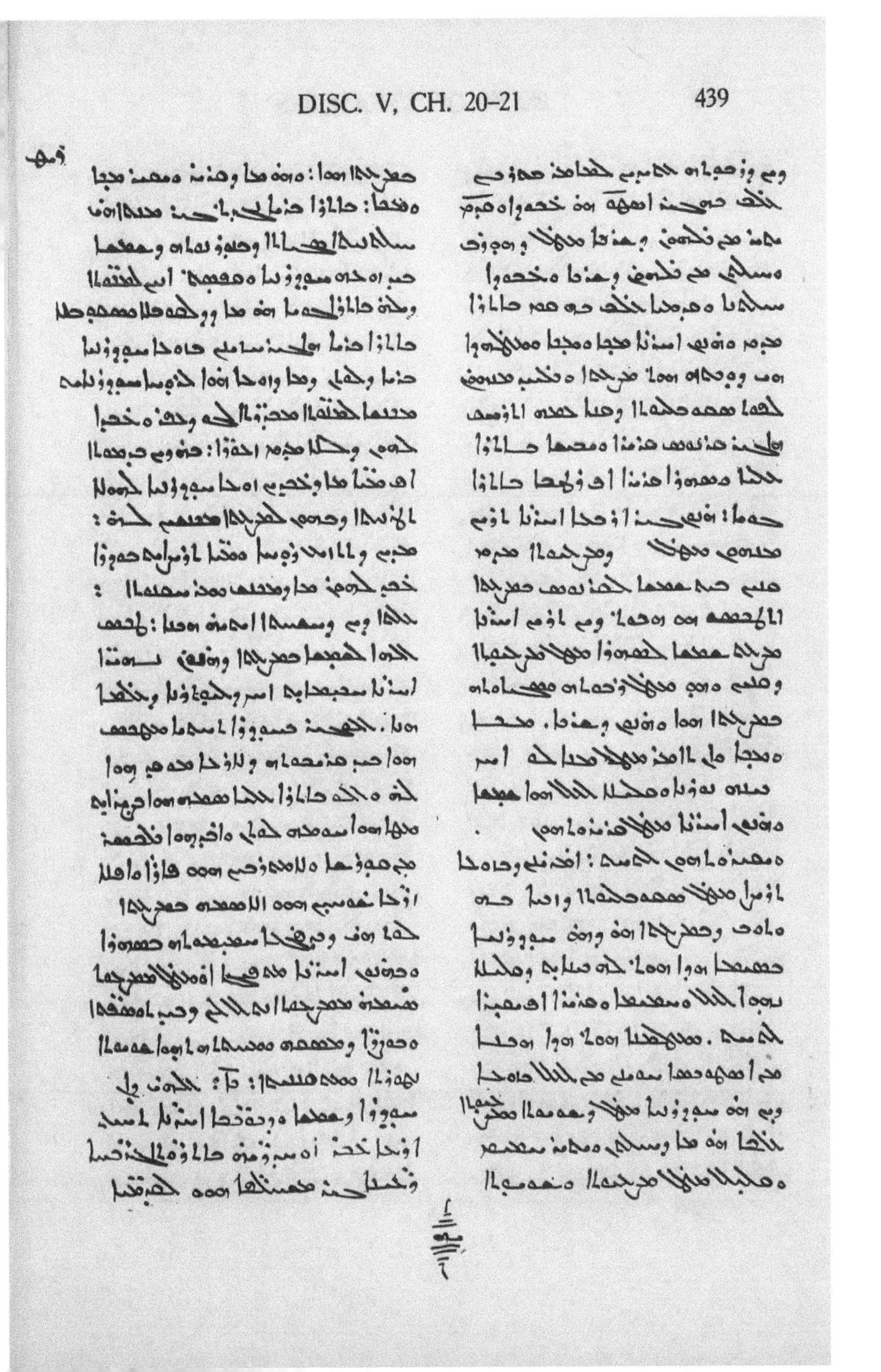

[illegible]

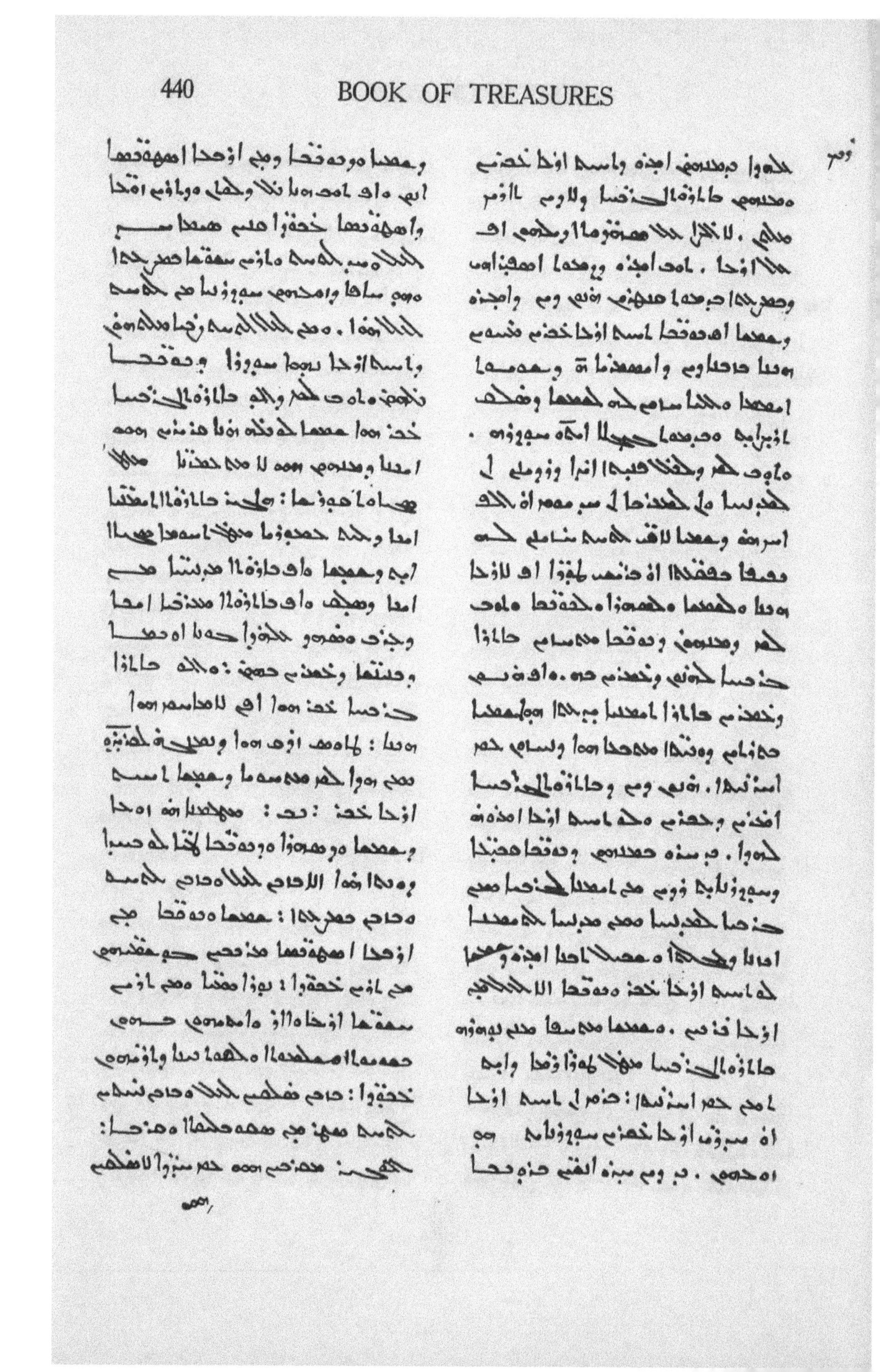

[illegible]

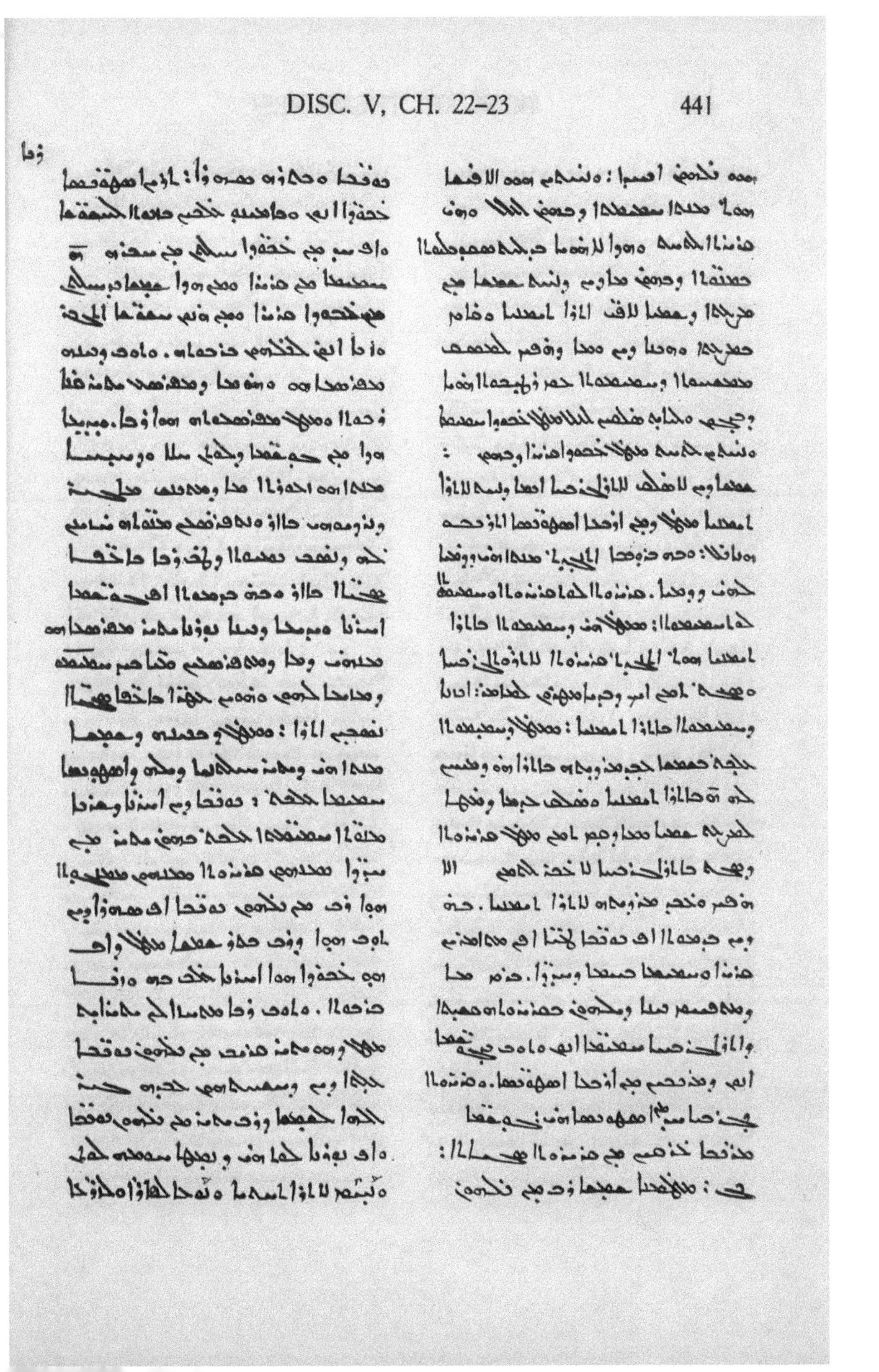

[illegible]

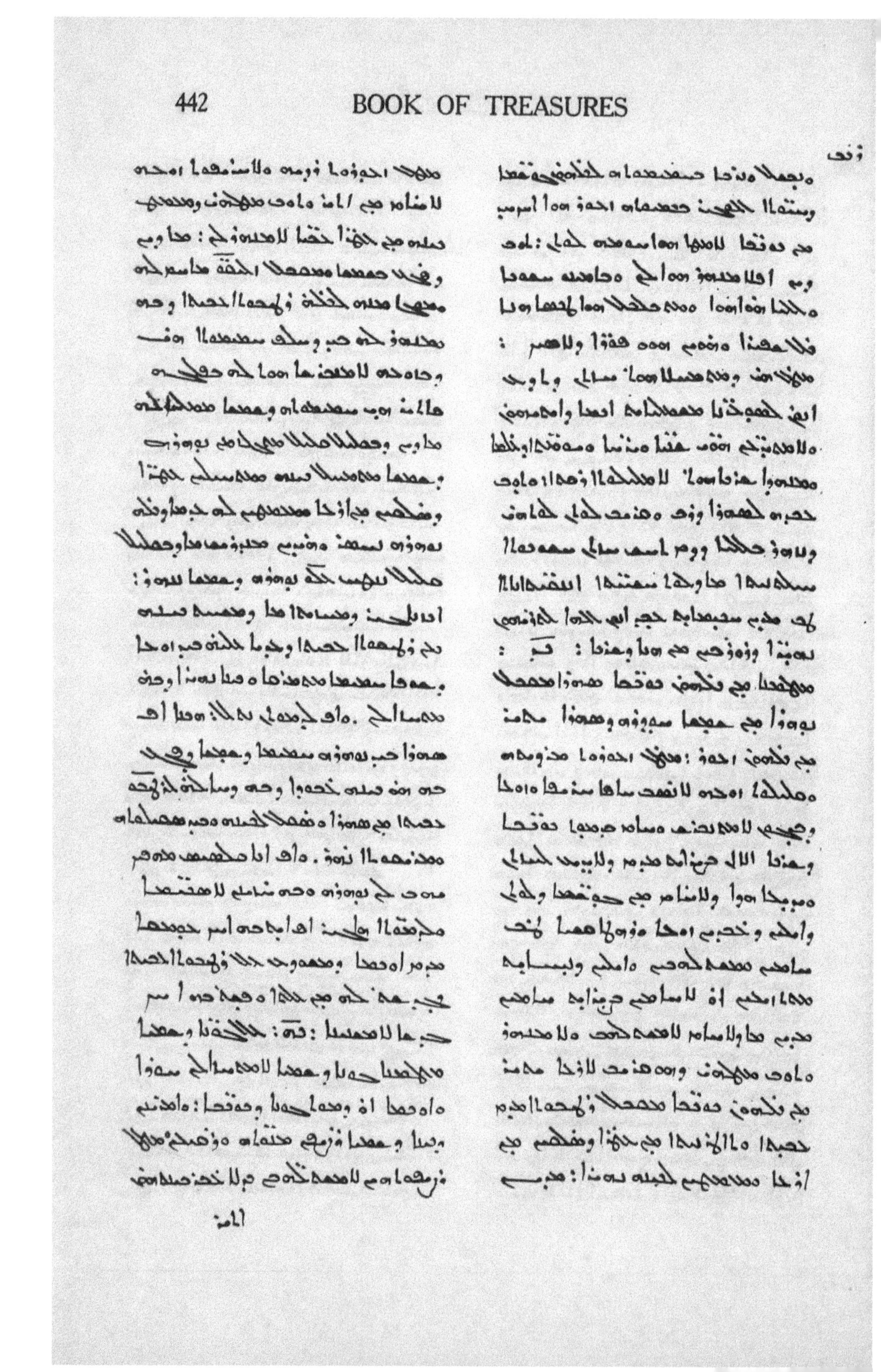

[illegible]

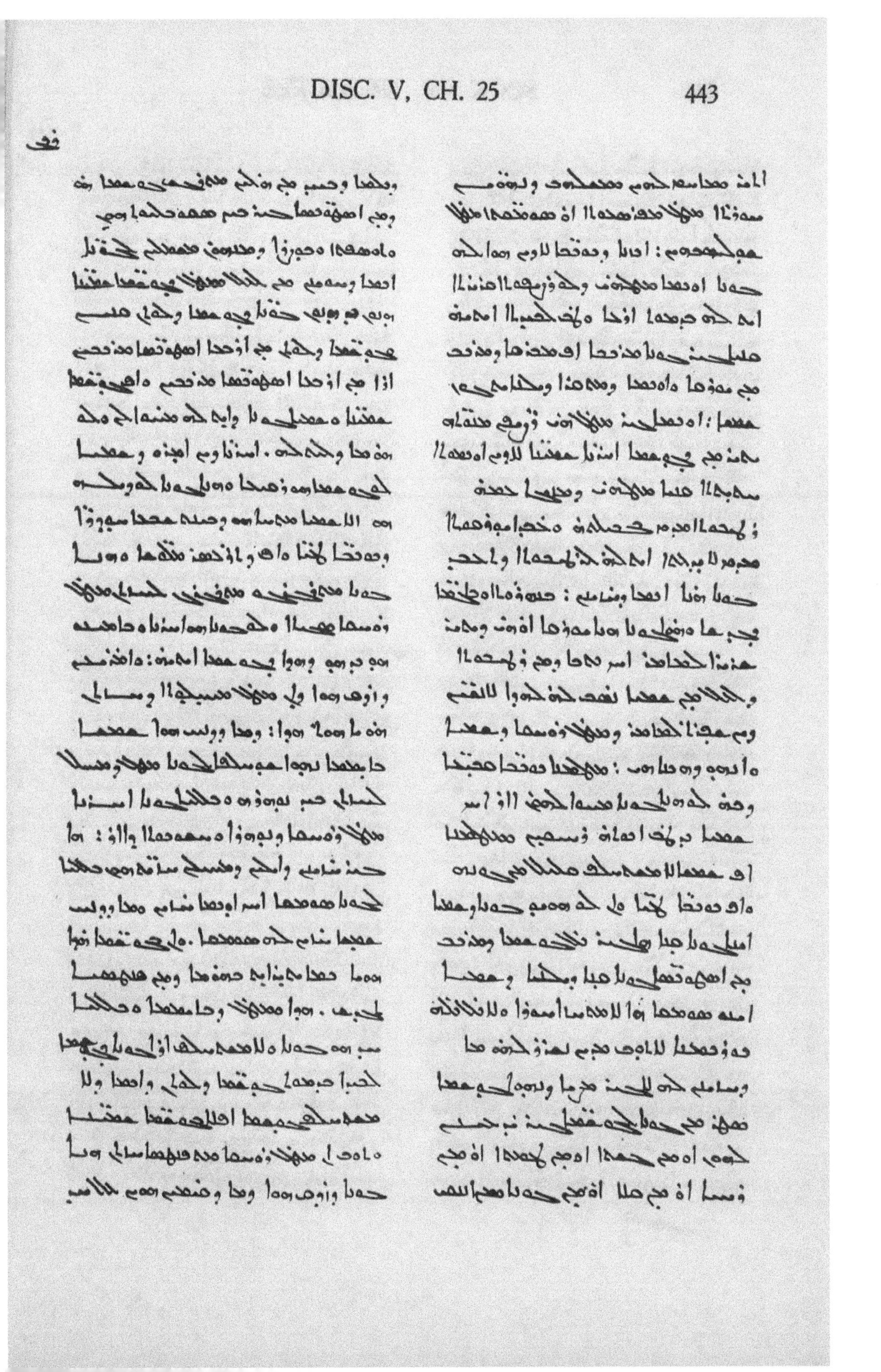

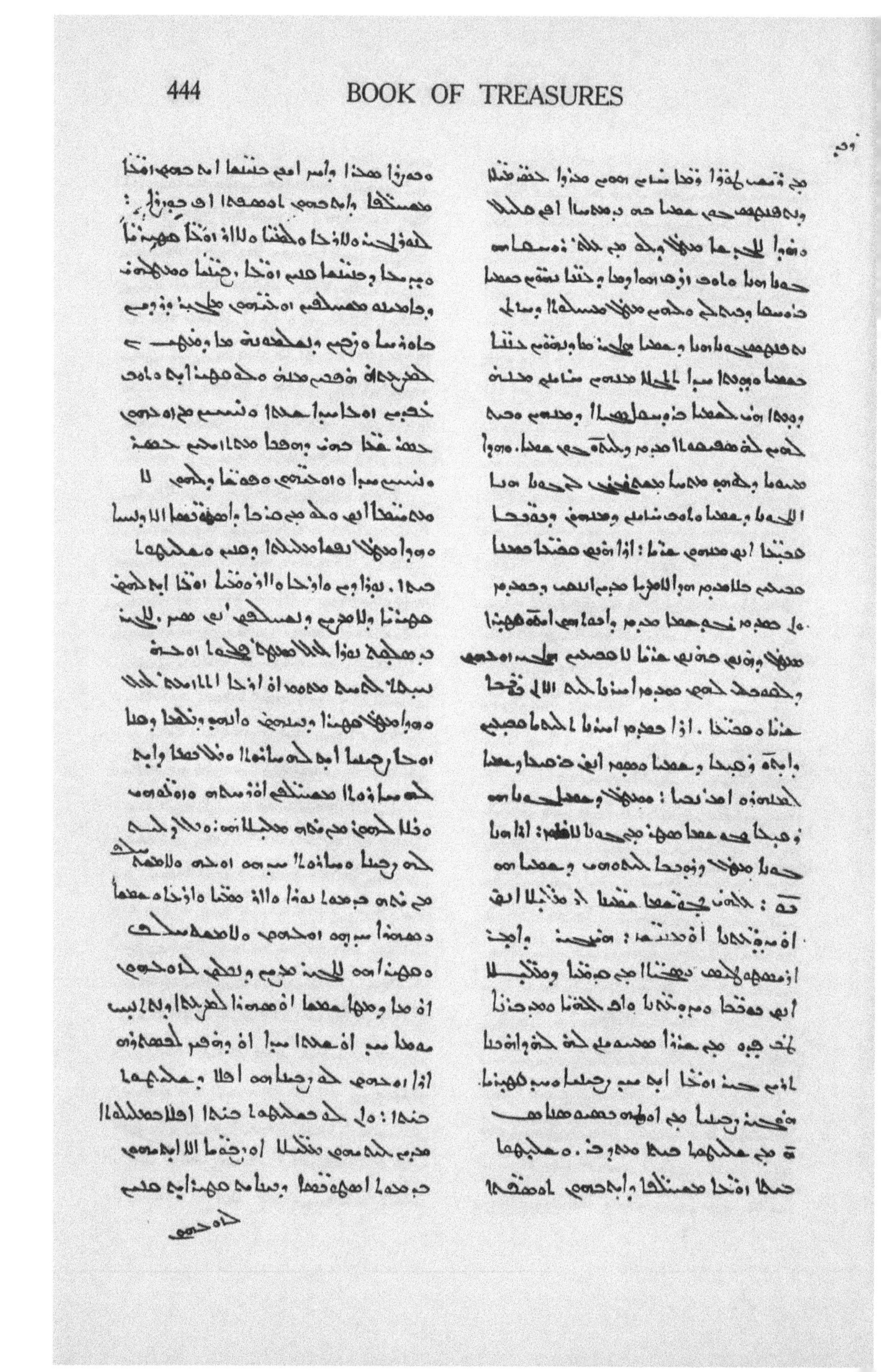

ܠܗܘܢ ܗܘܐ ܕܗܘܝܐ ܗܢܐ ܚܝܠܐ ܡܚܠܘܦ
ܕܡܫܝܚܝ̈ܐ ܐܝܬܝܗܘܢ ܒܫܡܐ ܢܘܗܪܐ. ܐܝܟܢܐ
ܕܫܡܫܐ ܕܐܙܠܩܘ ܘܫܪܝܪܐ ܐܝܠܝܢ ܕܫܪܝܪܘܬ
ܘܢܘܗܪܐ ܠܐܦܝ̈ܢ ܚܝܠܐ ܐܢܫܝܐ ܐܝܬ ܗܘܐ ܠܗܘܢ
ܣܓܝ ܡܢ ܕܐܣܛܘܟܣ̈ܐ. ܡܛܠ ܐܝܬܘܗܝ ܗܘܐ ܟܠܠ
ܡܠܟܘܬܗܘܢ ܦܫܝܛܐ. ܘܐܦ ܡܛܠ ܐܡܪ ܚܕ ܡܢܗܘܢ
ܘܒܘܢ̈ܝܐ ܕܢܦܩ ܠܡܥܡܘܕܝܬܐ ܐܘ ܢܦܩ ܡܢ ܡܫܟܢ ܥܠܡܐ
ܘܡܫܪܐ ܠܒܫ ܗܘܐ ܐܝܟܢܐ ܕܦܓܪ̈ܐ ܕܒܪܝܫ ܒܗ
ܢܘܪܐ ܘܢܘܪܐ ܡܢ ܦܓܪ̈ܐ ܘܢܦܩܝܢ ܠܡܥܡܘܕܝܬܐ
ܘܒܪ ܡܫܬܟܚܝܢ ܠܗܘܢ ܐܘ ܪܓܠ̈ܐ ܐܘ ܫܟܝܢ
ܟܕ ܐܢܫܐ ܡܬܒܪܝܢ ܠܗܘܢ ܟܠܗ ܘܡܬܚܒܝܢ
ܫܢܝܐ ܡܬܚܘܝܐ ܘܒܗ ܡܫܬܡܠܝܐ ܐܝܟ ܠܗܠ
ܘܡܬܒܪܝܢ ܡܢ ܦܓܪ̈ܐ ܚܝ̈ܐ ܠܗܘܢ ܕܒܪܝܫ
ܠܗܘܪܝܐ ܐܘ ܠܗܘܢ ܡܬܩܛܠܝܢ ܐܘ ܠܗܘܢ ܡܬܒܪܝܢ
ܐܪܐ ܗܘ ܕܢܣܒܘܢ ܡܫܬܒܩ ܫܘ ܫܠܡܐ ܐܘ ܐܫܬܟܚ
ܐܘ ܒܥܘܕܪܐ ܐܘ ܡܟܒܪ ܐܘ ܡܫܬܡܫ ܐܝܟܢܐ ܠܐܝܠܝܢ
ܘܪܚܝܩܝܢ ܡܢܗ ܢܦܫܗ ܘܒܗ ܠܐܠܗܘܢ ܗܘ ܡܕܡ
ܘܠܐ ܦܠܚ ܐܘ ܢܟܪܙ ܐܘ ܢܫܦܘܩ ܐܠܐ ܡܟܝܠ ܐܝܟ
ܟܕ ܡܫܬܡܥܝܢ ܐܝܟܢܐ ܘܐܣܛܘܟܣ̈ܐ. ܢܘܦܠܘܢ
ܡܬܚܠܐ ܠܗܘܢܐ ܘܡܢܗܝܢܐ. ܘܡܬܚܐܪܐ ܘܡܫܬܡܥܐ
ܘܡܢ ܟܠ ܕܡܬܚܪܐ ܘܟܘܡܪܐ ܡܢܝܢܐ ܡܬܚܐܪܐ
ܟܠ ܕܡܬܥܒܕ ܠܗܘܢ ܘܡܣܬܪܩܝܢ ܟܠ ܕܡܢܟܝܢ
ܘܐܪܟܐ ܟܠ ܕܡܬܚܫܒ ܠܗ ܟܠܗ ܘܡܬܥܢܝܢ
ܟܠܗ ܡܬܟܬܒܝܢ ܡܢܗ ܢܟܪܝܐ ܘܪ̈ܓܠܐ
ܘܫܦܝܪܝܢ ܢܟܪ̈ܝܐ ܟܠܗ ܘܢܦܠܘ ܠܟܠܗ ܘܦܠܚܘ
ܡܦܩܢܐ ܘܡܫܬܡܥܐ ܗܘܝܐ ܘܡܢ ܐܦ ܗܘ ܟܕ ܗܝ
ܘܠܠܐ ܫܡܥܐ ܟܠ ܚܝܝܢ ܡܟܠ ܕܠܐ ܐܪ̈ܐ ܘܡܢܗ
ܘܢܦܩܝܢ ܡܢܗܘܢ ܐܦ ܡܫܬܡܥ ܘܢܘܚ
ܐܢܫܐ ܟܢ ܒܫܡܐ ܗܢܝܢ ܘܠܡܬܚܫܒ
ܠܡܬܚܫܒܘ ܟܕ ܡܬܚܙܝܢ ܘܐܦ ܕܟܪܝܢ

ܠܟܢܫܐ ܡܫܪܐ ܐܘ ܡܫܡܫܐ: ܘܐܦ ܕܢܡܫܚܘܢ ܘܢܐܣܘ
ܦܠܚܝܢ ܟܒܝܐ ܕܗܘ ܡܟܝܠܐ ܘܫܡܫܐ ܕܘܟܗ
ܟܪܝܗܐ ܕܐܣܛܘܟܣ̈ܐ ܡܫܬܚܠܦܘܗܝ ܟܒܝܐ ܘܡܟܝܠܐ
ܘܡܬܚܠܦܐ ܘܢܦ̈ܠܠܐ ܘܡܬܕܟ̈ܝܢܐ ܠܟܠܗܘܢ: ܐܪܐ
ܟܕ ܐܦ ܕܢܗ ܡܦܫܩܝܢ ܘܦܪܣܝܢ ܡܦܠܚܝܢ
ܘܡܫܬܡܥܝܢ ܡܫܬܕܪܝܢ ܘܡܫܬܡܫܝܢ ܐܠܐ ܕܡܕܡ
ܐܢܫܐ ܕܐܝܬܘܗܝ ܠܟܠܗ ܒܥܒܕܐ ܕܝܠܗ ܘܟܠܐ
ܕܗܘ ܫܡܫܬܗ ܕܠܐ ܡܫܬܕܪ ܠܗܘܢ ܢܦܫܗ
ܡܬܚܙܝܐ ܘܠܐܢܫܐ ܟܕܪܐ ܘܠܐ ܢܫܬܐ ܡܫܡܫܢܘܬܐ
ܐܢܫܐ ܕܟܠ ܕܒܪܐ: ܐܝܟܢܐ ܗܘ ܟܠ ܡܬܠܟܝܢ
ܐܝܟܢܐ ܡܬܥܕܪܝܢ ܦܠܚܝܢ ܫܪܝܪܝܢ ܕܡܫܬܡܠܐ
ܡܢܗܘܢ ܕܢܦ̈ܫܬܐ ܡܫܬܠܚܝܢ ܘܡܢܗܘܢ ܕܒܥܪ̈ܐ
ܟܬܪܝܢ ܘܡܢܗܘܢ ܕܐܢܫ̈ܐ ܦܪܣܝܢ ܘܫܪܪܐ
ܕܡܫܬܟܚܐ ܕܦܫܝܛ ܟܪܝܗܐ ܟܕܟܪܐ. ܐܪܐ ܒ
ܦܠܚ ܟܫܝܪܐ ܡܫܬܡܥܝܢ ܘܟܠ ܐܪܟܐ ܠܐ ܐܝܬܝܗܘܢ
ܒܟܠܗ ܡܫܬܐ ܐܝܟ ܐܣܛܘܟܣ̈ܐ. ܐܝܟܢܐ ܠܗܘܢ
ܢܦܠܚܘܢ ܘܟܠܗ ܕܡܬܪܨ ܦܠܚܝܢ ܫܪܝܪܝܢ
ܘܠܗܘܢ ܐܝܟܐ ܘܡܫܬܡܥ ܟܒܝܪܐ ܘܟܫܝܪܐ ܐܘ ܕܡܬܚܪܐ
ܟܠܗܘ ܕܫܘܝܐ ܘܒܟܠܗ ܫܡܝܐ ܘܡܢ ܕܗܘ
ܠܗܘ ܡܦܪܢܟ ܠܗܘܢ ܦܠܚܝܢ ܡܟܠܐ ܐܝܟ
ܒܪܓܐ ܕܢܗܪ ܠܐܠܗ ܦܩܕ ܕܢܟܠ ܠܟܠ ܟܫܝܪܐ
ܐܘ ܫܪܝܐ ܐܢܫܝܐ ܘܗܘܐ ܦܩܕ ܟܕ ܠܟܘܢܐ
ܕܡܢ ܕܝܢ ܠܒܝܬ ܕܡܫܬܡܠܐ ܒܪܝܐ ܘܥܒܪ̈ܐ
ܘܡܠܟܘܢ ܢܗܘܐ ܡܢ ܢܦܫܐ ܐܝܟܢܐ ܕܟܫܝܪܐ
ܡܫܬܡܠܐ ܐܝܟ ܗܘܐ ܘܢܫܬܟܚ ܕܠܐ ܐܢܫܝܢ
ܠܦܠܚܝܢ ܟܫܝܪܐ ܘܟܠܗ ܐܝܟܢܐ
ܘܗܘ ܗܘܐ ܟܕ ܗܘܐ ܡܕܡ ܘܡܫܬܡܫܢܐ
ܡܬܚܫܒ ܗܘܐ ܠܗ ܘܡܠܟܘܬܗ ܗܘ ܐܠܐ ܕܗܘ
ܕܟܫܝܪܐ ܗܘ ܘܟܠܗ ܡܢ ܗܘ: ܡܕܡ
ܘܗܘܐ ܠܐ ܡܫܬܡܥܝܢ ܡܕܡ ܕܡܢ ܟܠܐ ܕܐܣܛܘ

ܪܦܘ

[illegible]

ܬܘܒ ܥܡ ܐܠܗܐ ܢܫܪܐ ܡܐܡܪܐ
ܫܬܝܬܝܐ

ܐ ܢܐܡܪ ܥܠ ܡܠܐܟܐ

[illegible]

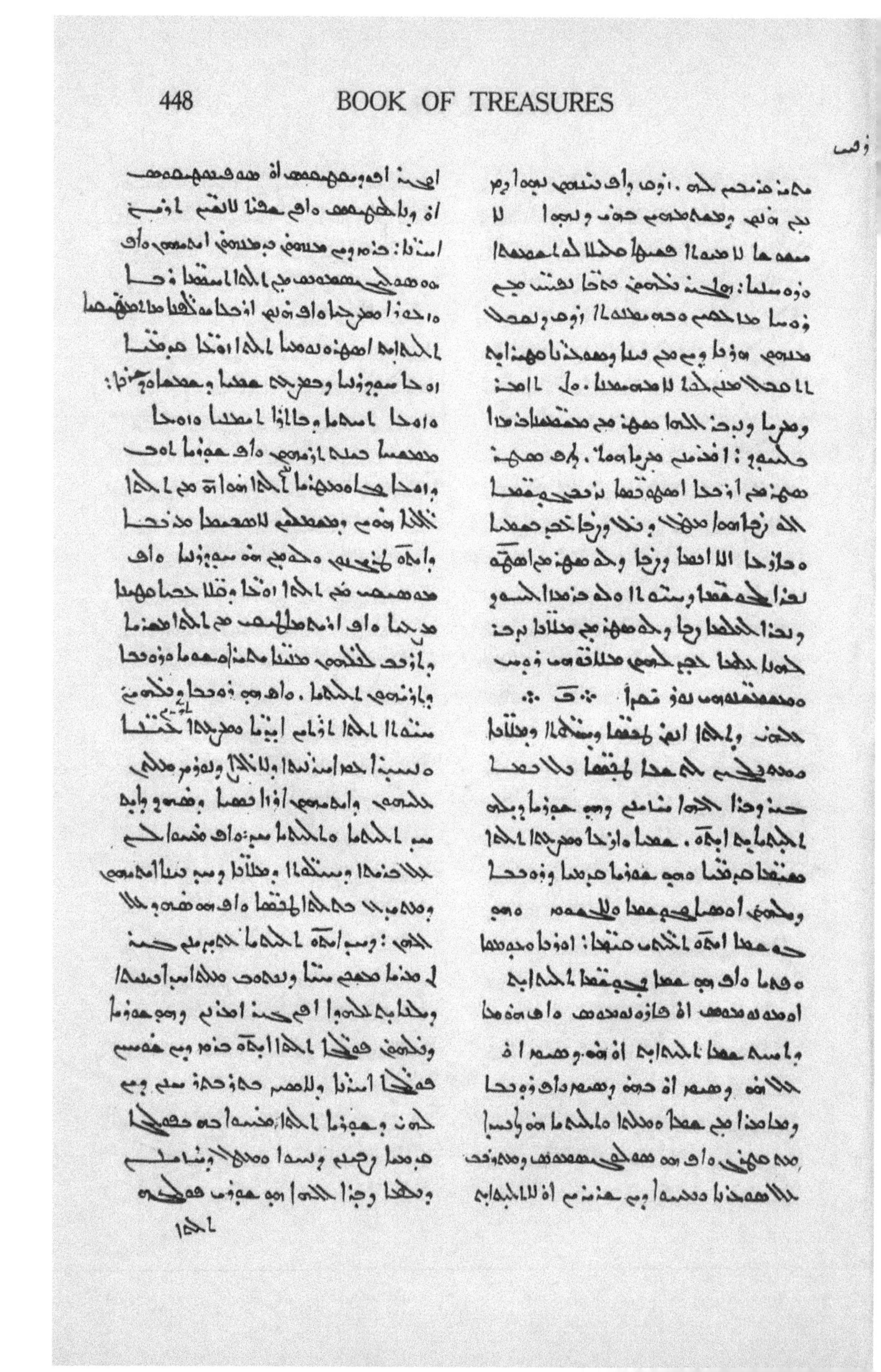

[illegible]

ܐܠܗܐ ܡܫܡܠܝܐ ܡܢ ܐܢܫ ܘܐܦ ܚܘܪܐ ܘܕܢܚ.
ܘܡܠܐܟܐ ܐܠܗܐ ܐܠܦܐ ܐܡܪܗ ܘܡܢ ܦܘܡ ܐܠܗܐ
ܘܡܢ ܡܠܟܘܬܗ ܕܒܫܡܝܐ ܐܡܪ ܕܢܐ. ܗܘ
ܗܝ ܕܐܠܗܐ ܕܢܛܪ ܐܠܗܐ ܐܠܗܐ ܗܘ ܕܢܗܝ
ܐܠܗܐ ܐܡܪܗ ܕܠܐ ܚܘܪܐ ܘܕܠܐ ܡܫܠܡܢܘܬܐ. ܘܡܛܠ
ܗܘܐ ܐܠܗܐ ܗܘ ܠܗ ܕܠܐ ܐܠܗܐ ܠܗ ܦܓܪܐ. ܗܢܐ ܕܝܢ
ܗܘ ܚܘܪܐ ܘܠܟܠܗܘܢ ܢܦܫܬܐ ܕܢܗܘܐ ܐܠܗܐ ܢܚܘܐ
ܘܡܛܠ ܕܢܦܫܗ ܚܝܐ ܕܠܗ ܐܠܗܐ ܗܘܝܐ ܠܐ ܥܒܕ
ܗܘ ܕܢܝ ܘܐܠܗܐ ܐܝܟ ܗܘ ܦܓܪܐ ܐܡܪܬ
ܘܐܡܪܝܢ ܕܒܐܠܗܘܬܐ ܗܝ ܡܠܬܐ ܗܘ ܐܠܗܐ ܐܠܗܘܬܐ
ܘܐܝܟܢܐ ܕܦܓܪܐ ܠܗ ܐܠܗܘܬܐ ܠܐ
ܗܘ ܕܝܢ ܡܠܟܘܬܐ ܐܝܟ ܕܐܠܗܐ: ܘܡܫܝܚܐ ܡܪܢ
ܥܒܕܐ ܠܡܫܡܠܝܘ ܡܫܝܚܐ ܡܠܟܐ ܕܡܫܝܚܐ
ܡܫܝܚܐ ܡܢ ܕܢܚܐ ܒܗ ܐܠܗܐ ܘܡܫܝܚܐ ܐܡܪܗ
ܡܫܝܚܐ ܗܘ ܠܗ ܠܗܘ ܗܘܐ ܘܠܗ ܐܡܪ ܘܡܫܝܚܐ
ܘܡܢ ܐܠܗܐ ܕܒܗ ܡܠܬܐ ܗܘ ܡܠܬܐ ܡܫܝܚܐ
ܒܗ ܡܠܬܐ ܐܝܟ ܡܢ ܐܠܗܘܬܐ ܘܐܢܫܘܬܐ:
ܡܢ ܠܗ ܒܪܢܫܐ ܐܠܗܐ ܒܪܗ
ܦܓܪܐ ܕܡܫܝܚܐ ܕܠܗ ܕܡܢ ܐܢܫܘܬܐ
ܒܐܠܗܘܬܐ ܘܡܢ ܐܠܗܘܬܐ ܠܗ ܒܪܢܫܐ
ܐܦܢ ܕܡܫܝܚܐ ܕܢܐ ܘܡܢ ܐܠܗܐ ܕܢܚܐ ܒܗ
ܫܡܐ ܐܢܫܐ ܘܐܫܬܡܠܝ ܘܚܘܪܐ ܘܡܠܐܟܐ ܘܠܐ
ܫܡܐ ܘܐܫܬܡܠܝ ܘܚܘܪܐ: ܡܫܝܚܐ ܡܢ
ܐܢܫܘܬܐ ܘܐܠܗܘܬܐ ܕܢܚܐ ܠܗ ܐܠܗܘܬܐ
ܗܝ ܡܠܬܐ ܗܘ ܠܐܢܫܐ ܘܐܫܬܡܠܝ ܘܚܘܪܐ
ܠܒܪܢܫܐ ܡܢ ܗܘ ܕܐܠܗܘܬܐ ܘܡܢ ܐܢܫܘܬܐ
ܐܦ ܚܝܐ ܗܘ ܕܐܠܗܘܬܐ ܕܠܐܠܗܘܬܐ ܐܠܗܐ.
ܗܘ ܕܡܫܝܚܐ ܗܘ ܕܐܠܗܘܬܐ ܕܢܚܐ
ܡܢ ܒܥܘܬܐ ܐܠܗܐ ܡܢ ܗܘ ܘܐܠܗܐ ܗܘ
ܘܗܘ ܐܠܗܐ ܘܠܐ ܡܫܬܡܠܝ: ܒܗ ܘܐܠܗܘܬܐ

ܐܠܗܐ ܐܡܪ: ܕܢܐ ܐܝܟ ܡܫܝܚܐ ܘܚܝܐ ܡܠܟܘܬܐ
ܘܫܡܐ ܘܐܫܬܡܠܝ ܐܢܫܐ. ܒܗ ܐܠܗܐ ܡܢ ܕܢܚܐ
ܠܗ ܐܠܗܘܬܐ ܠܗ ܡܫܝܚܐ ܐܠܗܐ ܡܫܝܚܐ
ܠܗ ܐܢܫܘܬܐ. ܐܠܗܐ ܡܫܝܚܐ ܠܗ ܐܢܫܘܬܐ
ܘܡܢ ܗܢܐ ܘܡܫܝܚܐ ܘܡܠܟܐ ܐܠܗܐ ܕܡܫܝܚܐ
ܗܘ ܕܢܐ ܐܝܟ ܐܠܗܐ ܕܡܫܝܚܐ ܕܐܠܗܘܬܐ
ܗܘ ܘܐܠܗܐ ܗܘ ܠܐܠܗܐ ܗܘܐ ܠܗ ܡܫܝܚܐ
ܘܡܫܝܚܐ ܐܠܗܐ ܗܘ ܡܢ ܐܠܗܐ ܕܐܠܗܘܬܐ ܠܐܠܗܐ
ܗܘ ܐܝܟ ܐܢܫܐ ܡܫܝܚܐ ܕܠܗ ܐܠܗܘܬܐ
ܡܢ ܚܘܪܐ ܘܡܠܟܐ ܐܠܗܐ ܗܘ ܡܢ ܡܠܬܐ
ܕܡܫܝܚܐ ܠܗ ܕܐܠܗܐ ܘܡܠܟܐ ܐܝܟ ܚܘܪܐ
ܠܐ ܡܫܬܡܠܝ ܠܐܢܫܘܬܐ ܡܫܝܚܐ ܘܐܠܗܘܬܐ
ܘܡܢ ܐܝܟ ܐܢܫܐ ܘܡܫܝܚܐ ܐܝܟ ܐܢܫܘܬܐ
ܡܫܝܚܐ ܘܐܢܫܐ ܘܐܠܗܐ ܘܐܠܗܐ ܐܝܟ ܡܫܝܚܐ
ܡܢ ܡܫܝܚܐ ܘܐܢܫܘܬܐ ܘܐܠܗܐ ܕܐܠܗܘܬܐ
ܠܗ ܡܫܝܚܐ ܘܐܠܗܐ ܘܡܢ ܐܠܗܐ ܕܐܠܗܘܬܐ
ܠܡܫܝܚܐ ܘܡܫܝܚܐ ܘܐܠܗܐ ܕܢܚܐ ܐܝܟ ܡܫܝܚܐ
ܘܡܫܝܚܐ ܘܐܠܗܐ ܘܐܢܫܐ ܐܝܟ ܕܡܫܝܚܐ
ܡܫܝܚܐ ܘܐܠܗܐ ܐܢܫܐ ܘܐܠܗܐ ܐܝܟ ܕܡܫܝܚܐ
ܠܐܠܗܐ ܡܫܝܚܐ ܡܢ ܐܢܫܐ ܗܘ ܘܡܫܝܚܐ ܠܗ
ܐܢܫܐ ܘܐܠܗܘܬܐ ܡܫܝܚܐ ܘܠܗ. ܘܐܠܗܘܬܐ
ܠܗ ܐܠܗܘܬܐ ܐܢܫܘܬܐ ܘܡܢ ܐܢܫܘܬܐ
ܘܡܫܝܚܐ ܠܐ ܐܢܫܐ: ܘܡܫܝܚܐ ܕܢܚܐ ܘܡܫܝܚܐ
ܘܐܠܗܘܬܐ ܘܐܢܫܘܬܐ ܠܐ ܐܠܗܐ ܡܫܝܚܐ ܕܢܚܐ
ܗܘ ܡܫܝܚܐ ܘܐܠܗܘܬܐ ܐܠܗܐ ܐܝܟ ܐܢܫܘܬܐ
ܘܡܫܝܚܐ ܐܠܗܐ ܡܠܬܐ. ܐܠܗܐ ܡܫܝܚܐ
ܡܠܬܐ ܡܢ ܐܠܗܐ ܐܝܟ ܕܢܚܐ ܠܡܫܝܚܐ ܠܗ
ܗܘ ܕܐܠܗܐ ܘܡܫܝܚܐ ܘܡܫܝܚܐ ܘܠܗ. ܕܢܚܐ
ܚܝܐ ܕܐܠܗܐ ܠܗ ܡܫܝܚܐ ܡܢ ܡܠܬܐ

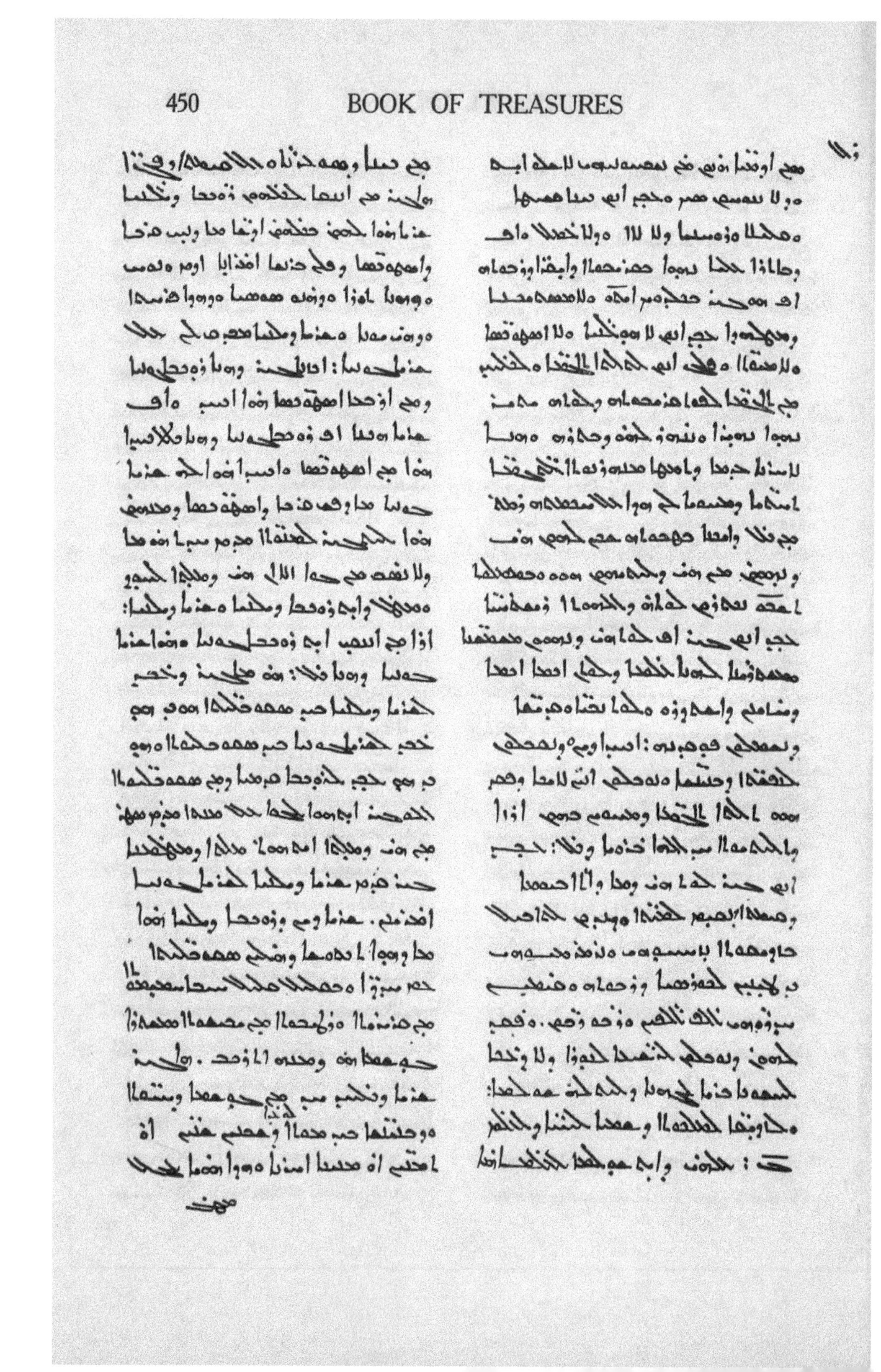

[illegible]

[illegible]

ܡܛܠ ܟܪܘܙܐ ܕܫܘܡܠܝܐ ܕܐܝܬ ܒܗ̇ ܒܥܘܡܩ
ܘܡܛܠ ܐܚܪܢܐ ܗܘܐ ܕܝܢ ܫܪܝܟܘܬܐ ܕܒܝ̈ܫܬܐ
ܘܚܛ̈ܝܐ: ܡܛܠ ܟܝܢܐ ܕܫܘܠܡܢܐ ܘܣܟܠܘܬܐ
ܘܫܘܡܠܝܐ ܡܫܬܟܚ ܚܕ ܐܚܝܐ ܕܫܘܠܡܢܐ
ܘܚܕ ܙܒܢܐ ܕܦܘܪܫܢܐ. ܘܐܦ ܕܪܘܚܢܝܘܬܐ
ܐܝܟܢܐ ܗܘܐ ܡܫܬܡܠܝܐ ܪܘܚܢܐ ܘܢܦܫܐ ܒܟܠܗ
ܟܠܗ: ܗܟܢܐ ܘܫܘܠܡܢܐ ܐܝܟܢܐ ܗܘܐ ܡܫܬܡܠܝܐ
ܘܫܪܝܐ ܘܚܝܐ ܒܟܠܗ ܟܠܗ ܠܘܬ: ܡܕܝܢ ܕܗܢܘܢ
ܡܫܬܢܐ ܡܢܗ ܕܡܛܠܗ ܕܗܢܐ ܕܡܢ ܙܒܢ
ܒܙܒܢܐ ܡܫܬܢܝܐ ܠܗܠܝܢ ܕܡܢ ܚܘܝܐ
ܘܠܐ ܡܫܬܟܚܐ ܗܘܐ. ܘܐܝܟ ܕܡܕܡ ܘܠܐ ܡܫܬܟܚܐ
ܘܐܦ ܡܫܬܢܝܐ: ܡܛܠ ܕܡܢ ܐܝܟ ܟܠܐ ܒܗܘ
ܡܫܬܟܚܐ ܡܬܚܙܝܐ ܘܡܫܬܢܐ ܡܢ ܡܫܬܟܚܐ
ܗܘ ܡܢ ܘܠܐ ܡܫܬܢܝܐ ܐܠܐ ܕܝܢܐ ܕܠܗ
ܡܫܬܟܚܐ ܘܚܝܐ ܗܘ ܦܫܝܛܐ ܘܠܗ ܒܗ
ܡܫܬܟܚܐ ܕܫܘܡܠܝܐ: ܡܛܠ ܕܚܢܢ ܗܟܢܐ
ܗܘ ܒܗ ܦܫܝܛܐ ܗܘ ܐܠܐ ܡܢ ܫܘܡܠܝܗ ܐܡܪܢܢ
ܐܬܪ ܡܢ ܐܝܟܢܐ ܡܫܬܡܪܐ ܘܦܠܚܝܢ. ܘܐܦ ܐܝܟ
ܕܚܘܪܐ ܕܫܘܡܠܝܐ ܕܡܫܬܟܚܐ ܗܘ ܚܘܪܐ
ܘܗܘ ܚܘܪܐ ܡܫܬܟܚܐ ܘܡܫܬܪܝܐ ܘܡܫܬܡܠܝܢܐ
ܗܘܐ ܐܬܪ ܡܢ ܐܝܟ ܕܡܫܬܡܠܝܢܐ ܕܗܘܐ ܘܫܪܝܐ
ܐܝܟܢܐ ܕܐܦ ܘܚܝܐ: ܘܐܦ ܐܝܟ ܕܚܝܐ
ܘܚܒܝܫܐ ܗܘܐ ܡܫܘܪܝܐ ܕܪܘܚܢܐ ܒܚܝܐ
ܘܡܢ ܐܣܛܘܟܣ̈ܐ ܘܫܪܝܐ ܡܫܬܡܠܝܢܘܬܐ
ܐܬܡܠܝܢ ܕܐܣܛܘܟܣ̈ܐ ܠܗ ܒܪܘܚܐ ܘܐܦ ܐܬܐ
ܘܐܦ ܡܫܬܡܠܝܐ ܗܘܝܐ ܠܗ ܒܪܘܚܐ ܘܚܝܠܗ
ܘܠܗ ܡܫܬܢܝܐ ܢܦܫ ܗܘܐ ܘܡܫܬܚܠܦܐ ܕܡܫܬܥܒܕܐ
ܢܦܫ ܢܘܚܐ ܘܫܠܝܐ. ܐܬܪܐ ܢܦܫܐ ܡܫܬܥܒܕܐ
ܠܐܣܛܘܟܣ̈ܐ. ܘܢܘܚܐ ܘܫܠܝܐ ܘܐܚܕܐ ܕܐܢܬܘܢ
ܫܪܝܐ ܘܚܛܝܐ ܘܫܘܠܡܐ. ܥܠ ܐܬܪ ܕܐܝܟ

ܕܪܘܚܢܝܘܬܐ ܕܡܢ ܐܣܛܘܟܣ̈ܐ ܡܬܬܪܐ ܡܢ
ܐܝܟ ܕܐܦ ܗܘ ܪܘܚܢܐ ܕܐܣܛܘܟܣ̈ܐ ܡܪ̈ܟܒܐ
ܡܬܬܪܐ. ܢܘܪܐ ܘܐܐܪ ܘܡ̈ܝܐ ܘܐܪܥܐ ܦܢܝܢ
ܠܚܕܕܐ ܐܣܛܘܟܣ̈ܐ ܦܫܝ̈ܛܐ: ܘܐܝܟ ܕܗܘ ܕܐܝܟܢܐ
ܐܡܬܝ ܕܢܬܦܫܛܘܢ ܐܣܛܘܟܣ̈ܐ ܦܫܝ̈ܛܐ ܡܬܝܕܥܝܢ
ܘܠܐ ܚܫܝܢ. ܘܐܦܢ ܡܪ̈ܟܒܐ ܘܡܢ ܐܣܛܘܟ̈ܣ
ܡܪ̈ܟܒܐ ܗܘܘ ܘܐܬܪܟܒܘ ܐܣܛܘܟܣ̈ܐ ܡܪ̈ܟܒܐ
ܗܘܘ: ܘܟܕ ܐܢܘܢ ܗܦܟܝܢ ܐܣܛܘܟܣ̈ܐ ܡܪ̈ܟܒܐ
ܠܦܫܝ̈ܛܐ ܐܣܛܘܟܣ̈ܐ ܕܡܢ ܦܫܝ̈ܛܐ ܠܐ ܗܦܟܝܢ
ܠܡܪ̈ܟܒܐ. ܡܛܠ ܕܦܫܝܛܝܢ ܒܟܝܢܗܘܢ. ܐܠܐ ܠܗ ܒܣܘܦ
ܡܫܬܘܕܥܐ ܠܗ ܠܟܠܗܘܢ ܕܡܪܟܒܘܬܐ ܘܐܦ ܦܫܝܛܘܬܐ
ܒܟܠܗ. ܐܠܐ ܠܗ ܒܠܚܘܕ ܠܦܫܝܛܘܬܐ: ܡܛܠ ܕܗܘܝܢ
ܘܠܟܠܗܘܢ ܠܗܘܢ ܐܣܛܘܟܣ̈ܐ ܦܫܝ̈ܛܐ ܐܠܐ ܠܗ
ܠܡܪ̈ܟܒܐ ܘܐܝܟ ܐܢܘܢ. ܡܫܬܢܝܢ ܟܕ ܠܚܘܕ
ܠܡܠܝܢ ܡܢ ܟܠܗܝܢ ܐܬܪ̈ܘܬܐ ܡܫܬܪܝܢ ܐܣܛܘܟܣ̈ܐ
ܡܪ̈ܟܒܐ ܠܦܫܝ̈ܛܐ ܗܢܘܢ ܘܟܠܗܘܢ ܡܬܦܫܛܝܢ
ܐܝܟܢܐ ܕܒܝܕ ܦܘܫܛܐ ܠܐܣܛܘܟܣ̈ܐ ܡܪ̈ܟܒܐ
ܘܦܫܝ̈ܛܐ ܠܟܝܢܗ ܦܫܝܛܐ ܠܗ ܠܗ ܗܢܘܢ ܕܢܫܬܠܡܘܢ
ܘܢܬܦܫܛܘܢ ܡܢ ܗܢܐ ܡܟܝܠ ܘܐܝܬܝܗܘܢ ܐܠܐ ܕܢܬܦܫܛܘܢ
ܠܡܘܣܦܐ ܡܢ ܗܢܘܢ ܘܐܝܬܝܗܘܢ ܠܟܠ ܡܕܡ
ܐܡܪܢܐ. ܘܗܟܢܐ ܕܡܢ ܐܦܝ ܟܝ̈ܢܐ ܡܫܚܠܦܐ ܗܘܐ
ܠܟܠܗܝܢ ܡܘܣܦܐ ܡܢ ܗܢܐ ܟܠܐ ܘܐܝܬܝܗܘܢ ܠܟܠ
ܡܕܡ ܐܡܪܢܐ ܕܢܦܫܗ ܘܐܝܟܢܐ ܡܘܫܚܬܐ ܘܗܢܐ
ܟܠܐ ܘܐܝܬܝܗܘܢ ܘܡܫܚܬܐ ܘܗܢܐ ܟܠܐ ܘܠܗ
ܡܬܡܣܟܢ ܘܐܬܐܡܪ ܦܫܝܛܘܬܐ ܠܟܠܗܝܢ ܦܫܝܛܝܢ
ܐܣܛܘܟܣ̈ܐ ܡܢ ܗܘܢ ܕܫܡܝ̈ܢܐ ܘܐܪܥܢܝܐ
ܘܡܫܘܚܬܐ. ܗܢܘܢ ܕܝܢ ܡܢܗܘܢ ܐܝܟܢܐ ܕܗܘܝܘ
ܥܠ ܗܘܐ ܦܘܠܓܐ ܡܢܗܘܢ ܦܫܝܛܝܢ ܘܠܐ ܡܟܝܠ:
ܡܕܝܢ ܢܗܘܪܐ ܕܡܫܡܫܢܘܬܐ ܘܡܫܡܫܢܘܬܐ ܘܦܘܪܫܢܘܬܐ
ܘܡܫܡܫܢܘܬܐ ܕܐܣܛܘܟܣ̈ܐ. ܒܬܪܘܢ ܦܫܝܛ
ܡܫܡܫܐ

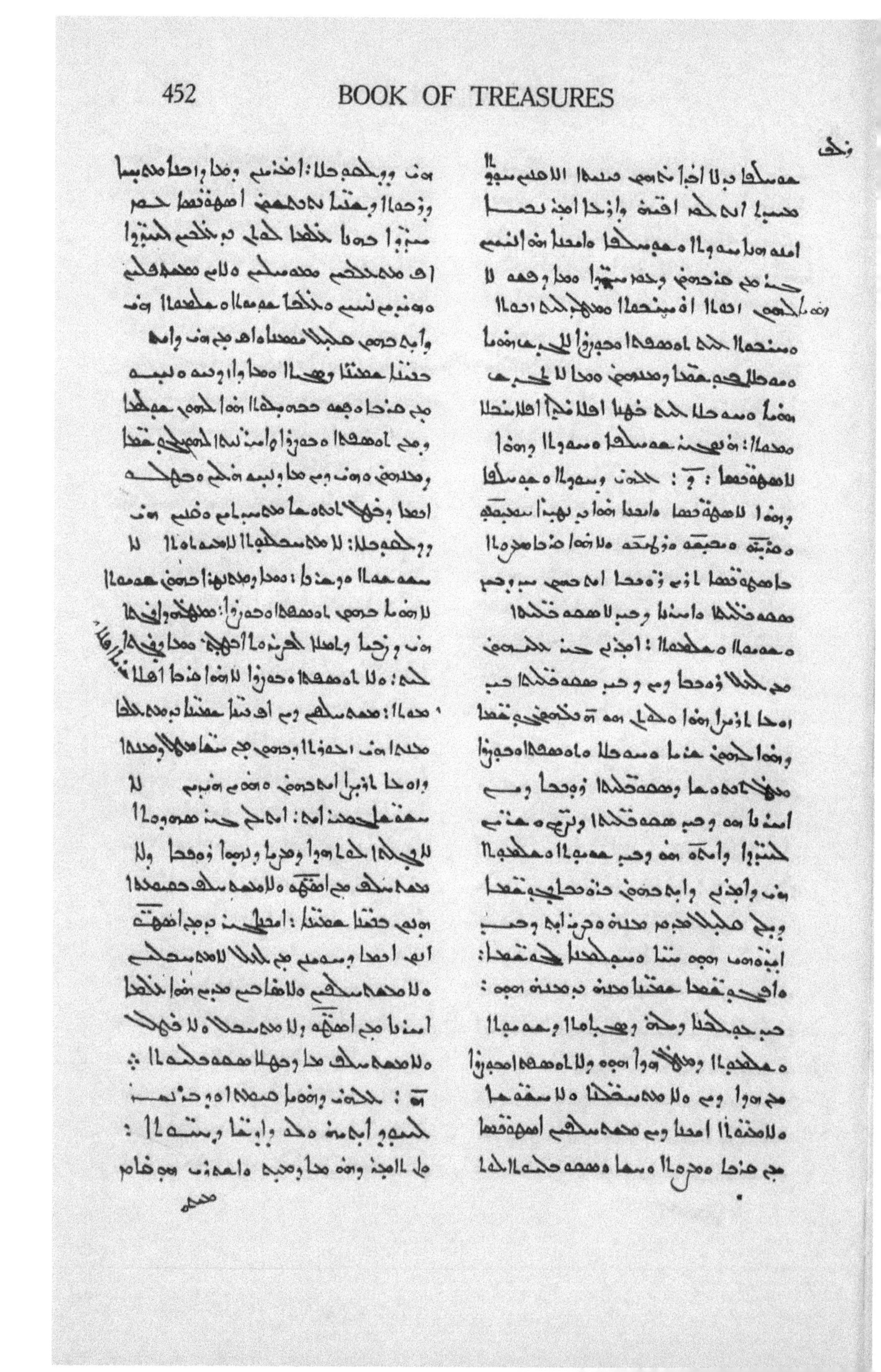

[illegible]

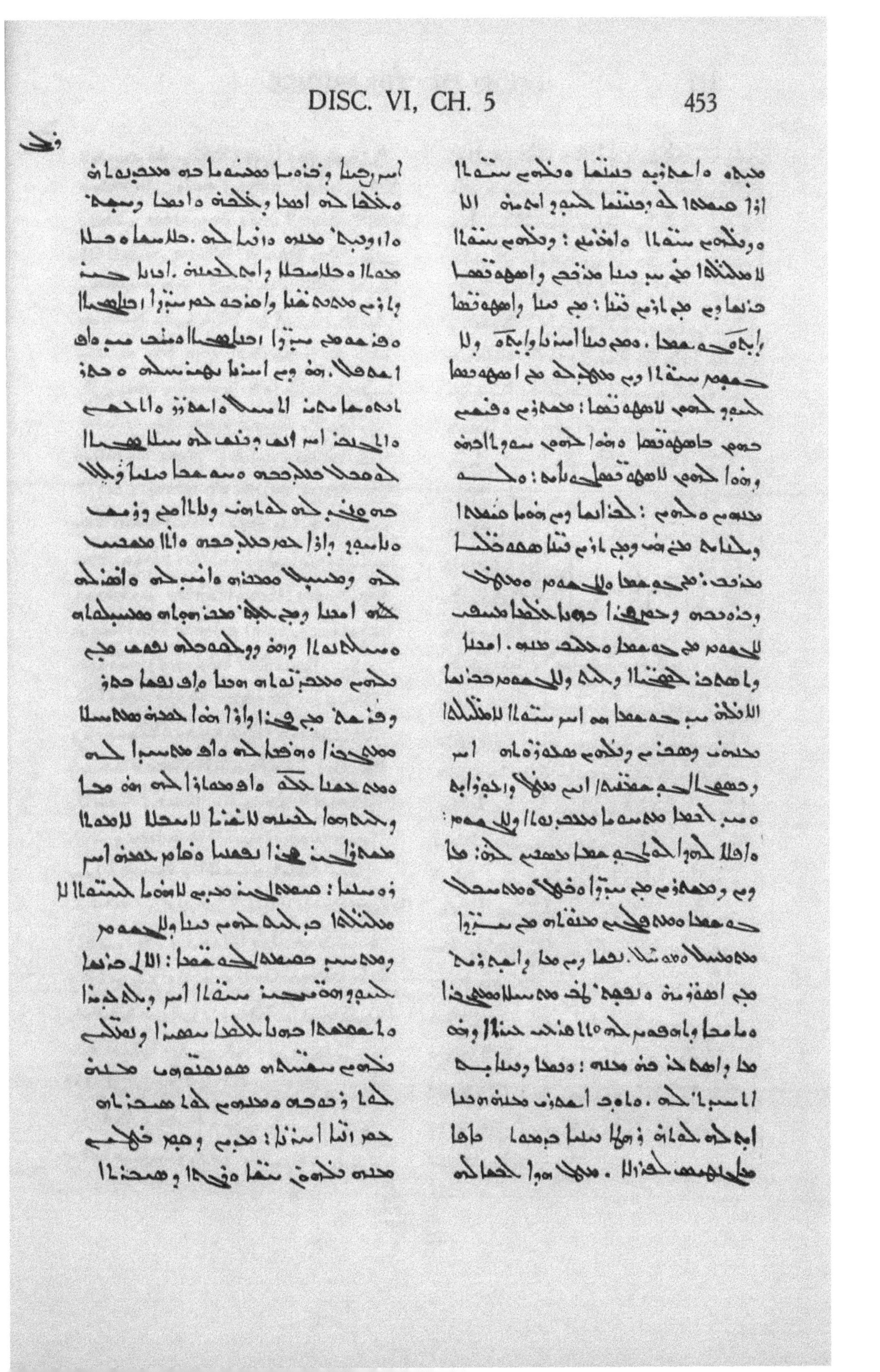

[illegible]

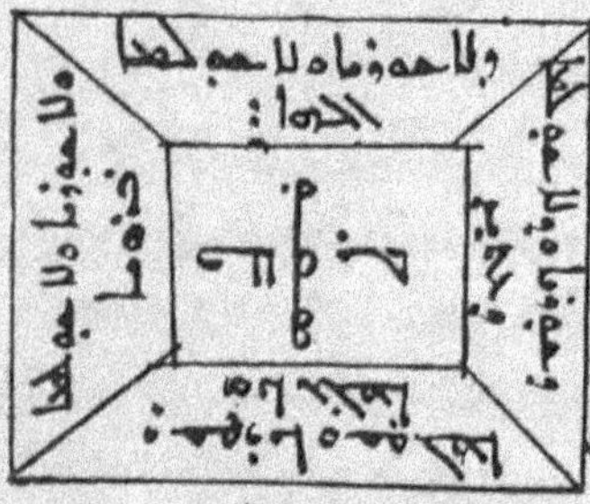

[illegible]

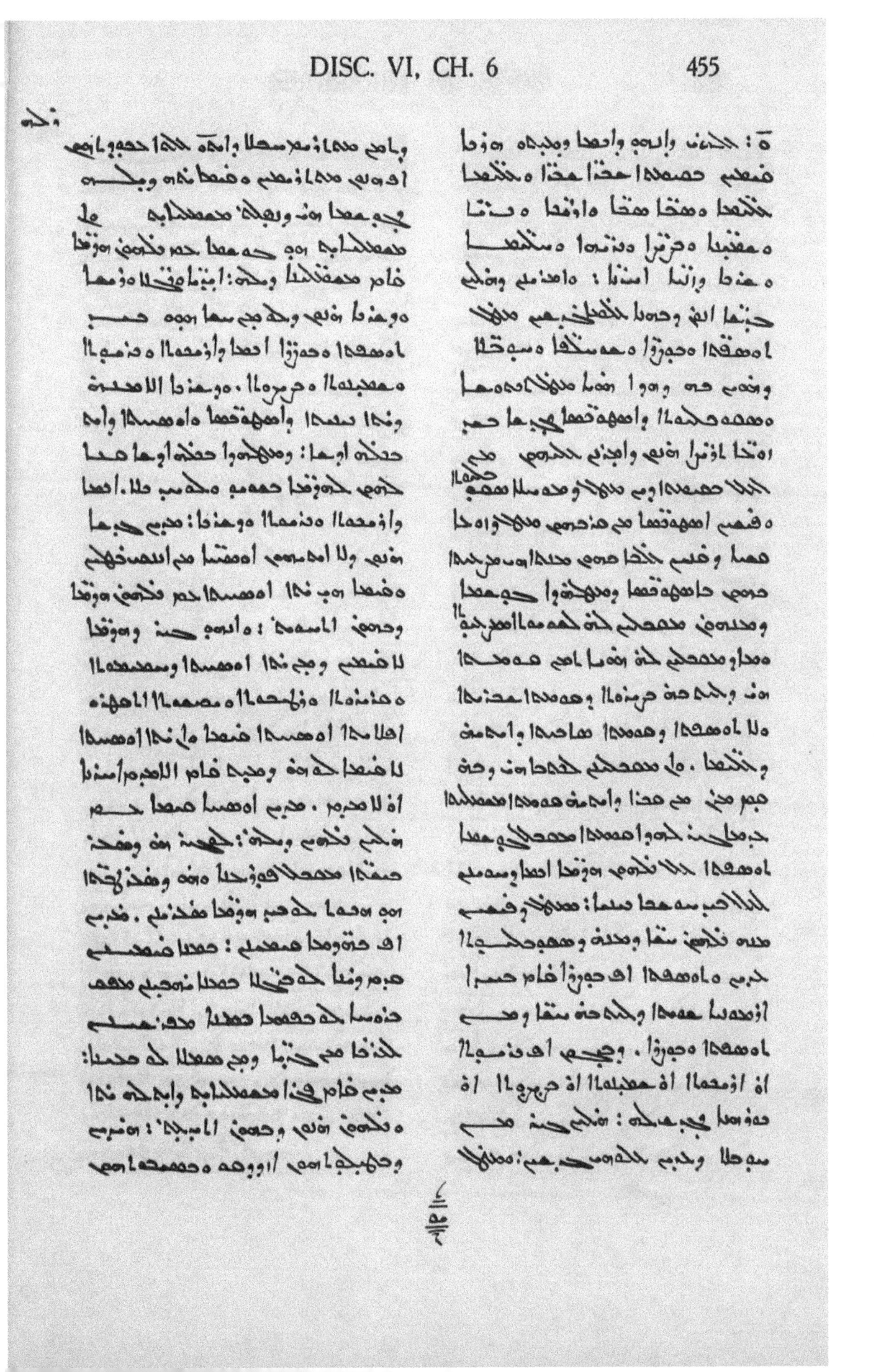

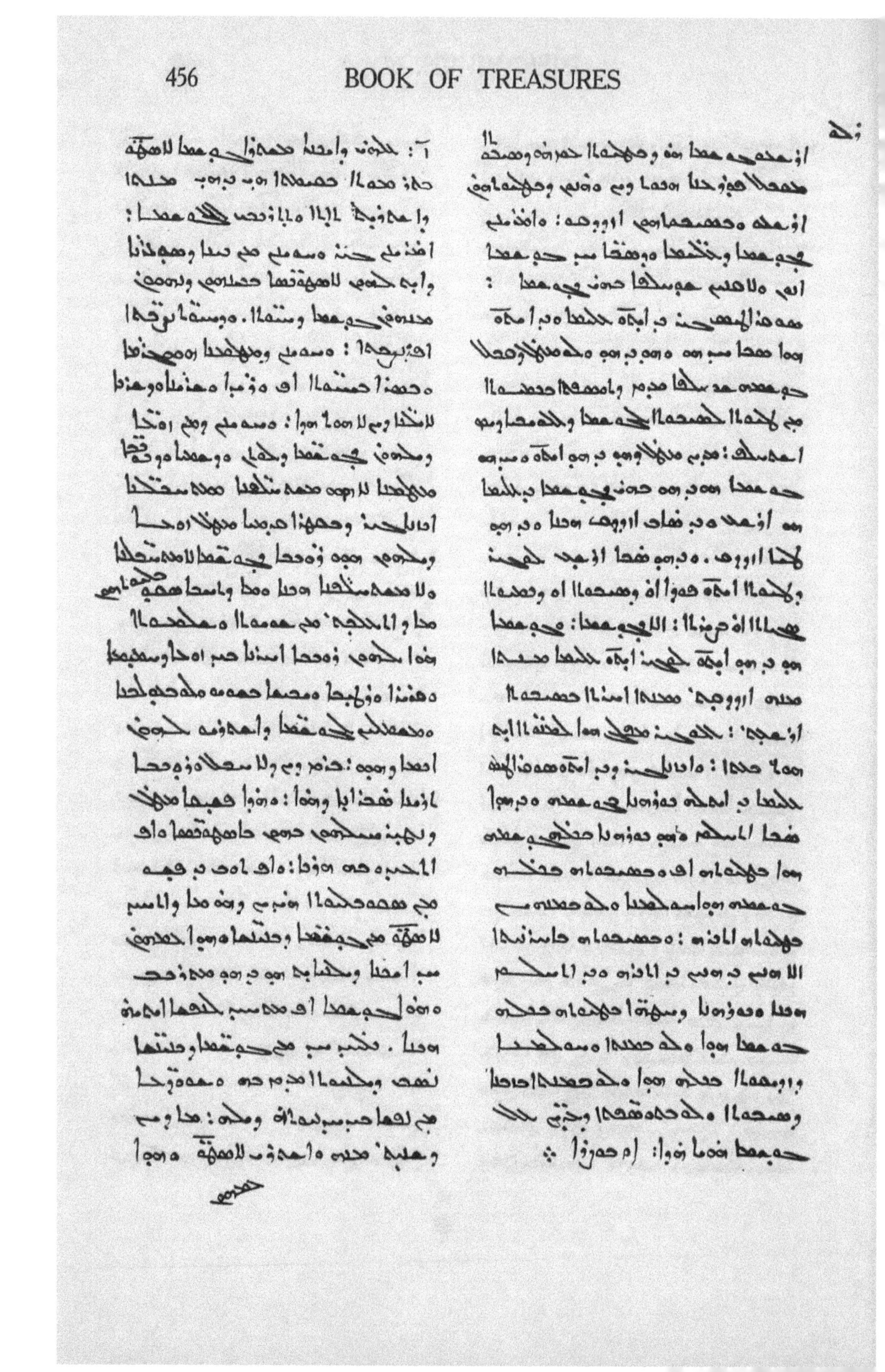

[illegible]

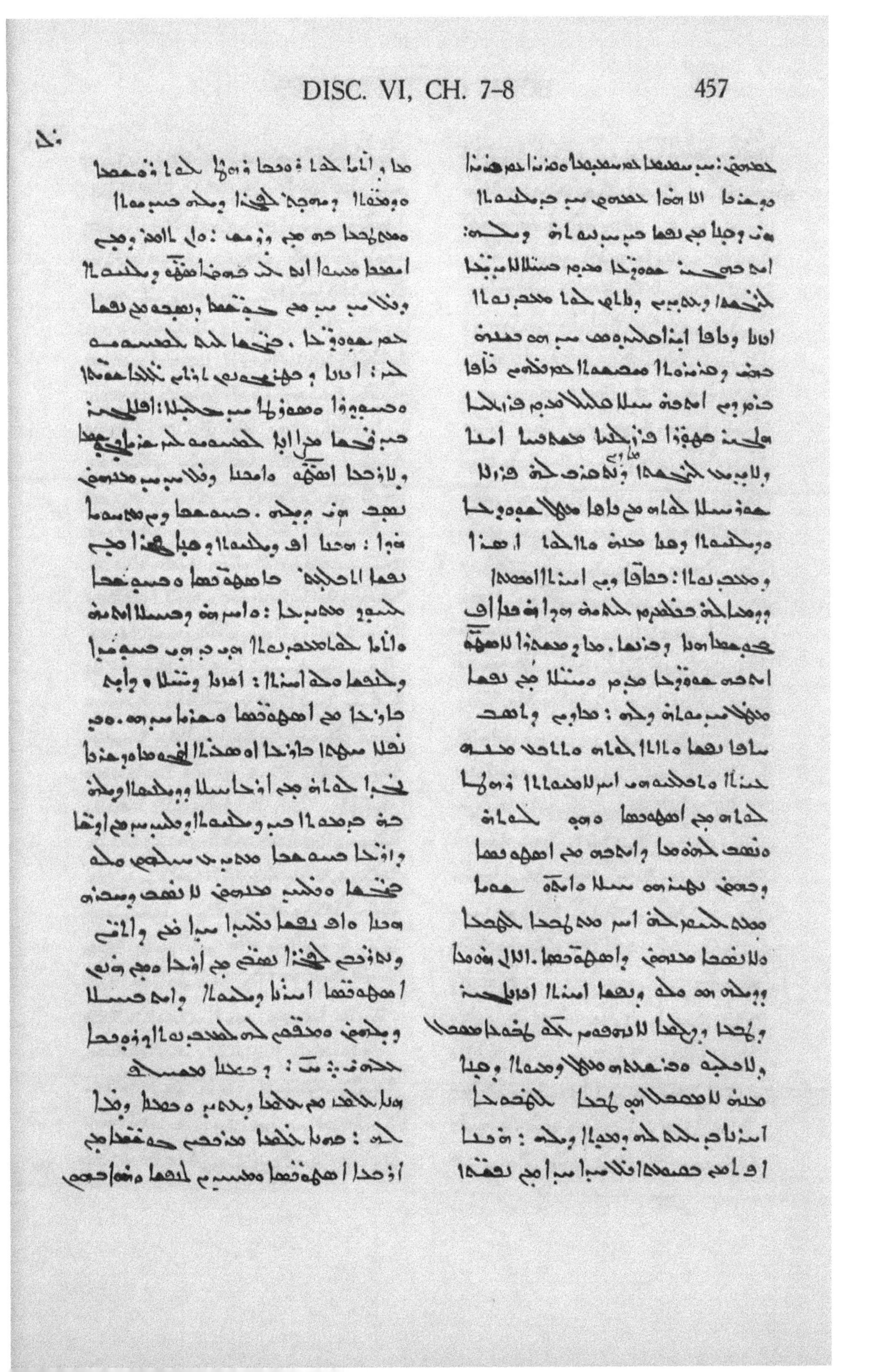

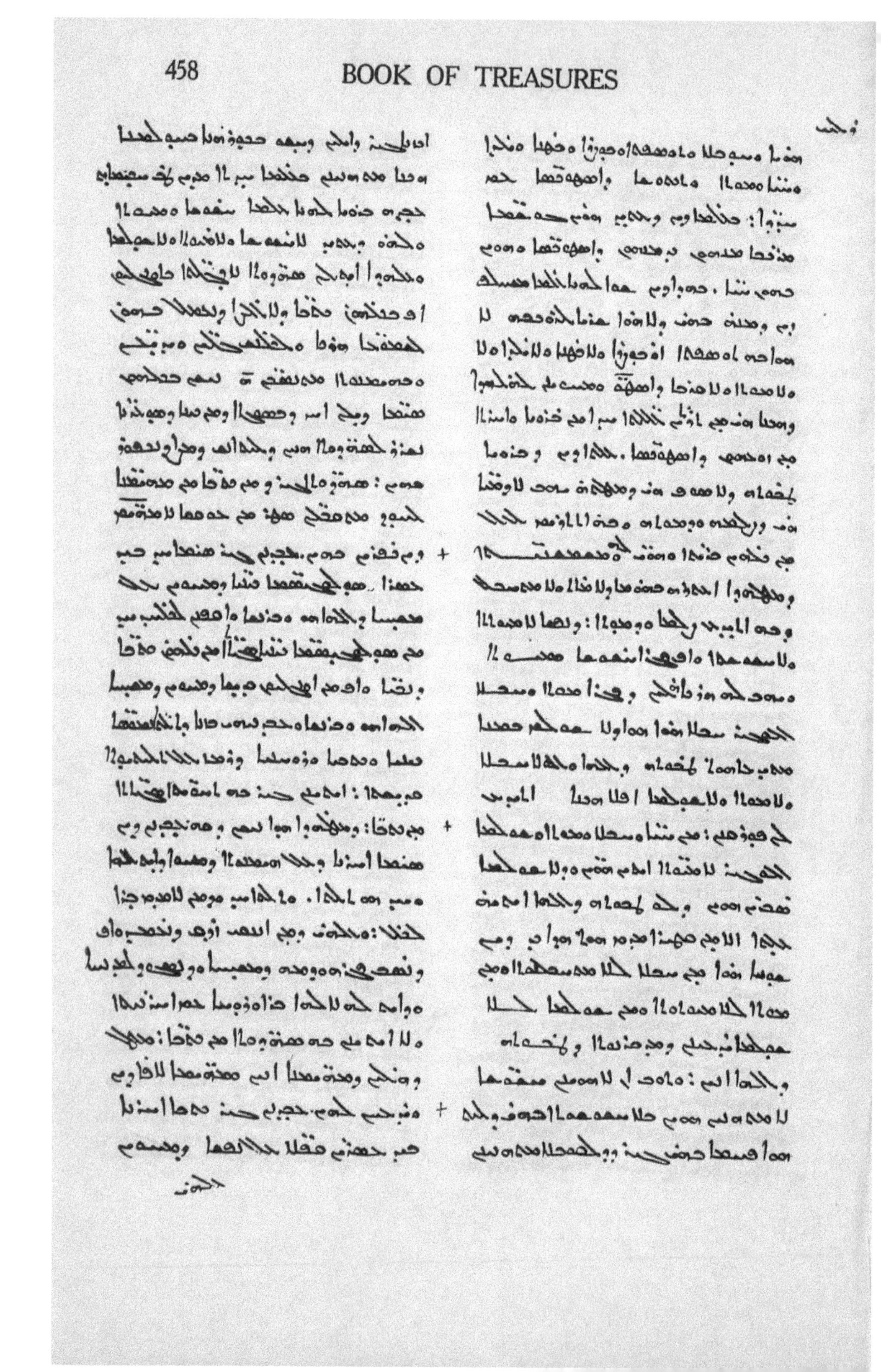

[illegible]

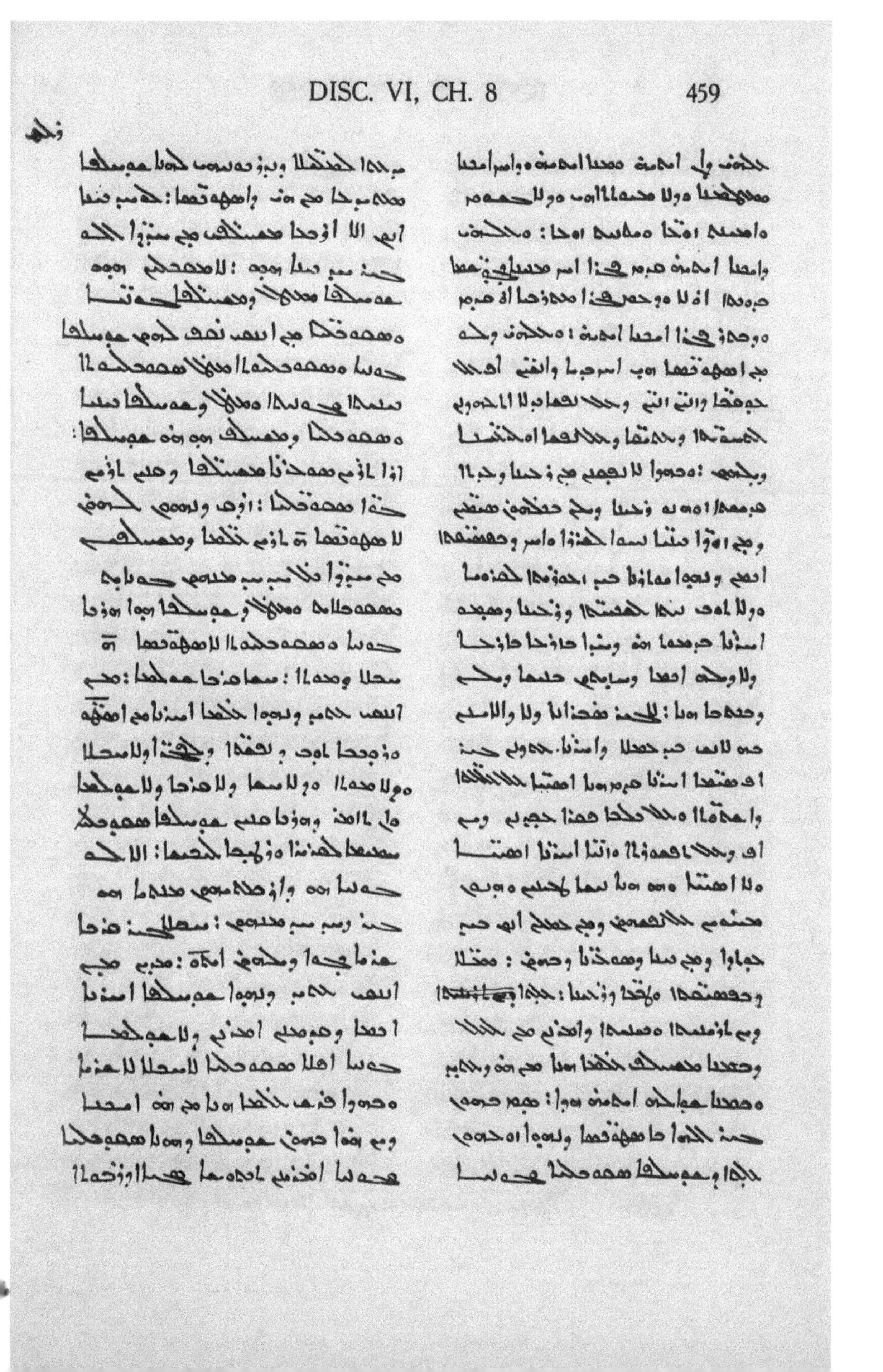

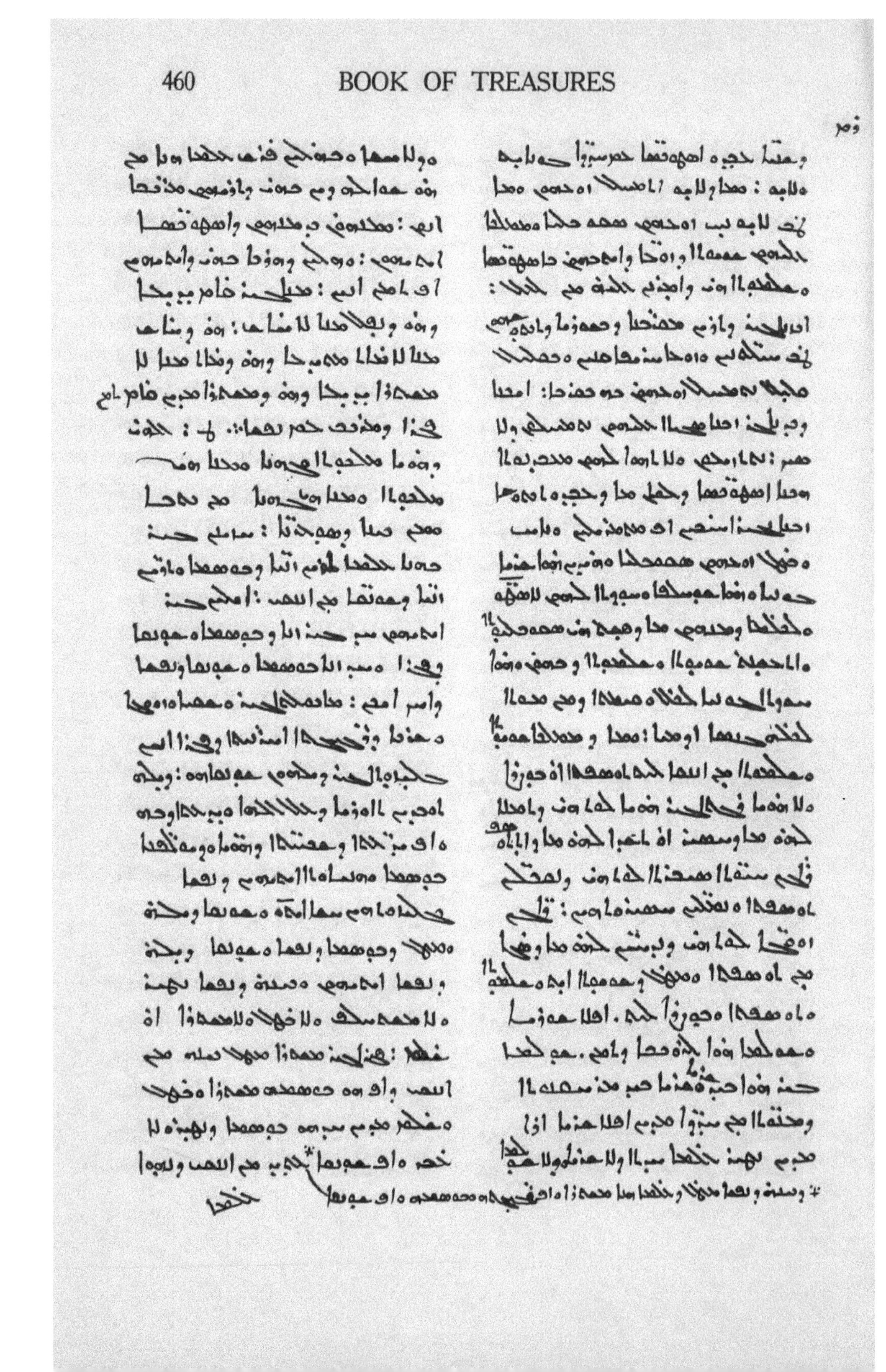

[illegible]

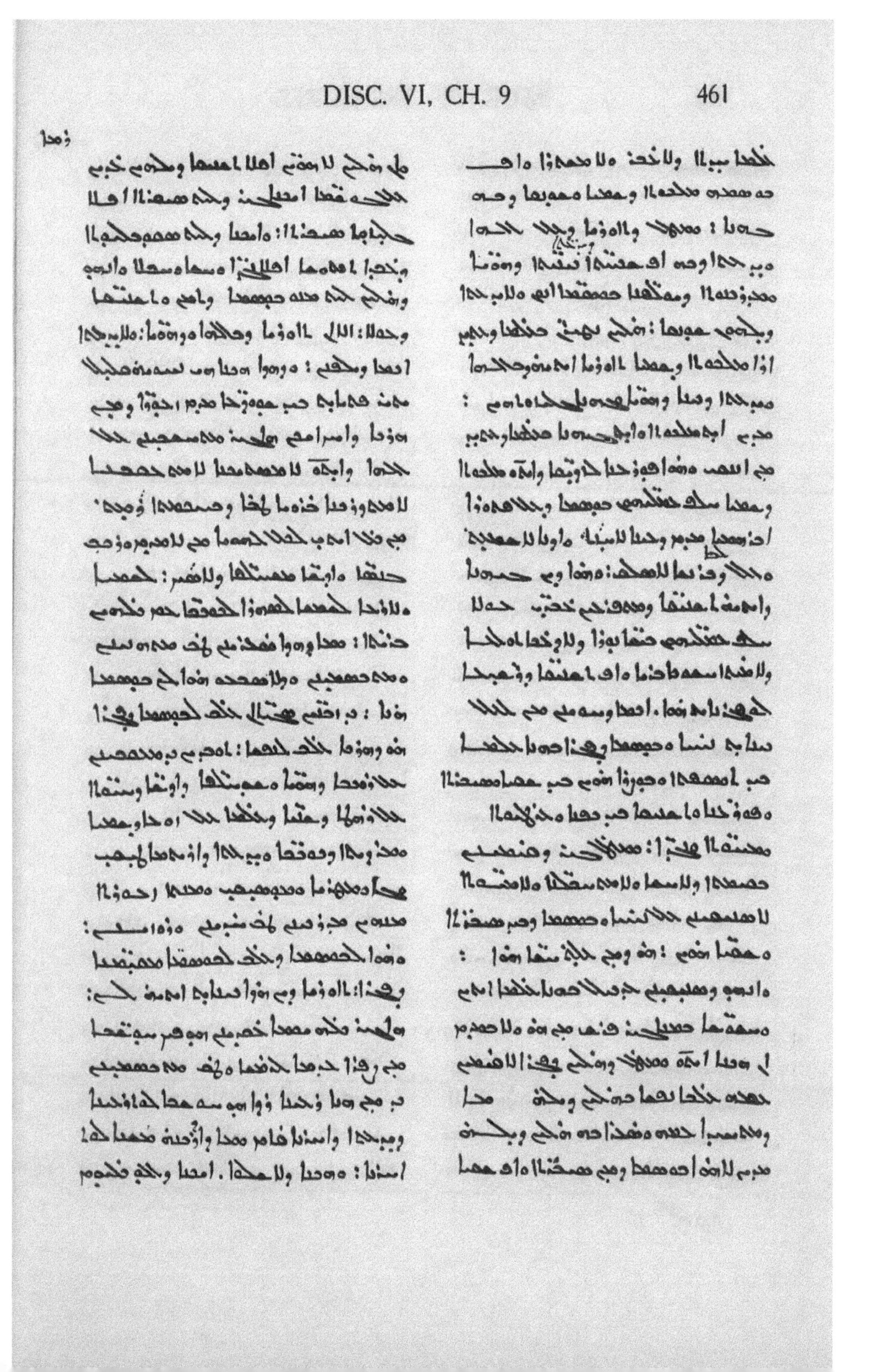

[illegible]

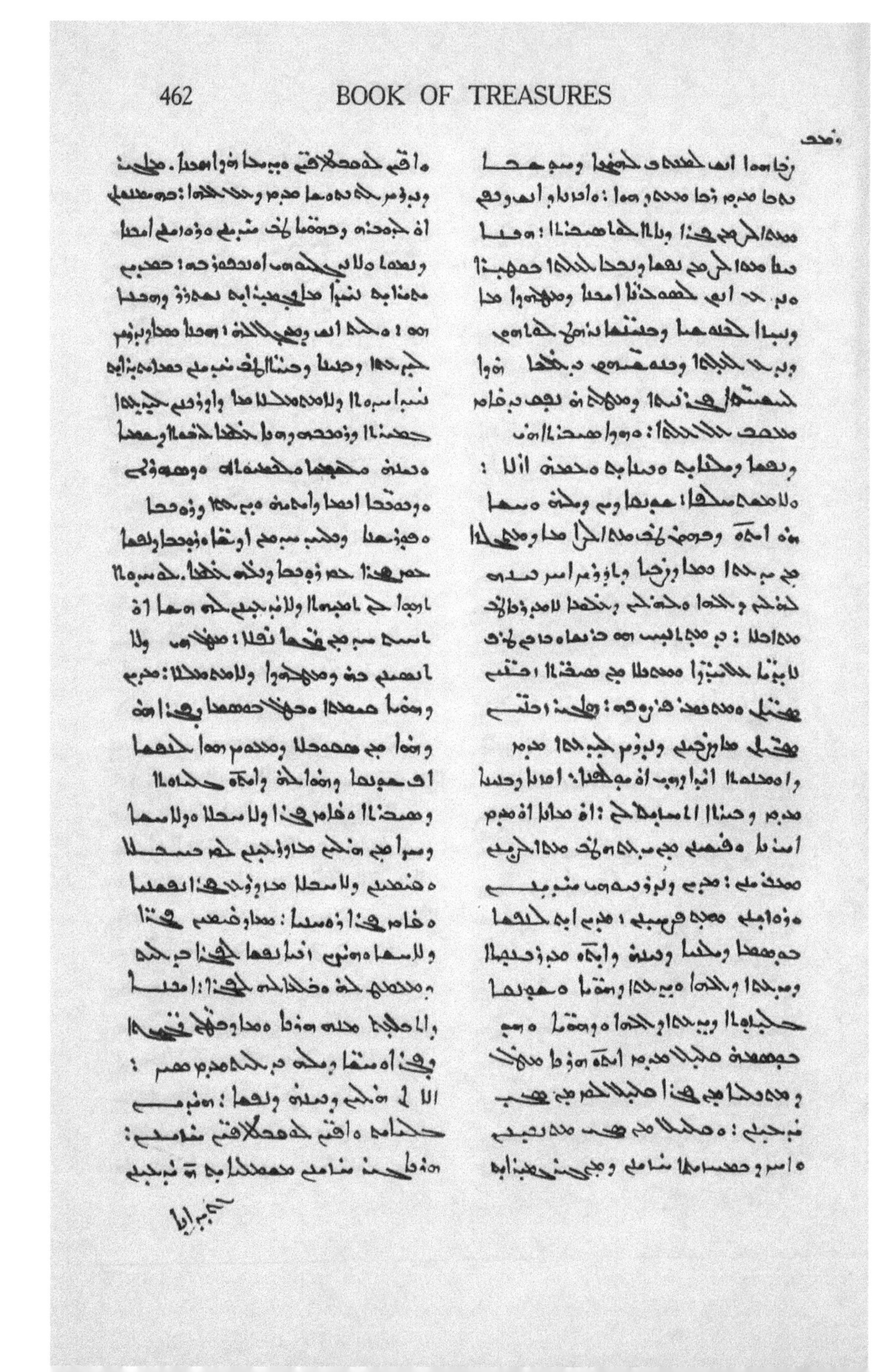

[illegible]

ܡܛܝܒܐ ܠܗ ܠܡܫܝܚܐ ܟܝܢܐ ܘܐܝܬܝܗ ܡܫܝܚܐ
ܘܡܫܘܚܐ ܗܘ ܘܐܦ ܡܫܝܚܐ: ܗܘ ܡܫܝܚܐ
ܘܟܝܢܐ ܗܘ ܘܟܝܢܗ ܐܝܟܢܐ: ܡܫܝܚܐ
ܟܝܢܐ ܐܦܠܐ ܠܐܠܗܘܬܐ ܐܝܟܢܐ ܘܠܐ
ܡܕܡ ܐܚܪܢܐ ܐܠܐ ܐܠܗܐ ܘܒܪܢܫܐ ܘܐܝܟ
ܒܚܝܐ ܘܢܦܫܐ ܡܬܒܣܡܐ ܠܐܠܗܐ ܘܡܗܝܡܢܘܬܐ
ܘܡܬܝܕܥ ܒܝܕܥܬܐ ܘܠܗܘ: ܡܛܝܒܐ ܠܗ ܟܝܢܐ
ܘܠܐܝܢܐ ܕܝܠܗ. ܘܠܐ ܐܢܫ ܡܫܢܐ
ܢܦܫܗ ܘܠܐ ܐܫܬܢܝܬ ܘܗܘܬ ܐܝܟ ܟܝܢܗ ܐܝܟ
ܟܠ ܠܐ ܡܫܢܝܐ ܡܢ ܐܚܪܢܐ ܘܐܦܢ ܗܘܐ ܡܫܡܫܢܐ
ܘܒܗܝܢ ܟܝܢܐ: ܐܢܫܐ ܘܡܢ ܒܬܪ ܐܠܗܐ
ܘܡܫܡܫܢܐ ܘܐܠܗܐ ܘܗܘܐ ܡܫܡܫܢܐ ܘܒܗܘܢ
ܘܡܛܠܗܕܐ ܘܗܘܐ ܐܝܟ ܡܫܝܚܐ ܟܪܝܐ ܘܩܐܡ
ܡܕܡ ܕܐܝܬܘܗܝ ܘܢܦܫܐ ܡܫܬܠܚܐ ܘܠܐ ܐܬܡܫܚ
ܟܕ ܗܘܐ ܡܫܝܚܐ ܘܠܐ ܡܫܬܚܠܦ ܘܐܬܚܠܦ ܗܘܐ
ܗܝܕܝܢ ܐܝܟ ܐܠܗܐ: ܡܫܝܚܐ ܡܫܝܚܐ
ܘܟܝܢܗ ܘܡܬܢܗ ܟܝܢܐ ܡܕܡ ܟܝܢܗ ܘܢܦܫܐ
ܡܫܡܫܢܐ ܟܠܗ ܘܐܢܫܐ: ܗܝܕܝܢ ܐܝܬܘܗܝ ܦܫܝܛܐ
ܫ ܟܝܢܘܬܐ ܐܝܬ ܐܚܪܢܐ ܟܕ ܗܘܝܐ ܘܠܐ ܡܫܬܢܝܐ
ܘܦܫܝܛ ܗܟܢܐ ܕܝܢ ܠܐ ܟܝܢܐ ܘܐܦܢ
ܘܟܠܗ ܘܠܐ ܢܬܚܫܒ ܗܘܐ ܘܗܘ ܠܐܢܫ
ܘܡܬܚܫܒ ܟܝܢܐ ܘܗܘ ܘܐܝܟܢܐ ܡܫܡܫܢܐ
ܟܪܝܐ ܘܐܢܫܐ ܘܠܡܫܝܚܐ ܘܗܘܝܐ ܕܡܫܡܫܢܐ
ܡܢ ܒܪܢܫܐ ܡܕܡ ܕܐܝܬܘܗܝ ܚܝܐ ܚܝܐ ܘܐܦ
ܡܫܡܫܢܐ ܒܪܢܫܐ ܘܚܝܐ ܢܬܚܙܐ ܠܗ ܟܝܢܘܗܝ
ܟܝܢܐ ܐܠܐ ܐܦ ܟܝܢܗ ܗܘܐ ܘܦܫܝܛܐ: ܡܫܡܫ ܡܢܗ
ܢܦܫ ܐܝܟ ܚܝܐ ܘܠܐ ܡܫܬܢܝܐ ܘܠܐ ܡܫܡܫܢܐ:
ܡܛܠ ܢܦܫܗ ܚܝܐ ܘܢܦܫܐ ܗܘ ܡܬܚܙܝܐ: ܟܪܝܐ
ܡܕܡ ܘܟܠܐ ܐܘ ܡܕܡ: ܗܝܕܝܢ ܐܝܟ ܚܝܐ
ܟܪܝܐ ܘܡܫܝܚܐ ܐܝܟܢܐ ܘܢܦܫܗ ܡܢ ܡܫܝܚܐ

ܦܩܕ

ܘܗܟܢܐ ܗܘܐ ܟܝܢܐ ܢܦܫܐ ܐܝܟ ܐܢܫܐ ܡܢ ܟܪܝܐ ܐܝܟ
ܟܝܢܐ ܠܐ ܗܘ ܟܝܢܐ ܘܠܐ ܟܝܢܐ ܘܐܝܠܐ ܗܢܘܢ ܡܢ
ܢܦܫܐ ܟܕ ܚܘܒܐ ܘܐܠܗܐ ܐܝܟ ܗܘ ܐܠܐ ܟܝܢܐ
ܘܡܫܡܫܢܐ ܟܠܗ ܘܡܫܡܫܢܐ ܘܠܐ ܐܢܫ ܐܢܫܐ ܘܠܐ
ܡܫܡܫܢܐ ܘܟܠܗ ܡܬܚܙܝܐ ܘܐܦ ܟܝܢܐ ܡܫܝܚܐ
ܫܘܝܐ: ܟܡܫܝܚܐ ܘܗܢܐ ܟܝܢܐ. ܡܛܠ ܡܢ ܚܝܐ
ܘܐܝܟܢܐ ܟܡܫܝܚܐ ܘܗܢܐ ܟܝܢܐ ܡܢ ܡܫܝܚܐ
ܘܠܐ ܡܫܡܫܢܐ ܠܗ ܠܟܝܢܐ: ܘܟܠܗ ܐܦ ܗܘܐ
ܘܡܫܡܫܢܐ ܘܟܝܢܐ ܗܘ: ܗܘ ܕܝܢ ܐܦ ܫܘܝܐ
ܐܦ ܟܡܫܡܫܢܐ: ܗܘ ܕܝܢ ܡܫܡܫܢܐ ܟܕ ܘܟܝܢܐ
ܫܪܝܪܐ ܗܘ ܘܠܗ ܟܕ ܫܘܝܐ ܠܗ ܚܝܐ:
ܗܟܢܐ ܘܠܐ ܗܘܐ ܐܠܗ ܫܘܝܐ ܠܗ ܚܝܐ ܗܘ
ܟܝܢܐ ܘܟܠ ܡܫܡܫܢܐ: ܐܘܟܝܬ ܘܡܢ ܘܡܫܝܚܐ
ܘܫܡܫܐ ܘܡܫܡܫܢܐ ܟܕ ܟܝܢܗ ܘܚܝܐ ܡܫܡܫܢܐ
ܟܕ ܡܫܡܫܢܐ ܐܝܟ ܡܫܡܫܢܐ: ܢܦܫܗ ܟܝܢܐ
ܐܝܟ ܡܫܡܫܢܐ ܟܝܢܐ ܫܘܝܐ ܗܘ ܠܐ ܟܝܢܐ ܘܡܫܡܫܢܐ
ܫܘܝܐ ܗܘ ܘܡܢ ܐܦܢ ܠܗ ܡܫܡܫܢܐ ܡܢ ܟܝܢܘܬܐ
ܘܠܐ ܡܫܡܫܢܐ ܘܠܐ ܢܬܚܫܒ ܐܠܐ ܡܛܠܗ ܕܡܫܝܚܐ
ܡܢ ܟܝܢܐ ܐܚܪܢܐ ܡܫܡܫܢܐ ܟܝܢܐ ܘܢܦܫܐ ܘܐܦ ܗܘ
ܘܐܝܟ ܡܫܡܫܢܐ ܟܝܢܐ ܗܟܢܐ ܘܟܠܗ ܘܐܝܟ ܟܝܢܐ
ܫܘܝܐ ܟܝܢܐ: ܗܝܕܝܢ ܐܝܟ ܫܘܝܐ ܘܟܠܗܘܢ ܘܢܦܫܗ
ܘܐܝܟ ܐܝܬܝܗ ܐܠܗܐ ܕܡܫܡܫܢ ܘ ܟܝܢܐ
ܘܡܫܡܫܢܐ ܘܚܝܐ ܫܘܝܐ ܐܝܟ ܢܦܫܘܬܐ ܡܫܡܫܢܐ
ܡܢ ܫܘܝܐ ܡܫܡܫܢܐ ܟܝܢܗ ܘܢܦܫܗ ܘܐܝܬܝܐ
ܠܐܠܗܐ ܘܐܝܟ ܡܫܡܫܢܐ ܗܘ: ܠܗ ܡܫܡܫܢ ܟܠܗ
ܠܐܢܫ ܘܟܠ ܡܫܡܫܢܐ ܘܗܘܝܐ ܢܦܫܗ ܠܐܠܗܐ
ܡܢ ܟܝܢܐ ܘܐܠܗܐ ܫܘܝܐ ܐܝܟܢܐ: ܘܫܘܝܐ ܡܫܡܫܢܘܬܐ
ܘܡܫܡܫܢܐ: ܡܫܡܫܢ ܟܠܗ ܐܢܫܝܐ ܘܐܦܢܐ
ܘܢܦܫܗ ܢܦܫܐ ܟܝܢܐ ܘܟܠܗܘܢ ܘܢܦܫܗ ܟܝܢܐ
ܘܟܠܗܘܢ: ܡܕܡ ܡܢ ܟܝܢܐ ܘܐܠܗܐ ܐܝܬܝܗܘܢ ܫܘܝܐ

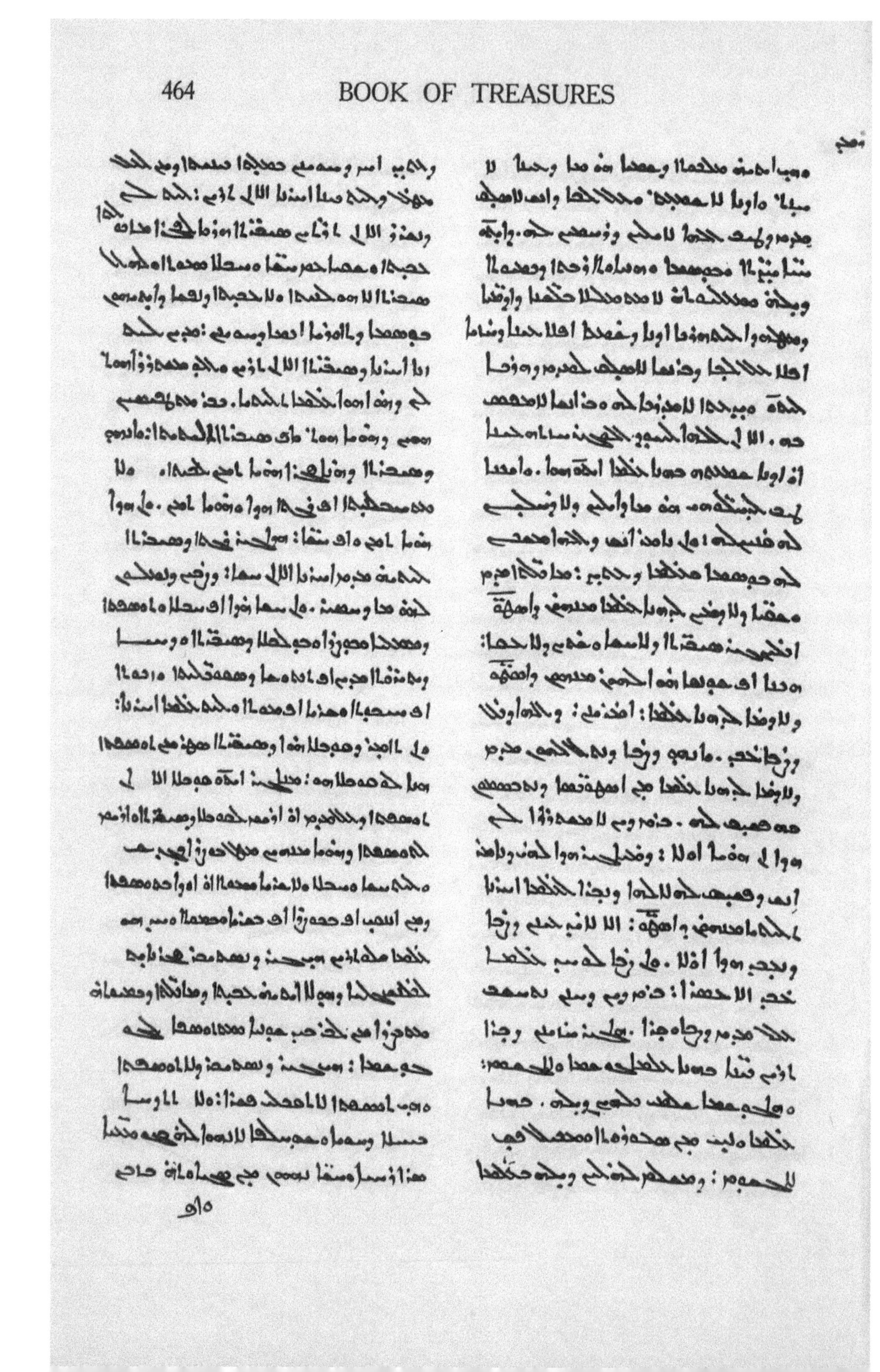

[illegible]

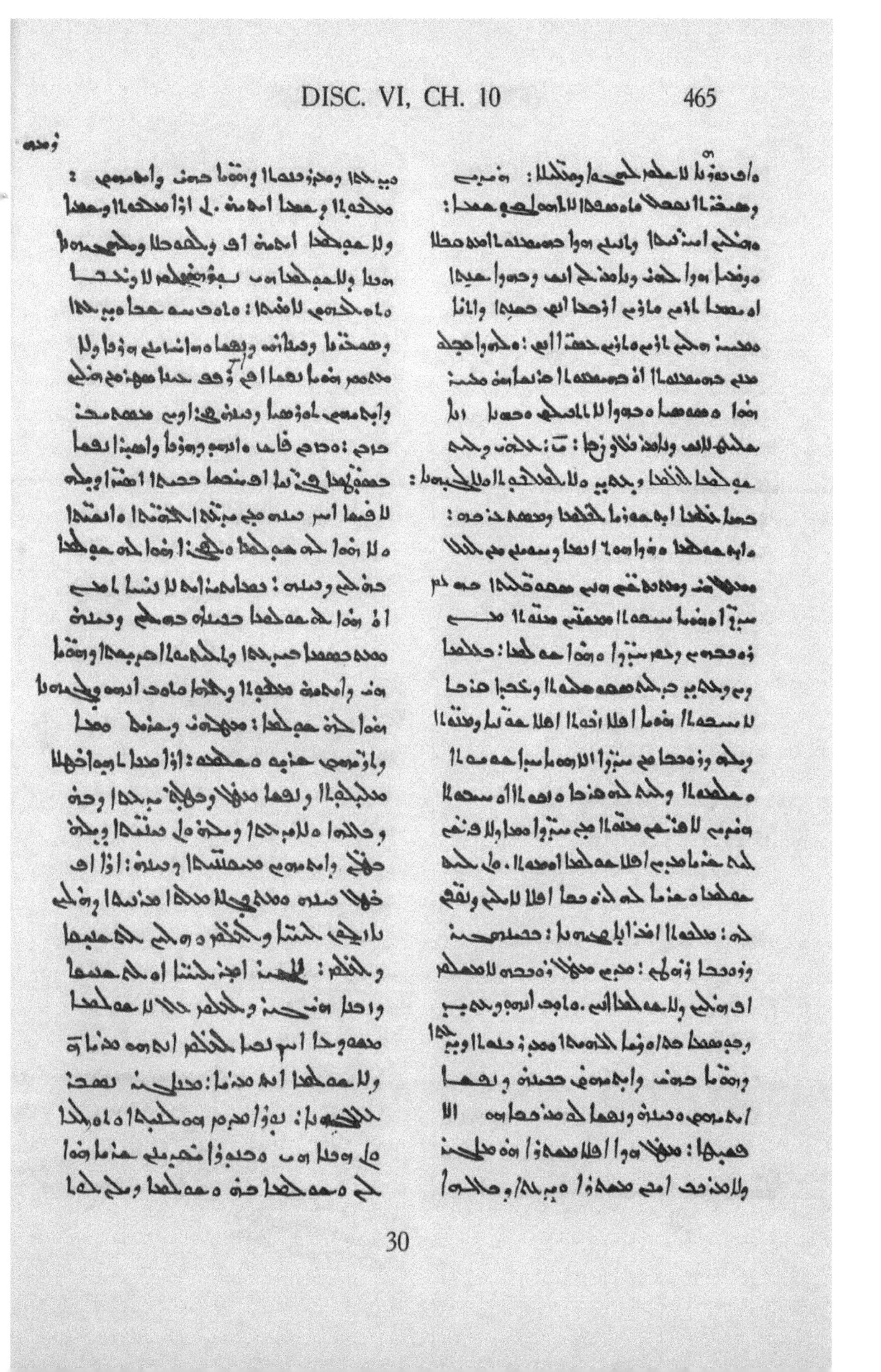

[illegible]

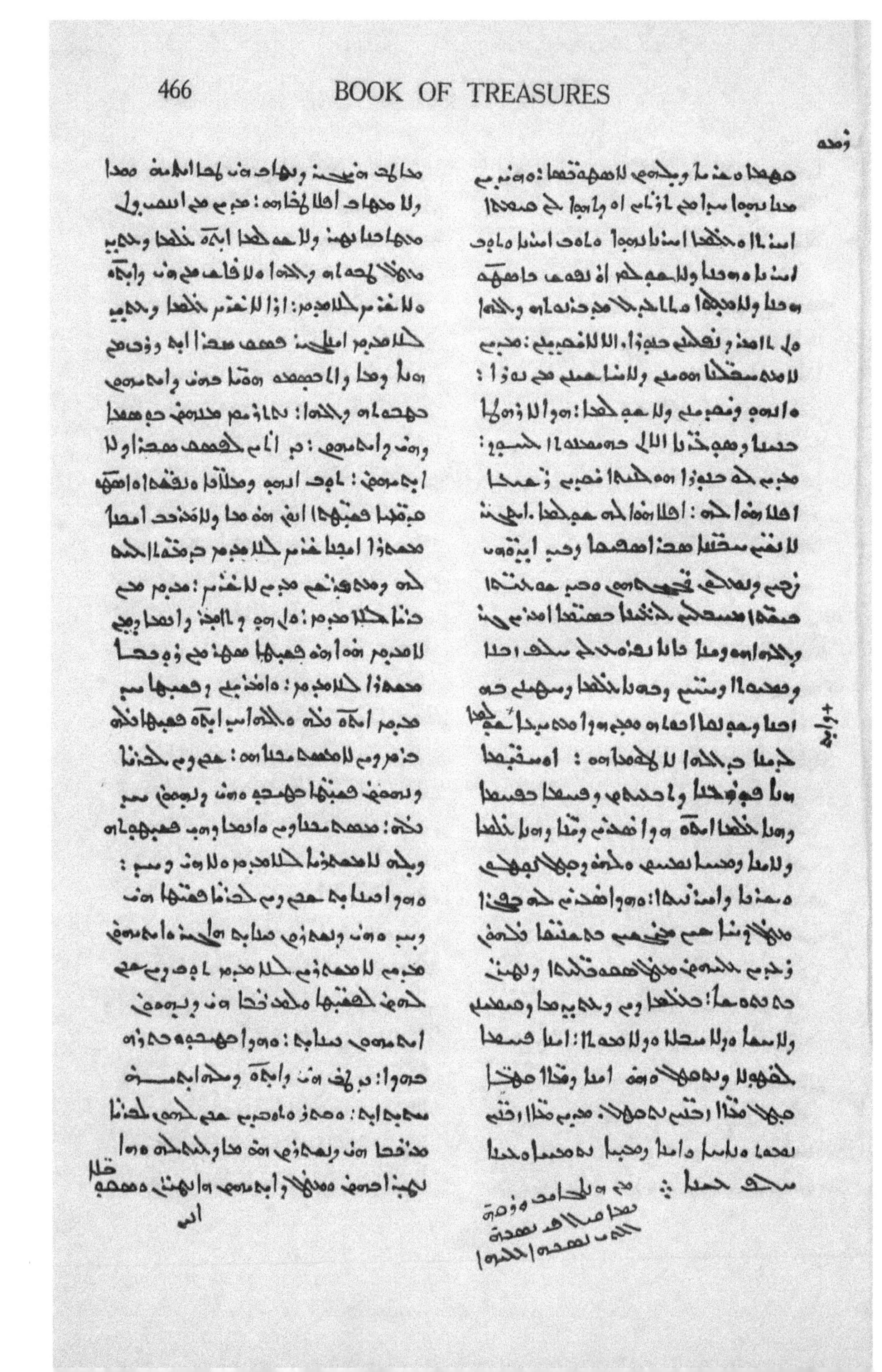

[illegible]

[illegible]

ܐܝܟ ܕܡܬܝܕܥܐ: ܐܘܦ ܦܫܝܛܬܐ ܐܝܟ ܡܢܗܝܢ ܡܢ ܦܣܝܩܬܐ ܘܠܐ
ܡܛܠ ܕܡܢ ܟܠܗ ܡܟܝܠ ܡܬܪܟܒܐ ܡܛܠ ܕܡܢ ܦܫܝܛܬܐ
ܘܠܐ ܐܝܟ ܕܪܘܚܐ ܒܪܘܚܢܝܐ ܗܘܐ ܦܫܝܛܐ ܘܡܪܟܒܐ
ܡܐ: ܒܠܘܬ ܕܡܬܚܙܝܢܐ ܒܝܬ ܐܠܗܐ ܡܫܝܚܐ ܠܐܠܗܐ
ܫܪܪܐ ܘܒܠܘܬ ܕܒܗ ܒܝܕ ܠܐ ܫܪܪܐ ܠܗ ܡܫܝܚܐ ܕܐܠܗܐ
ܚܒܝܫ ܐܝܬܘܗܝ ܦܓܪܐ ܘܡܬܪܟܒܢܘܬܐ ܘܚܕ ܗܘ ܗܘ ܐܝܬܘܗܝ
ܘܡܬܚܙܐ ܐܠܗܐ ܒܪܢܫܐ ܡܫܝܚܐ ܗܘܐ ܘܡܢ ܩܕܝܡ
ܡܢ ܕܒܗ ܒܝܕ ܣܝܡܬܐ ܘܐܘܟܝܬ ܗܘ ܕܟܝܢܗ
ܒܡܠܘܗܝ ܘܒܕܡܘܬܗ ܘܡܬܚܙܐ ܒܟܢܐ: ܦܓܪܢܝܐ
ܒܝܫ ܒܟܠܗ ܫܘܪܝܐ ܘܐܘܟܡܐ ܘܡܬܚܙܐ ܦܓܪܢܐܝܬ ܠܗ ܗܘ
ܡܬܪܟܒܢܘܬܗ ܗܘ ܚܒܫ ܒܟܠܗ ܗܘ ܡܪܟܐ ܘܠܐ ܦܓܪ
ܡܢ ܒܟܠܗ ܫܘܪܝܐ ܘܒܗܘܢܐ ܕܦܬܓܡܐ ܒܢܒ
ܚܢܢ ܘܡܫܝܚܐ ܐܠܗܐ ܗܘܐ ܐܠܗܝܘܬܐ ܘܠܐ
ܡܫܝܚܐ ܗܘܐ ܗܘܐ ܒܡܬܚܙܝܢܘܬ ܐܠܗܝܘܬܐ
ܐܘܡܐ ܘܠܐ ܡܫܝܚܐ: ܘܡܬܚܙܐ ܗܘܐ ܒܝܬ ܡܫܝܚܐ
ܐܠܗܝܘܬܐ ܘܡܢܗ ܗܘܐ ܒܪܢܫܐ ܘܡܬܚܙܝܢܘܬܗ ܗܘܐ ܘܡܢܐ
ܘܗܘ ܠܡܣܪ ܒܪܢܫܐ ܘܡܢܗ ܡܢ ܚܛܝܬܐ ܘܟܝܢܐ
ܚܒ ܫܡܝܐ ܘܡܢܗܐ ܐܠܗ ܡܬܚܙܝܢܐ ܘܡܢܗ
ܢܦܫ ܡܬܚܙܝܢܘܬܗ ܘܠܐ ܡܬܚܙܝܐ ܐܟܠܐ ܡܫܝܚܐ
ܟܕ ܒܗ ܐܠܗܝܘܬܐ ܐܝܬܝܐ ܘܒܗ ܗܘ ܫܡܝܐ
ܘܠܐ ܒܟܝܢܐ ܐܦܝܐ ܘܚܝܐ ܘܠܐ ܡܫܝܚܐ: ܐܠܗܐ
ܚܕ ܘܡܬܚܙܝܢܘܬܗ ܘܚܣܝܢ ܐܝܬܘܗܝ
ܒܝܬ ܐܠܗܘܬܐ ܗܘܐ ܡܫܝܚܐ ܘܗܘܐ ܡܢ ܒܟܝܢܐ
ܘܡܢܐ ܘܗܘܐ ܡܬܚܙܝܢܘܬܗ ܒܝܬ ܐܠܗܘܬܐ
ܐܢܫܐ ܘܟܠܗ ܒܟܝܢܐ ܘܡܟܠ ܘܦܓܪܐ
ܘܒܝܬܗ ܘܠܐ ܡܫܝܚܐ ܘܐܢܫ ܒܟܠܗܘܢ ܟܠܗ ܡܟܠ
ܘܠܐ ܡܫܐ ܘܠܐ ܡܫܟܚܐ ܘܡܢ ܟܠܗ ܦܓܪܐ ܘܠܐ
ܗܘ ܡܬܠܝܢ ܐܦ ܒܟܝܢܐ ܘܪܘܚܐ ܡܢ ܟܠܗ
ܒܟܝܢܐ ܘܒܪܗ ܐܠܗܐ ܕܡܫܝܚܐ ܘܒܗ ܒܝܬܗ
ܒܪܘܚܐ ܐܠܗܐ ܡܫܝܚܐ ܘܒܗ ܒܝܬ ܠܐ ܒܝܬ ܠܗ

ܪ̈ܡܙܐ
ܡܫܝܚܐ: ܘܒܗ ܗܘܐ ܐܠܗܐ ܒܝܬ ܒܗ ܡܫܝܚܐ
ܒܪܘܚܐ ܕܡܢ ܒܪ ܐܠܗܐ ܗܘ ܡܫܝܚ ܦܓܪܐ
ܐܝܟ ܕܒܐܠܗܘܬܐ ܘܒܪܢܫܐ ܒܝܬ ܐܝܟ ܕܡܬܝܕܥ
ܐܠܗܐ ܒܗܘܢܐ ܡܫܝܚܐ ܘܒܪܢܫ ܘܠܐ ܡܫܝܚܐ:
ܘܐܦ ܕܠܐ ܒܪܢܫ ܟܠܗ ܡܟܝܠ ܐܠܐ ܕܟܠܗ ܡܬܚܙܐ ܡܢ ܟܠܗܘܢ
ܒܗ ܒܪܢܫܐ ܚܕ ܗܘܐ ܐܠܗܐ ܒܪܢܫܐ ܘܐܠܗܘܬܗ ܘܠܐ
ܡܢ ܡܬܚܙܐ ܘܐܠܗܘܬܐ ܐܝܬܝܐ ܚܕ ܐܠܗܐ ܒܪܢܫܐ
ܘܒܗ ܐܝܬ ܚܕ ܡܫܝܚܐ ܘܒܪܢܫܐ ܘܗܘ ܒܗ ܒܗ ܐܠܐ ܐܝܟ
ܕܒܣܪܝܬܐ ܘܠܐ ܒܣܪܢܝܬܐ: ܒܗ ܘܐܦ ܐܦܢ ܟܠܗ ܡܬܚܙܐ
ܐܦܢ: ܘܡܬܚܙܝܢܘܬܐ ܐܝܟܢܐ ܘܡܢ ܟܠܗ ܦܓܪܢܐ. ܘܗܘ ܗܘܐ
ܒܟܠ ܡܫܝܚܐ ܐܠܗܘܬܐ ܗܘܐ ܘܡܬܚܙܐ ܒܟܠܗ
ܐܠܗܘܬܐ ܕܡܢ ܡܢ ܟܠܗܘܢ ܡܬܚܙܝܢ ܘܒܟܠ
ܡܢ ܟܠܗ ܒܟܠܗ ܡܢ ܠܐ ܡܬܚܙܐ ܡܫܝܚܐ ܘܠܐ ܡܫܝܚ
ܡܢ ܐܢܫܐ ܘܡܢ ܕܟܝܢܐ ܘܡܢ ܐܠܗܘܬܐ ܘܡܢ
ܡܫܝܚܐ ܘܡܢ ܐܠܗܐ ܡܢ ܒܪܢܫܐ ܘܡܬܚܙܐ ܡܫܝܚܐ
ܡܬܚܙܝܢ ܐܝܟ ܐܢܫܐ ܒܡܬܚܙܝܢܐ ܐܝܟ ܐܢܫܐ ܘܒܟܠܗ
ܘܡܬܚܙܝܢܐ ܘܡܬܚܙܝܢ ܘܡܢ ܢܦܫ ܘܦܓܪܐ ܘܒܫܪܪܐ
ܘܦܓܪܐ ܘܡܢ ܘܒܗ ܘܠܐ ܡܫܝܚܐ ܘܡܬܚܙܐ ܐܠܗܐ
ܠܟܠ ܡܢܗ ܘܡܬܚܙܐ ܡܢ ܒܟܠܗ ܗܘܐ ܘܐܠܗܘܬܐ
ܘܡܢ ܘܗܘܐ ܒܢܦܫ ܢܦܫ ܡܬܚܙܝܢ ܐܝܟ ܘܐܦܢ
ܘܗܘܐ ܐܝܬܘܗܝ ܒܝܬ ܒܗ ܘܡܫܝܚܐ ܡܢ ܦܓܪܐ:
ܘܐܝܟ ܕܡܬܚܙܝܢܘܬܐ ܐܝܬܘܗܝ ܒܟܠ ܟܠܗ ܘܡܢ ܒܗ
ܘܐܦ ܒܝܬ ܒܗ ܘܡܬܚܙܐ ܘܐܦ ܠܐ ܡܬܚܙܝܢ ܐܝܟ
ܒܪܢܫܐ ܘܐܝܟ ܕܡܬܚܙܝܢܘܬܐ ܐܝܬܘܗܝ ܐܠܐ ܡܢ
ܒܝܬ ܒܗ ܡܬܚܙܝܢܐ ܘܐܦ ܒܗ ܡܬܚܙܝܢ ܐܝܟ
ܒܗ ܘܡܬܚܙܝܢ ܘܐܝܬܘܗܝ ܡܬܚܙܝܢܘܬܐ ܐܝܬܘܗܝ
ܘܟܠܗ ܡܢ ܒܪܢܫܐ ܕܠܐ ܐܢܫ ܒܗ ܘܠܐ ܦܓܪ
ܘܒܝܬ ܒܗ ܘܡܢ ܒܗ ܒܗ ܡܬܚܙܝܢܐ ܐܠܗܐ
ܒܡܬܚܙܝܢܘܬܐ ܘܐܢܫܐ ܘܡܢܫܐ ܘܠܐ ܡܬܚܙܝܢ ܘܠܐ ܡܫܝܚܐ
ܘܒܢܫܐ ܗܘ ܘܡܫܝܚܐ ܟܠܗ ܘܡܬܚܙܝܢܐ

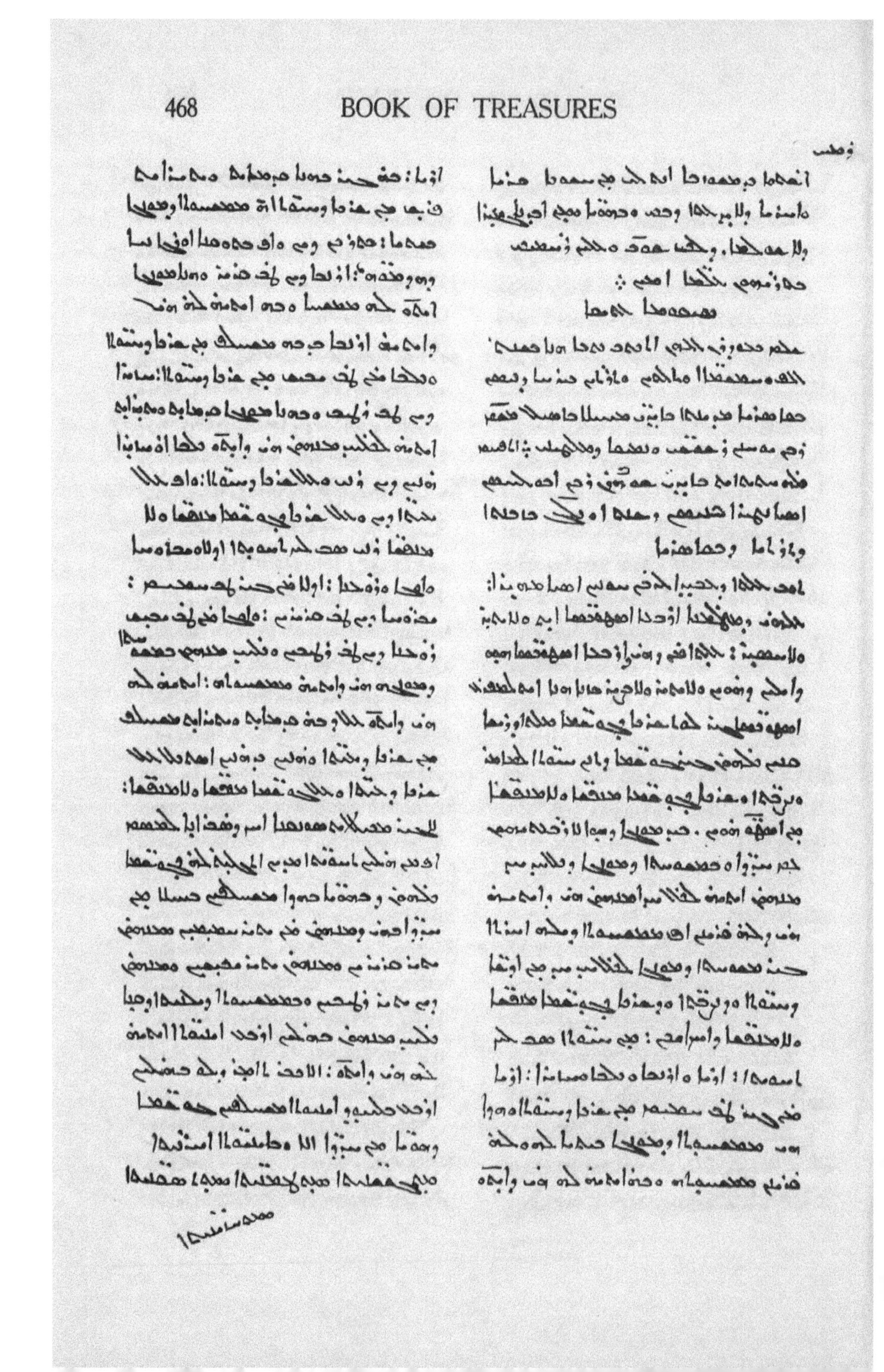

[illegible]

[illegible]

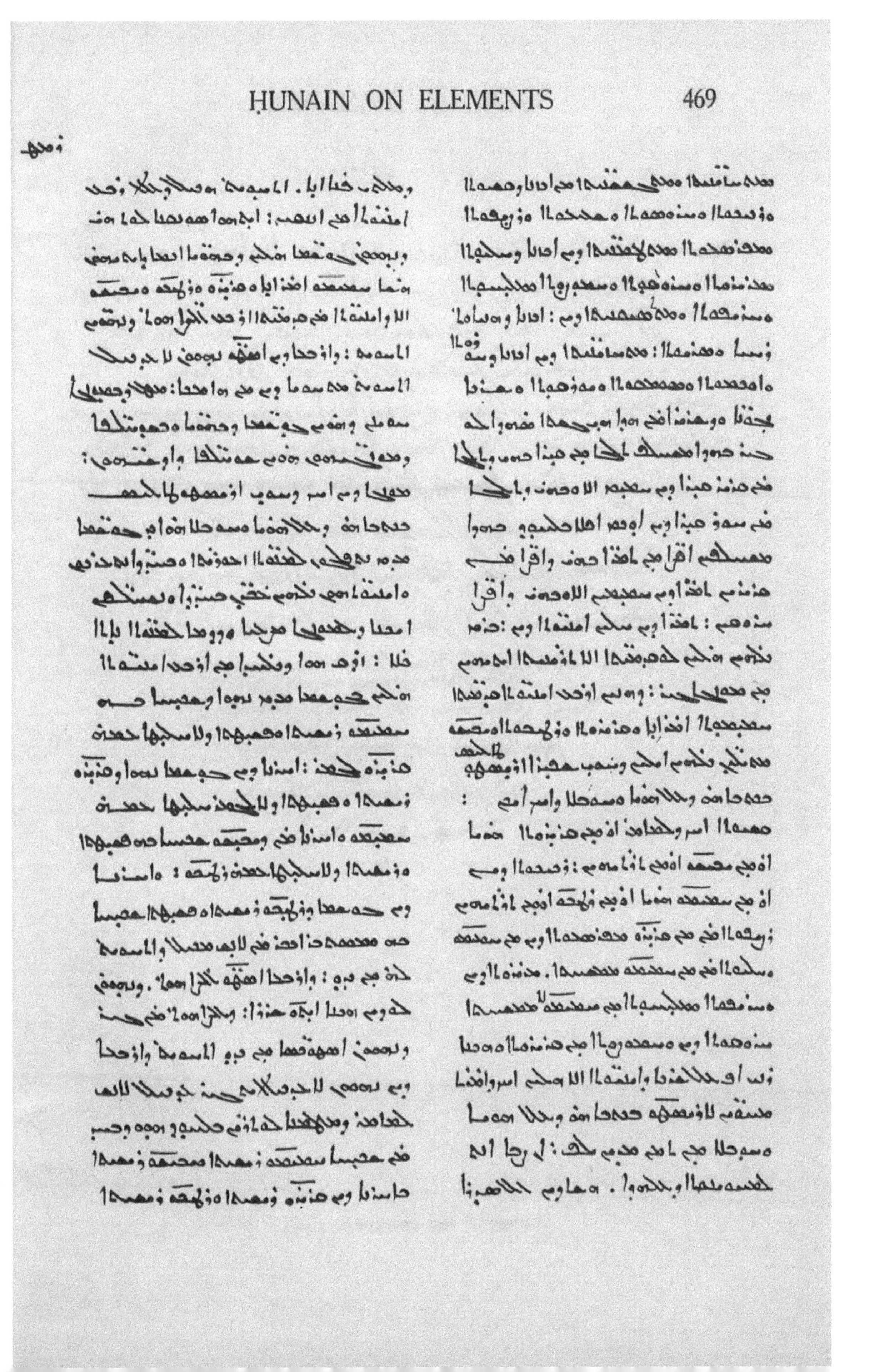

[illegible]

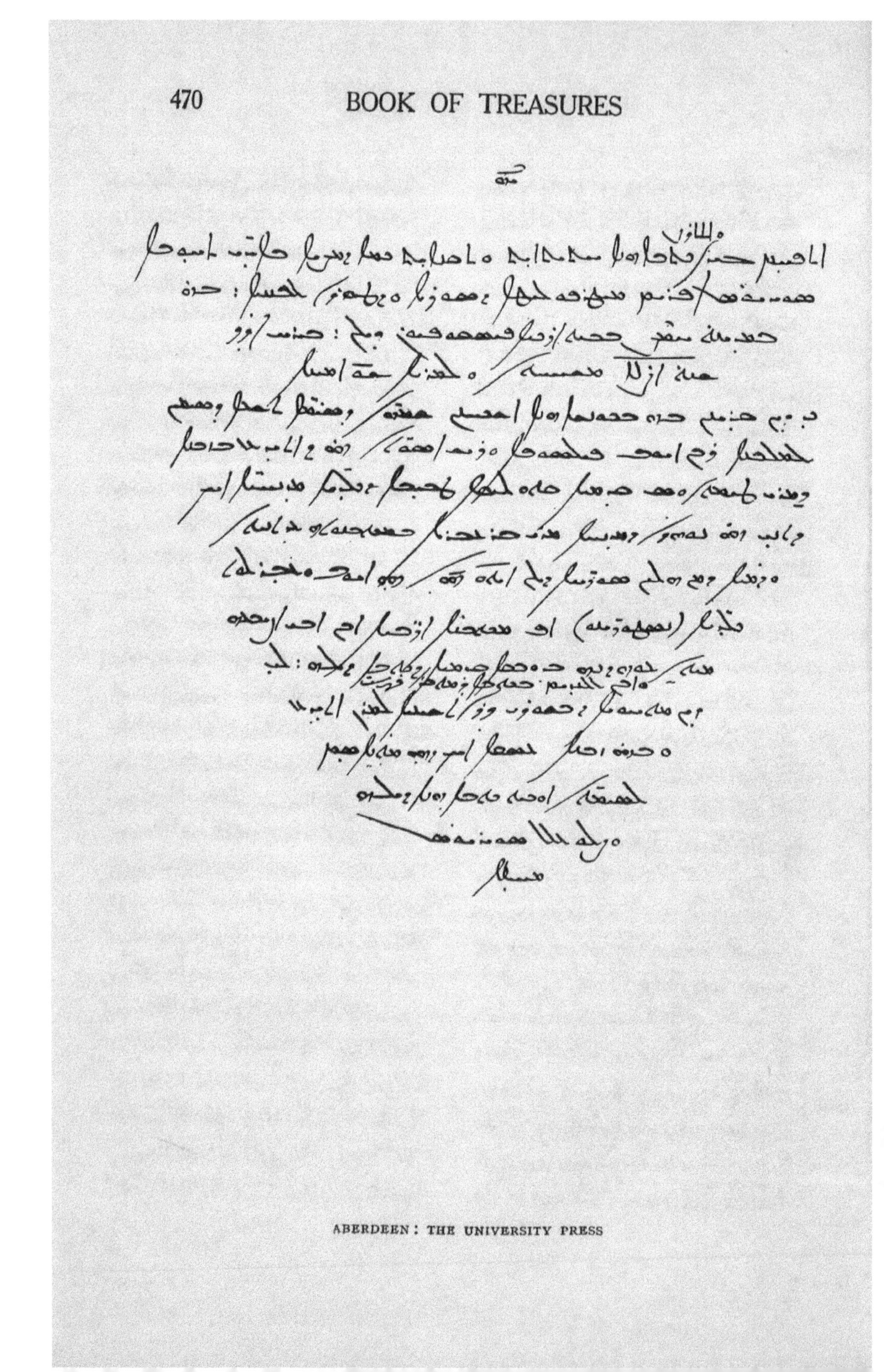

[illegible]

ABERDEEN : THE UNIVERSITY PRESS

www.ingramcontent.com/pod-product-compliance
Lightning Source LLC
LaVergne TN
LVHW062244200926
848484LV00006B/1607